# Synthesis and Applications of New Spin Crossover Compounds

# Synthesis and Applications of New Spin Crossover Compounds

Special Issue Editor
**Takafumi Kitazawa**

MDPI • Basel • Beijing • Wuhan • Barcelona • Belgrade

*Special Issue Editor*
Takafumi Kitazawa
Toho University
Japan


*Editorial Office*
MDPI
St. Alban-Anlage 66
4052 Basel, Switzerland


This is a reprint of articles from the Special Issue published online in the open access journal *Crystals* (ISSN 2073-4352) from 2018 to 2019 (available at: https://www.mdpi.com/journal/crystals/special_issues/SpinCrossover_Compounds).


For citation purposes, cite each article independently as indicated on the article page online and as indicated below:

LastName, A.A.; LastName, B.B.; LastName, C.C. Article Title. *Journal Name* **Year**, *Article Number*, Page Range.


**ISBN 978-3-03921-361-0 (Pbk)**
**ISBN 978-3-03921-362-7 (PDF)**


© 2019 by the authors. Articles in this book are Open Access and distributed under the Creative Commons Attribution (CC BY) license, which allows users to download, copy and build upon published articles, as long as the author and publisher are properly credited, which ensures maximum dissemination and a wider impact of our publications.
The book as a whole is distributed by MDPI under the terms and conditions of the Creative Commons license CC BY-NC-ND.

# Contents

**About the Special Issue Editor** ........................................... vii

**Takafumi Kitazawa**
Synthesis and Applications of New Spin Crossover Compounds
Reprinted from: *Crystals* **2019**, *9*, 382, doi:10.3390/cryst9080382 ..................... 1

**José Alberto Rodríguez-Velamazán, Kosuke Kitase, Elías Palacios, Miguel Castro, Ángel Fernández-Blanco, Ramón Burriel and Takafumi Kitazawa**
Structural Insights into the Two-Step Spin-Crossover Compound
Fe(3,4-dimethyl-pyridine)$_2$[Ag(CN)$_2$]$_2$
Reprinted from: *Crystals* **2019**, *9*, 316, doi:10.3390/cryst9060316 ..................... 5

**Natalia Artiukhova, Galina Romanenko, Gleb Letyagin, Artem Bogomyakov, Sergey Veber, Olga Minakova, Marina Petrova, Vitaliy Morozov and Victor Ovcharenko**
Spin Transition in the Cu(hfac)$_2$ Complex with (4-Ethylpyridin-3-yl)-Substituted Nitronyl Nitroxide Caused by the "Asymmetric" Structural Rearrangement of Exchange Clusters in the Heterospin Molecule
Reprinted from: *Crystals* **2019**, *9*, 285, doi:10.3390/cryst9060285 ..................... 13

**Tomoe Matsuyama, Keishi Nakata, Hiroaki Hagiwara and Taro Udagawa**
Iron(II) Spin Crossover Complex with the 1,2,3-Triazole-Containing Linear Pentadentate Schiff-Base Ligand and the MeCN Monodentate Ligand
Reprinted from: *Crystals* **2019**, *9*, 276, doi:10.3390/cryst9060276 ..................... 26

**Darunee Sertphon, Phimphaka Harding, Keith S. Murray, Boujemaa Moubaraki, Suzanne M. Neville, Lujia Liu, Shane G. Telfer and David J. Harding**
Solvent Effects on the Spin Crossover Properties of Iron(II) Imidazolylimine Complexes
Reprinted from: *Crystals* **2019**, *9*, 116, doi:10.3390/cryst9020116 ..................... 42

**Satoshi Tokinobu, Haruka Dote and Satoru Nakashima**
Threefold Spiral Structure Constructed by 1D Chains of [[M(NCS)$_2$(bpa)$_2$]·biphenyl]$_n$ (M = Fe, Co; bpa = 1,2-bis(4-pyridyl)ethane)
Reprinted from: *Crystals* **2019**, *9*, 97, doi:10.3390/cryst9020097 ..................... 54

**Kazuyuki Takahashi, Kaoru Yamamoto, Takashi Yamamoto, Yasuaki Einaga, Yoshihito Shiota, Kazunari Yoshizawa and Hatsumi Mori**
High-Temperature Cooperative Spin Crossover Transitions and Single-Crystal Reflection Spectra of [Fe$^{III}$(qsal)$_2$](CH$_3$OSO$_3$) and Related Compounds
Reprinted from: *Crystals* **2019**, *9*, 81, doi:10.3390/cryst9020081 ..................... 63

**Houcem Fourati, Guillaume Bouchez, Miguel Paez-Espejo, Smail Triki and Kamel Boukheddaden**
Spatio-temporal Investigations of the Incomplete Spin Transition in a Single Crystal of [Fe(2-pytrz)$_2${Pt(CN)$_4$}]·3H$_2$O: Experiment and Theory
Reprinted from: *Crystals* **2019**, *9*, 46, doi:10.3390/cryst9010046 ..................... 85

**Sriram Sundaresan, Irina A. Kühne, Conor T. Kelly, Andrew Barker, Daniel Salley, Helge Müller-Bunz, Annie K. Powell and Grace G. Morgan**
Anion Influence on Spin State in Two Novel Fe(III) Compounds: [Fe(5F-sal$_2$333)]X
Reprinted from: *Crystals* **2019**, *9*, 19, doi:10.3390/cryst9010019 ..................... 103

**Masaya Enomoto, Hiromichi Ida, Atsushi Okazawa and Norimichi Kojima**
Effect of Transition Metal Substitution on the Charge-Transfer Phase Transition and Ferromagnetism of Dithiooxalato-Bridged Hetero Metal Complexes, $(n\text{-}C_3H_7)_4N[Fe^{II}_{1-x}Mn^{II}_xFe^{III}(dto)_3]$
Reprinted from: *Crystals* 2018, *8*, 446, doi:10.3390/cryst8120446 . . . . . . . . . . . . . . . . . . . . . 115

**Verónica Jornet-Mollá, Carlos Giménez-Saiz and Francisco M. Romero**
Synthesis, Structure, and Photomagnetic Properties of a Hydrogen-Bonded Lattice of $[Fe(bpp)_2]^{2+}$ Spin-Crossover Complexes and Nicotinate Anions
Reprinted from: *Crystals* 2018, *8*, 439, doi:10.3390/cryst8110439 . . . . . . . . . . . . . . . . . . . . . 132

**Ahmed Yousef Mohamed, Minji Lee, Kosuke Kitase, Takafumi Kitazawa, Jae-Young Kim and Deok-Yong Cho**
Soft X-ray Absorption Spectroscopy Study of Spin Crossover Fe-Compounds: Persistent High Spin Configurations under Soft X-ray Irradiation
Reprinted from: *Crystals* 2018, *8*, 433, doi:10.3390/cryst8110433 . . . . . . . . . . . . . . . . . . . . . 145

**Takashi Kosone, Itaru Tomori, Daisuke Akahoshi, Toshiaki Saito and Takafumi Kitazawa**
New Iron(II) Spin Crossover Complexes with Unique Supramolecular Networks Assembled by Hydrogen Bonding and Intermetallic Bonding
Reprinted from: *Crystals* 2018, *8*, 415, doi:10.3390/cryst8110415 . . . . . . . . . . . . . . . . . . . . . 154

**Ichiro Terasaki, Masamichi Ikuta, Takafumi D. Yamamoto and Hiroki Taniguchi**
Impurity-Induced Spin-State Crossover in $La_{0.8}Sr_{0.2}Co_{1-x}Al_xO_3$
Reprinted from: *Crystals* 2018, *8*, 411, doi:10.3390/cryst8110411 . . . . . . . . . . . . . . . . . . . . . 163

**Taous Houari, Emmelyne Cuza, Dawid Pinkowicz, Mathieu Marchivie, Said Yefsah and Smail Triki**
Iron(II) Spin Crossover (SCO) Materials Based on Dipyridyl-*N*-Alkylamine
Reprinted from: *Crystals* 2018, *8*, 401, doi:10.3390/cryst8110401 . . . . . . . . . . . . . . . . . . . . . 176

**Nataliya G. Spitsyna, Yuri N. Shvachko, Denis V. Starichenko, Erkki Lahderanta, Anton A. Komlev, Leokadiya V. Zorina, Sergey V. Simonov, Maksim A. Blagov and Eduard B. Yagubskii**
Evolution of Spin-Crossover Transition in Hybrid Crystals Involving Cationic Iron Complexes $[Fe(III)(3\text{-}OMesal_2\text{-}trien)]^+$ and Anionic Gold Bis(dithiolene) Complexes $Au(dmit)_2$ and $Au(dddt)_2$
Reprinted from: *Crystals* 2018, *8*, 382, doi:10.3390/cryst8100382 . . . . . . . . . . . . . . . . . . . . . 190

**Alexander R. Craze, Mohan M. Bhadbhade, Cameron J. Kepert, Leonard F. Lindoy, Christopher E. Marjo and Feng Li**
Solvent Effects on the Spin-Transition in a Series of Fe(II) Dinuclear Triple Helicate Compounds
Reprinted from: *Crystals* 2018, *8*, 376, doi:10.3390/cryst8100376 . . . . . . . . . . . . . . . . . . . . . 206

**Sabine Lakhloufi, Elodie Tailleur, Wenbin Guo, Frédéric Le Gac, Mathieu Marchivie, Marie-Hélène Lemée-Cailleau, Guillaume Chastanet and Philippe Guionneau**
Mosaicity of Spin-Crossover Crystals
Reprinted from: *Crystals* 2018, *8*, 363, doi:10.3390/cryst8090363 . . . . . . . . . . . . . . . . . . . . . 222

**Akihiro Ondo and Takayuki Ishida**
Cobalt(II) Terpyridin-4′-yl Nitroxide Complex as an Exchange-Coupled Spin-Crossover Material
Reprinted from: *Crystals* 2018, *8*, 155, doi:10.3390/cryst8040155 . . . . . . . . . . . . . . . . . . . . . 232

# About the Special Issue Editor

**Takafumi Kitazawa**, Dr., Full Professor, is a chemist and received his doctoral degree in 1992 from the University of Tokyo for his thesis "Mineralomimetic Inclusion Structures Built of Cadmium Cyanide System" (Supervisor: Prof. Toschitake IWAMOTO). During his doctorate, Kitazawa was a JSPS Fellow (DC). In 1992, he became Research Associate, Department of Chemistry, Toho University, where he was promoted to Lecturer in 1995, to Associate Professor in 2000, and to Full Professor in 2010.

Kitazawa received the 1997 Young Chemist Award (Japan Society of Coordination Chemistry) for his work on "Silica-Mimetic Polymorphism of the $Cd(CN)_2$ Host Lattice Depending on the Guest G in $Cd(CN)_2 \cdot xG$ Clathrates". In 1996, Kitazawa, along with his co-workers, published the first Hofmann-like spin crossover (SCO) coordination polymer $\{Fe(py)_2[Ni(CN)_4]\}_n$ (py = pyridine). Kitazawa has also served as Editorial Board Member of the Journal of Nuclear and Radiochemical Sciences in 1999–2002 and as a Project Leader of the JSPS-PAN (Poland) Research Project "High Pressure Synthesis for Mineralomimetic Metal Coordination Compounds" in 2000 and 2001. Kitazawa is also recipient of the JSPS-CISC (Spain) Fellowship, CSIC en Aragón, Zaragoza University.

Kitazawa has published over 100 scientific papers. His current research interests include spin crossover materials related to Hofmann-like coordination polymers, and highly organized and self-assembled materials with cadmium cyanide and polycyanopolycadmate systems. His research interests also encompass radiochemistry and nuclear chemistry, including Mössbauer spectroscopy and actinide coordination chemistry.

*Editorial*

# Synthesis and Applications of New Spin Crossover Compounds

Takafumi Kitazawa [1,2]

1. Department of Chemistry, Toho University, Chiba 274-8510, Japan; kitazawa@chem.sci.toho-u.ac.jp
2. Research Centre for Materials with Integrated Properties, Toho University, Chiba 274-8510, Japan

Received: 17 July 2019; Accepted: 23 July 2019; Published: 25 July 2019

---

The spin crossover (SCO) between multi-stable states in transition metal material is one of the attractive molecular switching phenomena which is responsive to various external stimuli such as temperature, pressure, light, electromagnetic field, radiation, nuclear decay, soft-X-ray, guest molecule inclusion, chemical environments and so forth. The light induced excited spin state trapping (LIESST) effect, the nuclear decay induced excited spin state trapping (NIESST) effect and the soft X-ray induced excited spin state trapping (SOXIESST) effect are associated with the SCO phenomena.

The crystal chemistry of SCO behavior in inorganic crystal materials might be able to be potentially associated with smart materials and promising materials for applications as components of memory devices, displays, sensors and mechanical devices and, especially, actuators such as artificial muscles. This is possible after Cambi and colleagues' pioneering research on the anomalous magnetic behaviors of mononuclear.

The Fe(III) coordination complexes [1] was first demonstrated as SCO phenomena in the early 1930s. Further, significant and fundamental scientific attention has been focused on the SCO phenomena in a wide research range of fields of fundamental chemical and physical and related sciences [2]. The interdisciplinary regions of chemical and physical sciences related to the SCO phenomena are also important.

The Special Issue is devoted to various aspects of the SCO and related research containing 18 interesting original papers on valuable and important SCO topics.

Regarding the interdisciplinary regions related to SCO research, impurity-induced spin-state crossover in $La_{0.8}Sr_{0.2}Co_{1-x}Al_xO$ was reported. However, the spin-state crossover also semi-quantitatively explained the enhanced thermopower and the anomalously large coercive field induced by the substituted Al ion [3]. The classic SCO is impossible in Cu(II) complexes with diamagnetic ligands, including the diamagnetic structural analogs to nitroxides which link to solid SCO-like phenomena of heterospin coordination compounds based on copper hexafluoroacetylacetonate $[Cu(hfac)_2]$ with nitronylnitroxide radicals. This was described because they can undergo structural transformations accompanied by spin transitions induced by external effects [4]. The SCO behavior of cobalt(II) terpyridin-4'-yl nitroxide complex as an exchange-coupled SCO material was reported as a successful example of multifunctional SCO materials with combining magnetic exchange coupling interactions [5]. A charge-transfer phase transition (CTPT) accompanied by an electron transfer between adjacent $Fe^{II}$ and $Fe^{III}$ sites was also reported in relation to the dithiooxalato-bridged iron mixed-valence complex [6].

The octahedral Fe(II) SCO systems with $3d^6$, which can be transited between the diamagnetic $(t_{2g})^6$ and the paramagnetic $(t_{2g})^4(e_g)^2$ configuration, are able to be widely and deeply investigated as potentially smart materials. The temperature dependence of the mosaicity for 5 thermo-induced iron(II) SCO compounds were investigated using X-ray diffraction, as the volume of high-spin (HS) and low-spin (LS) crystal packings are known to be very different [7]. Regarding the solvent effects of the SCO, the effects of lattice solvent on the solid-state SCO of a dinuclear Fe(II) triple helicate complex series [8] and SCO Fe(II) imidazolylimine complexes [9] were reported in supramolecular crystal systems with delicate and subtle host-guest interactions. The synthesis, crystal chemistry, and photomagnetic properties of the SCO complexes with [Fe(bpp)$_2$]$^{2+}$ were researched.in a 3D supramolecular architecture, including hydrogen bonds between iron(II) complexes, nicotinate anions, and water molecules [10]. The 1,2,3-triazole-containing polydentate ligand iron(II) SCO family into a linear pentadentate ligand system was reported. This was shown in an abrupt and incomplete HT SCO at approximately 400 K while the SCO transition was irreversible due to the crystal-to-amorphous transformation in association with the loss of the lattice MeCN solvent molecules [11]. A series of SCO Fe(II) complexes based on dipyridyl-N-alkylamine and thiocyanate ligands were investigated, and the higher SCO transition temperature explained the more pronounced linearity of the Fe–N–C angles in the crystal recently indicated by experimental and theoretical magneto-structural research [12].

A particularly successful and potentially developing synthetic kingdom for SCO iorn(II) polymeric complexes with valuable and sophisticated functional crystal properties are the SCO Hofmann-type coordination polymers. The first compound of this type Fe(pyridine)$_2$Ni(CN)$_4$ reported in 1996 [13], opened various roads to a number of Hofmann-like SCO compounds with a large display of functional properties. The special issue contains 4 original research articles which are devoted to the synthesis and characterizations of various Hofmann-like polymeric systems. The optical microscopy technique to investigate the thermal and the spatio-temporal properties of the Hofmann-related SCO single crystal [Fe(2-pytrz)$_2${Pt(CN)$_4$}]·3H$_2$O was described to show a first-order SCO behavior from a full high-spin (HS) state at high temperatures to intermediate, high-spin low-spin (HS-LS) states [14]. The precise crystallographic investigation on the polymeric SCO Hofmann-like compound Fe(3,4-dimethyl-pyridine)$_2$[Ag(CN)$_2$]$_2$ was reported, and its temperature dependence was followed by the means of a single-crystal and powder X-ray diffraction [15]. These very important article reported in the special issue demonstrate a soft X-ray–induced excited spin state trapping (SOXEISST) effect in Hofmann-like SCO coordination polymers of Fe$^{II}$(4-methylpyrimidine)$_2$[Au(CN)$_2$]$_2$ and Fe$^{II}$(pyridine)$_2$[Ni(CN)$_4$] [16]. The emission Mössbauer spectra of $^{57}$Co-labelled Co(pyridine)$_2$Ni(CN)$_4$ indicated that $^{57}$Fe atoms were assumed to be trapped in the excited electronic state ($^5$T) by the nuclear decay induced excited spin state trapping (NIESST) effect [17]. The two SCO coordination polymers built up by the Hofmann-like frameworks combining Fe$^{II}$ octahedral ions, 4-cyanopyridine and [Au(CN)$_2$]$^-$ liner units were described exhibiting ferromagnetic interaction [18]. A single crystal X-ray structural analysis showed that polymeric [[Fe(NCS)$_2$(bpa)$_2$]·biphenyl]$_n$ and [[Co(NCS)$_2$(bpa)$_2$]·biphenyl]$_n$ had a chiral propeller structure of pyridines around the central metal, which was associated with crystal chemistry and their SCO phenomena for the [[Fe(NCX)$_2$(bpa)$_2$]_(guest)]$_n$ family [19].

The Fe(III) SCO compounds are also important and attractive compounds as smart materials with multifunctional properties. The influence of geometry and counterion effects in determining the spin states in an iron (III) complex [Fe(5F-sal$_2$333)]X was investigated using a crystal analysis, UV-Vis spectroscopy, SQUID and EPR spectroscopy. The R-sal$_2$333 ligands promoting SCO in Fe(III) sites both in the solid state and in solution was established [20]. The hybrid ion-pair crystals containing hexadentate [Fe(III)(3-OMesal$_2$-trien)]$^+$ SCO cationic coordination units and anionic gold complex units [Au(dmit)$_2$]$^-$ and [Au(dddt)$_2$]$^-$ were investigated by a single-crystal X-ray diffraction method, P-XRD, and SQUID measurements [21]. Fe(III) SCO compounds from qsal ligand (Hqsal = N-(8-quinolyl)salicylaldimine) were described. The optical conductivity spectra were calculated from the single-crystal reflection spectra, which were, to the best of their knowledge, the first optical conductivity spectra of SCO complexes [22].

Finally, the contribution of all the authors for sending their works is greatly appreciated. The Special Issue, "Synthesis and Applications of New Spin Crossover Compounds" presents a comprehensive report on the current work on SCO materials and will be interesting for the readers. The author is also deeply grateful to all the anonymous reviewers for their valuable suggestions and very dedicated evaluations, which have been very helpful for improving the quality of the Special Issue. The author thanks the editorial staff for their valuable efforts in the planning, review processes and publication of this Special Issue.

In addition, the readers' submission of their valuable papers to the Special Issue "Synthesis and Applications of New Spin Crossover Compounds (Volume II)" would be further appreciated.

**Conflicts of Interest:** The author declares no conflict of interest.

## References

1. Cambi, L.; Szegö, L. Uber die magnetische Susceptibilitat der komplexen Verbindungen. *Ber. Dtsch. Chem. Ges.* **1931**, *64*, 2591–2598. [CrossRef]
2. Takahashi, K. Spin-Crossover Complexes. *Inorganics* **2018**, *6*, 32. [CrossRef]
3. Terasaki, I.; Ikuta, M.; Yamamoto, T.D.; Taniguchi, H. Impurity-Induced Spin-State Crossover in $La_{0.8}Sr_{0.2}Co_{1-x}Al_xO_3$. *Crystals* **2018**, *8*, 411. [CrossRef]
4. Artiukhova, N.; Romanenko, G.; Letyagin, G.; Bogomyakov, A.; Veber, S.; Minakova, O.; Petrova, M.; Morozov, V.; Ovcharenko, V. Spin Transition in the $Cu(hfac)_2$ Complex with (4-Ethylpyridin-3-yl)-Substituted Nitronyl Nitroxide Caused by the "Asymmetric" Structural Rearrangement of Exchange Clusters in the Heterospin Molecule. *Crystals* **2019**, *9*, 285. [CrossRef]
5. Ondo, A.; Ishida, T. Cobalt(II) Terpyridin-4′-yl Nitroxide Complex as an Exchange-Coupled Spin-Crossover Material. *Crystals* **2018**, *8*, 155. [CrossRef]
6. Enomoto, M.; Ida, H.; Okazawa, A.; Kojima, N. Effect of Transition Metal Substitution on the Charge-Transfer Phase Transition and Ferromagnetism of Dithiooxalato-Bridged Hetero Metal Complexes, $(n-C_3H_7)_4N[Fe^{II}_{1-x}Mn^{II}_xFe^{III}(dto)_3]$. *Crystals* **2018**, *8*, 446. [CrossRef]
7. Lakhloufi, S.; Tailleur, E.; Guo, W.; Le Gac, F.; Marchivie, M.; Lemée-Cailleau, M.-H.; Chastanet, G.; Guionneau, P. Mosaicity of Spin-Crossover Crystals. *Crystals* **2018**, *8*, 363. [CrossRef]
8. Craze, A.R.; Bhadbhade, M.M.; Kepert, C.J.; Lindoy, L.F.; Marjo, C.E.; Li, F. Solvent Effects on the Spin-Transition in a Series of Fe(II) Dinuclear Triple Helicate Compounds. *Crystals* **2018**, *8*, 376. [CrossRef]
9. Sertphon, D.; Harding, P.; Murray, K.S.; Moubaraki, B.; Neville, S.M.; Liu, L.; Telfer, S.G.; Harding, D.J. Solvent Effects on the Spin Crossover Properties of Iron(II) Imidazolylimine Complexes. *Crystals* **2019**, *9*, 116. [CrossRef]
10. Jornet-Mollá, V.; Giménez-Saiz, C.; Romero, F.M. Synthesis, Structure, and Photomagnetic Properties of a Hydrogen-Bonded Lattice of $[Fe(bpp)_2]^{2+}$ Spin-Crossover Complexes and Nicotinate Anions. *Crystals* **2018**, *8*, 439. [CrossRef]
11. Matsuyama, T.; Nakata, K.; Hagiwara, H.; Udagawa, T. Iron(II) Spin Crossover Complex with the 1,2,3-Triazole-Containing Linear Pentadentate Schiff-Base Ligand and the MeCN Monodentate Ligand. *Crystals* **2019**, *9*, 276. [CrossRef]
12. Houari, T.; Cuza, E.; Pinkowicz, D.; Marchivie, M.; Yefsah, S.; Triki, S. Iron(II) Spin Crossover (SCO) Materials Based on Dipyridyl-N-Alkylamine. *Crystals* **2018**, *8*, 401. [CrossRef]
13. Kitazawa, T.; Gomi, Y.; Takahasi, M.; Takeda, M.; Enomoto, M.; Miyazaki, A.; Enoki, T. Spin-crossover behaviour of the coordination polymer $Fe^{II}(C_5H_5N)_2Ni^{II}(CN)_4$. *J. Mater. Chem.* **1996**, *6*, 119–121. [CrossRef]
14. Fourati, H.; Bouchez, G.; Paez-Espejo, M.; Triki, S.; Boukheddaden, K. Spatio-temporal Investigations of the Incomplete Spin Transition in a Single Crystal of $[Fe(2\text{-pytrz})_2\{Pt(CN)_4\}]\cdot3H_2O$: Experiment and Theory. *Crystals* **2019**, *9*, 46. [CrossRef]
15. Rodríguez-Velamazán, J.A.; Kitase, K.; Palacios, E.; Castro, M.; Fernández-Blanco, Á.; Burriel, R.; Kitazawa, T. Structural Insights into the Two-Step Spin-Crossover Compound $Fe(3,4\text{-dimethyl-pyridine})_2[Ag(CN)_2]_2$. *Crystals* **2019**, *9*, 316. [CrossRef]
16. Mohamed, A.Y.; Lee, M.; Kitase, K.; Kitazawa, T.; Kim, J.-Y.; Cho, D.-Y. Soft X-ray Absorption Spectroscopy Study of Spin Crossover Fe-Compounds: Persistent High Spin Configurations under Soft X-ray Irradiation. *Crystals* **2018**, *8*, 433. [CrossRef]
17. Sato, T.; Ambe, F.; Kitazawa, T.; Sano, H.; Takeda, M. Conversion of the Valence States of $^{57}Fe$ Atoms Produced in $^{57}Co$-labelled $[Co(pyridine)_2Ni(CN)_4]$. *Chem. Lett.* **1997**, *26*, 1287. [CrossRef]
18. Kosone, T.; Tomori, I.; Akahoshi, D.; Saito, T.; Kitazawa, T. New Iron(II) Spin Crossover Complexes with Unique Supramolecular Networks Assembled by Hydrogen Bonding and Intermetallic Bonding. *Crystals* **2018**, *8*, 415. [CrossRef]

19. Tokinobu, S.; Dote, H.; Nakashima, S. Threefold Spiral Structure Constructed by 1D Chains of [[M(NCS)$_2$(bpa)$_2$]·biphenyl]$_n$ (M = Fe, Co; bpa = 1,2-bis(4-pyridyl)ethane). *Crystals* **2019**, *9*, 97. [CrossRef]
20. Sundaresan, S.; Kühne, I.A.; Kelly, C.T.; Barker, A.; Salley, D.; Müller-Bunz, H.; Powell, A.K.; Morgan, G.G. Anion Influence on Spin State in Two Novel Fe(III) Compounds: [Fe(5F-sal$_2$333)]X. *Crystals* **2019**, *9*, 19. [CrossRef]
21. Spitsyna, N.G.; Shvachko, Y.N.; Starichenko, D.V.; Lahderanta, E.; Komlev, A.A.; Zorina, L.V.; Simonov, S.V.; Blagov, M.A.; Yagubskii, E.B. Evolution of Spin-Crossover Transition in Hybrid Crystals Involving Cationic Iron Complexes [Fe(III)(3-OMesal$_2$-trien)]$^+$ and Anionic Gold Bis(dithiolene) Complexes Au(dmit)$_2$ and Au(dddt)$_2$. *Crystals* **2018**, *8*, 382. [CrossRef]
22. Takahashi, K.; Yamamoto, K.; Yamamoto, T.; Einaga, Y.; Shiota, Y.; Yoshizawa, K.; Mori, H. High-Temperature Cooperative Spin Crossover Transitions and Single-Crystal Reflection Spectra of [Fe$^{III}$(qsal)$_2$](CH$_3$OSO$_3$) and Related Compounds. *Crystals* **2019**, *9*, 81. [CrossRef]

© 2019 by the author. Licensee MDPI, Basel, Switzerland. This article is an open access article distributed under the terms and conditions of the Creative Commons Attribution (CC BY) license (http://creativecommons.org/licenses/by/4.0/).

*Article*

# Structural Insights into the Two-Step Spin-Crossover Compound Fe(3,4-dimethyl-pyridine)$_2$[Ag(CN)$_2$]$_2$

José Alberto Rodríguez-Velamazán [1,*], Kosuke Kitase [2], Elías Palacios [3], Miguel Castro [3], Ángel Fernández-Blanco [1], Ramón Burriel [3] and Takafumi Kitazawa [2]

1. Institut Laue-Langevin, 71 Avenue des Martyrs, CS 20156-38042 Grenoble, France; fernandez-blanco@ill.eu
2. Department of Chemistry, Faculty of Science, Toho University, Miyama, Funabashi, Chiba 274-8510, Japan; 6117004k@st.toho-u.ac.jp (K.K.); kitazawa@chem.sci.toho-u.ac.jp (T.K.)
3. Instituto de Ciencia de Materiales de Aragón (ICMA), CSIC—Universidad de Zaragoza, 50009 Zaragoza, Spain; elias@unizar.es (E.P.); mcastro@unizar.es (M.C.); burriel@unizar.es (R.B.)
* Correspondence: velamazan@ill.eu; Tel.: +33-(0)47-620-7260

Received: 24 May 2019; Accepted: 16 June 2019; Published: 19 June 2019

**Abstract:** The crystal structure of the polymeric spin crossover compound Fe(3,4-dimethyl-pyridine)$_2$[Ag(CN)$_2$]$_2$ has been solved and its temperature dependence followed by means of single-crystal and powder X-ray diffraction. This compound presents a two-step spin transition with relatively abrupt steps centred at ca. 170 K and 145 K and a plateau at around 155 K. The origin of the two-step transition is discussed in light of these structural studies. The observations are compatible with a mostly disordered state between the two steps, consisting of mixing of high-spin and low-spin species, while weak substructure reflections in the mixed phase could indicate some degree of long-range order of the high-spin and low-spin sites.

**Keywords:** iron (II), spin crossover; X-ray diffraction; coordination polymers

## 1. Introduction

The quest for switchable magnetic compounds represents one of the major topics in the field of molecular magnetism. One of the most typical examples of molecular switching in the solid state is that of spin-crossover (SCO) compounds. These systems can interconvert between low-spin (LS) and high-spin (HS) states as a result of different stimuli, like a change in temperature, pressure, the application of electric and magnetic fields, by light irradiation, or by adsorption/desorption of guest molecules [1–5].

It is well-known that the SCO phenomenon is explained from ligand-field theory considerations [6]. The LS state is the ground-state of SCO compounds, being the stable state at low temperatures and involving high ligand-field strength, while the HS state becomes stable at high temperatures and involves low ligand-field strength [7]. Usually, the LS to HS spin transition encompasses an increase of the metal–ligand bond lengths and, indeed, the interplay between the ligand field strength dependence on the metal–ligand distance and the electron–electron repulsion can be considered the mainspring for the SCO [6]. Most commonly, spin crossover compounds contain Fe(II) in a 3d$^6$ electronic configuration, the spin transition occurring between a diamagnetic LS state (S = 0) and a paramagnetic HS state (S = 2), and a number of them present an N$_6$ first coordination sphere. For these compounds, the change in the Fe–N distances from the LS to the HS state is typically ~0.2 Å [8,9]. The structure around the central ions has therefore a particular relevance for understanding the properties of these compounds [10].

A particularly successful synthetic route for spin-crossover compounds with interesting functional properties is that of the SCO Hofmann-type coordination polymers, where cyanometalate complexes connect the SCO centres forming extended networks with different dimensionalities and topologies [11,12]. Since the first compound of this type reported by Prof. Kitazawa in

1996 [13], a number of SCO compounds with a large display of functional properties have been described [3,14]. Some time ago we reported the synthesis and characterisation of the 2D spin-crossover compound Fe(py)$_2$[Ag(CN)$_2$]$_2$ (py = pyridine) [15], and later we described the magnetic, photomagnetic, calorimetric, Mössbauer and reflectivity studies of the derived Fe(X-py)$_2$[Ag(CN)$_2$)]$_2$ family (X = H; 3-methyl; 4-methyl; 3,4-dimethyl; 3-Cl) [16]. The main characteristic of this series is the two-step character of the transitions in most of the compounds, which is also observed in the relaxation of the photoinduced HS metastable state. The compounds derived by halogen substitution in the position 3 of the pyridine, Fe(3-Xpy)$_2$[M(CN)$_2$]$_2$ (X = F, Cl, Br, I and M = Ag, Au) have been reported by the group of Prof. J. A. Real [17,18]. The 3-Fpy derivatives also present two-step transitions (at ambient pressure for the Ag compound and at 0.18–0.26 GPa for the Au one). In turn, the use of the bicyclic ligand 1,6-naphthyridine, has allowed a wide thermal hysteresis centred near room temperature to be obtained [19]. In other closely related systems, also derived from [M$^I$(CN)$_2$]$^-$ (M$^I$ = Ag, Au) building blocks, like Fe(3-F-4-methyl-py)$_2$[Au(CN)$_2$]$_2$ [20] or Fe(4-(3-pentyl)py)$_2$[Au(CN)$_2$]$_2$ [21], two-step transitions are also observed. Finally, opening the focus to the Fe(II) Hofmann-like coordination polymers constituted of [M$^{II}$(CN)$_4$]$^{2-}$ (MII = Ni, Pd, Pt) building blocks, multistep spin transitions are also found [22,23]. Therefore, it is worth trying to shed light on the origin of this frequent feature in this type of compounds. In some of the cited cases, a structural inequivalence of different SCO sites in the intermediate phases is at the origin of the multistep behaviour [22]. But, unfortunately, there is a lack of precise structural information of some of these systems due to the difficulties in obtaining single crystals. In this work we report on the structure of Fe(3,4-dimethyl-pyridine)$_2$[Ag(CN)$_2$]$_2$. This compound shows a clear two-step behaviour in the magnetic susceptibility and heat capacity curves in the region of the spin transition [16], which starts at ca. 200 K and spans in a temperature range of around 90 K with a plateau at around 155 K (Figure 1a). The relatively abrupt steps are centred at ca. 170 K and 145 K, with around 47% and 53% of spin conversion, respectively, and with practical absence of residual fraction of high-spin species at low temperature [16]. In this work, the crystal structure of Fe(3,4-dimethyl-pyridine)$_2$[Ag(CN)$_2$]$_2$ has been solved and its temperature dependence followed by means of single-crystal and powder X-ray diffraction.

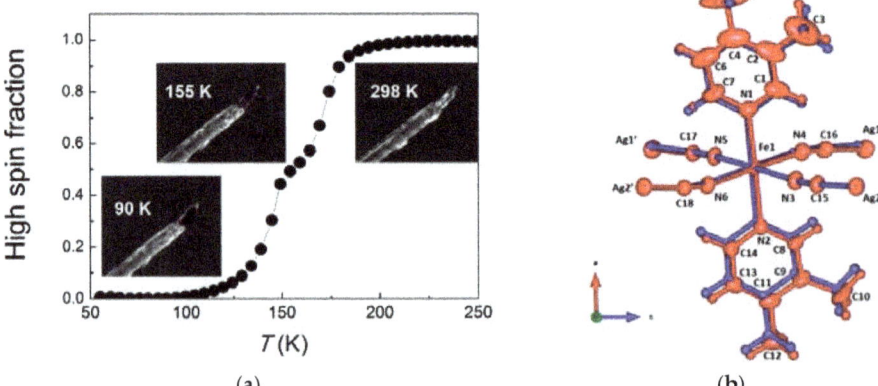

**Figure 1.** (a) Temperature dependence of the fraction of centres in the high-spin state for Fe(3,4-dimethyl-pyridine)$_2$[Ag(CN)$_2$]$_2$ deduced from magnetic susceptibility measurements (from Reference [16]) and photographs of the crystals at the temperatures of the structural determination. (b) Representation of a fragment of the structure of Fe(3,4-dimethyl-pyridine)$_2$[Ag(CN)$_2$]$_2$ at room temperature (red) and 90 K (blue) containing its asymmetric unit and atom numbering (excepting H atoms, for clarity). Thermal ellipsoids of non-hydrogen atoms are represented at 40% probability (H atoms are represented as spheres with arbitrary radius).

## 2. Materials and Methods

Fe(3,4-dimethyl-pyridine)$_2$[Ag(CN)$_2$]$_2$ was synthesized as described previously [15,16]. Single crystals were formed by a slow diffusion method after more than 2 days at room temperature from an aqueous mixture of (i) FeSO$_4$(NH$_4$)$_2$SO$_4$·6H$_2$O and 3,4-dimethyl-pyridine in 10 mL of water and (ii) K[Ag(CN)$_2$] in 10 mL of water. The colour of the samples changed from white (powder)/colourless (crystals) at room temperature (HS) to purple at low temperature (LS). The elemental analysis of the samples confirmed the organic contents.

The crystal structures were determined by single-crystal X-ray diffraction using a BRUKER APEX SMART CCD (Bruker, Billerica, MA, USA) area-detection diffractometer with monochromatized Mo $K\alpha$ radiation ($\lambda$ = 0.71 Å) at 298 K, 155 K and 90 K. The diffraction data were treated using SMART and SAINT (Bruker (2012). Bruker AXS Inc., Madison, WI, USA), while the absorption corrections were performed using SADABS [24]. The structures were solved by direct methods with SHELXTL [25]. All non-hydrogen atoms were refined anisotropically, and the hydrogen atoms were generated geometrically. Graphical representations of the structures were produced using the program VESTA [26]. COD 3000244 and 3000245 contain the Supplementary Materials crystallographic data for this paper. These data can be obtained free of charge via http://www.crystallography.net/search.html.

Powder X-ray diffraction patterns were recorded at room temperature, 210 K, 155 K and 80 K using a D-max Rigaku system, with a Cu rotating anode generator operated at 35 kV, 80 mA and a graphite monochromator to select the CuK$\alpha$ radiation. Step-scanned patterns were measured between $2\Theta$ = 5° and 60° (in steps of 0.03°). Data have been analysed using the FULLPROF program [27].

## 3. Results and Discussion

The crystal structure of Fe(3,4-dimethyl-pyridine)$_2$[Ag(CN)$_2$]$_2$ in both the HS and LS states was determined by single-crystal X-ray diffraction at room temperature and 90 K, respectively. At these temperatures the crystals of this compound are respectively colourless and purple (Figure 1a), which is an indication of the spin transition. An additional structural determination was performed in the plateau region, at 155 K (see below). The compound crystallizes in the monoclinic $P2_1/c$ space group, which is maintained after the spin transition. There is only one crystallographically independent Fe(II) site in both high-temperature and low-temperature structures. The Fe(II) atom is surrounded by six N atoms corresponding to four CN groups belonging to two crystallographically independent [Ag(CN)$_2$]$^-$ units in the equatorial positions and two 3,4-dimethyl-pyridine ligands in the axial ones (Figure 1b).

At room temperature, we distinguish two types of Fe–N bond distances, the axial ones [Fe–N(1) = 2.211(4) Å; Fe–N(2) = 2.208(4) Å] being longer than the equatorial ones [Fe–N(3) = 2.155(4) Å; Fe–N(4) = 2.162(4) Å; Fe–N(5) = 2.162(4) Å; Fe–N(6) = 2.180(4) Å] (Figure 1). These Fe–N distances were consistent with a HS state. The two independent [Ag(CN)$_2$]$^-$ units were very similar and display an almost linear geometry; the C–Ag–C bond angle is in the 175.7(6)–177.9(2)° range. These groups connect two iron atoms defining chains parallel to the diagonals of the $bc$-plane, thus forming two-dimensional corrugated layers (Figure 2). The layers are organized in pairs, held together by strong argentophilic interactions (Figure 2), which are attractive interactions occurring between silver centres when two or more low-coordinated silver cations appear in pairs or groups in molecular structures with distances between them lower than the sum of van der Waals radii of Ag atoms (3.44 Å) [28]. The Ag–Ag distance in the double layer is 2.9853(7) Å, while the distance between double layers is 7.1273(8) Å. The two-dimensional mesh is penetrated by the 3,4-dimethyl-pyridine ligands from the upper and lower layers. The stacking of the pyridine rings was held by $\pi$–$\pi$ interactions. Some of the displacement ellipsoids of C atoms, in particular those of the methyl groups, are significantly larger than in the rest of the structure (Figure 1), probably a consequence of some degree of disorder (static or dynamic) in these groups involving uncertainty in the position of H atoms.

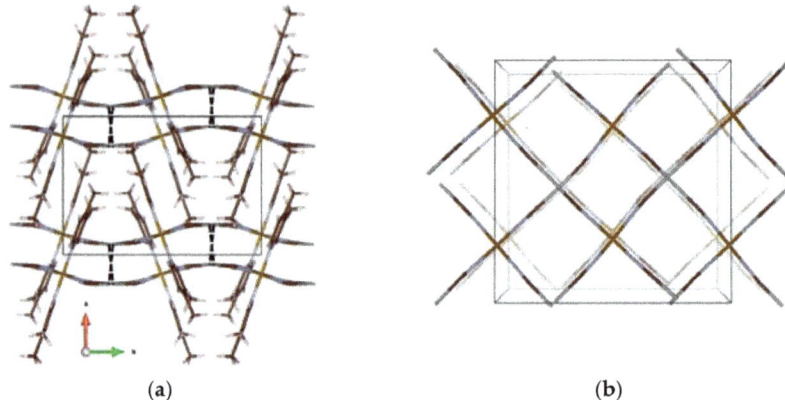

(a) (b)

**Figure 2.** View of the relative disposition of the layers of Fe(3,4-dimethyl-pyridine)$_2$[Ag(CN)$_2$]$_2$. (a) Stacking of four consecutive layers along the [001] direction. The dashed lines show the argentophilic interactions. (b) View of the relative disposition of two grids along the [100] direction. Colour code: iron: yellow; silver: grey; nitrogen: light blue; carbon: brown; hydrogen: light pink.

At 90 K, the structure is isostructural with that at room temperature, but with significant changes related with the spin transition. An overall contraction of the cell is observed (Table 1), and is more relevant in the *bc*-plane (parallel to the 2D layers) than in the *a*-direction. The most significant changes are those associated with the contraction produced in the [FeN$_6$] coordination octahedron associated with the spin transition. The average Fe–N bond distance shortens by 0.219 Å, as expected for a complete SCO in an iron(II) complex. Again we distinguish two types of Fe–N bond distances, the axial ones [Fe–N(1) = 1.997(5) Å; Fe–N(2) = 2.001(4) Å], longer than the equatorial ones [Fe–N(3) = 1.939(4) Å; Fe–N(4) = 1.943(4) Å; Fe–N(5) = 1.942(4) Å; Fe–N(6) = 1.940(3) Å] (Figure 1). The contraction in the *bc*-plane results in slightly more corrugated 2D layers. The Ag–Ag distance in the double layer is reduced to 2.9263(6) Å, while the distance between double layers is slightly increased to 7.1300(11) Å as a consequence of the transition. The larger displacement ellipsoids observed at high temperature are no more present at low temperature; however, the methyl groups make a slight turn to accommodate in the LS structure (Figure 1).

**Table 1.** Crystallographic data for Fe(3,4-dimethyl-pyridine)$_2$[Ag(CN)$_2$]$_2$ at different temperatures.

| Formula | C$_{18}$ H$_{18}$ Ag$_2$ Fe N$_6$ | | |
|---|---|---|---|
| Formula weight | | 589.97 | |
| T (K) | 298 K | 155 K | 90 K |
| Crystal system | monoclinic | monoclinic | monoclinic |
| Space Group | P 2$_1$/c | P 2$_1$/c | P 2$_1$/c |
| Z | 4 | 4 | 4 |
| a (Å) | 10.0093(8) | 9.898(8) | 9.9550(13) |
| b (Å) | 15.0039(11) | 14.609(12) | 14.2900(19) |
| c (Å) | 14.8435(11) | 14.573(12) | 14.3979(19) |
| β (°) | 91.4390(10) | 91.356(10) | 91.090(2) |
| V (Å$^3$) | 2228.5(3) | 2107(3) | 2047.8(5) |
| Fe–N(1) (Å) | 2.211(4) | 2.09(2) | 1.997(5) |
| Fe–N(2) (Å) | 2.208(4) | 2.08(2) | 2.001(4) |
| Fe–N(3) (Å) | 2.155(4) | 2.01(2) | 1.939(4) |
| Fe–N(4) (Å) | 2.162(4) | 2.08(3) | 1.943(4) |
| Fe–N(5) (Å) | 2.162(4) | 2.03(3) | 1.942(4) |
| Fe–N(6) (Å) | 2.180(4) | 2.04(3) | 1.940(3) |
| No. of measured, independent and observed [$I > 2\sigma(I)$] reflections ($R_{int}$) | 10391, 3404, 2672 (0.023) | 9951, 3900, 2788 (0.084) | 13965, 4884, 4043 (0.030) |
| $R[F^2 > 2\sigma(F^2)]$, $wR(F^2)$, S | 0.030, 0.067, 1.04 | 0.109, 0.313, 1.85 | 0.040, 0.141, 1.01 |

The evolution with temperature of the crystal structure of Fe(3,4-dimethyl-pyridine)$_2$[Ag(CN)$_2$]$_2$ across the spin transition was followed by powder X-ray diffraction. The pattern at room temperature perfectly agrees with the structure determined by single-crystal diffraction (Figure 3) without any modification of the structural parameters. On lowering the temperature, the pattern at 210 K (just above the transition) closely resembles the diffractogram at room temperature. Changes in some intensities are observed at 155 K (in the plateau of the spin transition), in particular in the 25° < 2Θ < 30° region, with the pattern appearing as a mixture of those at room temperature and 80 K, where the diffractogram again corresponds with the structure determined by single-crystal diffraction in the LS state at 90 K. The cell volume decreases continuously with temperature, with the contraction being more pronounced in the region of the spin transition. The volume decrease is mainly due to the contraction in the *bc*-plane, with the *b* and *c* axes showing little variation between room temperature and 210 K, and decreasing significantly with the spin transition (ca. 4%). The *a*-axis shows the opposite behaviour, with a relatively smaller overall contraction (ca. 0.6 % in the whole temperature range) and only a slight variation in the region of the spin transition; in particular, an anomalous increase is observed in the *a*-axis in the low temperature step of the transition (the axial Fe–N bonds are approximately in this direction). The evolution of the lattice parameters can be observed by following the position of the reflections (0 2 2)—*bc*-plane—and (2 0 0) —*a*-axis—in the diffraction patterns shown in the inset of Figure 3-left, while the lattice parameters and cell volume obtained from the fit of the diffractograms are shown in Figure 3-middle and right, respectively.

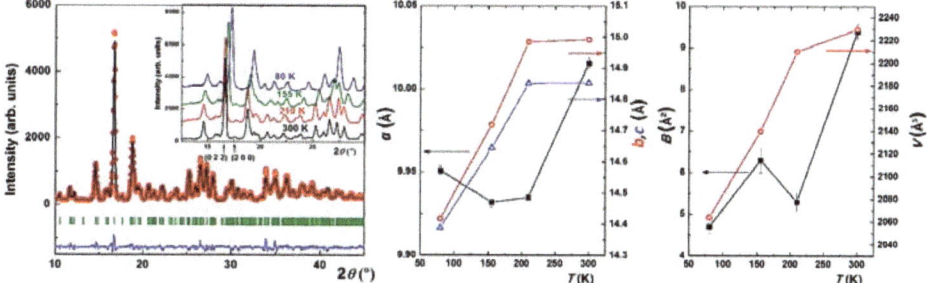

**Figure 3.** **Left**: Powder X-ray diffraction pattern of Fe(3,4-dimethyl-pyridine)$_2$[Ag(CN)$_2$]$_2$ at room temperature. Experimental (red symbols) and calculated patterns (black line), and Bragg peak positions (vertical green marks). Inset: detail of the experimental patterns at room temperature, 210 K, 155 K and 80 K; the arrows indicate the positions of the (0 2 2) and (2 0 0) reflections. **Middle**: lattice parameters obtained by powder X-ray diffraction vs. temperature. **Right**: thermal displacement parameter and cell volume vs. temperature.

From the magnetic measurements, we can infer that the proportion of iron centres in the HS and LS states in the plateau region is around one-half/one-half (see Figure 1a and Reference [16]). Nothing in the powder diffraction patterns suggests that these are distributed in an ordered way, which, together with the fact that a single crystallographically independent Fe site is present in both the HS and LS structures, rather supports a scenario with an equal mixture of HS and LS species not ordered at long range, in the same way as in the Fe(pyridine)$_2$[Ag(CN)$_2$]$_2$ compound [15]. In line with this scenario, if we refine an overall "thermal" displacement factor, *B*, as a function of temperature (together with the cell parameters, but keeping invariant the rest of the structural parameters determined by single-crystal diffraction), we observe an anomalous increase in the region of the plateau of the spin transition (Figure 3-right), which is consistent with having an average of Fe(II) ions in two different spin states. The spin transition involves a significant change in the Fe–N distance, and therefore the anomalous "thermal" factor is in this case due to positional disorder caused by the presence of HS and LS species in the same crystallographic position.

The single-crystal data at 155 K can be interpreted as well as a structure described in the monoclinic $P2_1/c$ space group, with only one crystallographically independent Fe(II) site. However, the quality of the refinement is poorer as a result of the disorder related with the coexistence of HS and LS states and the presence of twinning. In particular, some atomic displacement ellipsoids present an abnormal shape, due to the uncertainty in the distances, specifically those that are more affected by the change in spin state. In agreement with the powder diffraction results, the $a$-axis appears anomalously more contracted than at 90 K, although the cell volume follows a continuous decrease with temperature. In the single-crystal data, there are some weak but observed reflections at 155 K of type (h0l) with l = odd, i.e., forbidden by the $P2_1/c$ space group (unobserved in the powder diffraction data). These reflections would indicate two non-equivalent Fe sites, and some degree of order in the position of the atoms in HS and LS state, although crystal twinning could account (at least in part) for these forbidden reflections. A refinement with the triclinic space group $P$-1 was also tested. The refined cell has alpha = 90.11(2)°, beta = 91.53(1)°, gamma = 91.22(2)°. The results in the $P$-1 group show very slightly shorter Fe–N distances in one of the two inequivalent Fe sites, being in all cases intermediate between those at room temperature and at 90 K. This can be interpreted as though, most probably, a symmetry breaking occurs in the plateau region implying some order of HS and LS species, but cannot be clearly resolved due to a low degree of long-range ordering (disorder) and twinning. Refining the data in the $P$-1 group gives very similar results as for the $P2_1/c$ group, and the use of the less symmetric group is therefore not clearly justified. Then, the most plausible situation is a distribution of HS and LS species with a low degree of long-range order (and probably with some short-range order, but the present diffraction experiments are not sensitive to it). This kind of mostly disordered state has been explained for previously reported examples [15,29] as a result of competing long and short-range ferro- and antiferro-elastic interactions, respectively, between SCO centres, transmitted via the bonds of the polymeric network.

## 4. Conclusions

We have investigated the origin of the two-step spin-crossover transition of Fe(3,4-dimethyl-pyridine)$_2$[Ag(CN)$_2$]$_2$ by means of single-crystal and powder X-ray diffraction. The structure has been solved in both high-spin (room temperature) and low-spin states (90 K). The structure is described in the monoclinic $P2_1/c$ space group, which is maintained across the spin transition, with a single crystallographically independent Fe(II) site. The observations in the plateau region between the two steps of the spin transition (155 K) are compatible with a state between the two steps consisting of an equal mixture of high-spin and low-spin species with a low degree of long-range order.

**Supplementary Materials:** Supplementary crystallographic data at 155 K are available online at http://www.mdpi.com/2073-4352/9/6/316/s1.

**Author Contributions:** Conceptualization, J.A.R.-V.; Data curation, K.K., Á.F.-B. and T.K.; Formal analysis, K.K. and E.P.; Funding acquisition, M.C. and R.B.; Investigation, J.A.R.-V., K.K., E.P. and T.K.; Supervision, J.A.R.-V., M.C., R.B. and T.K.; Visualization, Á.F.-B.; Writing—original draft, J.A.R.-V.; Writing—review & editing, E.P., M.C., Á.F.-B. and R.B.

**Funding:** This research was funded by the Spanish Ministerio de Ciencia, Innovación y Universidades (grant numbers MAT2015-68200-C02-2-P and MAT2017-86019-R) and Diputación General de Aragón (Project E11-17R).

**Acknowledgments:** The authors acknowledge the use of Servicios Generales de Apoyo a la Investigación from Universidad de Zaragoza and thank Takayuki Suda in Department of Chemistry, Toho University for his contribution in the preparation of the compound.

**Conflicts of Interest:** The authors declare no conflict of interest. The funders had no role in the design of the study; in the collection, analyses, or interpretation of data; in the writing of the manuscript, or in the decision to publish the results.

## References

1. Linert, W.; Verdaguer, M. *Molecular Magnets: Recent Highlights*; Springer: Wien, Austria, 2003.
2. Gutlich, P.; Goodwin, H.A. Topics in Current Chemistry. In *Spin-Crossover in Transition Metal Compounds*; Springer: Berlin, Germany, 2004.
3. Muñoz, M.C.; Real, J.A. Thermo-, piezo-, photo- and chemoswitchable spin crossover iron(II) metallocyanate based coordination polymers. *Coord. Chem. Rev.* **2011**, *255*, 2068–2093. [CrossRef]
4. Halcrow, M.A. *Spin-Crossover Materials: Properties and Applications*; John Wiley & Sons: Chichester, UK, 2013.
5. Senthil Kumar, K.; Ruben, M. Emerging trends in spin crossover (SCO) based functional materials and devices. *Coord. Chem. Rev.* **2017**, *346*, 176–205. [CrossRef]
6. Hauser, A. Ligand Field Theoretical Considerations. *Top. Curr. Chem.* **2004**, *233*, 49–58.
7. Gütlich, P.; Goodwin, H. Spin Crossover—An Overall Perspective. *Top. Curr. Chem.* **2004**, *233*, 1–47.
8. König, E. Nature and dynamics of the spin-state interconversion in metal complexes. *Struct. Bond.* **1991**, *76*, 51–152.
9. Collet, E.; Guionneau, P. Structural analysis of spin-crossover materials: From molecules to materials. *C. R. Chim.* **2018**, *21*, 1133–1151. [CrossRef]
10. Guionneau, P.; Marchivie, M.; Bravic, G.; Létard, J.-F.; Chasseau, D. Structural Aspects of Spin Crossover. Example of the [Fe$^{II}$Ln(NCS)$_2$] Complexes. *Top. Curr. Chem.* **2004**, *234*, 97–128.
11. Garcia, Y.; Niel, V.; Muñoz, M.C.; Real, J.A. Spin crossover in 1D, 2D and 3D polymeric Fe(II) networks. *Top. Curr. Chem.* **2004**, *233*, 229–257.
12. Sciortino, N.F.; Neville, S.M. Two-Dimensional Coordination Polymers with Spin Crossover Functionality. *Aust. J. Chem.* **2014**, *67*, 1553–1562. [CrossRef]
13. Kitazawa, T.; Gomi, Y.; Takahasi, M.; Takeda, M.; Enomoto, M.; Miyazaki, A.; Enoki, T. Spin-crossover behaviour of the coordination polymer Fe$^{II}$(C$_5$H$_5$N)$_2$Ni$^{II}$(CN)$_4$. *J. Mater. Chem.* **1996**, *6*, 119–121. [CrossRef]
14. Ni, Z.-P.; Liu, J.-L.; Hogue, N.; Liu, W.; Li, J.-Y.; Chen, Y.-C.; Tong, M.-L. Recent advances in guest effects on spin-crossover behavior in Hofmann-type metal-organic frameworks. *Coord. Chem. Rev.* **2017**, *335*, 28–43. [CrossRef]
15. Rodríguez-Velamazán, J.A.; Castro, M.; Palacios, E.; Burriel, R.; Kitazawa, T.; Kawasaki, T. A Two-Step Spin Transition with a Disordered Intermediate State in a New Two-Dimensional Coordination Polymer. *J. Phys. Chem. B* **2007**, *111*, 1256–1261. [CrossRef] [PubMed]
16. Rodríguez-Velamazán, J.A.; Carbonera, C.; Castro, M.; Palacios, E.; Kitazawa, T.; Létard, J.F.; Burriel, R. Two-Step Thermal Spin Transition and LIESST Relaxation of the Polymeric Spin-Crossover Compounds Fe(X-py)$_2$[Ag(CN)$_2$]$_2$ (X=H, 3-methyl, 4-methyl, 3,4-dimethyl, 3-Cl). *Chem. Eur. J.* **2010**, *16*, 8785–8796. [CrossRef] [PubMed]
17. Muñoz, M.C.; Gaspar, A.B.; Galet, A.; Real, J.A. Spin-Crossover Behavior in Cyanide-Bridged Iron(II)-Silver(I) Bimetallic 2D Hofmann-like Metal-Organic Frameworks. *Inorg. Chem.* **2007**, *46*, 8182–8192. [CrossRef] [PubMed]
18. Agustí, C.; Muñoz, M.C.; Gaspar, A.B.; Real, J.A. Spin-Crossover Behavior in Cyanide-bridged Iron(II)-Gold(I) Bimetallic 2D Hofmann-like Metal-Organic Frameworks. *Inorg. Chem.* **2008**, *47*, 2552–2561. [CrossRef] [PubMed]
19. Hiiuk, V.M.; Shova, S.; Rotaru, A.; Ksenofontov, V.; Fritsky, I.O.; Gural'skiy, I.A. Room temperature hysteretic spin crossover in a new cyanoheterometallic framework. *Chem. Commun.* **2019**, *55*, 3359–3362. [CrossRef] [PubMed]
20. Kosone, T.; Kawasaki, T.; Tomori, I.; Okabayashi, J.; Kitazawa, T. Modification of Cooperativity and Critical Temperatures on a Hofmann-Like Template Structure by Modular Substituent. *Inorganics* **2017**, *5*, 55. [CrossRef]
21. Kosone, T.; Kitazawa, T. Guest-dependent spin transition with long range intermediate state for 2-dimensional Hofmann-like coordination polymer. *Inorg. Chim. Acta* **2016**, *439*, 159–163. [CrossRef]
22. Sciortino, N.F.; Scherl-Gruenwald, K.R.; Chastanet, G.; Halder, G.J.; Chapman, K.W.; Letard, J.-F.; Kepert, C.J. Hysteretic Three-Step Spin Crossover in a Thermo- and Photochromic 3D Pillared Hofmann-type Metal-Organic Framework. *Angew. Chem. Int. Ed.* **2012**, *51*, 10154–10158. [CrossRef]

23. Kucheriv, O.I.; Shylin, S.I.; Ksenofontov, V.; Dechert, S.; Haukka, M.; Fritsky, I.O.; Gural'skiy, I.A. Spin Crossover in Fe(II)–M(II) Cyanoheterobimetallic Frameworks (M = Ni, Pd, Pt) with 2-Substituted Pyrazines. *Inorg. Chem.* **2016**, *55*, 4906–4914. [CrossRef]
24. Sheldrick, G.M. *SADABS Program for Empirical Absorption Correction for Area Detector Data*; University of Göttingen: Göttingen, Germany, 1996.
25. Sheldrick, G.M. *SHELXTL Program for the Solution of Crystal Structure*; University of Göttingen: Göttingen, Germany, 1997.
26. Momma, K.; Izumi, F. VESTA 3 for three-dimensional visualization of crystal, volumetric and morphology data *J. Appl. Crystallogr.* **2011**, *44*, 1272–1276. [CrossRef]
27. Rodríguez-Carvajal, J. Recent advances in magnetic structure determination by neutron powder diffraction. *Phys. B* **1993**, *192*, 55–69. [CrossRef]
28. Schmidbaur, H.; Schier, A. Argentophilic Interactions. *Angew. Chem. Int. Ed.* **2015**, *54*, 746–784. [CrossRef] [PubMed]
29. Romstedt, H.; Hauser, A.; Spiering, H.; Gutlich, P. Modelling of two step high spin⇌ low spin transitions using the cluster variation method. *J. Phys. Chem. Solids* **1998**, *59*, 1353–1362. [CrossRef]

© 2019 by the authors. Licensee MDPI, Basel, Switzerland. This article is an open access article distributed under the terms and conditions of the Creative Commons Attribution (CC BY) license (http://creativecommons.org/licenses/by/4.0/).

Article

# Spin Transition in the Cu(hfac)$_2$ Complex with (4-Ethylpyridin-3-yl)-Substituted Nitronyl Nitroxide Caused by the "Asymmetric" Structural Rearrangement of Exchange Clusters in the Heterospin Molecule

Natalia Artiukhova [1,2,*], Galina Romanenko [1], Gleb Letyagin [1,2], Artem Bogomyakov [1,2], Sergey Veber [1,2], Olga Minakova [1,2], Marina Petrova [1], Vitaliy Morozov [1,2] and Victor Ovcharenko [1]

1. International Tomography Center, SB RAS, Institutskaya Str., 3A, Novosibirsk 630090, Russia; romanenko@tomo.nsc.ru (G.R.); gl@tomo.nsc.ru (G.L.); bus@tomo.nsc.ru (A.B.); sergey.veber@tomo.nsc.ru (S.V.); o.minakova@g.nsu.ru (O.M.); petrovamv@tomo.nsc.ru (M.P.); moroz@tomo.nsc.ru (V.M.); Victor.Ovcharenko@tomo.nsc.ru (V.O.)
2. Novosibirsk State University, Pirogova St., 1, Novosibirsk 630090, Russia
* Correspondence: natalya.artyukhova@tomo.nsc.ru

Received: 29 April 2019; Accepted: 24 May 2019; Published: 1 June 2019

**Abstract:** Methods for the synthesis of binuclear [Cu(hfac)$_2$L$^{Et}$]$_2$ and tetranuclear [[Cu(hfac)$_2$]$_4$(L$^{Et}$)$_2$] heterospin compounds based on copper hexafluoroacetylacetonate [Cu(hfac)$_2$] and 2-(4-ethylpyridin-3-yl)-4,5-bis(spirocyclopentyl)-4,5-dihydro-1$H$-imidazole-3-oxide-1-oxyl (L$^{Et}$), were developed. The crystals of the complexes are elastic and do not crash during repeated cooling–heating cycles. It was found that a singlet–triplet conversion occurred in all of the {Cu(II)–O•–N<} exchange clusters in the molecules of the binuclear [Cu(hfac)$_2$L$^{Et}$]$_2$ which led to spin coupling with cooling. The transition occurred in a wide temperature range with a maximum gradient $\Delta\chi$T at ≈180 K. The structural transformation of the crystals takes place at T < 200 K and is accompanied by the lowering of symmetry from monoclinic to triclinic, twinning, and a considerable shortening of the Cu–O$_{NO}$ distance (2.19 and 1.97 Å at 295 and 50 K, respectively). For the tetranuclear [[Cu(hfac)$_2$]$_4$(L$^{Et}$)$_2$], two structural transitions were recorded (at ≈154 K and ≈118 K), which led to a considerable change in the spatial position of the Et substituent in the nitronyl nitroxyl fragment. The low-temperature process was accompanied by a spin transition recorded as a hysteresis loop on the $\chi$T(T) curve during the repeated cooling–heating cycles (T$\frac{1}{2}\uparrow$ = 122 K, T$\frac{1}{2}\downarrow$ = 115 K). This transition is unusual because it causes spin coupling in half of all of the {>N–•O–Cu$^{2+}$} terminal exchange clusters, leading to spin compensation for only two paramagnetic centers of the six centers in the molecule.

**Keywords:** spin crossover; Cu(II) complexes; nitroxides; phase transitions; magnetostructural correlations

## 1. Introduction

Solid phases of heterospin complexes based on copper hexafluoroacetylacetonate [Cu(hfac)$_2$] with nitronylnitroxide radicals are of interest because they can undergo structural transformations accompanied by spin transitions induced by external effects [1]. Since the classic spin crossover is impossible in Cu(II) complexes with diamagnetic ligands, including the diamagnetic structural analogs of nitroxides [2], an essential condition for this effect is the coordination of an additional paramagnetic center, that is, the formation of at least a two-center exchange cluster [3]. A spin-crossover-like phenomenon is observed when the external effect changes the mutual orientation of the paramagnetic centers and consequently the character of interaction of odd electrons. The thermally induced change in the distances between the paramagnetic centers in the {>N–•O–Cu(II)} or {>N–•O–Cu(II) –O•–N<} exchange clusters generally leads to an abrupt change in the energy and/or sign of the exchange interaction, which just gives rise to an anomaly on the curve of the temperature dependence of the magnetic susceptibility $\chi T(T)$.

Pyridyl-substituted nitronyl and imino nitroxides (Figure 1) were the first stable nitroxides for which spin transitions were recorded in their heterospin Cu(hfac)$_2$ complexes [4–8]. These transitions are generally reversible and often occur without destruction of the crystal (Single Crystal to Single Crystal transformations), due to which it is possible to trace the temperature dynamics of the structure and compare the structural features of the high- and low-temperature phases with the magnetic properties of the compound [9–21]. In the majority of these heterospin compounds, the magnetic anomalies are caused by strong antiferromagnetic exchange interactions in the {>N–•O–Cu(II)} exchange clusters, which appear at low temperatures and lead to full compensation of spins.

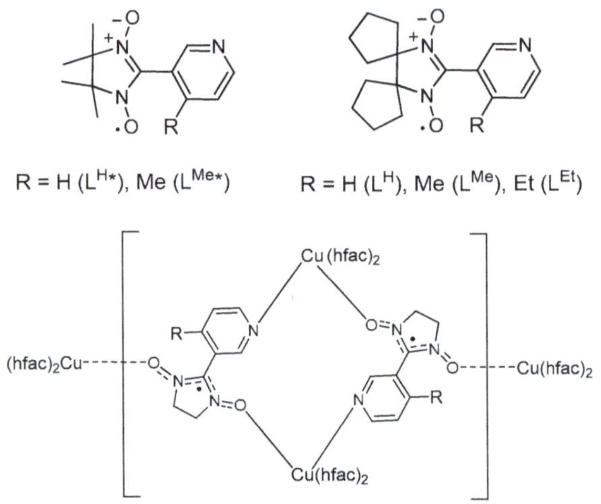

R = H (L$^{H*}$), Me (L$^{Me*}$)    R = H (L$^H$), Me (L$^{Me}$), Et (L$^{Et}$)

**Figure 1.** L$^R$ and L$^{R*}$ radicals; structure of the [[Cu(hfac)$_2$]$_4$(L$^R$)$_2$] tetranuclear complex. The fragment in square brackets is the [Cu(hfac)$_2$L]$_2$ binuclear fragment.

Earlier, it was shown that Cu(hfac)$_2$ forms binuclear and tetranuclear complexes with both L$^{R*}$ and L$^R$ (Figure 1); it also forms chain polymers {[[Cu(hfac)$_2$]$_2$L$_2$][Cu(hfac)$_2$]}$_\infty$, in which the binuclear fragments are linked into chains via additional Cu(hfac)$_2$ fragments. The spin transitions were recorded for [Cu(hfac)$_2$L$^H$]$_2$ [18], [Cu(hfac)$_2$L$^{Me}$]$_2$•Solv (Solv = n-C$_6$H$_{14}$, n-C$_{10}$H$_{22}$, n-C$_{16}$H$_{34}$) [9], [[Cu(hfac)$_2$]$_4$(L$^H$)$_2$] [18], and [[Cu(hfac)$_2$]$_4$(L$^{H*}$)$_2$] complexes [4,16]. We synthesized the binuclear [Cu(hfac)$_2$L$^{Et}$]$_2$ and tetranuclear [[Cu(hfac)$_2$]$_4$(L$^{Et}$)$_2$] and studied their structure and magnetic properties. The present paper reports the results of this study, namely, the spin transition found for [Cu(hfac)$_2$L$^{Et}$]$_2$ in a wide temperature range and an unusual spin transition for [[Cu(hfac)$_2$]$_4$(L$^{Et}$)$_2$] that is due to the coupling of only one third of the total number of spins in the molecule.

## 2. Materials and Methods

### 2.1. Chemical Materials

Cu(hfac)$_2$ [22], 1,1′-dihydroxylamino-bis-cyclopentyl sulfate monohydrate [18], and 4-ethyl nicotinaldehyde [23] used in the present study were synthesized by the known procedures.Commercial reagents and solvents were used without any additional purification. For chromatographic procedures,

TLC plates Silica Gel 60 $F_{254}$, aluminum sheets (Macherey-Nagel, Düren, Germany), and silica gel "0.063–0.200 mm for column chromatography" (Merck KGaA, Darmstadt, Germany) were used. The IR spectra of the samples pelletized with KBr were recorded on a VECTOR-22 spectrometer (Bruker, Karlsruhe, Germany). The melting points were determined on a heating table (VEB kombinat Nagema, Radebeul, DDR). The microanalyses were performed on an EuroEA-3000 CHNS analyser (HEKAtech, Wegberg, Germany).

*2.2. Synthesis of Compounds*

2-(4-Ethylpyridin-3-yl)-4,5-bis(spirocyclopentyl)-4,5-dihydro-1*H*-imidazole-3-oxide-1-oxyl ($L^{Et}$). 4-Ethylnicotinaldehyde (0.135 g, 1 mmol) was added to a solution of 1,1'-dihydroxylamino biscyclopentyl sulfate monohydrate (0.3 g, 1 mmol) in $H_2O$ (3 mL) at room temperature. The reaction mixture was stirred at 50 °C for 1 h and then kept at 4 °C for 1 day. The solution was neutralized with $NaHCO_3$ until the gas ceased to evolve. The resulting precipitate was filtered off, washed on an $H_2O$ filter, and dried in a vacuum box. This gave 2-(4-ethylpyridin-3-yl)-4,5-bis(spirocyclopentyl)-1,3-dihydroxy-imidazoline (0.250 g) in the form of a white powder, which was then used without any additional purification.

$MnO_2$ (1 g) was added to a suspension of the resulting adduct in MeOH (3 mL) while cooling it on a water bath, and the mixture was stirred for 1 h. The solution was filtered, the residue washed with MeOH, and the combined filtrates were evaporated. The residue was dissolved in EtOAc and filtered through a $SiO_2$ layer (2 ×5 cm). The eluate was evaporated on a rotor evaporator. The product was recrystallized from the $CH_2Cl_2$–hexane mixture, keeping it at 4 °C. The yield of $L^{Et}$ was 0.4 g (13%), as dark blue prismatic crystals. M.p. 90–92 °C (decomp.). IR spectrum $\nu/cm^{-1}$: 3553, 3416, 3237, 2968, 2944, 1638, 1618, 1590, 1392, 1320, 1177, 1131, 1032, 953, 844, 623, 481. Found (%): C, 68.7; H, 7.5; N, 13.5. $C_{18}H_{24}N_3O_2$. Calculated (%): C, 68.8; H, 7.7; N, 13.4.

$[Cu(hfac)_2L^{Et}]_2$. A mixture of $Cu(hfac)_2$ (0.048 g, 0.1 mmol) and $L^{Et}$ (0.030 g, 0.1 mmol) was dissolved in acetone (1 mL). Then, hexane (4 mL) was added and the solution was kept in a refrigerator at 4 °C for 1 day. The resulting brown prismatic crystals were filtered off, washed with cold hexane, and dried in air. Yield 56 mg (72%). Found (%):C, 42.7; H, 3.2; N, 5.4; F, 28.6. $C_{56}H_{52}Cu_4F_{24}N_6O_{12}$. Calculated (%): C, 42.5; H, 3.3; N, 5.3; F, 28.8.

$[[Cu(hfac)_2]_4(L^{Et})_2]$. A mixture of $Cu(hfac)_2$ (0.076 g, 0.160 mmol) and $L^{Et}$ (0.024 g, 0.08 mmol) was dissolved in $CH_2Cl_2$ (1 mL). Then, hexane (3 mL) was added and the solution was kept in an open vessel at 4 °C for 1 day. The brownish red crystalline precipitate in the form of square prisms was filtered off, washed with cold hexane, and dried in air. Yield 68 mg (68%). Found (%): C, 36.0; H, 2.2; N, 3.2; F, 36.1. $C_{76}H_{56}Cu_4F_{48}N_6O_{20}$. Calculated (%): C, 36.0; H, 2.2; N, 3.3; F, 35.9.

*2.3. Magnetic Measurements*

The magnetic susceptibility of the polycrystalline samples was measured on a MPMSXL SQUID magnetometer (Quantum Design, San Diego, CA, USA) in the temperature range 2–360 K in a magnetic field of up to 5 kOe. The paramagnetic components of the magnetic susceptibility χ were determined with allowance for the diamagnetic contribution evaluated from the Pascal constants. The magnetic properties were analyzed using spin-Hamiltonian in general form of $H = -2\sum J_{ij}S_iS_j$.

*2.4. Crystal Structure Determinations*

The arrays of reflections from single crystals were collected on Bruker (Bruker AXS GmbH, Karlsruhe, Germany) AXS-Smart Apex II (with a Helix low-temperature accessory, Oxford Cryosystems, Oxford, United Kingdom) and Apex Duo diffractometers (the absorption was included using the SADABS, version 2.10 program, Bruker AXS Inc, Karlsruhe, Germany.). The structures were solved by direct methods and refined by the full-matrix least-squares method in an anisotropic approximation for all non-hydrogen atoms. The positions of H atoms were calculated geometrically and refined in the riding model. All calculations on structure solution and refinement were performed with

SHELXL-2014/6 and SHELXL-2017/1 programs (Shelx, Göttingen, Germany) [24]. The selected bond lengths and crystal data for the compounds are listed in Tables S1 and S2, Supplementary Materials. Full information on the structures was deposited at CCDC (1907461-1907475) and can be requested at the site www/ccdc.cam.ac.uk/data_request/cif.

*2.5. IR-Spectrum Experiments*

The spectra were recorded over the range of 4000–700 cm$^{-1}$ using a HYPERION 2000 IR microscope (Bruker Optics, Ettlingen, Germany) equipped with a D316 MCT detector and coupled to a Bruker Vertex 80v FTIR spectrometer (Bruker Optics, Ettlingen, Germany). The spectral resolution was 1 cm$^{-1}$. A Linkam FTIR600 sample stage (Linkam Scientific Instruments, Surrey, United Kingdom) equipped with BaF$_2$ windows was used to control the temperature of the single crystal. The variable temperature FTIR (VT-FTIR) spectra of a thin single crystal of [[Cu(hfac)$_2$]$_4$(L$^{Et}$)$_2$] were recorded in the mid-IR range at 80–274 K at a step of 2 K. The probing area of the crystal was≈0.2×0.2 mm$^2$, which was slightly smaller than the crystal size.

*2.6. Density Functional Theory (DFT) Calculations*

The quantum-chemical calculations were performed in the Quantum Espresso 5.3 program package (MaX Centre of Excellence, Modena, Italy) for DFT band structure calculations in the plane wave basis by the pseudopotential method. Ultrasoft pseudopotentials with nonlinear core corrections (NLCC) and the Perdew–Burke–Ernzerhof (PBE) exchange correlation potential were used [25]. The cut off energy for the expansion of the electronic wave functions in plane waves was taken to be 35 Ry; for the charge density, the cutoff energy was 280 Ry. When integrating in the reciprocal space, we used 2 × 2 × 2 k-points in the Monkhorst–Pack grid [26] in the first Brillouin zone; the Gaussian broadening was chosen to be 0.136 eV. The Hubbard corrections on the copper and oxygen atoms were applied within the framework of the GGA+U scheme with the parameters Ud(Cu) = 9.8 eV and Up(O) = 5.0 eV, respectively [27,28]. In the band DFT approach, combined with the conventional BS method to obtain exchange integrals, we calculated the set of unit cell energies for a complete number of spin polarization states of the crystal unit cell. Such an approach allows one to cover the entire net of exchange interactions of the system under study, including both the exchange couplings between the spins inside the unit cell and the exchange couplings involving spins from adjacent unit cells. Then, we mapped the selected spin Hamiltonian (the form of spin Hamiltonians are discussed in Sections 3.6 and 3.7) onto the set of energy levels obtained above. Solving the resulting system of linear equations was the final step to find the desired set of exchange integrals, characterizing the spin Hamiltonian. A comparison of the theoretical and experimental χT(*T*) dependences shows in all cases that the calculated exchange integrals are slightly exaggerated (approximately twofold) because of the exaggerated spin density delocalization in the DFT calculation.

## 3. Results

*3.1. Synthesis of L$^{Et}$*

Nitronylnitroxide L$^{Et}$ was prepared by condensation of 4-ethylnicotinaldehyde with 1,1′-dihydroxylamino biscyclopentyl sulfate monohydrate followed by the oxidation of the resulting adduct with MnO$_2$ in EtOH (Scheme 1). The L$^{Et}$ single crystals were grown from a CH$_2$Cl$_2$–hexane mixture.

## 3.2. Magnetic and X-Ray Investigations for $L^{Et}$

The XRD study showed that in the $L^{Et}$ molecule, the angle between the planes of the pyridine ring and the O•–N–C=N→O fragment of the 2-imidazoline ring (∠Py–CN$_2$O$_2$) is 71.3°, and the ethyl group is almost perpendicular to the pyridine ring (Figure 2): the angle between the plane of the pyridine ring and the ethyl group (∠Py–Et) is 84.7°. The bond lengths in the N–O groups are 1.288(6) and 1.268(6) Å; the shortest distances between the paramagnetic centers in the structure exceed 4.5 Å. This agrees with $\chi$T of 0.336 K·cm$^3$/mol, which is almost constant at 30–300 K and close to the theoretical pure spin value of 0.375 K·cm$^3$/mol for one paramagnetic center with spin S = $\frac{1}{2}$ at g = 2. Antiferromagnetic exchange interactions between spins of the $L^{Et}$ molecules are negligibly small and cause small decreasing of $\chi$T only below 20 K. The exchange coupling parameter $J$ may be estimated as −1.37 cm$^{-1}$, using a dimer model (spin Hamiltonian $H = -2J·S_1 S_2$) for the analysis of the $\chi$T(T) dependence (Figure 2).

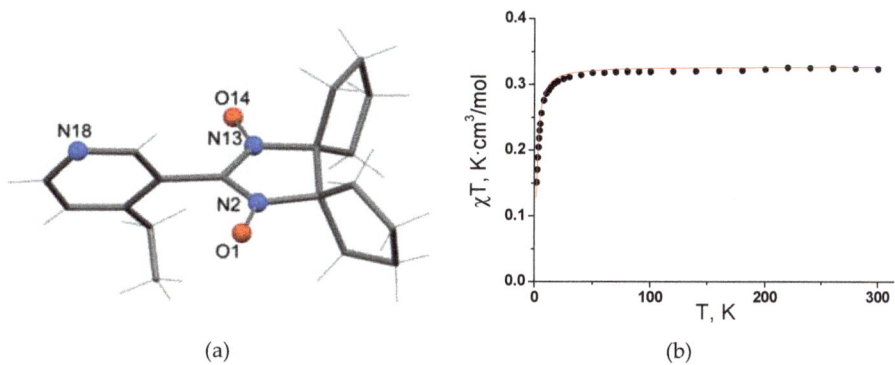

**Figure 2.** Structure of $L^{Et}$(**a**) and temperature dependence of $\chi$T for $L^{Et}$(**b**).

Variation of the Cu(hfac)$_2$/$L^{Et}$ ratio allowed us to obtain two heterospin complexes: binuclear [Cu(hfac)$_2 L^{Et}$]$_2$ and tetranuclear [[Cu(hfac)$_2$]$_4$($L^{Et}$)$_2$], whose solid phases exhibit thermally induced spin transitions, as shown below.

## 3.3. Magnetic and X-Ray Investigations for [Cu(hfac)$_2 L^{Et}$]$_2$

The binuclear [Cu(hfac)$_2 L^{Et}$]$_2$ complex is formed by the centrosymmetric molecules (Figure 3), and the statistically averaged environment of the Cu atom at room temperature can be described as a flattened octahedron. In the octahedron, the axial positions are occupied by the pyridine N(18R) atom (1.996(2) Å) and one of the O$_{hfac}$ atoms (O2) (1.965(2) Å), whereas the O(14R) atom of the NO group (2.193(2) Å) and the other three O$_{hfac}$ atoms (2.074(2), 2.105(2), and 2.204(4) Å, Table 1) lie in the equatorial plane. When the crystal is cooled to 240 K, the Cu–O distances in the equatorial plane in the coordination unit become almost equal (2.122(4)–2.152(4) Å), which actually reflects that approximately half of all molecules have the configuration of the high-temperature phase, and the other half have the configuration of the low-temperature phase. Below 200 K, the crystal symmetry lowers from monoclinic to triclinic (see Table S1, Supplementary Materials), and the twinning effect takes place. At 50 K, the structure contains two centrosymmetric crystallographically independent molecules, in which Cu is surrounded by an elongated octahedron with axial distances of 2.270(10)–2.338(10) Å; equatorial Cu–N distances of 1.940(11) and 1.977(10); Cu–O$_{NO}$ 1.963(9) and 1.979(8) Å; Cu–O$_{hfac}$ 1.944(9), 1.987(9), 1.961(9), and 2.035(9) Å for Cu1 and Cu2, respectively (Table 1).

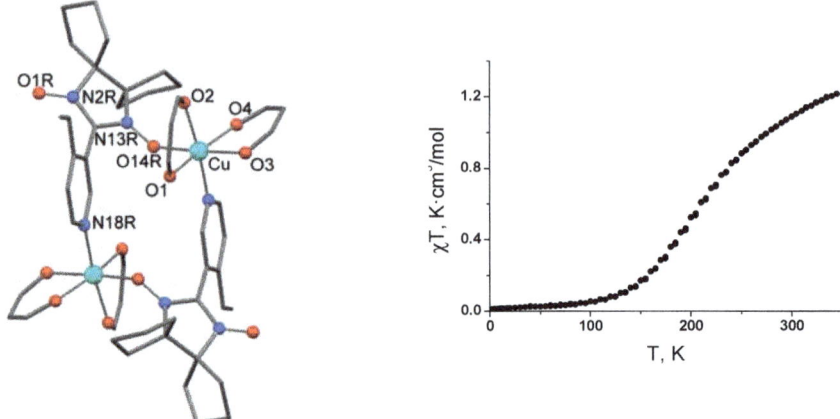

**Figure 3.** Molecular structure and temperature dependence of χT for [Cu(hfac)$_2$L$^{Et}$]$_2$ (H, CF$_3$ are omitted for clarity).

**Table 1.** Selected bond lengths (Å) and angles (°) in [Cu(hfac)$_2$L$^{Et}$]$_2$.

| T (K) | 295 | 240 | 200 | | 50 |
|---|---|---|---|---|---|
| Cu–O$_{NO}$ | 2.193(2) | 2.128(2) | 2.062(5) | 1.963(9) | 1.979(8) |
| ∠CuO$_R$N | 127.8(2) | 126.53(2) | 126.2(4) | 122.0(7) | 124.9(7) |
| Cu–O$_{hfac}$ | 1.965(2), 2.074(2) 2.105(2), 2.204(4) | 1.951(3), 2.123(3) 2.122(4), 2.152(3) | 1.921(5), 2.056(8) 2.182(7), 2.217(6) | 1.944(9), 1.987(9) 2.302(10), 2.338(10) | 1.961(9), 2.035(9) 2.270(10), 2.279(10) |
| Cu–N | 1.996(2) | 1.989(3) | 1.991(6) | 1.940(11) | 1.977(10) |
| ∠CN$_2$O$_2$–Py | 54.7 | 53.8 | 53.0 | 50.1 | 53.6 |
| ∠Py–Et | 6.7(5) | 9.8(5) | 7(2) | 16(1) | 12(2) |

In [Cu(hfac)$_2$L$^{Et}$]$_2$, the interplanar angle ∠Py-CN$_2$O$_2$ is considerably smaller (≈52.0°) than in free L$^{Et}$, and the ethyl group lies in the plane of the pyridine ring (Table 1).

For [Cu(hfac)$_2$L$^{Et}$]$_2$, χT is 1.214 K·cm$^3$/mol at 315 K, which is lower than the theoretical pure spin value of 1.50 K·cm$^3$/mol for four paramagnetic centers with S = $\frac{1}{2}$ and g = 2. When the compound is cooled, χT smoothly decreases to ≈0.01 K·cm$^3$/mol, indicating that the spins almost completely vanish in the system, that is the result of the strong antiferromagnetic exchange interactions, which is typical for the equatorial coordination of the O•–N< group in Cu(II) complexes [3,5]. The character of the χT(T) dependence is consistent with the XRD data on the shortening of the Cu–O$_{NO}$ distances with lowering temperature (Table 1).

### 3.4. X-Ray Investigations for[[Cu(hfac)$_2$]$_4$(L$^{Et}$)$_2$]

The solid [[Cu(hfac)$_2$]$_4$(L$^{Et}$)$_2$] is formed by the tetranuclear molecules (Figure 4). The structure of their cyclic fragment (Figure 4) is similar to that of [Cu(hfac)$_2$L$^{Et}$]$_2$ (Figure 3). The difference is that all four O atoms of the bridging L$^{Et}$ are coordinated by the Cu atoms.

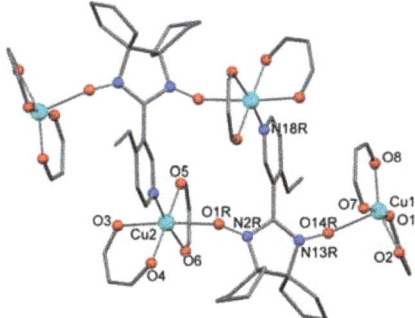

**Figure 4.** Molecular structure of [[Cu(hfac)$_2$]$_4$(L$^{Et}$)$_2$] at 295 K and atomic numbering in its crystallographically independent part (H, CF$_3$ are omitted for clarity).

In the [[Cu(hfac)$_2$]$_4$(L$^{Et}$)$_2$] molecule, the Cu1 atoms is in a square pyramid environment formed by the O$_{NO}$ atom lying at the apex (Cu1–O14R 2.332(2) Å) and four O$_{hfac}$ atoms (Cu–O$_{hfac}$ 1.915(4)–1.930(4) Å) lying at the base (Figure 4). For Cu1, the degree of distortion τ of the coordination polyhedron of the Cu atom (τ = 1 for the trigonal bipyramid and 0 for the square pyramid) is 0.0603 [29]. The O$_{NO}$ and O$_{hfac}$ atoms lie on the elongated axis of the square bipyramid at the intracyclic Cu2 atom (Cu2–O1R 2.553(4) Å and Cu2–O3 2.285(4) Å). The interplanar angles ∠Py–CN$_2$O$_2$ and ∠Py–Et are 56.3 and 16.2°, respectively. When the crystal was cooled from 295 to 154 K, the distances in the coordination units slightly shortened (Table 2), and the angle ∠Py–Et markedly increased, from 16.2 to 36.5°. When the sample was cooled by another 4 K (i.e., to 150 K), the structure transformed in a nontrivial way: the $b$ unit cell parameter increased three fold, and so did the crystallographically independent part, which now contained half of the centrosymmetric molecule with the Cu5 and Cu6 atoms and the complete molecule with the Cu1–Cu4 atoms (Figure 5a). The square pyramidal surrounding of the terminal Cu atoms also became much more distorted: the τ parameter increased to 0.228–0.266 (Table 2). The ∠Py–Et angle increased to 41.6° in the centrosymmetric molecule and to 53.2 and 56.3° in the non-centrosymmetric one. In the latter (the molecule in the right part of Figure 5a), the Et groups, magenta-colored in the figure, are orientated in the same direction relative to the {Cu$_2$(O$_{NO}$)$_2$} square plane in such a way that the terminal atom of the group is directed toward the terminal Cu(hfac)$_2$ fragment in one group and toward the plane of the hfac ligand from the cyclic dimer in the other.

**Table 2.** Selected bond lengths and angles in the [[Cu(hfac)$_2$]$_4$(L$^{Et}$)$_2$] molecules (* – after heating from 50 K).

| T, K | 295 | 240 | 200 | 154 | 150 | 100 | 50 | 240 * |
|---|---|---|---|---|---|---|---|---|
| Cu–O$_R$ (inn.) | 2.553(4) | 2.498(2) | 2.497(2) | 2.480(3) | 2.526(2), 2.413(2), 2.490(2) | 2.428(3), 2.424(3) | 2.406(3), 2.425(3) | 2.511(3) |
| Cu–O$_R$ (term.) | 2.276(4) | 2.287(2) | 2.287(2) | 2.285(4) | 2.308(2), 2.290(2), 2.290(2) | 1.954(3), 2.281(3) | 1.961(3), 2.270(3) | 2.276(3) |
| τ | 0.060 | 0.045 | 0.04650 | 0.042 | 0.228, 0.266, 0.239 | 0.375, 0.259 | 0.346, 0.253 | 0.263 |
| Cu–N | 1.998(4) | 2.016(2) | 2.016(2) | 2.009(4) | 2.008(3), 2.018(3), 2.016(3) | 2.015(4), 2.017(4) | 2.022(4), 2.021(4) | 2.005(3) |
| ∠CN$_2$O$_2$–Py | 56.3 | 56.7 | 56.7 | 57.0 | 53.5, 54.7, 60.3 | 54.9, 57.0 | 54.4, 56.6 | 56.2 |
| ∠Py–Et | 16.2 | 24.3 | 24.3 | 36.5 | 53.2, 41.6, 56.3 | 60.7, 63.6 | 61.1, 63.4 | 19.4 |

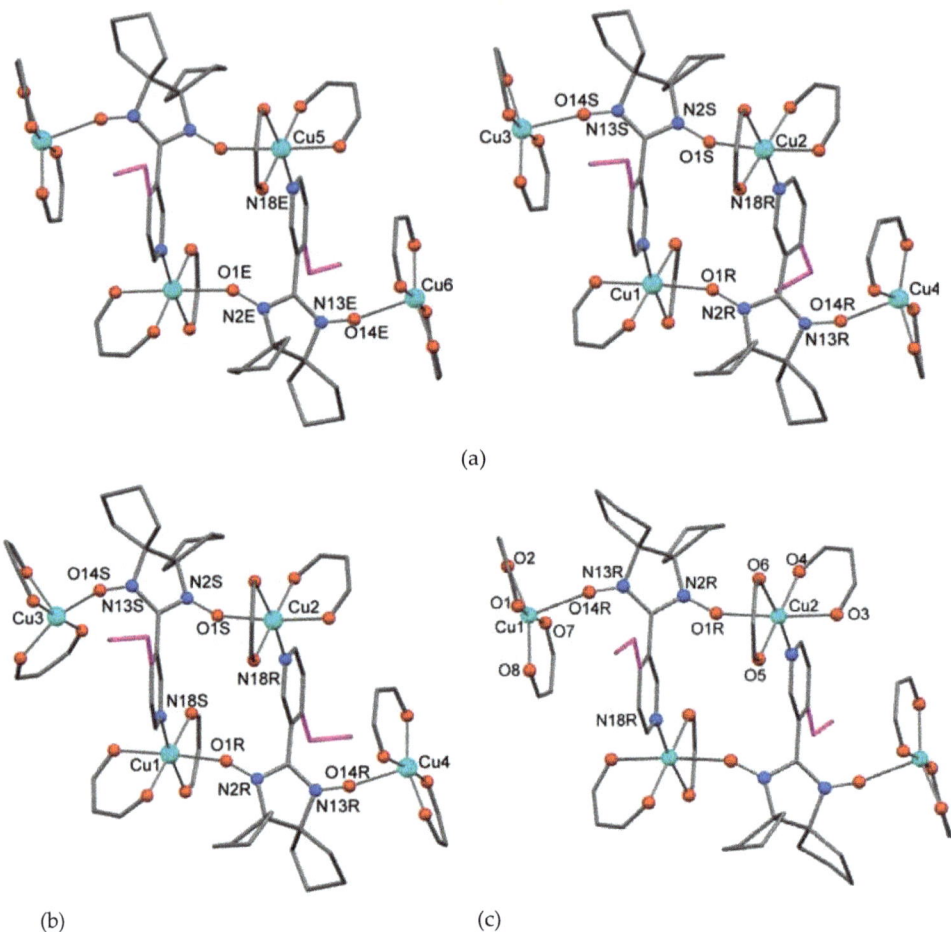

**Figure 5.** [[Cu(hfac)$_2$]$_4$(L$^{Et}$)$_2$]: molecular structure and atomic numbering in the crystallographically independent part (**a**) at 150 K, (**b**) at 100 K, and (**c**) after return to 240 K; the Et groups are magenta-colored (H and CF$_3$ are omitted for clarity).

When the crystal was further cooled to 100 K, the $b$ parameter became close to the initial value, whereas the $a$ parameter doubled. The crystallographically independent part of the structure at 100 K was a complete [[Cu(hfac)$_2$]$_4$(L$^{Et}$)$_2$] molecule. According to Figure 5b, the Et groups were again orientated in different directions (the ∠Py–Et angle is 60.7 and 63.3°). The surrounding of the terminal Cu3 and Cu4 atoms changed dramatically. For Cu3, the square pyramidal environment was preserved, with the O$_{NO}$ atom at the apex (Cu–O 2.281(3) Å). For Cu4, however, the apex of the square pyramid was occupied by the O$_{hfac}$ atom (Cu–O 2.158(3) Å), while O$_{NO}$ shifted to the base (Cu–O$_{NO}$ 1.954(3), Cu–O$_{hfac}$ 1.918(4)–1.957(3) Å) and τ increased to 0.375 (Table 2). Further cooling of the crystal to 50 K did not reveal any significant changes in the structure. After the crystal was further heated to 240 K, the high-temperature phase was recovered (Table S2 (Supplementary Materials) and Table 2, Figure 5c) except that the square pyramidal environment of the terminal Cu atom remained rather strongly distorted: τ = 0.263, whereas its initial value was 0.045.

## 3.5. IR Spectroscopy of [[Cu(hfac)$_2$]$_4$(L$^{Et}$)$_2$]

Earlier, VT-FTIR was shown to be highly sensitive to the temperature-induced structural transitions in copper–nitroxide complexes [30]. Figure 6a presents the temperature dependence of a fragment of the absorption spectrum of the [[Cu(hfac)$_2$]$_4$(L$^{Et}$)$_2$] single crystal at 80–274 K and shows the regions for which integration was performed and the temperature dependences calculated (Figure 6b). When the crystal was cooled from 274 to 158 K, the intensity of the IR absorption bands changed insignificantly and monotonically. Below 158 K, some of the IR absorption lines significantly changed their intensity and position, confirming the XRD data considered above. These changes were most dramatic at energies of 849–841 cm$^{-1}$, at which a new absorption band centered at 847 cm$^{-1}$ appeared. Further cooling to 80 K revealed only one more structural transition, which strongly affected all the absorption bands being considered (Figure 6b). The temperature of this transition agreed well with the temperature of the magnetic transition (Figure 6c) and also demonstrated a hysteresis of a few K. Thus, VT-FTIR of [[Cu(hfac)$_2$]$_4$(L$^{Et}$)$_2$] reliably confirmed both transitions observed by XRD and SQUID.

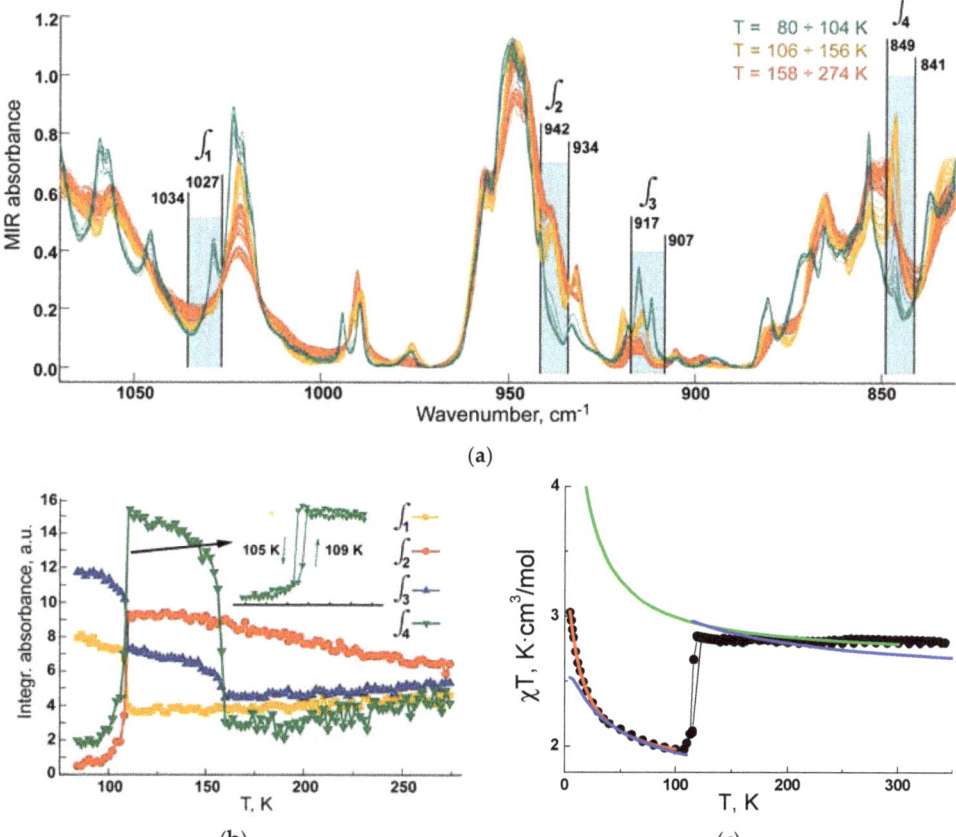

**Figure 6.** (a) Variable temperature FTIR (VT-FTIR) spectra of the [[Cu(hfac)$_2$]$_4$(L$^{Et}$)$_2$] single crystal in the mid-IR range measured at different temperatures; (b) temperature dependencies of the integrated intensity of the absorption bands in the ranges marked in (a); (c) dependence χT(T). The solid lines on the χT(T) dependence are the theoretical curves (see the text for explanations).

## 3.6. Magnetic Investigations for [[Cu(hfac)$_2$]$_4$(L$^{Et}$)$_2$]

Figure 6c shows the temperature dependence of $\chi T$ for [[Cu(hfac)$_2$]$_4$(L$^{Et}$)$_2$]. The $\chi T$ value does not change in the range of 122–345 K and equals 2.80 K cm$^3$/mol, which is slightly higher than the theoretical pure spin value of 2.25 K·cm$^3$/mol for six non-interacting paramagnetic centers with spin $S = \frac{1}{2}$ and g = 2. The exaggerated $\chi T$ value is caused by g$_{Cu}$ > 2. Below 122 K, $\chi T$ drastically decreases, reaching 1.97 K·cm$^3$/mol at 107 K, which agrees with the theoretical value for four non-interacting unpaired electrons, and then increases to 3.03 K cm$^3$/mol at 5 K. As mentioned above, at 150–100 K the environment of one of the terminal atoms in the structure transforms with a drastic shortening of the Cu–O$_{NO}$ distance, giving rise to strong antiferromagnetic exchange interactions in the >N–•O–Cu$^{2+}$ exchange clusters, which lead to complete compensation of their spins. When the sample is further cooled, $\chi T$ increases, which confirms the intramolecular ferromagnetic exchange between the remaining paramagnetic centers. The $\chi T(T)$ dependence in the range of 5–100 K was analyzed in terms of the following spin Hamiltonian:

$$H = -2J_1 \cdot S_{Cu3}S_{R1} - 2J_2 \cdot (S_{Cu2}S_{R1} + S_{Cu1}S_{R2}) - 2J_3 \cdot (S_{Cu2}S_{R2} + S_{Cu1}S_{R1}) - 2J_4 \cdot S_{Cu4}S_{R2} \tag{1}$$

using the PHI program [31]. The optimum values of the $J_1$, $J_2$, $J_3$, and $J_4$ parameters are 15.3, 8.75, 2.12, and −461 cm$^{-1}$, respectively, at g$_{Cu}$ = 2.27 and g$_R$ = 2.0 (fixed) (Figure 6c, red line). If we set that $J_4 = J_1 = 15.3$ cm$^{-1}$, then the theoretical curve describes well the experimental data in the high-temperature range of 150–300 K (Figure 6c, green line).

## 3.7. Quantum-Chemical Calculations for [[Cu(hfac)$_2$]$_4$(L$^{Et}$)$_2$]

The quantum-chemical calculations of the exchange interaction parameters for [[Cu(hfac)$_2$]$_4$(L$^{Et}$)$_2$] were performed using the following spin Hamiltonian:

$$H = -2J_1 \cdot S_{Cu3}S_{R1} - 2J_{21} \cdot S_{Cu2}S_{R1} - 2J_{22} \cdot S_{Cu1}S_{R2} - 2J_4\, S_{Cu4}S_{R2} \tag{2}$$

where $S_{R1}$ and $S_{R2}$ are the spins of the nitronylnitroxides lying between Cu1 and Cu3 (R1) and Cu2 and Cu4 (R2), respectively (Figure 5b). The exchange integrals of the spin Hamiltonian were calculated by the broken symmetry method [32] from the energies of the spin configurations determined by DFT quantum-chemical band structure calculations. The resulting exchange integrals are shown in Table 3.

Table 3. Exchange integrals (cm$^{-1}$) in the [[Cu(hfac)$_2$]$_4$(L$^{Et}$)$_2$] molecules.

| T, K | $J_{21}$ | $J_1$ | $J_{22}$ | $J_4$ |
|---|---|---|---|---|
| 50  | 37.7 | 15.2 | 49.9 | −719.0 |
| 100 | 37.7 | 14.9 | 49.0 | −720.0 |
| 153 | 37.1 | 12.9 | 37.1 | 12.9 |
| 295 | 32.7 | 9.6  | 32.7 | 9.6 |

Our calculations confirmed the appearance of strong antiferromagnetic exchange in the terminal >N–•O–Cu$^{2+}$ fragment when the position of the coordinated O$_{NO}$ atom changed from axial to equatorial at T < 122 K. The exchange integral calculated for this case, $J_4 = -720$ cm$^{-1}$, as well as the positive exchange integrals at T > 154 K, qualitatively agree with the exchange parameters obtained by fitting the experimental curve $\chi T(T)$ (Figure 6c), but are slightly overestimated, what is typical for quantum-chemical calculations [33]. The temperature dependences $\chi T(T)$ calculated using the exchange integrals from Table 3 are shown in Figure 6c (blue lines). At T < 100 K, the exchange integrals were used for T = 100 K; at T > 153 K, for T = 295 K, g$_{Cu}$ = 2.15, g$_R$ = 2.05. The quantum-chemical calculations also show that other exchange integrals in [[Cu(hfac)$_2$]$_4$(L$^{Et}$)$_2$] are positive, which explains the fast growth of the $\chi T$ when the temperature lowers at T < 100 K. It was also assumed that the exchange interaction between the spins of the intracyclic copper Cu1 and Cu2 ions with nitroxides R1 and R2 via the pyridine ring is small. For this reason, it was not calculated, although it can contribute to the growth of the $\chi T$ to 3.03 K cm$^3$/mol at T < 10 K for [[Cu(hfac)$_2$]$_4$(L$^{Et}$)$_2$].

## 4. Conclusions

New molecular heterospin complexes [Cu(hfac)$_2$L$^{Et}$]$_2$ and [[Cu(hfac)$_2$]$_4$(L$^{Et}$)$_2$] exhibiting reversible spin transitions were obtained as a result of this study. The crystals of the complexes are mechanically stable and do not crash during the repeated cooling–heating cycles, which allowed us to study the structure of both the high- and low-temperature phases. The thermally induced structural rearrangement in the binuclear complex was accompanied by the change in its symmetry. When the tetranuclear complex was cooled, two structural phase transitions were observed, the low-temperature transition provoking the spin transition. This transition is unusual, since it causes spin coupling in half of all terminal {>N–•O–Cu$^{2+}$} exchange clusters, which leads to spin compensation of only two paramagnetic centers of the six centers present in the molecule. This "asymmetric" coupling of the electrons of the paramagnetic centers inside one molecule, that has not been observed earlier for multinuclear Cu(II) complexes with nitroxides [4,5,18]. This effect can evidently be responsible for the appearance of stepwise spin transitions in multinuclear compounds with several exchange clusters in the molecule. For [[Cu(hfac)$_2$]$_4$(L$^{Et}$)$_2$] molecules, however, the transition to the low-spin state in the second terminal {>N–•O–Cu$^{2+}$} fragment was not recorded after the cooling to 2 K.

**Supplementary Materials:** The following are available online at http://www.mdpi.com/2073-4352/9/6/285/s1, Table S1: Crystallographic data and experimentdetails for [Cu(hfac)$_2$(L$^{Et}$)]$_2$; Table S2: Crystallographic data and experiment details for [[Cu(hfac)$_2$]$_4$(L$^{Et}$)$_2$].

**Author Contributions:** Conceptualization, V.O. and G.R.; methodology, N.A.; software, V.M.; validation, N.A., G.R., S.V., V.M. and A.B.; formal analysis, N.A.; investigation, all authors; resources, V.O.; data curation, V.M. and M.P.; writing—original draft preparation, N.A. and G.R.; writing—review and editing, V.O.; visualization, N.A.; supervision, V.O.; project administration, V.O.; funding acquisition, V.O., G.R. and N.A.

**Funding:** This study was financially supported by the Russian Scientific Foundation (grant no. 17-13-01022, magnetic measurements and XRD study), the Russian Foundation for Basic Research (grant no. 18-33-00491, synthesis of the ligand), MK-1970.2018.3 (synthesis of the complexes) and the Ministry of Science and Education of the Russian Federation (IR measurements and quantum-chemical calculations).

**Conflicts of Interest:** The authors declare no conflict of interest.

## References

1. Ovcharenko, V.I.; Bagryanskaya, E.G. *Spin-Crossover Materials: Properties and Applications*; Halcrow, M., Ed.; John Wiley & Sons Ltd.: Chichester, UK, 2013.
2. Tretyakov, E.V.; Tolstikov, S.E.; Suvorova, A.O.; Polushkin, A.V.; Romanenko, G.V.; Bogomyakov, A.S.; Veber, S.L.; Fedin, M.V.; Stass, D.V. Crucial role of paramagnetic ligands for magnetostructural anomalies in "breathing crystals". *Inorg. Chem.* **2012**, *51*, 9385–9394. [CrossRef] [PubMed]
3. Ovcharenko, V.I.; Romanenko, G.V.; Maryunina, K.Y.; Bogomyakov, A.S.; Gorelik, E.V. Thermally Induced Magnetic Anomalies in Solvates of the Bis(hexafluoroacetylacetonate) copper (II) Complex with Pyrazolyl-Substituted NitronylNitroxide. *Inorg. Chem.* **2008**, *47*, 9537–9552. [CrossRef]
4. Lanfranc de Panthou, F.; Belorizky, E.; Calemczuk, R.; Luneau, D.; Marcenat, C.; Ressouche, E.; Turek, P.; Rey, P. A new type of thermally-induced spin transition associated with an equatorial. dblarw. axial conversion in a copper(II)-nitroxide cluster. *J. Am. Chem. Soc.* **1995**, *117*, 11247–11253.
5. Lanfranc de Panthou, F.; Luneau, D.; Musin, R.N.; Öhrström, L.; Grand, A.; Turek, P.; Rey, P. Spin-Transition and Ferromagnetic Interactions in Copper(II) Complexes of a 3-Pyridyl-Substituted IminoNitroxide. Dependence of the Magnetic Properties upon Crystal Packing. *Inorg. Chem.* **1996**, *35*, 3484–3491. [CrossRef]
6. Rey, P.; Ovcharenko, V.I. Copper(II) Nitroxide Molecular Spin-transition Complexes. In *Magnetism: Molecules to Materials: 5 Volumes Set*; Miller, J.S., Drillon, M., Eds.; Wiley-VCH: New York, NY, USA, 2003; pp. 41–63.
7. Ovcharenko, V.I. Metal-nitroxide complexes: synthesis and Magnetostructural Correlations. In *Stable Radicals: Fundamentals and Applied Aspects of Odd-Electron Compounds*; Hicks, R., Ed.; Wiley-VCH: New York, NY, USA, 2010; pp. 461–507.

8. Halcrow, M. *Spin-Crossover Materials: Properties and Applications*; John Wiley & Sons Ltd.: Chichester, UK, 2013.
9. Tolstikov, S.E.; Artiukhova, N.A.; Romanenko, G.V.; Bogomyakov, A.S.; Zueva, E.M.; Barskaya, I.Y.; Fedin, M.V.; Maryunina, K.Y.; Tretyakov, E.V.; Sagdeev, R.Z. Heterospin complex showing spin transition at room temperature. *Polyhedron* **2015**, *100*, 132–138. [CrossRef]
10. Wang, Y.-L.; Gao, Y.-Y.; Yang, M.-F.; Gao, T.; Ma, Y.; Wang, Q.-L.; Liao, D.-Z. Four new Cu-coordination compounds based on different nitronylnitroxide radicals: Structural design and magneto-structural correlations. *Polyhedron* **2013**, *61*, 105–111. [CrossRef]
11. Yang, M.; Sun, J.; Guo, J.; Sun, G.; Li, L. Cu-Ln compounds based on nitronylnitroxide radicals: Synthesis, structure, and magnetic and fluorescence properties. *Cryst. Eng. Comm.* **2016**, *19*, 9345–9356. [CrossRef]
12. Souza, D.A.; Florencio, A.S.; Soriano, S.; Calvo, R.; Sartoris, R.P.; Carneiro, J.W.d.M.; Sangregorio, C.; Novak, M.A.; Vaz, M.G.F. New copper(II)-radical one dimensional chain: Synthesis, crystal structure, EPR, magnetic properties and DFT calculations. *Dalton Trans.* **2009**, *34*, 6816–6824. [CrossRef]
13. Souza, D.A.; Moreno, Y.; Ponzio, E.A.; Resende, J.A.; Jordao, A.K.; Cunha, A.C.; Ferreira, V.F.; Novak, M.A.; Vaz, M.G.F. Synthesis, crystal structure, magnetism and electrochemical properties of two copper(II) furoyltrifluoroacetonate complexes with nitroxide radical. *Inorg. Chim. Acta* **2011**, *370*, 469–473. [CrossRef]
14. Caneschi, A.; Ferraro, F.; Gatteschi, D.; Rey, P.; Sessoli, R. Structure and magnetic properties of a chain compound formed by copper (II) and a tridentate nitronylnitroxide radical. *Inorg. Chem.* **1991**, *30*, 3162–3166. [CrossRef]
15. Yeltsov, I.; Ovcharenko, V.; Ikorskii, V.; Romanenko, G.; Vasilevsky, S. Copper(II) Thenoyltrifluoroacetonate as Acceptor Matrix in Design of Heterospin Complexes. *Polyhedron* **2001**, *20*, 1215–1222. [CrossRef]
16. Hirel, C.; Li, L.; Brough, P.; Vostrikova, K.; Pecaut, J.; Mehdaoui, B.; Bernard, M.; Turek, P.; Rey, P. New spin-transition-like-nitroxide species. *Inorg. Chem.* **2007**, *46*, 7545–7552. [CrossRef]
17. Wang, K.-M.; Du, L.; Fang, R.-B.; Zhao, Q.-H. Synthesis and Crystal Structure of Copper(II)-Hexafluoro-Acetylacetonate Complexes with Pyridyl-Substituted Nitronyl and Imino-Nitroxide Radicals. *J. Chem. Cryst.* **2010**, *40*, 472–475. [CrossRef]
18. Artiukhova, N.A.; Maryunina, K.Y.; Fokin, S.V.; Tretyakov, E.V.; Romanenko, G.V.; Polushkin, A.V.; Bogomyakov, A.S.; Sagdeev, R.Z.; Ovcharenko, V.I. Spirocyclic Derivatives of NitronylNitroxides in the Design Of Heterospin Cu(II) Complexes Manifesting Spin Transitions. *Russ. Chem. Bull.* **2013**, *62*, 2132–2140. [CrossRef]
19. Wang, X.-F.; Licun Li, P.H.; Sutter, J.-P. [(Cu-Radical)$_2$-Ln]: Structure and Magnetic Properties of a Hetero-tri-spin Chain of Rings (Ln = Y$^{III}$, Gd$^{III}$, Tb$^{III}$, Dy$^{III}$). *Inorg. Chem.* **2015**, *54*, 9664–9669. [CrossRef]
20. Wang, X.-F.; Hu, P.; Li, Y.-G.; Li, L.-C. Construction of NitronylNitroxide-Based 3d–4f Clusters: Structure and Magnetism. *Chem. Asian J.* **2015**, *10*, 325–328. [CrossRef]
21. Artiukhova, N.A.; Romanenko, G.V.; Letyagin, G.A.; Bogomyakov, A.S.; Tolstikov, S.E.; Ovcharenko, V.I. Spin transition characteristics of molecular solvates of Cu$^{II}$ complexes with nitroxides: Sensitivity to the packing type. *Russ. Chem. Bull.* **2019**, *68*, 732–742. [CrossRef]
22. Bertrand, J.A.; Kaplan, R.I. A Study of Bis(hexafluoroacetylacetonato)copper(II). *Inorg. Chem.* **1966**, *5*, 489–491. [CrossRef]
23. Gueritte, F.; Guillou, C.; Husson, H.-P.; Kozielski, F.; Labriere, C.; Skoufias, D.; Tcherniuk, S.; Thal, C. Use of Derivatives of Indoles for the Treatment of Cancer. EP2266562A1, 29 December 2010.
24. Sheldrick, G.M. Crystal structure refinement with SHELXL. *Acta Cryst.* **2015**, *C71*, 3–8.
25. Giannozzi, P.; Baroni, S.; Bonini, N.; Calandra, M.; Car, R.; Cavazzoni, C.; Ceresoli, D.; Chiarotti, G.L.; Cococcioni, M.; Dabo, I. QUANTUM ESPRESSO: A modular and open-source software project for quantum simulations of materials. *J. Phys. Condens. Matter.* **2009**, *21*, 395502–395520. [CrossRef]
26. Monkhorst, H.J.; Pack, J.D. Special points for Brillouin-zone integrations. *Phys. Rev. B* **1976**, *13*, 5188–5192. [CrossRef]
27. Streltsov, S.V.; Petrova, M.V.; Morozov, V.A.; Romanenko, G.V.; Anisimov, V.I.; Lukzen, N.N. Interplay between lattice, orbital, and magnetic degrees of freedom in the chain-polymer Cu(II) breathing crystals. *Phys. Rev. B* **2013**, *87*, 024425–024425/6. [CrossRef]
28. Morozov, V.A.; Petrova, M.V.; Lukzen, N.N. Exchange coupling transformations in Cu (II) heterospin complexes of "breathing crystals" under structural phase transitions. *AIP Adv.* **2015**, *5*, 087161. [CrossRef]

29. Addison, A.W.; Rao, T.N.; Reedijk, J.; van Rijn, J.; Verschoor, G.C. Synthesis, structure, and spectroscopic properties of copper(II) compounds containing nitrogen–sulphur donor ligands; the crystal and molecular structure of aqua [1,7-bis(N-methylbenzimidazol-2′-yl)-2,6-dithiaheptane]copper(II) perchlorate. *J. Chem. Soc. Dalton Trans.* **1984**, *7*, 1349–1356. [CrossRef]
30. Veber, S.L.; Suturina, E.A.; Fedin, M.V.; Boldyrev, K.N.; Maryunina, K.Y.; Sagdeev, R.Z.; Ovcharenko, V.I.; Gritsan, N.P.; Bagryanskaya, E.G. FTIR study of thermally induced magnetostructural transitions in breathing crystals. *Inorg Chem.* **2015**, *54*, 3446–3455. [CrossRef]
31. Chilton, N.F.; Anderson, R.P.; Turner, L.D.; Soncini, A.; Murray, K.S. PHI: A powerful new program for the analysis of anisotropic monomeric and exchange-coupled polynuclear d- and f-block complexes. *J. Comput. Chem.* **2013**, *34*, 1164–1175. [CrossRef]
32. Yamaguchi, K.; Fukui, H.; Fueno, T. Molecular Orbital (MO) Theory for Magnetically Interacting Organic Compounds. Ab-Initio MO Calculations of the Effective Exchange Integrals For Cyclophane-Type Carbene Dimers. *Chem. Lett.* **1986**, *15*, 625–628. [CrossRef]
33. Cho, D.; Ko, K.C.; Ikabata, Y.; Wakayama, K.; Yoshikawa, T.; Nakai, H.; Lee, J.Y. Effect of Hartree-Fock exact exchange on intramolecular magnetic coupling constants of organic diradicals. *J. Chem. Phys.* **2015**, *142*, 024318. [CrossRef]

© 2019 by the authors. Licensee MDPI, Basel, Switzerland. This article is an open access article distributed under the terms and conditions of the Creative Commons Attribution (CC BY) license (http://creativecommons.org/licenses/by/4.0/).

Article

# Iron(II) Spin Crossover Complex with the 1,2,3-Triazole-Containing Linear Pentadentate Schiff-Base Ligand and the MeCN Monodentate Ligand

Tomoe Matsuyama [1], Keishi Nakata [2], Hiroaki Hagiwara [1,*] and Taro Udagawa [3]

1. Department of Chemistry, Faculty of Education, Gifu University, Yanagido 1-1, Gifu 501-1193, Japan; v1008407@edu.gifu-u.ac.jp
2. Graduate School of Education, Gifu University, Yanagido 1-1, Gifu 501-1193, Japan; x1131019@edu.gifu-u.ac.jp
3. Department of Chemistry and Biomolecular Science, Faculty of Engineering, Gifu University, Yanagido 1-1, Gifu 501-1193, Japan; udagawa@gifu-u.ac.jp
* Correspondence: hagiwara@gifu-u.ac.jp; Tel.: +81-58-293-2253

Received: 16 May 2019; Accepted: 25 May 2019; Published: 28 May 2019

**Abstract:** A mononuclear iron(II) complex bearing the linear pentadentate $N_5$ Schiff-base ligand containing two 1,2,3-triazole moieties and the MeCN monodentate ligand, [$Fe^{II}$MeCN($L_{3\text{-Me-3}}^{Ph}$)](BPh$_4$)$_2$·MeCN·H$_2$O (**1**), have been prepared ($L_{3\text{-Me-3}}^{Ph}$ = bis(N,N'-1-Phenyl-1H-1,2,3-triazol-4-yl-methylideneaminopropyl)methylamine). Variable-temperature magnetic susceptibility measurements revealed an incomplete one-step spin crossover (SCO) from the room-temperature low-spin (LS, $S = 0$) state to a mixture of the LS and high-spin (HS, $S = 2$) species at the higher temperature of around 400 K upon first heating, which is irreversible on the consecutive cooling mode. The magnetic modulation at around 400 K was induced by the crystal-to-amorphous transformation accompanied by the loss of lattice MeCN solvent, which was evident from powder X-ray diffraction (PXRD) studies and themogravimetry. The single-crystal X-ray diffraction studies showed that the complex is in the LS state ($S = 0$) between 296 and 387 K. In the crystal lattice, the complex-cations and B(1)Ph$_4^-$ ions are alternately connected by intermolecular CH···π interactions between the methyl group of the MeCN ligand and phenyl groups of B(1)Ph$_4^-$ ions, forming a 1D chain structure. The 1D chains are further connected by P4AE (parallel fourfold aryl embrace) interactions between two neighboring complex-cations, constructing a 2D extended structure. B(2)Ph$_4^-$ ions and MeCN lattice solvents exist in the spaces of the 2D layer. DFT calculations verified that the 1,2,3-triazole-containing ligand $L_{3\text{-Me-3}}^{Ph}$ gives a stronger ligand field around the octahedral coordination environment of the iron(II) ion than the analogous imidazole-containing ligand $H_2L^{2Me}$ (= bis(N,N'-2-methylimidazol-4-yl-methylideneaminopropyl)methylamine) of the known compound [$Fe^{II}$MeCN($H_2L^{2Me}$)](BPh$_4$)$_{1.5}$·Cl$_{0.5}$·0.5MeCN (**2**) reported by Matsumoto et al. (Nishi, K.; Fujinami, T.; Kitabayashi, A.; Matsumoto, N. Tetrameric spin crossover iron(II) complex constructed by imidazole···chloride hydrogen bonds. *Inorg. Chem. Commun.* **2011**, *14*, 1073–1076), resulting in the much higher spin transition temperature of **1** than that of **2**.

**Keywords:** spin crossover; linear pentadentate ligand; iron(II); mononuclear; 1,2,3-triazole; crystal structure; magnetic properties; DFT calculation; intermolecular interactions; amorphous

## 1. Introduction

Spin crossover (SCO) materials exhibiting spin state-interconversion between the high-spin (HS) and the low-spin (LS) states have attracted much attention since they have a great potential for applications in data storage, display, switching device, sensors for temperature, pressure, gas, and solvent, and also multi-modal sensing technology [1–5]. For the real-world applications, SCO compounds must fulfill demands such as a proper room temperature (RT) response, abrupt spin transition, wide thermal hysteresis, and high durability at least [6,7]. Multistep [8–11] and high-temperature (HT) [12–18] SCO complexes also provide demands for the development of multinary memories and for the investigation of the thermal stability of spin transition under extreme conditions, respectively. These properties are complicatedly affected by the ligand field strength of each SCO molecule and cooperativity between SCO metal sites. Generally, the important factors are the ligand backbone, molecular structure of each SCO molecule, accompanying components (counter ions and lattice solvents), crystal packing, and cooperative interactions through intermolecular interactions [19,20] and/or bridging ligands [21]. The most essential need, namely RT operation, is achievable by controlling the spin transition temperature ($T_{1/2}$). Thus, the various molecular design and modifications have been reported for the precise tuning of $T_{1/2}$ [22–29].

Octahedral metal complexes with pentadentate ligand are of interest since their ligand field strength can be systematically controlled by modifying the additional monodentate ligand in line with the spectrochemical series [30–32], and such a characteristic is useful not only for exhibiting SCO but also for the possible tuning of $T_{1/2}$. These complexes are also beneficial for the construction of polynuclear materials such as di [33–42], tri [43], tetra [44], penta [45], hepta [46–48], nona [48] and dodecanuclear [49] complexes by using bridging ligands instead of monodentate ligands for exhibiting multi-step spin transition. In the pentadentate ligand system, although iron(III) SCO compounds are well known [50–57], iron(II) spin transition systems are quite rare [58,59]. One of the reasons for this is the difficulty of the crystallization of iron(II) complexes with pentadentate ligand, as pointed out by Matsumoto et al. [60]. To the best of our knowledge, only one SCO iron(II) complex with linear pentadentate ligand, which shows the gradual and partial SCO below RT, has been reported so far [59]. So, our ongoing interest for exploring RT and HT SCO systems by using 1,2,3-triazole-containing multidentate Schiff-base ligands spontaneously extends to the next project for the synthesis of a SCO iron(II) complex bearing linear pentadentate ligand, which is missing in our tridentate to hexadentate ligand family [15,18,26,61–63].

In this study, we have synthesized an iron(II) complex with $N_6$ donor atoms from the 1,2,3-triazole-containing pentadentate ligand $L_{3\text{-Me-}3}{}^{Ph}$ and the monodentate neutral ligand MeCN, [$Fe^{II}$MeCN($L_{3\text{-Me-}3}{}^{Ph}$)](BPh$_4$)$_2$·MeCN·H$_2$O (**1**) ($L_{3\text{-Me-}3}{}^{Ph}$ = bis($N,N'$-1-Phenyl-1$H$-1,2,3-triazol-4-yl-methylideneaminopropyl)methylamine, Scheme 1) inspired by the above-mentioned SCO iron(II) complex having imidazole-containing linear pentadentate ligand, [$Fe^{II}$MeCN(H$_2$L$^{2Me}$)](BPh$_4$)$_{1.5}$·Cl$_{0.5}$·0.5MeCN (**2**)(H$_2$L$^{2Me}$ = bis($N,N'$-2-methylimidazol-4-yl-methylideneaminopropyl)methylamine [59]. We report here the synthesis, crystal structures, and thermal and magnetic properties of **1** with the first theoretical comparison of the 1,2,3-triazole- and imidazole-containing multidentate ligand system.

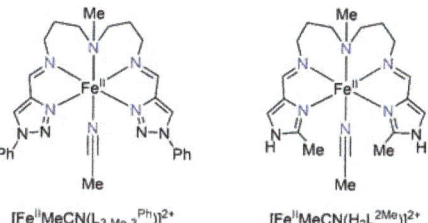

**Scheme 1.** Schematic structures of [$Fe^{II}$MeCN($L_{3\text{-Me-}3}{}^{Ph}$)]$^{2+}$ and [$Fe^{II}$MeCN(H$_2$L$^{2Me}$)]$^{2+}$.

## 2. Materials and Methods

### 2.1. Synthesis of $Fe^{II}$ Complex

#### 2.1.1. General

All reagents and solvents were purchased from commercial sources and used for the syntheses without further purification. The 1-phenyl-1$H$-1,2,3-triazole-4-carbaldehyde was synthesized according to the reported procedures [62,64–66]. Complexation and crystallization of 1 were performed under nitrogen atmosphere using standard Schlenk techniques. Other synthetic procedures were carried out in air.

#### 2.1.2. Synthesis of the Linear Pentadentate $N_5$ Ligand $L_{3\text{-Me-3}}^{Ph}$ = bis(N,N'-1-Phenyl-1H-1,2,3-triazol-4-yl-methylideneaminopropyl)methylamine

The ligand $L_{3\text{-Me-3}}^{Ph}$ was prepared by mixing 3,3′-diamino-$N$-methyldipropylamine and 1-phenyl-1$H$-1,2,3-triazole-4-carbaldehyde with a 1:2 molar ratio in MeCN. The ligand solution thus prepared was used for the synthesis of $Fe^{II}$ complex without further purification and isolation.

#### 2.1.3. Preparation of [$Fe^{II}$MeCN($L_{3\text{-Me-3}}^{Ph}$)]($BPh_4$)$_2$·MeCN·$H_2O$ (1)

3,3′-diamino-$N$-methyldipropylamine (0.109 g, 0.75 mmol) in MeCN (3 mL) was added to a solution of 1-phenyl-1$H$-1,2,3-triazole-4-carbaldehyde (0.260 g, 1.5 mmol) in MeCN (7.5 mL). The resulting pale-yellow solution was stirred at RT for 1 h. A solution of $NaBPh_4$ (0.513 g, 1.5 mmol) in MeOH (4 mL) was added to a solution of $Fe^{II}Cl_2·4H_2O$ (0.149 g, 0.75 mmol) in MeOH (4 mL), and the resulting pale-yellow solution was stirred at RT for 5 min. Both reaction mixtures were filtered, and they were mixed under nitrogen atmosphere. The resulting mixture was allowed to stand for a week in a fridge, during which time the precipitated dark red-brown prismatic crystals were collected by suction filtration. Yield: 0.203 g (21%). Anal. Calcd for [$Fe^{II}$MeCN($L_{3\text{-Me-3}}^{Ph}$)]($BPh_4$)$_2$·MeCN·$H_2O$ (1) = $C_{77}H_{77}B_2FeN_{11}O$: C, 73.99; H, 6.21; N, 12.33. Found: C, 73.78; H, 6.03; N, 12.03%. IR (KBr): $\nu_{C\equiv N}$ 2271, 2251, $\nu_{C=N}$ 1616, 1593, $\nu_{BPh_4}$ 734, 704 cm$^{-1}$. A weight loss of 3.4% corresponding to the MeCN (3.3%) and a subsequent 1.2% loss of water (1.4%) were observed at 408 and 444 K, respectively, by TG measurement (Figure 1).

### 2.2. Physical Measurements

Elemental C, H, and N analyses were performed on a J-Science Lab (Kyoto, Japan) MICRO CORDER JM-10. Infrared (IR) spectra were recorded at RT using a JASCO (Tokyo, Japan) FT/IR 460Plus spectrophotometer with the samples prepared as KBr disks. Thermogravimetric (TG) data was collected on a Rigaku (Tokyo, Japan) Thermo plus EVO2 TG-DTA8122 instrument in the temperature range of 19–359 °C (292–632 K) at a sweep rate of 10 K min$^{-1}$ under a nitrogen atmosphere (200 mL min$^{-1}$). Real-time sample images during TG analysis were recorded under an optional direct monitoring system of the TG-DTA instrument. Magnetic susceptibilities were measured in the temperature range of 5–400 K at a sweep rate of 2 K min$^{-1}$ under an applied magnetic field of 1 T using a Quantum Design (San Diego, CA, USA) MPMS-XL7 SQUID magnetometer. The sample was wrapped in an aluminum foil and was then inserted into a quartz glass tube with a small amount of glass wool filler. Corrections for diamagnetism of the sample were made using Pascal's constants [67,68]. A background correction for the sample holder was also applied. Powder X-ray diffraction (PXRD) patterns were recorded at RT on a portion of polycrystalline powders placed on a non-reflecting silicon plate, using a Rigaku MiniFlex600 diffractometer with Cu K$\alpha$ radiation ($\lambda$ = 1.5418 Å) operated at 0.4 kW power (40 kV, 10 mA).

*2.3. Crystallographic Data Collection and Structure Analyses*

X-ray diffraction data were collected by a Rigaku (Tokyo, Japan) AFC7R Mercury CCD diffractometer using graphite monochromated Mo Kα radiation (λ = 0.71075 Å) operated at 5 kW power (50 kV, 100 mA). A single crystal was mounted on a glass fiber and the diffraction data were collected at 296 K. Following the measurement at 296 K, the crystal was warmed and the subsequent measurements were performed at 350, 375, 387, and 400 K. The temperature of the crystal was maintained at the selected value by means of a Rigaku cooling device with nitrogen flow to within an accuracy of ± 2 K. Data reductions and empirical absorption correction using spherical harmonics, implemented in a SCALE3 ABSPACK scaling algorithm (multi-scan method) [69] were performed using the CrysAlis$^{Pro}$ software package (version 1.171.39.46) [70]. The structures were solved by the direct method using SHELXT [71] and refined on $F^2$ data using the full-matrix least-squares algorithm using SHELXL [72], both of which were implemented in the program OLEX2 (version 1.2.10) [73] with anisotropic displacement parameters for all non-hydrogen atoms. Hydrogen atoms were placed in calculated positions with idealized geometries and refined by using a riding model and isotropic displacement parameters. The continuous shape measures (CShMs) of the Fe$^{II}$ centers relative to the ideal octahedron, S(Oh) was calculated by the program SHAPE 2.1 [74]. The octahedral volumes of the Fe$^{II}$ centers were calculated using OLEX2 [73]. CCDC 1911292–1911295 contains the supplementary crystallographic data for this paper. These data can be obtained free of charge via http://www.ccdc.cam.ac.uk/conts/retrieving.html or from the CCDC (12 Union Road, Cambridge CB2 1EZ, UK; Fax: +44 1223 336033; E-mail: deposit@ccdc.cam.ac.uk).

*2.4. Computational Details*

In the present study, electronic energy was evaluated by (U)M06L [75] DFT method with a combination of 6-311G(d) electronic basis set (for all atoms except for Fe) and LANL2DZ pseudo potential (for Fe) in gas phase. The M06-L functional is a local functional (i.e., 0% Hartree-Fock exchange) and is known as one of the good exchange-correlation density functionals for transition metal chemistry. The single crystal X-ray crystallography structures were used as the initial geometries for DFT geometry optimization. We confirmed that all DFT optimized structures have no imaginary frequencies. All calculations were performed with the aid of the GAUSSIAN09 program package [76].

## 3. Results and Discussion

*3.1. Synthesis and Characterization*

The linear pentadentate N$_5$ ligand L$_{3\text{-Me-3}}^{Ph}$ was prepared by the condensation reaction of 1-phenyl-1$H$-1,2,3-triazole-4-carbaldehyde and 3,3'-diamino-$N$-methyldipropylamine with a 2:1 molar ratio in MeCN. The iron(II) complex **1** was prepared by mixing the ligand solution in MeCN, methanolic solutions of FeCl$_2$·4H$_2$O, and NaBPh$_4$ with a 1:1:2 molar ratio under an inert nitrogen atmosphere at ambient temperature. Dark red-brown prismatic crystals were precipitated in a week in a fridge, which are stable in the air with no efflorescence. The chemical formula of [Fe$^{II}$MeCN(L$_{3\text{-Me-3}}^{Ph}$)](BPh$_4$)$_2$·MeCN·H$_2$O was confirmed by the elemental analysis and thermogravimetry (TG; Figure 1). As shown in Figure 1, when the powdery sample was heated from 19 °C (292 K) at a sweep rate of 10 °C min$^{-1}$ under a nitrogen atmosphere (200 mL min$^{-1}$), the sample weight decreased gradually and a 3.4% weight loss was observed at 135 °C (408 K), which corresponds to the calculated weight percentage of one MeCN molecule per [Fe$^{II}$MeCN(L$_{3\text{-Me-3}}^{Ph}$)](BPh$_4$)$_2$·MeCN·H$_2$O (3.3%). Above this temperature, an additional gradual weight loss of 1.2%, corresponding to one H$_2$O molecule (1.4%) was detected at 171 °C (444 K). Finally, above 171 °C (444 K), the weight loss became more and more abrupt. During the TG measurement, the real-time sample images were also recorded (Figure 1). Initial orange-brown color of the grinding samples at 20 °C (293 K) was retained until ca. 124 °C (397 K) upon heating, and then slightly darkened around 125 °C (398 K). Upon further increasing the temperature, the samples were gradually shrinking from ca. 127 °C (400 K) with darkening, and were then melting from ca. 137 °C (410 K), and finally melted over ca. 147 °C (420 K). These changes agree with two broad endothermic peaks detected in the DTA curve. These observations and corresponding TG/DTA profiles indicated that the compound **1** showed some sort of structural modification at around 127 °C (400 K) associated with the loss of the MeCN lattice solvent, and further heating above this temperature caused melting. Thus, the magnetic susceptibilities described later were measured below 400 K. The PXRD pattern at RT showed no apparent extra reflections compared to the simulated pattern from the structure of single-crystal X-ray diffraction analysis at 296 K, ensuring the phase purity of **1** (Figure S1). The IR spectrum of **1** showed characteristic bands at ca. 1616 and 1593 cm$^{-1}$, corresponding to the C=N stretching vibration of the coordinated Schiff-base ligand, ca. 734 and 704 cm$^{-1}$, corresponding to the BPh$_4^-$ ion, and ca. 2271 and 2251 cm$^{-1}$, corresponding to the C≡N stretching vibration of the MeCN molecules (Figure S2) [53,77].

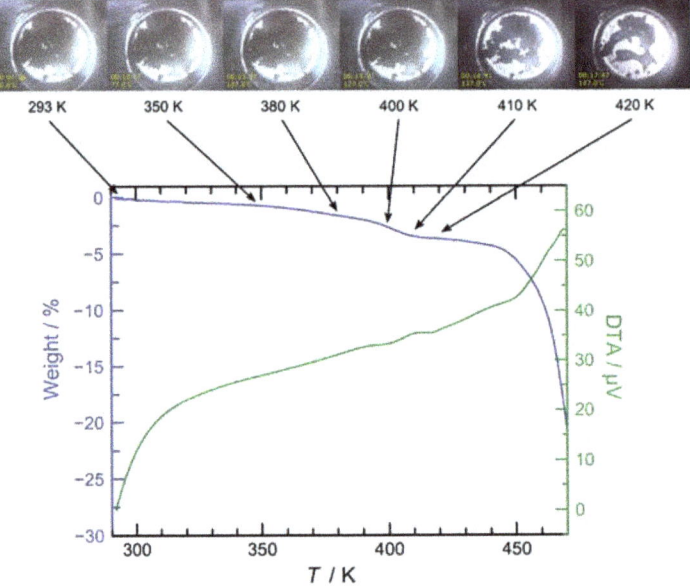

**Figure 1.** TG/DTA curve of **1** with the selected real-time sample images (from 293 to 420 K) during the TG measurement.

## 3.2. Magnetic Properties

The magnetic susceptibilities of **1** were measured between 5 to 400 K at a sweep rate of 2 K min$^{-1}$ under an applied magnetic field of 1 T using a MPMS-XL7 SQUID magnetometer (Quantum Design, San Diego, CA, USA). The $\chi_M T$ vs. $T$ plots are shown in Figure 2, where $\chi_M$ is the molar magnetic susceptibility and $T$ is the absolute temperature. On first cooling, the $\chi_M T$ value of **1** is 0.2 cm$^3$ K mol$^{-1}$ at 300 K and decreases moderately to reach 0.0 cm$^3$ K mol$^{-1}$ at 5 K, indicating that **1** is a LS Fe$^{II}$ ($S = 0$) complex. On subsequent heating, the $\chi_M T$ value increases slightly from 0.0 cm$^3$ K mol$^{-1}$ at 5 K to ca. 0.5 cm$^3$ K mol$^{-1}$ at 389 K, and then increases abruptly to reach ca. 2.2 cm$^3$ K mol$^{-1}$ at 400 K. When the temperature is held at 400 K for 30 min, the $\chi_M T$ value further increases to reach 2.7 cm$^3$ K mol$^{-1}$ as a saturated value, indicating that ca. 90% of LS species show spin transition to the HS state. On further cooling, the $\chi_M T$ value decreases gradually from 2.7 cm$^3$ K mol$^{-1}$ at 400 K to 2.4 cm$^3$ K mol$^{-1}$ at 374 K, then decreases more smoothly to 1.6 cm$^3$ K mol$^{-1}$ at 20 K, and finally decreases abruptly to ca. 1.1 cm$^3$ K mol$^{-1}$ at 5 K, revealing the coexistence of HS and LS species in the whole temperature region after the first heating. The decreasing of the $\chi_M T$ value below 20 K is due to the zero-field splitting of the HS Fe$^{II}$ complex.

To prove the structural modulation before and after the initial spin transition upon first heating, we took PXRD data for **1** after SQUID measurements (Figure S1). As clearly apparent from Figure S1, the crystalline phase of **1** was converted to an amorphous form after SQUID measurements. The IR spectrum for this amorphous sample was also measured (Figure S2), and the spectrum showed the additional characteristic band at ca. 1637 cm$^{-1}$, corresponding to the C=N stretching vibration of the coordinated Schiff-base ligand of the HS complex [26,62]. This result indicated the existence of both HS (albeit being not fully characterized) and LS species in the amorphous phase at RT, and was consistent with the magnetic data in the second cycle. On the other hand, the characteristic bands of the C≡N stretching vibration of MeCN (2271 and 2251 cm$^{-1}$) were weakened (but not perfectly disappeared) due mainly to the loss of the lattice MeCN solvent molecule. To sum it all up, these results demonstrate that the irreversible spin conversion at around 400 K in the first heating process is related to the crystal-to-amorphous transformation associated with the loss of lattice MeCN solvent. Desolvation effects are reported in a variety of SCO systems in both positive (occurrence of abrupt and/or hysteretic spin transition) and negative (disappearance of SCO) ways [78–84] but a concomitant crystal-to-amorphous transformation is rarely observed [85].

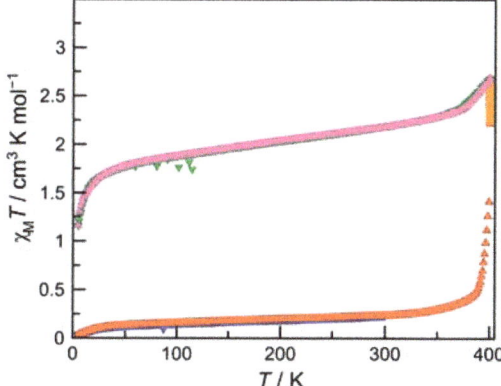

**Figure 2.** Temperature dependence of the $\chi_M T$ product for **1** at a sweep rate of 2 K min$^{-1}$. The sample was cooled from 300 to 5 K (blue inverted triangles) and then warmed from 5 to 400 K (red triangles) in the first cycle, and the temperature was held at 400 K for 30 min (orange triangles), and then the sample was cooled from 400 to 5 K (green inverted triangles) and then warmed from 5 to 400 K (magenta triangles) in the second cycle.

### 3.3. Crystal Structures

Single-crystal X-ray diffraction analyses were performed for **1** at 296, 350, 375, and 387 K. Structure determination at higher temperature, i.e., 400 K was unsuccessful since the single crystal became amorphous during the measurement. Table 1 summarizes the crystallographic data and Table 2 lists the relevant coordination bond lengths, angles, and additional structural parameters, such as Σ [86], Θ [87], S(Oh) [74], and octahedral volume. Since the crystal structures at the four temperatures are quite similar except for the subtle expansion of the cell volume and the FeN$_6$ coordination sphere from 296 to 387 K, we discuss below the structure at 296 K as a representative. The crystallographic unique unit consists of one complex-cation [Fe$^{II}$MeCN(L$_{3\text{-Me-3}}^{Ph}$)]$^{2+}$, two BPh$_4^-$ ions, and one MeCN molecule as the lattice solvent, which is disordered at two positions. The one H$_2$O molecule in each [Fe$^{II}$MeCN(L$_{3\text{-Me-3}}^{Ph}$)](BPh$_4$)$_2$ unit characterized by elemental analysis and TG measurement could not be found, while the Platon analysis [88] indicates that there are some voids which can involve water molecules.

Figure 3 shows the molecular structure of the complex-cation [Fe$^{II}$MeCN(L$_{3\text{-Me-3}}^{Ph}$)]$^{2+}$ at 296 K, in which the Fe$^{II}$ ion is coordinated by the N$_5$ donor atoms of the linear pentadentate Schiff-base ligand L$_{3\text{-Me-3}}^{\text{Ph}}$ and the nitrogen atom of the MeCN monodentate ligand to give an octahedral coordination environment. Two terminal triazole moieties take *cis*-positions, and one of two triazole moieties and the MeCN ligand at the sixth coordination position coordinate to the central Fe$^{II}$ ion from opposite directions. This coordination mode is same as that of the related imidazole-containing complex [59]. The bent angle of Fe–N10–C26 at Fe–NCMe is 172.2(1)°. The Fe–N lengths are in the range of 1.9580(13)–2.0901(13) Å, and the average Fe–N distance is 1.988 Å, typical for a LS Fe$^{II}$ complex with N$_6$ donors. It is noteworthy that the coordination bond length of Fe–N(amine) is longer than those of other Fe–N distances. In addition to the average Fe–N distance, the degree of both angular and trigonal distortion, i.e., Σ and Θ, and S(Oh) (Table 2) are lower than those of related imidazole-containing complex **2** (Average Fe–N distance, Σ, Θ, and S(Oh) at 296 K are = 2.085 Å, 79.7, 121.0, 1.004 for Fe1 site, and 2.155 Å, 89.1, 168.2, 1.541 for Fe2 site, respectively) [59]. These are consistent since the both the Fe1 and the Fe2 site of **2** at 296 K are mixtures of HS and LS species.

Figure 4 shows the selected intermolecular interactions of **1** at 296 K. Firstly, the MeCN ligand of the complex-cation is surrounded by four nearest phenyl rings of two B(1)Ph$_4^-$ ions via CH$\cdots\pi$ interactions between the methyl group of the MeCN ligand and phenyl groups of B(1)Ph$_4^-$ ions with the C27 (Me) to centroid (Ph) distances being in the range of 3.567–3.788 Å. Since the three hydrogen atoms of the methyl group of the MeCN can form only three CH$\cdots\pi$ interactions, the four CH$\cdots\pi$ interactions of each MeCN indicated in Figure 4 are averaged as one. Secondly, two neighboring complex-cations are connected by a P4AE (parallel fourfold aryl embrace) interaction [28,89], forming a dimeric structure with the C23 (Ph) to centroid (triazole) distance of CH$\cdots\pi$ and centroid (Ph) to centroid (Ph) distance of $\pi$–$\pi$ interactions are 3.690 and 3.663 Å, respectively. As a result, a 1D chain structure is constructed by alternately interacted complex-cations and B(1)Ph$_4^-$ ions via CH$\cdots\pi$ interactions (longitudinal direction in Figure 5), and further connections of the 1D chains through P4AE interactions (transverse direction in Figure 5) form a 2D extended layer structure. The remaining B(2)Ph$_4^-$ ions and MeCN lattice solvents exist in the spaces of the 2D layer with the intermolecular CH$\cdots$N interaction (C61$\cdots$N11 = 3.486 Å). Finally, there are additional CH$\cdots\pi$ interactions between the layers, resulting in the construction of a 3D supramolecular network in the whole crystal lattice. This molecular assembly is quite different from the tetrameric assembly through four intermolecular NH$\cdots$Cl interactions of **2** [59]. Therefore, this difference of molecular assembly is presumably responsible for the emergence of different SCO profiles between **1** and **2**.

**Table 1.** X-ray crystallographic data for **1**.

| Temperature/K | 296 | 350 | 375 | 387 |
| --- | --- | --- | --- | --- |
| Formula | | C$_{77}$H$_{75}$B$_2$FeN$_{11}$ | | |
| Formula weight | | 1231.95 | | |
| Crystal system | | monoclinic | | |
| Space group | | P2$_1$/n (No.14) | | |
| a/Å | 11.3072(2) | 11.3443(2) | 11.3962(3) | 11.4005(2) |
| b/Å | 41.0958(8) | 41.2866(9) | 41.3406(12) | 41.3938(10) |
| c/Å | 14.4788(4) | 14.4825(4) | 14.4308(5) | 14.4297(4) |
| β/deg | 92.977(2) | 92.796(2) | 92.513(3) | 92.473(2) |
| V/Å$^3$ | 6718.9(3) | 6775.1(3) | 6792.2(4) | 6803.2(3) |
| Z | 4 | 4 | 4 | 4 |
| d$_{calcd.}$/g cm$^{-3}$ | 1.218 | 1.208 | 1.205 | 1.203 |
| μ (Mo Kα)/mm$^{-1}$ | 0.277 | 0.275 | 0.274 | 0.273 |
| R$_1$ [a] (I>2sigma(I)) | 0.0458 | 0.0500 | 0.0779 | 0.0601 |
| wR$_2$ [b] (I>2sigma(I)) | 0.1062 | 0.1176 | 0.1944 | 0.1511 |
| R$_1$ [a] (all data) | 0.0678 | 0.0826 | 0.1220 | 0.1030 |
| wR$_2$ [b] (all data) | 0.1161 | 0.1328 | 0.2177 | 0.1738 |
| S | 1.029 | 1.022 | 1.065 | 1.022 |
| CCDC number | 1911292 | 1911293 | 1911294 | 1911295 |

[a] $R_1 = \Sigma||Fo| - |Fc||/\Sigma|Fo|$. [b] $wR_2 = [\Sigma w(|Fo|^2 - |Fc|^2)^2/\Sigma w|Fo^2|^2]^{1/2}$.

**Table 2.** Relevant coordination bond lengths (Å), angles (°) and structural parameters for **1**. Σ [86] and Θ [87] are angular indices characteristic for the spin state of the complex. S(Oh) is the continuous shape measures (CShMs) of the $Fe^{II}$ centers relative to the ideal octahedron [74].

| Temperature/K | 296 | 350 | 375 | 387 |
|---|---|---|---|---|
| Fe1–N3 | 1.9770(13) | 1.9788(15) | 1.989(3) | 1.986(2) |
| Fe1–N4 | 1.9955(11) | 1.9976(13) | 2.004(2) | 1.9985(18) |
| Fe1–N5 | 2.0901(13) | 2.0918(15) | 2.088(3) | 2.093(2) |
| Fe1–N6 | 1.9477(12) | 1.9494(14) | 1.947(2) | 1.9533(19) |
| Fe1–N7 | 1.9577(12) | 1.9619(14) | 1.966(2) | 1.9683(17) |
| Fe1–N10 | 1.9580(13) | 1.9599(15) | 1.961(3) | 1.967(2) |
| Average Fe–N | 1.988 | 1.990 | 1.993 | 1.994 |
| N3–Fe1–N4 | 80.25(5) | 80.04(6) | 79.94(10) | 79.85(8) |
| N3–Fe1–N5 | 177.92(5) | 177.74(6) | 177.53(10) | 177.45(8) |
| N3–Fe1–N6 | 95.80(5) | 95.89(6) | 96.07(11) | 96.13(8) |
| N3–Fe1–N7 | 83.58(5) | 83.46(6) | 83.33(10) | 83.42(7) |
| N3–Fe1–N10 | 88.57(5) | 88.49(6) | 88.51(10) | 88.08(8) |
| N4–Fe1–N5 | 97.68(5) | 97.70(6) | 97.59(11) | 97.61(8) |
| N4–Fe1–N6 | 173.48(5) | 173.33(6) | 173.26(10) | 173.32(8) |
| N4–Fe1–N7 | 94.11(5) | 94.16(5) | 94.04(9) | 94.05(7) |
| N4–Fe1–N10 | 90.57(5) | 90.45(6) | 90.48(10) | 90.40(7) |
| N5–Fe1–N6 | 86.23(5) | 86.34(6) | 86.38(11) | 86.39(8) |
| N5–Fe1–N7 | 96.37(5) | 96.56(6) | 96.82(10) | 96.67(8) |
| N5–Fe1–N10 | 91.69(5) | 91.72(6) | 91.60(10) | 92.09(8) |
| N6–Fe1–N7 | 80.23(5) | 80.06(6) | 80.04(10) | 80.13(8) |
| N6–Fe1–N10 | 94.53(5) | 94.74(6) | 94.87(10) | 94.82(8) |
| N7–Fe1–N10 | 170.05(5) | 169.89(6) | 169.83(11) | 169.56(8) |
| Σ | 61.89 | 62.83 | 63.27 | 63.90 |
| Θ | 93.47 | 95.21 | 96.77 | 95.69 |
| S(Oh) | 0.721 | 0.742 | 0.747 | 0.753 |
| Octahedral volume (Å$^3$) | 10.291 | 10.320 | 10.358 | 10.389 |

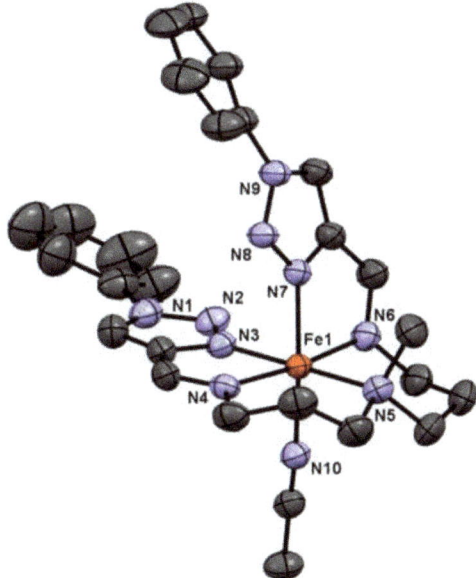

**Figure 3.** ORTEP drawing of the complex-cation $[Fe^{II}MeCN(L_{3\text{-Me-}3}{}^{Ph})]^{2+}$ of **1** at 296 K with the atom numbering scheme except for carbon and hydrogen atoms. The thermal ellipsoids are drawn with a 50% probability level. Hydrogen atoms have been omitted for clarity.

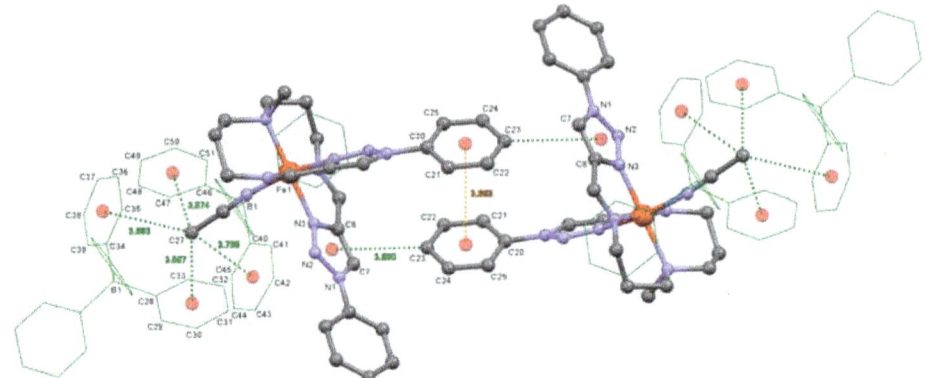

**Figure 4.** Representative intermolecular interactions of **1** at 296 K. Complex-cations are shown as a ball and stick model. B(1)Ph$_4^-$ ions are indicated as green wireframe. Centroids of aromatic rings are described as transparent-red balls. π–π (orange) and CH⋯π (light green) interactions are indicated as dotted lines. The MeCN ligand is surrounded by four nearest phenyl rings of two B(1)Ph$_4^-$ ions via CH⋯π interactions. Two neighboring complex-cations are connected by a P4AE interaction, forming a dimeric structure. Hydrogen atoms have been omitted for clarity.

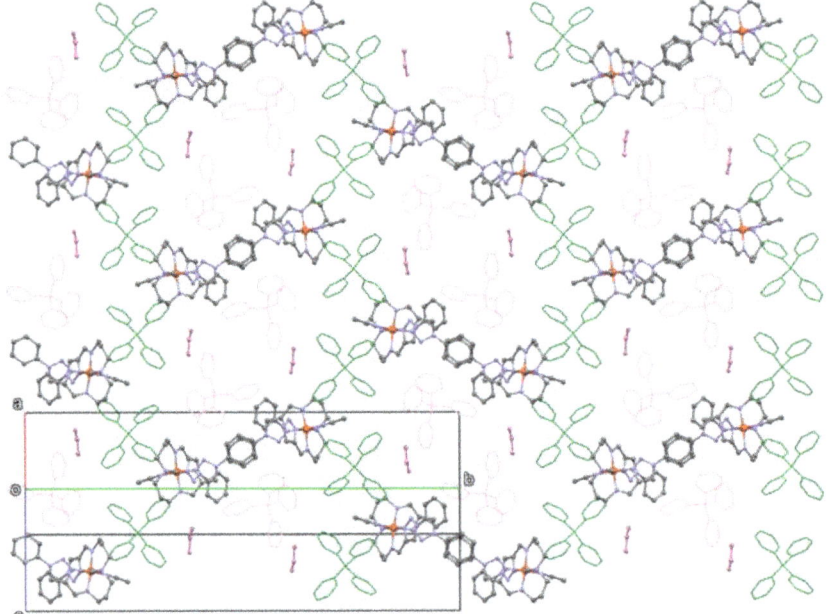

**Figure 5.** 2D layered structure of **1** at 296 K. Complex-cations and MeCN lattice solvents (violet) are shown as a ball and stick model. B(1)Ph$_4^-$ (green) and B(2)Ph$_4^-$ (pink) ions are indicated as wireframe. CH⋯π interactions connect complex-cations and B(1)Ph$_4^-$ ions alternately into a 1D chain (longitudinal direction). The 1D chains are further connected by P4AE interactions between two neighboring complex-cations (transverse direction), forming 2D extended structure. B(2)Ph$_4^-$ ions and MeCN lattice solvents exist in the spaces of the 2D layer. Hydrogen atoms have been omitted for clarity.

*3.4. DFT Calculations*

To explore the origin of the higher spin transition temperature of the 1,2,3-triazole-containing complex **1** than the imidazole-containing complex **2**, we performed DFT calculations. It should be noted here that only the experimental crystal structure data are available for LS **1** (vide supra) and for **2** corresponding to the mixture of HS and LS species at 296 K [59].

First, we performed DFT calculations to estimate HS–LS energy differences ($\Delta E$) at the experimental geometry (Table S1). Reflecting the spin state of the experimental crystal structures, large positive $\Delta E$ value (168.3 kJ mol$^{-1}$) and small $\Delta E$ values (45.5 kJ mol$^{-1}$ and −13.3 kJ mol$^{-1}$) were observed for **1** and **2**, respectively.

Next, we performed DFT geometry optimizations for both complexes in gas-phase. For the LS state of **1**, the structural parameters of DFT optimized structure indicated reasonable agreement with the crystal structure data of LS phase at 296 K (See, Table 2 and Table S2). Although the structural features (average Fe–N length, $\Sigma$, $\Theta$, S(Oh), and octahedral volume) of the DFT optimized LS **1** and **2** are similar to each other, the Fe–N(triazole) distances in **1** are shorter than the Fe–N(imidazole) distances in **2** by 0.05–0.07 Å. It should be mentioned here that the same tendency was observed in the comparison of the experimental crystal structures of the LS complex with the 1,2,3-triazole-containing linear hexadentate ligand and that of the imidazole-containing one [18]. Therefore, it can be speculated that 1,2,3-triazole-containing ligands form a stronger ligand field compared to imidazole-containing ones. For the HS state, the DFT optimized geometries of **1** and **2** are very similar to each other (Table S2). In addition, the structural parameters (average Fe–N distance, $\Sigma$, $\Theta$, S(Oh), and octahedral volume) of the DFT optimized HS **1** and **2** are also similar to those in the experimental geometry of HS phase of similar imidazole-containing Fe$^{II}$ complexes [60], while the monodentate ligand of them are different from **1** and **2** (average Fe–N distance, $\Sigma$, $\Theta$, S(Oh), and octahedral volume at 296 K are = 2.194 Å, 98.8, 187.1, 1.983, and 13.405 for NCS complex, and 2.198 Å, 99.3, 185.9, 1.986, and 13.474 for NCSe complex, respectively).

We also calculated $\Delta E$ values at the DFT optimized geometries to compare the strength of the ligand field of **1** and **2**. As shown in Table S3, the large positive $\Delta E$ values are found in LS geometry of both **1** and **2**. The $\Delta E$ value of **1** is 28.5 kJ mol$^{-1}$ larger than **2**, which implies that the strength of the ligand field is stronger in **1** rather than in **2**, as expected from the aforementioned structural features.

We also performed DFT geometry optimizations for model complexes **1'** and **2'**, in which two Ph groups in **1** and two Me groups in **2** were replaced by hydrogen atom to estimate the strength of the ligand field, excluding $\pi$ effects of the Ph ring and the steric effect of Me groups. The structural parameters of the DFT optimized structure are listed in Table S4. The structural parameters in **1'** were hardly affected by excluding Ph groups in **1**, which implies that the $\pi$ effects of the Ph ring are negligibly small, while the structural parameters in **2'** were slightly affected by excluding Me groups in **2**. It is worth mentioning that Fe–N(triazole) distances in **1'** are still shorter than the Fe–N(imidazole) ones in **2'** by 0.03 Å. Table S3 shows that the replacement of Ph groups in **1** also hardly affected the calculated $\Delta E$ value, and the $\Delta E$ value of **1'** (149.3 kJ mol$^{-1}$) is also still larger than **2'** (131.9 kJ mol$^{-1}$). These results demonstrated that the ligand field is stronger in the triazole-containing complex rather than in the imidazole-containing one, even if the $\pi$ effects of Ph rings and steric effect of Me groups were excluded.

From the above results, our DFT calculations elucidated that 1,2,3-triazole-containing ligands form the stronger ligand field compared to imidazole-containing ligands even for linear pentadentate ligand system, inducing the shift of the spin transition temperature from the lower temperature region in the imidazole-based complex to the higher region in the 1,2,3-triazole-based complex.

## 4. Conclusions

In conclusion, here we have expanded our 1,2,3-triazole-containing polydentate ligand iron(II) SCO family into a linear pentadentate ligand system. The newly synthesized complex **1** shows an abrupt and incomplete HT SCO at around 400 K while the spin transition is irreversible due to the crystal-to-amorphous transformation associated with the loss of the lattice MeCN solvent. Although the cooperativity through the molecular assembly for SCO profile of **1** was not directly compared to that of the imidazole-containing analogue **2**, the spin transition of **1** occurred more abruptly in the higher temperature region above RT than that of **2**. The 2D supramolecular structure based on the multiple CH···π interactions between MeCN ligand and two B(1)Ph$_4^-$ ions, and P4AE interactions between two neighboring complex-cations of **1** may have an important role for the emergence of cooperativity in the crystal lattice. DFT optimized HS and LS structures in the gas-phase of 1,2,3-triazole-containing system were compared to those of related imidazole-containing systems for the first time based on the experimental crystal structures of **1** (LS state) and **2** (mixture of LS and HS states), demonstrating that the 1,2,3-triazole-containing ligand L$_{3\text{-Me-}3}^{Ph}$ generates a stronger ligand field around the N$_6$ octahedral iron(II) core than its imidazole analogue ligand H$_2$L$^{2Me}$. Syntheses of analogues of **1** with different axial ligands are currently underway for the construction of the spectrochemical series and for fine tuning of $T_{1/2}$ of the present 1,2,3-triazole-based pentadentate ligand system.

**Supplementary Materials:** The following are available online at http://www.mdpi.com/2073-4352/9/6/276/s1, Figure S1: PXRD patterns of **1** at RT in different states: simulated from the single crystal X-ray data at 296 K; as-synthesized; after SQUID measurements, Figure S2: IR spectra (KBr) for **1** at RT in different states: as-synthesized; after SQUID measurements, Table S1: HS–LS energy differences of **1** and **2** for Experimental geometry, Table S2: Relevant coordination bond lengths (Å) and structural parameters in DFT optimized structures for **1** and **2**, Table S3: HS-LS energy differences of **1**, **2**, **1'**, and **2'** for DFT optimized geometry (LS), Table S4: Relevant coordination bond lengths (Å) and structural parameters in DFT optimized structures for model complexes **1'** and **2'**, Table S5: Cartesian coordinates (Å) of DFT optimized geometry of **1** (LS, in gas-phase), Table S6: Cartesian coordinates (Å) of DFT optimized geometry of **1** (HS, in gas-phase), Table S7: Cartesian coordinates (Å) of DFT optimized geometry of **2** (LS, in gas-phase), Table S8: Cartesian coordinates (Å) of DFT optimized geometry of **2** (HS, in gas-phase), Figure S3: DFT optimized structures of LS **1** (**a**), HS **1** (**b**), LS **2** (**c**), and HS **2** (**d**) in gas-phase.

**Author Contributions:** Conceptualization, H.H.; Methodology, H.H. and T.U.; validation, T.M., H.H., and T.U.; formal analysis, H.H. and T.U.; investigation, T.M., K.N., and H.H.; resources, H.H. and T.U.; data curation, H.H. and T.U.; writing—original draft preparation, H.H. and T.U.; writing—review and editing, H.H. and T.U.; visualization, H.H.; supervision, H.H.; project administration, H.H.

**Funding:** This work was supported in part by JSPS KAKENHI Grant numbers JP18K14240 (to H.H.) and JP18K05028 (to T.U.).

**Acknowledgments:** A part of this work was conducted in Institute for Molecular Science, supported by Nanotechnology Platform Program (Molecule and Material Synthesis) of the Ministry of Education, Culture, Sports, Science and Technology (MEXT), Japan.

**Conflicts of Interest:** The authors declare no conflict of interest.

## References

1. Gütlich, P.; Goodwin, H.A. (Eds.) *Spin Crossover in Transition Metal Compounds I-III*; Topics in Current Chemistry; Springer: Berlin, Germany, 2004; Volumes 233–235.
2. Halcrow, M.A. (Ed.) *Spin-Crossover Materials–Properties and Applications*; John Wiley & Sons: Chichester, UK, 2013.
3. Gütlich, P.; Gaspar, A.B.; Garcia, Y. Spin state switching in iron coordination compounds. *Beilstein J. Org. Chem.* **2013**, *9*, 342–391. [CrossRef] [PubMed]
4. Linares, J.; Codjovi, E.; Garcia, Y. Pressure and Temperature Spin Crossover Sensors with Optical Detection. *Sensors* **2012**, *12*, 4479–4492. [CrossRef] [PubMed]
5. Gentili, D.; Demitri, N.; Schäfer, B.; Liscio, F.; Bergenti, I.; Ruani, G.; Ruben, M.; Cavallini, M. Multi-modal sensing in spin crossover compounds. *J. Mater. Chem. C* **2015**, *3*, 7836–7844. [CrossRef]
6. Halcrow, M.A. Spin-crossover Compounds with Wide Thermal Hysteresis. *Chem. Lett.* **2014**, *43*, 1178–1188. [CrossRef]

7. Brooker, S. Spin crossover with thermal hysteresis: Practicalities and lessons learnt. *Chem. Soc. Rev.* **2015**, *44*, 2880–2892. [CrossRef] [PubMed]
8. Matsumoto, T.; Newton, G.N.; Shiga, T.; Hayami, S.; Matsui, Y.; Okamoto, H.; Kumai, R.; Murakami, Y.; Oshio, H. Programmable spin-state switching in a mixed-valence spin-crossover iron grid. *Nat. Commun.* **2014**, *5*, 1–8. [CrossRef] [PubMed]
9. Li, Z.-Y.; Ohtsu, H.; Kojima, T.; Dai, J.-W.; Yoshida, T.; Breedlove, B.K.; Zhang, W.-X.; Iguchi, H.; Sato, O.; Kawano, M.; et al. Direct Observation of Ordered High-Spin-Low-Spin Intermediate States of an Iron(III) Three-Step Spin-Crossover Complex. *Angew. Chem. Int. Ed.* **2016**, *55*, 5184–5189. [CrossRef]
10. Murphy, M.J.; Zenere, K.A.; Ragon, F.; Southon, P.D.; Kepert, C.J.; Neville, S.M. Guest programmable multistep spin crossover in a porous 2-D Hofmann-type material. *J. Am. Chem. Soc.* **2017**, *139*, 1330–1335. [CrossRef]
11. Zhang, D.; Trzop, E.; Valverde-Muñoz, F.J.; Piñeiro-López, L.; Muñoz, M.C.; Collet, E.; Real, J.A. Competing Phases Involving Spin-State and Ligand Structural Orderings in a Multistable Two-Dimensional Spin Crossover Coordination Polymer. *Cryst. Growth Des.* **2017**, *17*, 2736–2745. [CrossRef]
12. Bao, X.; Guo, P.-H.; Liu, W.; Tucek, J.; Zhang, W.-X.; Leng, J.-D.; Chen, X.-M.; Gural'skiy, I.; Salmon, L.; Bousseksou, A.; et al. Remarkably high-temperature spin transition exhibited by new 2D metal–organic frameworks. *Chem. Sci.* **2012**, *3*, 1629. [CrossRef]
13. Liu, W.; Bao, X.; Li, J.-Y.; Qin, Y.-L.; Chen, Y.-C.; Ni, Z.-P.; Tong, M.-L. High-Temperature Spin Crossover in Two Solvent-Free Coordination Polymers with Unusual High Thermal Stability. *Inorg. Chem.* **2015**, *54*, 3006–3011. [CrossRef]
14. Zheng, S.; Reintjens, N.R.M.; Siegler, M.A.; Roubeau, O.; Bouwman, E.; Rudavskyi, A.; Havenith, R.W.A.; Bonnet, S. Stabilization of the Low-Spin State in a Mononuclear Iron(II) Complex and High-Temperature Cooperative Spin Crossover Mediated by Hydrogen Bonding. *Chem. A Eur. J.* **2016**, *22*, 331–339. [CrossRef]
15. Hora, S.; Hagiwara, H. High-Temperature Wide Thermal Hysteresis of an Iron(II) Dinuclear Double Helicate. *Inorganics* **2017**, *5*, 49. [CrossRef]
16. Craze, A.R.; Howard-Smith, K.J.; Bhadbhade, M.M.; Mustonen, O.; Kepert, C.J.; Marjo, C.E.; Li, F. Investigation of the High-Temperature Spin-Transition of a Mononuclear Iron(II) Complex Using X-ray Photoelectron Spectroscopy. *Inorg. Chem.* **2018**, *57*, 6503–6510. [CrossRef] [PubMed]
17. Takahashi, K.; Yamamoto, K.; Yamamoto, T.; Einaga, Y.; Shiota, Y.; Yoshizawa, K.; Mori, H. High-Temperature Cooperative Spin Crossover Transitions and Single-Crystal Reflection Spectra of [$Fe^{III}$(qsal)$_2$](CH$_3$OSO$_3$) and Related Compounds. *Crystals* **2019**, *9*, 81. [CrossRef]
18. Hagiwara, H. High-temperature Spin Crossover of a Solvent-Free Iron(II) Complex with the Linear Hexadentate Ligand [Fe(L$_{2\text{-}3\text{-}2}^{Ph}$)](AsF$_6$)$_2$ (L$_{2\text{-}3\text{-}2}^{Ph}$ = bis[*N*-(1-Phenyl-1*H*-1,2,3-triazol-4-yl)methylidene-2-aminoethyl]-1,3- propanediamine). *Magnetochemistry* **2019**, *5*, 10. [CrossRef]
19. Weber, B.; Bauer, W.; Obel, J. An iron(II) spin-crossover complex with a 70 K wide thermal hysteresis loop. *Angew. Chem. Int. Ed.* **2008**, *47*, 10098–10101. [CrossRef]
20. Hayami, S.; Gu, Z.Z.; Yoshiki, H.; Fujishima, A.; Sato, O. Iron(III) spin-crossover compounds with a wide apparent thermal hysteresis around room temperature. *J. Am. Chem. Soc.* **2001**, *123*, 11644–11650. [CrossRef] [PubMed]
21. Kahn, O.; Martinez, C.J. Spin-transition polymers: From molecular materials toward memory devices. *Science* **1998**, *279*, 44–48. [CrossRef]
22. Southon, P.D.; Liu, L.; Fellows, E.A.; Price, D.J.; Halder, G.J.; Chapman, K.W.; Moubaraki, B.; Murray, K.S.; Létard, J.F.; Kepert, C.J. Dynamic interplay between spin-crossover and host-guest function in a nanoporous metal-organic framework material. *J. Am. Chem. Soc.* **2009**, *131*, 10998–11009. [CrossRef]
23. Arcis-Castíllo, Z.; Zheng, S.; Siegler, M.A.; Roubeau, O.; Bedoui, S.; Bonnet, S. Tuning the transition temperature and cooperativity of bapbpy-based mononuclear spin-crossover compounds: Interplay between molecular and crystal engineering. *Chem. A Eur. J.* **2011**, *17*, 14826–14836. [CrossRef] [PubMed]
24. Zhao, X.-H.; Zhang, S.-L.; Shao, D.; Wang, X.-Y. Spin Crossover in [Fe(2-Picolylamine)$_3$]$^{2+}$ Adjusted by Organosulfonate Anions. *Inorg. Chem.* **2015**, *54*, 7857–7867. [CrossRef] [PubMed]
25. Kershaw Cook, L.J.; Kulmaczewski, R.; Mohammed, R.; Dudley, S.; Barrett, S.A.; Little, M.A.; Deeth, R.J.; Halcrow, M.A. A Unified Treatment of the Relationship between Ligand Substituents and Spin State in a Family of Iron(II) Complexes. *Angew. Chem. Int. Ed.* **2016**, *55*, 4327–4331. [CrossRef]

26. Hagiwara, H.; Masuda, T.; Ohno, T.; Suzuki, M.; Udagawa, T.; Murai, K.I. Neutral Molecular Iron(II) Complexes Showing Tunable Bistability at Above, Below, and Just Room Temperature by a Crystal Engineering Approach: Ligand Mobility into a Three-Dimensional Flexible Supramolecular Network. *Cryst. Growth Des.* **2017**, *17*, 6006–6019. [CrossRef]
27. Nakanishi, T.; Okazawa, A.; Sato, O. Halogen Substituent Effect on the Spin-Transition Temperature in Spin-Crossover Fe(III) Compounds Bearing Salicylaldehyde 2-Pyridyl Hydrazone-Type Ligands and Dicarboxylic Acids. *Inorganics* **2017**, *5*, 53. [CrossRef]
28. Phonsri, W.; Macedo, D.S.; Vignesh, K.R.; Rajaraman, G.; Davies, C.G.; Jameson, G.N.L.; Moubaraki, B.; Ward, J.S.; Kruger, P.E.; Chastanet, G.; et al. Halogen Substitution Effects on N$_2$O Schiff Base Ligands in Unprecedented Abrupt Fe$^{II}$ Spin Crossover Complexes. *Chem. A Eur. J.* **2017**, *23*, 7052–7065. [CrossRef] [PubMed]
29. Kimura, A.; Ishida, T. Spin-Crossover Temperature Predictable from DFT Calculation for Iron(II) Complexes with 4-Substituted Pybox and Related Heteroaromatic Ligands. *ACS Omega* **2018**, *3*, 6737–6747. [CrossRef]
30. Nemec, I.; Herchel, R.; Boča, R.; Trávníček, Z.; Svoboda, I.; Fuess, H.; Linert, W. Tuning of spin crossover behaviour in iron(III) complexes involving pentadentate Schiff bases and pseudohalides. *Dalton Trans.* **2011**, *40*, 10090–10099. [CrossRef]
31. Tsuchida, R. Absorption Spectra of Co-ordination Compounds. *Bull. Chem. Soc. Jpn.* **1938**, *13*, 388–400. [CrossRef]
32. Shimura, Y. A Quantitative Scale of the Spectrochemical Series for the Mixed Ligand Complexes of d$^6$ Metals. *Bull. Chem. Soc. Jpn.* **1988**, *61*, 693–698. [CrossRef]
33. Ohta, S.; Yoshimura, C.; Matsumoto, N.; Okawa, H.; Ohyoshi, A. The Synthesis, Magnetic, and Spectroscopic Properties of Binuclear Iron(III) Complexes Bridged by Pyrazine, 1,1'-Tetramethylenebis(imidazol), or Bis(pyridine) Compounds Exhibiting a Spin-Equilibrium Behavior. *Bull. Chem. Soc. Jpn.* **1986**, *59*, 155–159. [CrossRef]
34. Hayami, S.; Inoue, K.; Osaki, S.; Maeda, Y. Synthesis and Magnetic Properties of Binuclear Iron(III) Complexes Containing Photoisomerization Ligand. *Chem. Lett.* **1998**, *27*, 987–988. [CrossRef]
35. Spiccia, L.; Fallon, G.D.; Grannas, M.J.; Nichols, P.J.; Tiekink, E.R.T. Synthesis and characterisation of mononuclear and binuclear iron(II) complexes of pentadentate and bis(pentadentate) ligands derived from 1,4,7-triazacyclononane. *Inorg. Chim. Acta* **1998**, *279*, 192–199. [CrossRef]
36. Boča, R.; Fukuda, Y.; Gembický, M.; Herchel, R.; Jaroščiak, R.; Linert, W.; Renz, F.; Yuzurihara, J. Spin crossover in mononuclear and binuclear iron(III) complexes with pentadentate Schiff-base ligands. *Chem. Phys. Lett.* **2000**, *325*, 411–419. [CrossRef]
37. Hayami, S.; Hosokoshi, Y.; Inoue, K.; Einaga, Y.; Sato, O.; Maeda, Y. Pressure-Stabilized Low-Spin State for Binuclear Iron(III) Spin-Crossover Compounds. *Bull. Chem. Soc. Jpn.* **2001**, *74*, 2361–2368. [CrossRef]
38. Kitashima, R.; Imatomi, S.; Yamada, M.; Matsumoto, N.; Maeda, Y. Gradual Two-step Spin Crossover Behavior of Binuclear Iron(III) Complex Bridged by trans -1,2-Bis(4-pyridyl)ethylene. *Chem. Lett.* **2005**, *34*, 1388–1389. [CrossRef]
39. Šalitroš, I.; Boča, R.; Dlháň, L.; Gembický, M.; Kožíšek, J.; Linares, J.; Moncol, J.; Nemec, I.; Perašínová, L.; Renz, F.; et al. Unconventional spin crossover in dinuclear and trinuclear iron(III) complexes with cyanido and metallacyanido bridges. *Eur. J. Inorg. Chem.* **2009**, 3141–3154. [CrossRef]
40. Nemec, I.; Boča, R.; Herchel, R.; Trávníček, Z.; Gembický, M.; Linert, W. Dinuclear Fe(III) complexes with spin crossover. *Monatsh. Chem.* **2009**, *140*, 815–828. [CrossRef]
41. Djukic, B.; Poddutoori, P.K.; Dube, P.A.; Seda, T.; Jenkins, H.A.; Lemaire, M.T. Bimetallic Iron(3+) Spin-Crossover Complexes Containing a 2,2'-Bithienyl Bridging bis-QsalH Ligand. *Inorg. Chem.* **2009**, *48*, 6109–6116. [CrossRef]
42. Fujinami, T.; Nishi, K.; Kitashima, R.; Murakami, K.; Matsumoto, N.; Iijima, S.; Toriumi, K. One-step and two-step spin crossover binuclear iron(III) complexes bridged by 4,4'-bipyridine. *Inorg. Chim. Acta* **2011**, *376*, 136–143. [CrossRef]
43. Boča, R.; Nemec, I.; Šalitroš, I.; Pavlik, J.; Herchel, R.; Renz, F. Interplay between spin crossover and exchange interaction in iron(III) complexes. *Pure Appl. Chem.* **2009**, *81*, 1357–1383. [CrossRef]
44. Herchel, R.; Boča, R.; Gembický, M.; Kožíšek, J.; Renz, F. Spin Crossover in a Tetranuclear Cr(III)−Fe(III)$_3$ Complex. *Inorg. Chem.* **2004**, *43*, 4103–4105. [CrossRef] [PubMed]

45. Renz, F.; Jung, S.; Klein, M.; Menzel, M.; Thünemann, A.F. Molecular switching complexes with iron and tin as central atom. *Polyhedron* **2009**, *28*, 1818–1821. [CrossRef]
46. Gembický, M.; Boča, R.; Renz, F. A heptanuclear Fe(II)–Fe(III)$_6$ system with twelve unpaired electrons. *Inorg. Chem. Commun.* **2000**, *3*, 662–665. [CrossRef]
47. Boča, R.; Šalitroš, I.; Kožíšek, J.; Linares, J.; Moncoľ, J.; Renz, F. Spin crossover in a heptanuclear mixed-valence iron complex. *Dalton Trans.* **2010**, *39*, 2198–2200. [CrossRef]
48. Renz, F.; Kerep, P. Unprecedented multiple electronic spin transition in hepta- and nonanuclear complex compounds observed by Mössbauer spectroscopy. *Polyhedron* **2005**, *24*, 2849–2851. [CrossRef]
49. Renz, F.; Hill, D.; Klein, M.; Hefner, J. Unprecedented multistability in dodecanuclear complex compound observed by Mössbauer spectroscopy. *Polyhedron* **2007**, *26*, 2325–2329. [CrossRef]
50. Ohyoshi, A.; Honbo, J.; Matsumoto, N.; Ohta, S.; Sakamoto, S. Spin-Equilibrium Behavior in Solution of an Iron(III) Complex [Bis[3-(3-methoxysalicylideneamine)propyl]amino-*O*,*N*,*N'*,*N''*,*O'*](pyridine)iron(III) Tetraphenylborate. *Bull. Chem. Soc. Jpn.* **1986**, *59*, 1611–1613. [CrossRef]
51. Maeda, Y.; Noda, Y.; Oshio, H.; Takashima, Y. $^{57}$Fe Mössbauer Spectra, Crystal Structure, and Spin-Crossover Behavior of [Fe(mbpN)(lut)]BPh$_4$. *Bull. Chem. Soc. Jpn.* **1992**, *65*, 1825–1831. [CrossRef]
52. Hirose, S.; Hayami, S.; Maeda, Y. Magnetic Properties of Iron(III) Complexes with Photoisomerizable Ligands. *Bull. Chem. Soc. Jpn.* **2000**, *73*, 2059–2066. [CrossRef]
53. Tanimura, K.; Kitashima, R.; Bréfuel, N.; Nakamura, M.; Matsumoto, N.; Shova, S.; Tuchagues, J.P. Infinite chain structure and steep spin crossover of a Fe$^{III}$ complex with a N$_3$O$_2$ pentadentate schiff-base ligand and 4-aminopyridine. *Bull. Chem. Soc. Jpn.* **2005**, *78*, 1279–1282. [CrossRef]
54. Bannwarth, A.; Schmidt, S.O.; Peters, G.; Sönnichsen, F.D.; Thimm, W.; Herges, R.; Tuczek, F. Fe$^{III}$ spin-crossover complexes with photoisomerizable ligands: Experimental and theoretical studies on the ligand-driven light-induced spin change effect. *Eur. J. Inorg. Chem.* **2012**, 2776–2783. [CrossRef]
55. Herchel, R.; Trávníček, Z. 5-Aminotetrazole induces spin crossover in iron(III) pentadentate Schiff base complexes: Experimental and theoretical investigations. *Dalton Trans.* **2013**, *42*, 16279–16288. [CrossRef] [PubMed]
56. Krüger, C.; Augustín, P.; Nemec, I.; Trávníček, Z.; Oshio, H.; Boča, R.; Renz, F. Spin crossover in iron(III) complexes with pentadentate schiff base ligands and pseudohalido coligands. *Eur. J. Inorg. Chem.* **2013**, 902–915. [CrossRef]
57. Krüger, C.; Augustín, P.; Dlháň, L.; Pavlik, J.; Moncoľ, J.; Nemec, I.; Boča, R.; Renz, F. Iron(III) complexes with pentadentate Schiff-base ligands: Influence of crystal packing change and pseudohalido coligand variations on spin crossover. *Polyhedron* **2015**, *87*, 194–201. [CrossRef]
58. Halcrow, M.A. The spin-states and spin-transitions of mononuclear iron(II) complexes of nitrogen-donor ligands. *Polyhedron* **2007**, *26*, 3523–3576. [CrossRef]
59. Nishi, K.; Fujinami, T.; Kitabayashi, A.; Matsumoto, N. Tetrameric spin crossover iron(II) complex constructed by imidazole···chloride hydrogen bonds. *Inorg. Chem. Commun.* **2011**, *14*, 1073–1076. [CrossRef]
60. Murakami, K.; Kitabayashi, A.; Yamauchi, S.; Nishi, K.; Fujinami, T.; Matsumoto, N.; Iijima, S.; Kojima, M. Iron(II) complexes with a linear pentadentate ligand H$_2$L$^1$ = bis(*N*,*N'*-2-methylimidazol-4-yl-methylideneaminopropyl)methylamine and a monodentate ligand X (X = N$_3^-$, NCS$^-$, NCSe$^-$). *Inorg. Chim. Acta* **2013**, *400*, 244–249. [CrossRef]
61. Hagiwara, H.; Tanaka, T.; Hora, S. Synthesis, structure, and spin crossover above room temperature of a mononuclear and related dinuclear double helicate iron(II) complexes. *Dalton Trans.* **2016**, *45*, 17132–17140. [CrossRef]
62. Hagiwara, H.; Okada, S. A polymorphism-dependent $T_{1/2}$ shift of 100 K in a hysteretic spin-crossover complex related to differences in intermolecular weak CH···X hydrogen bonds (X = S vs. S and N). *Chem. Commun.* **2016**, *52*, 815–818. [CrossRef]
63. Hagiwara, H.; Minoura, R.; Okada, S.; Sunatsuki, Y. Synthesis, Structure, and Magnetic Property of a New Mononuclear Iron(II) Spin Crossover Complex with a Tripodal Ligand Containing Three 1,2,3-Triazole Groups. *Chem. Lett.* **2014**, *43*, 950–952. [CrossRef]
64. Siddiki, A.A.; Takale, B.S.; Telvekar, V.N. One pot synthesis of aromatic azide using sodium nitrite and hydrazine hydrate. *Tetrahedron Lett.* **2013**, *54*, 1294–1297. [CrossRef]
65. Pathigoolla, A.; Pola, R.P.; Sureshan, K.M. A versatile solvent-free azide-alkyne click reaction catalyzed by in situ generated copper nanoparticles. *Appl. Catal. A Gen.* **2013**, *453*, 151–158. [CrossRef]

66. L'abbé, G.; Bruynseels, M.; Delbeke, P.; Toppet, S. Molecular rearrangements of 4-iminomethyl-1,2,3-triazoles. Replacement of 1-aryl substituents in 1 *H* -1,2,3-triazole-4-carbaldehydes. *J. Heterocycl. Chem.* **1990**, *27*, 2021–2027. [CrossRef]
67. Kahn, O. *Molecular Magnetism*; VCH: Weinheim, Germany, 1993.
68. Bain, G.A.; Berry, J.F. Diamagnetic Corrections and Pascal's Constants. *J. Chem. Educ.* **2008**, *85*, 532. [CrossRef]
69. *SCALE3 ABSPACK*, version 1.0.4; gui: 1.03; An oxford diffraction program; Oxford Diffraction Ltd.: Abingdon, UK, 2005.
70. *Rigaku Oxford Diffraction*, CrysAlisPro Software system, version 1.171.39.46; Rigaku Corporation: Oxford, UK, 2018.
71. Sheldrick, G.M. SHELXT—Integrated space-group and crystal-structure determination. *Acta Crystallogr. Sect. A Found. Adv.* **2015**, *71*, 3–8. [CrossRef] [PubMed]
72. Sheldrick, G.M. Crystal structure refinement with SHELXL. *Acta Crystallogr. Sect. C Struct. Chem.* **2015**, *71*, 3–8. [CrossRef]
73. Dolomanov, O.V.; Bourhis, L.J.; Gildea, R.J.; Howard, J.A.K.; Puschmann, H. OLEX2: A complete structure solution, refinement and analysis program. *J. Appl. Crystallogr.* **2009**, *42*, 339–341. [CrossRef]
74. Llunell, M.; Casanova, D.; Cirera, J.; Alemany, P.; Alvarez, S. *SHAPE2.1. Program for Calculating Continuous Shape Measures of Polyhedral Structures*; Universitat de Barcelona: Barcelona, Spain, 2013.
75. Zhao, Y.; Truhlar, D.G. The M06 suite of density functionals for main group thermochemistry, thermochemical kinetics, noncovalent interactions, excited states, and transition elements: Two new functionals and systematic testing of four M06-class functionals and 12 other functionals. *Theor. Chem. Acc.* **2008**, *120*, 215–241. [CrossRef]
76. Frisch, M.J.; Trucks, G.W.; Schlegel, H.B.; Scuseria, G.E.; Robb, M.A.; Cheeseman, J.R.; Scalmani, G.; Barone, V.; Mennucci, B.; Petersson, G.A.; et al. *Gaussian 09, Revision B.01*, Gaussian, Inc.: Wallingford, CT, USA, 2010.
77. Nakamoto, K. *Infrared and Raman Spectra of Inorganic and Coordination Compounds*, 6th ed.; John Wiley & Sons: Hoboken, NJ, USA, 2009.
78. Boča, R.; Baran, P.; Boča, M.; Dlháň, L.; Fuess, H.; Haase, W.; Linert, W.; Papánková, B.; Werner, R. Spin crossover in bis(2,6-bis(benzimidazol-2-yl)pyridine) iron(II) tetraphenylborate. *Inorg. Chim. Acta* **1998**, *278*, 190–196. [CrossRef]
79. Hagiwara, H.; Hashimoto, S.; Matsumoto, N.; Iijima, S. Two-Dimensional Iron(II) Spin Crossover Complex Constructed of Bifurcated NH···O$^-$ Hydrogen Bonds and π–π Interactions: [Fe$^{II}$(HL$^{H,Me}$)$_2$](ClO$_4$)$_2$·1.5MeCN (HL$^{H,Me}$ = Imidazol-4-yl-methylidene-8-amino-2-methylquinoline). *Inorg. Chem.* **2007**, *46*, 3136–3143. [CrossRef] [PubMed]
80. Costa, J.S.; Rodríguez-Jiménez, S.; Craig, G.A.; Barth, B.; Beavers, C.M.; Teat, S.J.; Aromí, G. Three-way crystal-to-crystal reversible transformation and controlled spin switching by a nonporous molecular material. *J. Am. Chem. Soc.* **2014**, *136*, 3869–3874. [CrossRef] [PubMed]
81. Yang, F.-L.; Chen, X.; Wu, W.-H.; Zhang, J.-H.; Zhao, X.-M.; Shi, Y.-H.; Shen, F. Spin switching in tris(8-aminoquinoline)iron(II)(BPh$_4$)$_2$: Quantitative guest-losing dependent spin crossover properties and single-crystal-to-single-crystal transformation. *Dalton Trans.* **2018**. [CrossRef] [PubMed]
82. Bushuev, M.B.; Vinogradova, K.A.; Gatilov, Y.V.; Korolkov, I.V.; Nikolaenkova, E.B.; Krivopalov, V.P. Spin crossover in iron(II) hexafluorophosphate complexes with 2-(pyridin-2-yl)-4-(3,5-di-R-1*H*-pyrazol-1-yl)-6-methylpyrimidines. *Inorg. Chim. Acta* **2017**, *467*, 238–243. [CrossRef]
83. Craze, A.R.; Bhadbhade, M.M.; Kepert, C.J.; Lindoy, L.F.; Marjo, C.E.; Li, F. Solvent Effects on the Spin-Transition in a Series of Fe(II) Dinuclear Triple Helicate Compounds. *Crystals* **2018**, *8*, 376. [CrossRef]
84. Sertphon, D.; Harding, P.; Murray, K.S.; Moubaraki, B.; Neville, S.M.; Liu, L.; Telfer, S.G.; Harding, D.J. Solvent Effects on the Spin Crossover Properties of Iron(II) Imidazolylimine Complexes. *Crystals* **2019**, *9*, 116. [CrossRef]
85. Jornet-Mollá, V.; Giménez-Saiz, C.; Romero, F.M. Synthesis, Structure, and Photomagnetic Properties of a Hydrogen-Bonded Lattice of [Fe(bpp)$_2$]$^{2+}$ Spin-Crossover Complexes and Nicotinate Anions. *Crystals* **2018**, *8*, 439. [CrossRef]
86. Guionneau, P.; Marchivie, M.; Bravic, G.; Létard, J.-F.; Chasseau, D. Structural aspects of spin crossover. Examples of the [Fe$^{II}$L$_n$(NCS)$_2$] complexes. *Top. Curr. Chem.* **2004**, *234*, 97–128.
87. Marchivie, M.; Guionneau, P.; Létard, J.-F.; Chasseau, D. Photo-induced spin-transition: The role of the iron(II) environment distortion. *Acta Crystallogr. Sect. B* **2005**, *61*, 25–28. [CrossRef]

88. Spek, A.L. Structure validation in chemical crystallography. *Acta Crystallogr. Sect. D Biol. Crystallogr.* **2009**, *65*, 148–155. [CrossRef]
89. Russell, V.; Scudder, M.; Dance, I. The crystal supramolecularity of metal phenanthroline complexes. *J. Chem. Soc. Dalton Trans.* **2001**, *0*, 789–799. [CrossRef]

© 2019 by the authors. Licensee MDPI, Basel, Switzerland. This article is an open access article distributed under the terms and conditions of the Creative Commons Attribution (CC BY) license (http://creativecommons.org/licenses/by/4.0/).

Article

# Solvent Effects on the Spin Crossover Properties of Iron(II) Imidazolylimine Complexes

**Darunee Sertphon** [1], **Phimphaka Harding** [1], **Keith S. Murray** [2], **Boujemaa Moubaraki** [2], **Suzanne M. Neville** [3], **Lujia Liu** [4], **Shane G. Telfer** [4] and **David J. Harding** [1,*]

1. Functional Materials and Nanotechnology Center of Excellence, Walailak University, Thasala, Nakhon Si Thammarat 80160, Thailand; sdarunee3@gmail.com (D.S.); kphimpha@mail.wu.ac.th (P.H.)
2. School of Chemistry, Monash University, Clayton, VIC 3800, Australia; keith.murray@monash.edu (K.S.M.); boujemaa.moubaraki@monash.edu (B.M.)
3. School of Chemistry, University of New South Wales, Sydney, NSW 2052, Australia; s.neville@unsw.edu.au
4. MacDiarmid Institute for Advanced Materials and Nanotechnology, Institute of Fundamental Sciences, Massey University, Palmerston North 4442, New Zealand; lujia.liu@northwestern.edu (L.L.); S.Telfer@massey.ac.nz (S.G.T.)
* Correspondence: hdavid@mail.wu.ac.th; Tel.: +66-75-672094

Received: 23 January 2019; Accepted: 20 February 2019; Published: 22 February 2019

**Abstract:** A series of Fe(II) complexes, *fac*-[Fe(4-ima-Bp)$_3$](Y)$_2$·sol (Y = ClO$_4$; sol = 3EtOH **1**, 3MeOH **2**; Y= BF$_4$; sol = EtOH·4H$_2$O **3**, 4H$_2$O **4** and 3.5MeCN **5**) have been prepared and structurally and magnetically characterized. The low temperature structures of **1**, **2** and **5** have been determined by X-ray crystallography with LS Fe(II) centres found in all cases. Extensive C–H···π interactions between the cations form 2D layers, which are linked to one another through N–H···O and O–H···O hydrogen bonds, resulting in high cooperativity. Despite **5** containing MeCN, N–H···O/F hydrogen bonds, and C–H···π and π-π interactions combine to give similar 2D layers. Magnetic measurements reveal moderately abrupt spin crossover for **1-4**; becoming more gradual and only 50% complete in **1** due to solvent loss. The MeCN solvate shows more gradual SCO and reinforces how subtle changes in packing can significantly influence SCO behaviour.

**Keywords:** spin crossover; iron(II) complexes; C–H···π interactions; magnetic properties; thermochromism

## 1. Introduction

Spin crossover (SCO) describes the interconversion between a high spin (HS) and low spin (LS) state induced by a range of external perturbations often temperature or light irradiation [1,2]. SCO in Fe(II) complexes with an octahedral geometry is dominant as it transforms a diamagnetic LS state (S = 0) to a paramagnetic HS state (S = 2) with a clear change in colour and lengthening of the Fe(II)-ligands bond distances [3–7]. Materials displaying SCO behaviour continue to be intensively studied due to their potential use as active components in memory, display and sensing devices [8–10] particularly photo-induced SCO complexes, or light-induced spin state trapping (LIESST) [11,12] which has been reported in many Fe(II) SCO systems.

Of the many iron(II) systems investigated those incorporating imidazole Schiff-base ligands have amongst the most varied SCO behaviours as exemplified in recent reviews by Kruger and Matsumoto [13,14]. In terms of mononuclear systems there are two basic designs one of which uses hexadentate ligands [15,16] and the other exploiting chelating imine ligands derived from an imidazolecarboxaldehyde. The ligand structures and abbreviations used in this article are shown in Chart 1. Amongst the first reports concerned [Fe{H$_3$(2-Me-im)$_3$-tren}]Cl·X (X = PF$_6$, AsF$_6$, SbF$_6$ and OTf) where the anion causes a change in magnetic behaviour from 50% SCO to abrupt and complete

SCO [17]. A feature of all the compounds are N-H···Cl hydrogen bonds that link the Fe(II) centres—this acts to enhance the communication pathways between SCO sites in the solid-state. Some years later, Seredyuk and co-workers studied [Fe{N-nBu-2-im)$_3$-tren}](PF$_6$)$_2$, which exhibits SCO behaviour with thermal hysteresis sensitive to scan rate (i.e., 14 K at 4 K min$^{-1}$ and 41 K at 0.1 K min$^{-1}$) [18,19]. This measurement scan rate dependency is due to the kinetically-driven formation of two distinct LS phases which differ in butyl group conformation.

**Chart 1.** Structure of the common ligands used in mononuclear imidazole SCO systems.

SCO systems with chelating imidazole ligands are also well described with [Fe(2-Me-4-ima-CH$_2$CH$_2$py)$_3$](X)$_2$ (X = PF$_6$, ClO$_4$, BF$_4$) all showing abrupt SCO due to N-H···N hydrogen bonds involving the imidazole and pyridine [20]. Surprisingly, despite their different shapes and sizes the anion has little effect on the spin transition temperature. In 2011, Matsumoto et al. examined *fac*-[Fe(2-Me-4-ima-R)$_3$]Cl·PF$_6$ {R= (Me), ethyl (Et), *n*-propyl (*n*-Pr), *n*-butyl (*n*-Bu), and *n*-pentyl (*n*-Pen)} [21]. Once again N-H···Cl hydrogen bonds link the spin centres but this time the different alkyl groups result in a variety of supramolecular motifs giving both gradual and abrupt SCO accompanied by hysteresis. Interestingly, *fac*-[Fe(2-Me-4-ima-*n*Pr)$_3$]Cl·PF$_6$ shows scan rate dependence of the hysteresis but unlike [Fe{N-*n*Bu-im)$_3$-tren}](PF$_6$)$_2$ there are no phase changes [22]. Kruger and co-workers reported [Fe(2-ima-*p*-C$_6$H$_4$OMe)$_3$](ClO$_4$)$_2$, a rare example of a *mer*-isomer [23]. In this case, π-π and C–H···π interactions and hydrogen bonds to the perchlorate anions link the Fe centres. However, the most interesting aspect of this complex is that it undergoes full switching under light irradiation [23]. Gu et al. have also investigated the impact of chirality on SCO in a series of complexes exemplified by *fac*-Λ-[Fe(*R*-N-Me-2-ima-CH(Me)Ph)$_3$](BF$_4$)$_2$·MeCN and *fac*-Δ-[Fe(*S*-N-Me-2-ima-CH(Me)Ph)$_3$](BF$_4$)$_2$·MeCN [24]. Racemisation of the stereogenic Fe(II) centre in the complexes is prevented by intramolecular π-π contacts between the imidazole and phenyl groups. In accordance with the identical packing arrangements both compounds exhibit moderately abrupt SCO. While the above shows there has been considerable research into imidazole based SCO systems aromatic groups remain poorly explored and in this work we report [Fe(4-ima-Bp)$_3$](Y)$_2$·Sol (Y = ClO$_4$; sol = EtOH **1**, MeOH **2**; Y= BF$_4$; sol = EtOH **3**, MeOH **4** and MeCN **5**) and investigate solvent and anion effects.

## 2. Materials and Methods

Perchlorate complexes are potentially explosive and should only be prepared in small quantities. 4-aminobiphenyl is a category 1 suspected carcinogen and facemasks and gloves must be used.

*2.1. General Remarks*

All manipulations were performed in air with reagent grade solvents. All chemicals were purchased from Aldrich Chemical Company (Singapore) or TCI Chemical Company (Tokyo, Japan) and used as received. Infrared spectra (as KBr discs) were recorded on a Perkin-Elmer Spectrum One infrared spectrophotometer in the range 400–4000 cm$^{-1}$. Electronic spectra were recorded in MeOH or MeCN at room temperature on a Shimadzu UV-1700 UV–VIS spectrophotometer (Kyoto, Japan). $^1$H NMR spectra were recorded on a Bruker 300 MHz FT-NMR spectrometer (Karlsruhe, Germany) at

25 °C in CDCl$_3$ with SiMe$_4$ added as an internal standard. Elemental analyses were carried out on a Eurovector EA3000 analyser (Pavia, Italy). ESI-MS were carried out on a Bruker Daltonics 7.0T Apex 4 FTICR mass spectrometer (Karlsruhe, Germany).

2.1.1. Synthesis of 4-ima-Bp

4-ima-Bp was prepared by mixing 4-imidazolecarboxaldehyde (0.480 g, 5.0 mmol) and 4-aminobiphenyl (0.846 g, 5.0 mmol) in methanol (15 cm$^3$). The mixture was warmed at ~50 °C under stirring for 1 h and then cooled to room temperature to give a pale yellow precipitate which was filtered. The pale yellow powder was dried in air, yield 1.194 g (97%). ν$_{max}$ (KBr)/cm$^{-1}$ 3123 w, 3054 w, 2948 w, 2787 m, 1621 s, 1585 s, 1485 m, 1456 m, 1329 m, 1119 s (Figure S1). λ$_{max}$/nm (DMF, ε/M$^{-1}$cm$^{-1}$) 340 (970). $^1$H NMR (CDCl$_3$, 295 K, 300 MHz) δ = 8.46 (s, 1H$_f$), 7.63 (s, 1H$_h$), 7.65-7.60 (m, 4H$_{2d, 2e}$), 7.48-7.31 (t, 2H$_{2c}$), 7.38-7.26 (m, 1H$_{a, 2b, g}$; Figure S2 Anal. Calc. for C$_{17}$H$_{14}$N$_2$O$_2$: C, 73.37; H, 5.07; N, 10.06. Found: C, 73.40; H, 5.12; N, 9.93%.

2.1.2. Synthesis of *fac*-[Fe(4-ima-Bp)$_3$](ClO$_4$)$_2$·3EtOH 1

The 4-ima-Bp ligand (0.148 g, 0.6 mmol) was dissolved in hot ethanol (3 cm$^3$) and a EtOH solution (2 cm$^3$) of Fe(ClO$_4$)$_2$·6H$_2$O (0.051 g, 0.2 mmol) was added dropwise with stirring to give a red solution. The mixture was stirred for 3 h and then cooled to room temperature to give dark red microcrystals of the product, which were dried in air, yield 0.106 g (53%). Red crystals suitable for single crystal X-ray diffraction were grow by slow evaporation of ethanol. m/z (ESI) 248.2 [4-ima-Bp]$^+$, 795.9 [Fe(4-ima-Bp)$_3$]$^+$, 895.5 [Fe(4-ima-Bp)$_3$][ClO$_4$]$^+$, 99 [ClO$_4$]$^-$. The microcrystals analyse for 2 equivalents of EtOH and 1 equivalent of water. Anal. Calc. for C$_{52}$H$_{53}$N$_9$O$_{11}$Cl$_2$Fe: C, 56.41; H, 4.83; N, 11.39. Found: C, 56.06; H, 4.54; N, 11.60%.

2.1.3. Synthesis of *fac*-[Fe(4-ima-Bp)$_3$](ClO$_4$)$_2$·3MeOH 2

The compound was made in a similar way to **1** using MeOH instead of EtOH giving dark red microcrystals, yield 0.217 g (67%). m/z (ESI) 248.2 [4-ima-Bp]$^+$, 557.2 [Fe(4-ima-Bp)$_2$]$^+$, 795.9 [Fe(4-ima-Bp)$_3$]$^+$, 99 [ClO$_4$]$^-$. Anal. Calc. for C$_{51}$H$_{51}$N$_9$O$_{11}$Cl$_2$Fe: C, 56.04; H, 4.71; N, 11.54. Found: C, 56.55; H, 4.65; N, 11.71%.

2.1.4. Synthesis of *fac*-[Fe(4-ima-Bp)$_3$](BF$_4$)$_2$·EtOH·4H$_2$O 3

The compound was made in a similar way to **1** using Fe(BF$_4$)$_2$·6H$_2$O instead of Fe(ClO$_4$)$_2$·6H$_2$O giving dark red microcrystals, yield 0.150 g (77%). m/z (ESI) 248.2 [4-ima-Bp]$^+$, 557.2 [Fe(4-ima-Bp)$_2$]$^+$, 795.9 [Fe(4-ima-Bp)$_3$]$^+$, 87 [BF$_4$]$^-$. Anal. Calc. for C$_{52}$H$_{47}$N$_9$B$_2$F$_8$O$_5$Fe: C, 56.36; H, 4.28; N, 11.38. Found: C, 56.11; H, 4.60; N, 11.56%.

2.1.5. Synthesis of *fac*-[Fe(4-ima-Bip)$_3$](BF$_4$)$_2$·4H$_2$O 4

The compound was made in a similar way to **3** using MeOH instead of EtOH giving red microcrystals yield 0.113 g (57%). m/z (ESI) 248.2 [4-ima-Bp]$^+$, 557.2 [Fe(4-ima-Bp)$_2$]$^+$, 795.9 [Fe(4-ima-Bp)$_3$]$^+$, 87 [BF$_4$]$^-$. Anal. Calc. for C$_{48}$H$_{47}$N$_9$F$_8$B$_2$O$_4$Fe: C, 55.21; H, 4.54; N, 12.08. Found: C, 54.93; H, 4.45; N, 11.90%.

2.1.6. Synthesis of *fac*-[Fe(4-ima-Bip)$_3$](BF$_4$)$_2$·3.5MeCN 5

Red crystals of the compound were made by dissolving 0.05 mmol of **3** in acetonitrile (5 cm$^3$) and allowing slow diffusion of Et$_2$O into the solution yielding red single crystals, 0.043 g (83%). m/z (ESI) 248.2 [4-ima-Bp]$^+$, 557.2 [Fe(4-ima-Bp)$_2$]$^+$, 87 [BF$_4$]$^-$. The compound analyses for 3 equivalents of MeCN. Anal. Calc. for C$_{54}$H$_{48}$N$_{12}$B$_2$F$_8$Fe: C, 59.22; H, 4.42; N, 15.56. Found: C, 59.11; H, 4.37; N, 15.40%.

## 2.2. VSM and SQUID Magnetometry Studies

Magnetic susceptibility data on **1-4** were collected on a Quantum Design Versalab Measurement System with a vibrating sample magnetometer (VSM) attachment within a small-bore hole cavity. Samples were contained within a polypropylene holder and held within a brass half-tube designed for VSM measurements. Measurements were taken continuously under an applied field of 0.3 T over the temperature range 300–50–300 K, at a ramp rate of 1 K min$^{-1}$ with no overshoot. Magnetic susceptibility data on **5** was collected on either a Quantum Design MPMS 5 or a MPMS XL-7 SQUID magnetometer (San Diego, USA) at a scan rate of 10 K·min$^{-1}$ being careful to allow long equilibrium times at each data point. All samples were taken freshly from the mother liquor in which the crystals were grown to limit any potential solvent loss. The raw data was corrected for the sample holder and diamagnetic contributions.

## 2.3. X-ray Crystallography

Crystal data and data processing parameters for the structures of **1**, **2** and **5** are given in Table 1. X-ray quality crystals of **1** and **2** were grown by slow evapouration of the solvent. Crystals were mounted on a glass fibre using perfluoropolyether oil and cooled rapidly to 100 K in a stream of cold nitrogen. The diffraction data of **1** and **2** were collected at 143 and 153 K on a Rigaku Spider diffractometer equipped with a MicroMax MM007 rotating anode generator, Cu$_\alpha$ radiation ($\lambda$ = 1.54178 Å), high-flux Osmic multilayer mirror optics, and a curved image-plate detector. The data were integrated, scaled and averaged with FS Process [25]. Diffraction data for **5** were collected at 123 K on a Bruker APEXII area detector with graphite monochromated MoK$\alpha$ ($\lambda$ = 0.71073 Å) [26]. After data collection, in each case an empirical absorption correction was applied [27]. The structures were then solved by direct methods and refined on all $F^2$ data using the SHELX suite of programs [28,29]. In all cases non-hydrogen atoms were refined with anisotropic thermal parameters; hydrogen atoms were included in calculated positions and refined with isotropic thermal parameters which were *ca.* 1.2 x (aromatic CH) or 1.5 x (CH$_2$, Me) the equivalent isotropic thermal parameters of their parent carbon atoms. All pictures were generated using Olex2 [30]. The CCDC numbers for the X-ray crystallographic data presented in this paper are 18925687-1892569 and can be obtained free of charge from the Cambridge Crystallographic Data Centre via www.ccdc.cam.ac.uk/data_request/cif.

Table 1. Crystallographic data and structure refinement for **1**, **2** and **5**.

| Compound | 1 | 2 | 5 |
|---|---|---|---|
| Formula | C$_{54}$H$_{57}$Cl$_2$FeN$_9$O$_{11}$ | C$_{51}$H$_{51}$Cl$_2$FeN$_9$O$_{11}$ | C$_{55.5}$H$_{49.5}$B$_2$FeN$_{12.5}$F$_8$ |
| Molecular weight/gmol$^{-1}$ | 1134.83 | 1092.76 | 1080.04 |
| Crystal system | Trigonal | Trigonal | Triclinic |
| Space group | $R\bar{3}$ | $R\bar{3}$ | $P\bar{1}$ |
| a/Å | 13.1242(9) | 12.9080(15) | 13.0204(7) |
| b/Å | 13.1242(9) | 12.9080(15) | 13.1932(8) |
| c/Å | 53.258(4) | 52.5240(4) | 18.4021(11) |
| $\alpha$/° | 90 | 90 | 73.540(4) |
| $\beta$/° | 90 | 90 | 86.375(4) |
| $\gamma$/° | 120 | 120 | 61.569(3) |
| T/K | 143(2) | 153(2) | 123(2) |
| Cell volume/Å$^3$ | 7944.4(12) | 7578.9(18) | 2656.3(3) |
| Z | 6 | 6 | 4 |
| Absorption coefficient/mm$^{-1}$ | 3.797 | 3.958 | 0.363 |
| Reflections collected | 23981 | 19730 | 39419 |
| Independent reflections, $R_{int}$ | 3413, 0.094 | 2875, 0.092 | 9354, 0.0825 |
| Max. and min. transmission | 0.561, 1.000 | 1.000 and 0.772 | - |
| Restraints/parameters | 2/206 | 0/224 | 0/878 |
| Final R indices [$I>2\sigma(I)$]: $R_1$, $wR_2$ | 0.1383, 0.3738 | 0.122, 0.392 | 0.1142, 0.2301 |
| CCDC no. | 1892568 | 1892567 | 1892569 |

## 3. Results

### 3.1. Synthesis and Characterization of fac-[Fe(4-ima-Bp)$_3$](Y)$_2$·sol Complexes

The synthesis of a family of *fac*-[Fe(4-ima-Bp)$_3$](Y)$_2$·sol complexes was achieved by a reaction between the 4-ima-Bp ligand in MeOH or EtOH and Fe(ClO$_4$)$_2$·XH$_2$O or Fe(BF$_4$)$_2$·6H$_2$O which affords dark red powders of the octahedral complexes *fac*-[Fe(4-ima-Bp)$_3$](Y)$_2$·sol (Y = ClO$_4$; sol = 3EtOH **1**, 3MeOH **2**; Y= BF$_4$; sol = EtOH·4H$_2$O **3**, 4H$_2$O **4** and 3.5MeCN **5**), Scheme 1. The acetonitrile solvate, **5** was prepared by recrystallization of **3** from MeCN/Et$_2$O.

**Scheme 1.** Synthesis of *fac*-[Fe(4-ima-Bp)$_3$](Y)$_2$·sol.

### 3.2. IR and UV–VIS Spectroscopy

IR spectroscopy of **1–5** shows an imine stretch at 1620 cm$^{-1}$ and at lower wavenumbers than the free ligand consistent with coordination to the metal (Table 2) [16,24,31]. Bands for the anions are also clearly visible in their expected positions (1083 and 1087 cm$^{-1}$). The presence of a further set of bands between 3362–3377 cm$^{-1}$ are consistent with O–H stretches suggesting that EtOH, MeOH or H$_2$O is present in the structures of these compounds.

**Table 2.** Physical and IR spectroscopic data for *fac*-[Fe(4-ima-Bp)$_3$](Y)$_2$·sol.

| Compound | %yield | Colour | IR (cm$^{-1}$) | | | | |
|---|---|---|---|---|---|---|---|
| | | | $\nu_{C=N}$ | $\nu_{C=C}$ | $\nu_{OH}$ | $\nu_{ArH}$ | $\nu_{anion}$ |
| **1** (ClO$_4$·3EtOH) | 53 | Dark red | 1620 | 1484 | 3362 | 3128 | 1087 |
| **2** (ClO$_4$·3EtOH) | 67 | Dark red | 1621 | 1484 | 3380 | 3135 | 1089 |
| **3** (BF$_4$·EtOH·4H$_2$O) | 77 | Dark red | 1620 | 1484 | 3377 | 3137 | 1083 |
| **4** (BF$_4$·4H$_2$O) | 57 | Dark red | 1620 | 1484 | 3377 | 3137 | 1083 |
| **5** (BF$_4$·3.5MeCN) | 63 | Red Orange | 1620 | 1484 | - | 3144 | 1051 |

In the visible region a DMF solution of 4-ima-Bp reveals an absorbance maximum at 340 nm ($\varepsilon$ = 24000 M$^{-1}$cm$^{-1}$) which arises from an intraligand $\pi\rightarrow\pi^*$ transition (Table 3) [32]. At room temperature, methanol solutions of **1-4** are orange and exhibit $\pi\rightarrow\pi^*$ transitions at approximately 330 and 280 nm (Figure 1). UV–VIS spectra of **1** and **3** in MeCN (in which **1-4** are more soluble), as representative examples of the compounds, reveal a possible band at *ca.* 840 nm consistent with the compounds being HS in solution (Figure S3).

**Table 3.** Wavelength maxima and extinction coefficients of *fac*-[Fe(4-ima-Bp)$_3$](Y)$_2$ in MeOH.

| Compound | $\lambda_{max}$/nm ($\varepsilon_{max}$/M$^{-1}$cm$^{-1}$) |
|---|---|
| 1 | 328 (65,000), 283 (72,000) |
| 2 | 326 (57,000), 283 (50,000) |
| 3 | 327 (66,200), 283 (61,000) |
| 4 | 328 (56,000), 283 (51,000) |

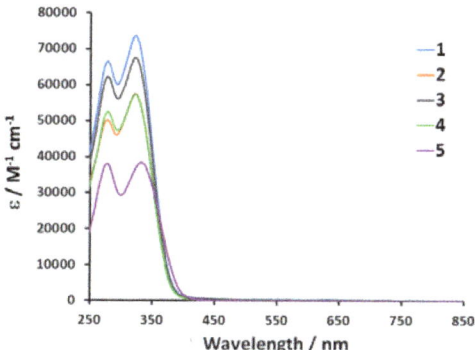

**Figure 1.** UV–VIS spectra of *fac*-[Fe(4-ima-Bp)$_2$](Y)$_2$·sol **1–5** in MeOH.

*3.3. Structural Studies of fac-[Fe(4-ima-Bp)$_3$](Y)$_2$·sol Complexes*

The structure of *fac*-[Fe(4-ima-Bp)$_3$](ClO$_4$)$_2$·3EtOH **1** determined by single-crystal X-ray diffraction at 143 K shows the compound reveals a trigonal symmetry (space group $R\bar{3}$, Figure 2). The methanol solvate **2**, is isostructural to **1** (collected at 153 K). The asymmetric units contain the Fe centre, a single 4-ima-Bp ligand, parts of two perchlorate anions and an EtOH (partially disordered in **1**) or MeOH molecule. The ligands coordinate with a facial (*fac*) disposition around the metal centre. In contrast, *fac*-[Fe(4-ima-Bp)$_3$](BF$_4$)$_2$·3.5MeCN **5** crystallizes in the triclinic $P\bar{1}$ space group. Despite the change in symmetry and the different anion the general features of **5** are remarkably similar to those of **1** and **2**. The Fe-ligand bond lengths and octahedral distortion parameters for **1**, **2** and **5** are given in Table 4. Comparison with [Fe(2-Me-4-ima-*n*Pr)$_3$]Cl·Y [22,31] and [Fe(*N*-Me-2-ima-CH(Me)Ph)$_3$](BF$_4$)$_2$·MeCN [24] indicates that the Fe(II) centres are LS at the low temperature used for X-ray data collection. Interestingly, in **1** and **2** the bond lengths are shorter by ca. 0.03 Å than the LS centre in [Fe(2-ima-*p*-C$_6$H$_4$OMe)$_3$](ClO$_4$)$_2$—the only other previously reported mononuclear system where the aromatic group is directly connected to the imine nitrogen [23]. The octahedral distortion parameters are also consistent with LS centres.

**Table 4.** Selected bond lengths, octahedral distortion parameters and hydrogen bonding distances (Å,°) for **1**, **2** and **5**.

| Bond lengths | 1-143 K | 2-153 K | | 5-123 K |
|---|---|---|---|---|
| Fe1–N1 | 1.962(6) | 1.964(8) | Fe1–N1 | 1.971(4) |
| | - | - | Fe1–N3 | 1.995(4) |
| | - | - | Fe1–N4 | 1.962(4) |
| Fe1–N3 | 2.005(5) | 1.983(8) | Fe1–N6 | 1.992(4) |
| | - | - | Fe1–N7 | 1.963(4) |
| | - | - | Fe1–N9 | 2.001(4) |
| Σ [33] | 68.7 | 55.7 | | 57.9 |
| Θ [34] | 196.4 | 127.3 | | 125.0 |
| N2-H2···O5 | 1.829(9) | 1.832(11) | N8-H8···F3 | 2.109(4) |
| O5-H5···O4 | - | 2.09(2) | N8-H8···F2 | 2.426(4) |
| | | | N5-H5···F1 | 1.939(4) |
| | | | N2-H2···N10 | 2.041(5) |

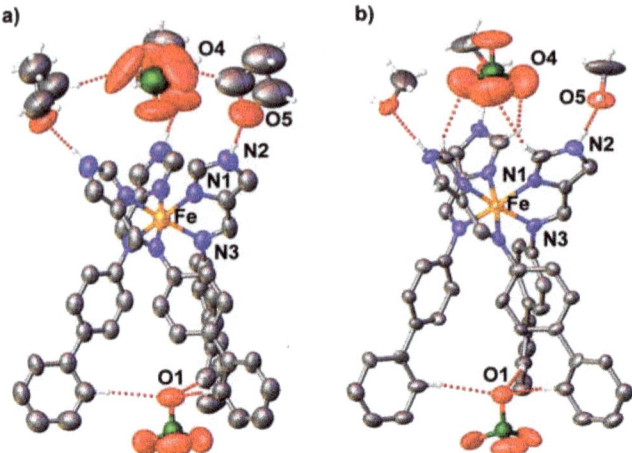

**Figure 2.** View of the molecular unit of (**a**) **1** and (**b**) **2**. Only selected hydrogen atoms and labels are shown in the interests of clarity. Ellipsoids are drawn at 50% probability.

A particular feature of the structures is that one of the anions sits in a pocket of biphenyl groups and is held in place by three C–H···O interactions (see Figure 2). The pocket is reinforced by three *intramolecular* C–H···π contacts, Figure 3. The imidazole hydrogens are involved in H-bonding to the solvent and not the anion as is seen in systems like [Fe(2-Me-4-ima-nPr)$_3$]Cl·Y. The second perchlorate anion instead forms H-bonds to the solvent molecules and weaker C–H···O interactions involving either ethanol or imidazole C–H groups. This change in packing at the second anion results in an inversion of the perchlorate in **1** compared with **2** (Figure 2).

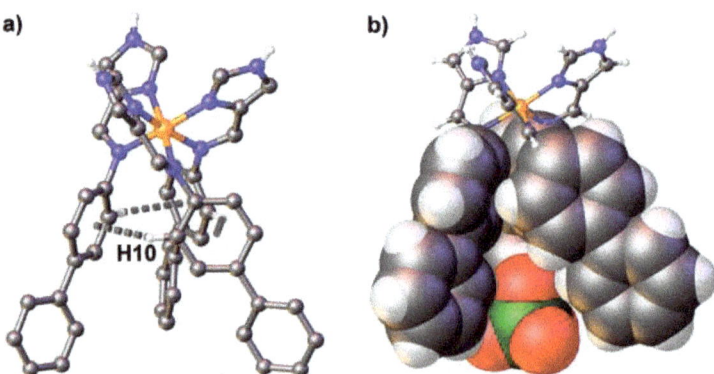

**Figure 3.** View of (**a**) the *intramolecular* C–H···π contacts in **1** and (**b**) spacefill diagram showing the perchlorate anion in the biphenyl pocket.

The overall packing for both compounds involves multiple C–H···π interactions of the propeller like biphenyl arms forming a triangular motif (Figure 4; Table S1). Within each 'triangle' of *fac*-[Fe(4-ima-Bp)$_3$]$^{2+}$ cations the Fe centres are chiral but as the intercalated triangle is of the opposite hand the overall structure is achiral, as expected. There are also C–H···π interactions between the Λ and Δ *fac*-[Fe(4-ima-Bp)$_3$]$^{2+}$ cations giving rise to a hexagonal motif (Figure 4b). As this is present in all the compounds in this series it is clearly very robust.

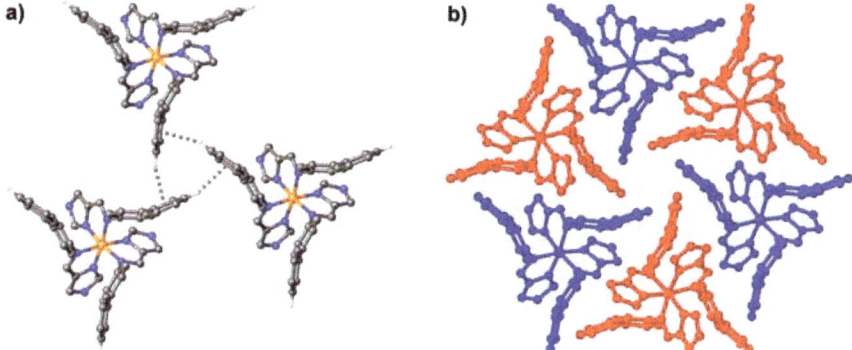

**Figure 4.** View of the (**a**) intermolecular C–H···π interactions that link the (**b**) Λ (blue) and Δ (red) *fac*-[Fe(4-ima-Bp)₃]²⁺ cations that form the hexagonal motif in **1**.

The hexagonal planes are approximately 12.5 Å thick and are separated from each other by an extensive network of perchlorate anions and the MeOH or EtOH solvent molecules held together by N-H···O and O-H···O hydrogen bonds (Figure 5; Figure S4). Similar 2D layers are present in the imidazolyl-imine dimers, [Fe(2-Me-ima-N-N-ima-2-Me)₃](ClO₄)₄ [32].

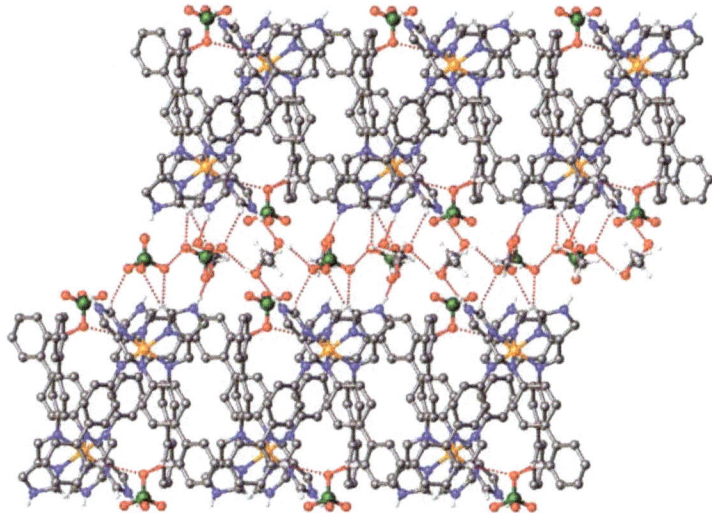

**Figure 5.** Side-on view of the 2D hexagonal planes and the ClO₄-MeOH layer in **2** that links the planes.

As noted above, the structure of the MeCN solvate **5** is very similar to **1** and **2**. A particular difference is that the BF₄⁻ anion directly links the 2D layers of the *fac*-[Fe(4-ima-Bp)₃]²⁺ cations through N-H···F hydrogen bonds. The remaining imidazole N-H group is hydrogen bonded to one of the acetonitrile molecules. A combination of C–H···N/F interactions hold the remaining MeCN molecules in the anion-solvent layer. The other subtle difference is that the 2D layers are no longer hexagonal, but are instead slightly distorted (Figures S5 and S6). This has a number of consequences including the loss of some C–H···π interactions and the concomitant formation of slightly angular π-π interactions (Figure 6). We also observe a reduction in the gap between the layers from ca. 13 Å in **1** and **2** to 10.8 Å in **5**.

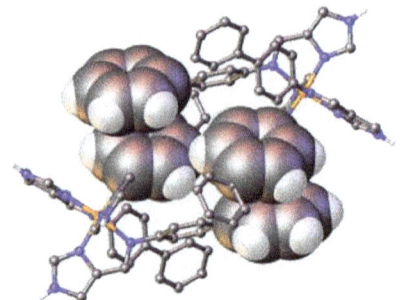

**Figure 6.** View of the π-π interactions in *fac*-[Fe(4-ima-Bp)$_3$](BF$_4$)$_2$·3.5MeCN **5**.

### 3.4. Magnetic studies of [Fe(4-ima-Bp)$_3$](Y)$_2$·sol Complexes

The magnetic properties of **1–5** have been studied by magnetic susceptibility (see $\chi_M T$ versus $T$ plots, Figures 7 and 8). All the compounds except **1** show a complete HS to LS transition around room temperature. The exception is **1** which shows a SCO profile that is more abrupt than **2** in the first warming (Figure S7) but thereafter exhibits a more gradual crossover with $\chi_M T$ going from 1.7 cm$^3$·mol$^{-1}$·K to 3.3 cm$^3$·mol$^{-1}$·K between 150 K and 350 K; indicating a 50% transition from the HS to LS state. Notably, the first measurement in **1** shows hysteresis, but this is only apparent, with the SCO profile changing upon subsequent cycles to finally give the profile shown in Figure 7. The change in SCO behaviour in **1** has been shown by TGA studies to be due to loss of one equivalent of EtOH (Figure S8). Similar solvent loss has been observed in *fac*-[Fe(*N*-Me-2-ima-CH(Me)Ph)$_3$](BF$_4$)$_2$·MeCN and also lowers the transition temperature [24]. Interestingly, despite the different solvents and anions **2-4** show very similar SCO profiles with $T_{1/2}$ varying slightly between 305 and 320 K. It is important to state that we cannot absolutely rule out solvent loss in the case of **2–5**, but the fact that we measured several thermal cycles with no change in the SCO profile seems to suggest that this is unlikely. We also note that the initial SCO profile in **5** is more gradual in the first 200 K, the exact reason for this behaviour is unclear but it is repeatable. Although we have been unable to obtain the structures of **3** and **4**, these results suggest that the hexagonal motif noted in Figure 4 is also present in **3** and **4** hence the almost identical SCO profiles observed. This is supported by the fact that in **5** the packing becomes *pseudo*-hexagonal and the SCO is now less abrupt and occurs at a slightly lower temperature. A comparison with [Fe(2-ima-*p*-C$_6$H$_4$OMe)$_3$](ClO$_4$)$_2$ which exhibits a complete spin crossover at 158 K [23] suggests that the biphenyl group, despite its size, stabilizes the LS state more than in [Fe(2-ima-*p*-C$_6$H$_4$OMe)$_3$](ClO$_4$)$_2$.

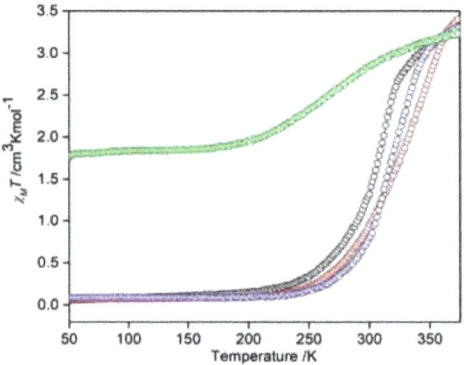

**Figure 7.** VSM profiles of *fac*-[Fe(4-ima-Bp)$_3$](Y)$_2$·sol as $\chi_M T$ vs. $T$ plots of a) **1** (green), b) **2** (red), c) **3** (blue) and d) **4** (black).

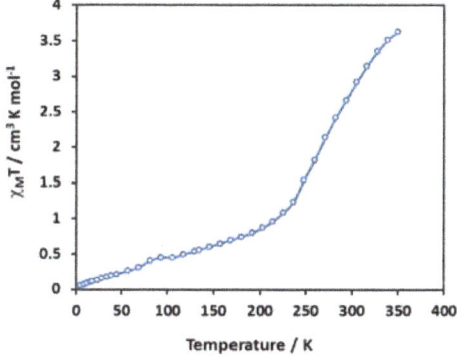

**Figure 8.** SQUID profile of *fac*-[Fe(4-ima-Bp)$_3$](BF$_4$)$_2$·3.5MeCN **5**.

*3.5. Thermochromism*

The complexes **1–4** all undergo a clear and reversible colour change from dark red to orange in the solid state with heating (Figure 9) associated with a LS to HS transition. Reports on thermochromism in Fe(II) imidazolyl complexes are rare and this colour change is different from the *fac*-[Fe(2-Me-4-ima-R)$_3$]Cl·PF$_6$ series where a change from yellow to orange/red is observed [35]. It follows that the R group on the imine nitrogen can be used to tune the thermochromic behaviour of such SCO systems. In addition, we have soaked filter paper in a solution of **2** and find that it reversibly changes colour from red to yellow between 30 and 60 °C (see supplementary video).

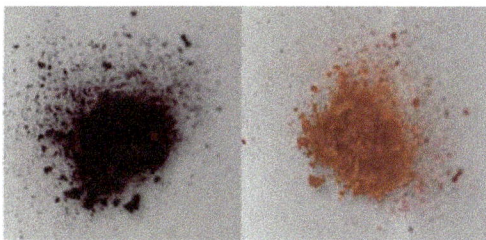

**Figure 9.** Colour change of *fac*-[Fe(4-ima-Bp)$_3$](ClO$_4$)$_2$·3MeOH **1** heating from 298 K (**left**) to 423 K (**right**).

## 4. Conclusions

In conclusion, we have prepared five Fe(II) complexes of the new 4-ima-Bp ligand and a variety of anions and solvent molecules. Structural studies show that all compounds crystallize as the *fac* isomer probably due to *intramolecular* C–H···π contacts involving the biphenyl groups and trapping of one of the anions. 2D hexagonal or *pseudo*-hexagonal layers of the *fac*-[Fe(4-ima-Bp)$_3$]$^{2+}$ cations form principally through C–H···π interactions. Strong hydrogen bonding between the layers is facilitated by the anions and solvent molecules giving rise to a high transition temperature, but moderately gradual SCO transitions. Notably, the solvent is found to influence SCO behaviour more than the anion. Moreover, the biphenyl group allows tuning of the spin transition temperature and represents a promising strategy in the design of more abrupt SCO systems that will operate at room temperature.

**Supplementary Materials:** The following are available online at http://www.mdpi.com/2073-4352/9/2/116/s1, Figure S1. IR spectra of **1–4**. Figure S2. $^1$H-NMR spectrum of 4-ima-Bp. Figure S3. UV–VIS of [Fe(4-ima-Bp)$_3$](ClO$_4$)$_2$ **1** and [Fe(4-ima-Bp)$_3$](BF$_4$)$_2$ **3** in MeCN in a 0.1 M solution. Figure S4. Side-on view of the packing in *fac*-[Fe(4-ima-Bp)$_3$](ClO$_4$)$_2$·3MeOH **2**. Figure S5. View of the *pseudo*-hexagonal packing motif in *fac*-[Fe(4-ima-Bp)$_3$](BF$_4$)$_2$·3.5MeCN **5**. Figure S6. Comparative view of the hexagonal and *pseudo*-hexagonal packing motifs found in **2** and **5**. Figure S7. SQUID profile of *fac*-[Fe(4-ima-Bp)$_3$](ClO$_4$)$_2$·3EtOH **1**. Figure S8.

TGA of *fac*-[Fe(4-ima-Bp)₃](ClO₄)₂·3EtOH **1**. Table S1. Geometric parameters of C–H···π and π–π interactions in **1-2** and **5**. A supplementary video showing the thermochromism in **2**.

**Author Contributions:** P.H. and D.J.H. designed, administered and supervised the project. P.H. and D.J.H. wrote the manuscript with input from all other authors. D.S. synthesized all compounds, and collected and analysed spectroscopic and crystallographic data; the latter for **5**. K.S.M. and B.M. conducted all SQUID magnetometry measurements. S.M.N. collected and analysed all VSM data. L.L. and S.G.T. collected and solved the structures for **1** and **2**.

**Funding:** We thank the Thailand Research Fund (grant nos. RSA5580028, BRG6180008) for funding this research and financial support from the Thailand Research Fund in the form of a Royal Golden Jubilee scholarship to DS (PHD/0135/2554).

**Conflicts of Interest:** The authors declare no conflict of interest.

### References

1. *Spin-Crossover Materials: Properties and Applications*; Halcrow, M.A. (Ed.) John Wiley & Sons Ltd.: Chichester, UK, 2013.
2. Gütlich, P.; Goodwin, H.A. Spin Crossover—An Overall Perspective. *Top. Curr. Chem.* **2004**, *233*, 1–47.
3. Boillot, M.L.; Weber, B. Mononuclear ferrous and ferric complexes. *Comptes Rendus Chim.* **2018**, *21*, 1196–1208. [CrossRef]
4. Collet, E.; Guionneau, P. Structural analysis of spin-crossover materials: From molecules to materials. *Comptes Rendus Chim.* **2018**, *21*, 1133–1151. [CrossRef]
5. Craig, G.A.; Roubeau, O.; Aromí, G. Spin state switching in 2,6-bis(pyrazol-3-yl)pyridine (3-bpp) based Fe(II) complexes. *Coord. Chem. Rev.* **2014**, *269*, 13–31. [CrossRef]
6. Brooker, S. Spin crossover with thermal hysteresis: practicalities and lessons learnt. *Chem. Soc. Rev.* **2015**, *44*, 2880–2892. [CrossRef] [PubMed]
7. Halcrow, M.A. Structure:function relationships in molecular spin-crossover complexes. *Chem. Soc. Rev.* **2011**, *40*, 4119–4142. [CrossRef] [PubMed]
8. Létard, J.-F.; Guionneau, P.; Goux-Capes, L. Towards Spin Crossover Applications. *Top. Curr. Chem.* **2004**, *235*, 221–249.
9. Senthil, K.; Ruben, M. Emerging trends in spin crossover (SCO) based functional materials and devices. *Coord. Chem. Rev.* **2017**, *346*, 176–205. [CrossRef]
10. Miller, R.G.; Brooker, S. Reversible quantitative guest sensing via spin crossover of an iron(ii) triazole. *Chem. Sci.* **2016**, *7*, 2501–2505. [CrossRef]
11. Létard, J.-F. Photomagnetism of iron(II) spin crossover complexes the T(LIESST) approach. *J. Mater. Chem.* **2006**, *16*, 2550–2559. [CrossRef]
12. Chastanet, G.; Desplanches, C.; Baldé, C.; Rosa, P.; Marchivie, M.; Guionneau, P. A critical review of the T(LIESST) temperature in spin crossover materials-What it is and what it is not. *Chem. Sq.* **2018**, *2*, 2. [CrossRef]
13. Scott, H.S.; Staniland, R.W.; Kruger, P.E. Spin crossover in homoleptic Fe(II) imidazolylimine complexes. *Coord. Chem. Rev.* **2018**, *362*, 24–43. [CrossRef]
14. Sunatsuki, Y.; Kawamoto, R.; Fujita, K.; Maruyama, H.; Suzuki, T.; Ishida, H.; Kojima, M.; Iijima, S.; Matsumoto, N. Structures and spin states of mono- and dinuclear iron(II) complexes of imidazole-4-carbaldehyde azine and its derivatives. *Coord. Chem. Rev.* **2010**, *254*, 1871–1881. [CrossRef]
15. Yamada, M.; Ooidemizu, M.; Ikuta, Y.; Osa, S.; Matsumoto, N.; Iijima, S.; Kojima, M.; Dahan, F.; Tuchagues, J.P. Interlayer Interaction of Two-Dimensional Layered Spin Crossover Complexes [Fe$^{II}$H$_3$LMe][Fe$^{II}$LMe]X (X = ClO$_4^-$, BF$_4^-$). *Inorg. Chem.* **2003**, *42*, 8406–8416. [CrossRef] [PubMed]
16. Bréfuel, N.; Watanabe, H.; Toupet, L.; Come, J.; Matsumoto, N.; Collet, E.; Tanaka, K.; Tuchagues, J.-P. Concerted spin crossover and symmetry breaking yield three thermally and one light-induced crystallographic phases of a molecular material. *Angew. Chem. Int. Ed.* **2009**, *48*, 9304–9307. [CrossRef] [PubMed]
17. Yamada, M.; Hagiwara, H.; Torigoe, H.; Matsumoto, N.; Kojima, M.; Dahan, F.; Tuchagues, J.P.; Re, N.; Iijima, S. A variety of spin-crossover behaviors depending on the counter anion: Two-dimensional complexes constructed by NH···Cl-hydrogen bonds, [Fe$^{II}$H$_3$LMe]Cl·X (X = PF$_6^-$, AsF$_6^-$, SbF$_6^-$, CF$_3$SO$_3^-$; H$_3$LMe = tris[2-{[(2methylimidazol-4-yl)methyl idene]amino}ethyl]amine). *Chem. Eur. J.* **2006**, *12*, 4536–4549. [CrossRef] [PubMed]

18. Seredyuk, M.; Muñoz, M.C.; Castro, M.; Romero-Morcillo, T.; Gaspar, A.B.; Real, J.A. Unprecedented multi-stable spin crossover molecular material with two thermal memory channels. *Chem. Eur. J.* **2013**, *19*, 6591–6596. [CrossRef] [PubMed]
19. Delgado, T.; Tissot, A.; Guénée, L.; Hauser, A.; Valverde-Muñoz, F.J.; Seredyuk, M.; Real, J.A.; Pillet, S.; Bendeif, E.-E.; Besnard, C. Very Long-Lived Photogenerated High-Spin Phase of a Multistable Spin-Crossover Molecular Material. *J. Am. Chem. Soc.* **2018**, *140*, 12870–12876. [CrossRef] [PubMed]
20. Nishi, K.; Arata, S.; Matsumoto, N.; Iijima, S.; Sunatsuki, Y.; Ishida, H.; Kojima, M. One-dimensional Spin-crossover iron(II) complexes bridged by intermolecular imidazole-pyridine NH···N hydrogen bonds, [Fe(HLMe)$_3$]X$_2$ (HLMe = (2-Methylimidazol-4-yl-methylideneamino-2-ethylpyridine; X = PF$_6^-$, ClO$_4^-$, BF$_4^-$). *Inorg. Chem.* **2010**, *49*, 1517–1523. [CrossRef]
21. Nishi, K.; Matsumoto, N.; Iijima, S.; Halcrow, M.A.; Sunatsuki, Y.; Kojima, M. A hydrogen bond motif giving a variety of supramolecular assembly structures and spin-crossover behaviors. *Inorg. Chem.* **2011**, *50*, 11303–11305. [CrossRef]
22. Fujinami, T.; Nishi, K.; Hamada, D.; Murakami, K.; Matsumoto, N.; Iijima, S.; Kojima, M.; Sunatsuki, Y. Scan Rate Dependent Spin Crossover Iron(II) Complex with Two Different Relaxations and Thermal Hysteresis fac-[Fe$^{II}$(HLn-Pr)$_3$]Cl·PF$_6$ (HLn-Pr = 2-Methylimidazol-4-yl-methylideneamino-n-propyl). *Inorg. Chem.* **2015**, *54*, 7291–7300. [CrossRef] [PubMed]
23. Thompson, J.R.; Archer, R.J.; Hawes, C.S.; Ferguson, A.; Wattiaux, A.; Mathonière, C.; Clérac, R.; Kruger, P.E. Thermally and photo-induced spin crossover behaviour in an Fe(II) imidazolylimine complex: [FeL$_3$](ClO$_4$)$_2$. *Dalton Trans.* **2012**, *41*, 12720–12725. [CrossRef] [PubMed]
24. Gu, Z.-G.; Pang, C.-Y.; Qiu, D.; Zhang, J.; Huang, J.-L.; Qin, L.-F.; Sun, A.-Q.; Li, Z. Homochiral iron(II) complexes based on imidazole Schiff-base ligands: Syntheses, structures, and spin-crossover properties. *Inorg. Chem. Commun.* **2013**, *35*, 164–168. [CrossRef]
25. Rigaku. *Rigaku XRD*; Rigaku Corporation: Tokyo, Japan, 1996.
26. *Bruker APEXII*; Bruker AXS Inc.: Madison, WI, USA, 2005.
27. *SAINT and SADABS*; Bruker AXS Inc.: Madison, WI, USA, 2003.
28. Sheldrick, G.M. Crystal structure refinement with SHELXL. *Acta Crystallogr. Sect. C Struct. Chem.* **2015**, *71*, 3–8. [CrossRef] [PubMed]
29. Sheldrick, G.M. SHELXT-Integrated space-group and crystal-structure determination. *Acta Crystallogr. Sect. A Found. Crystallogr.* **2015**, *71*, 3–8. [CrossRef] [PubMed]
30. Dolomanov, O.V.; Bourhis, L.J.; Gildea, R.J.; Howard, J.A.K.; Puschmann, H. OLEX2: a complete structure solution, refinement and analysis program. *J. Appl. Cryst.* **2009**, *42*, 339–342. [CrossRef]
31. Fujinami, T.; Nishi, K.; Matsumoto, N.; Iijima, S.; Halcrow, M.A.; Sunatsukid, Y.; Kojima, M. 1D and 2D assembly structures by imidazole···chloride hydrogen bonds of iron(II) complexes [Fe$^{II}$(HLn-Pr)$_3$]Cl·Y(HLn-Pr = 2-methylimidazol-4-yl-methylideneamino-n-propyl; Y = AsF$_6^-$, BF$_4^-$) and their spin states. *Dalton Trans.* **2011**, *40*, 12301–12309. [CrossRef]
32. Sunatsuki, Y.; Kawamoto, R.; Fujita, K.; Maruyama, H.; Suzuki, T.; Ishida, H.; Kojima, M.; Iijima, S.; Matsumoto, N. Structures and spin states of bis(tridentate)-type mononuclear and triple helicate dinuclear iron(II) complexes of imidazole-4-carbaldehyde azine. *Inorg. Chem.* **2009**, *48*, 8784–8795. [CrossRef]
33. McCusker, J.K.; Rheingold, A.L.; Hendrickson, D.N. Variable-Temperature Studies of Laser-Initiated $^5T_2$ to $^1A_1$ Intersystem Crossing in Spin-Crossover Complexes: Empirical Correlations between Activation Parameters and Ligand Structure in a Series of Polypyridyl Ferrous Complexes. *Inorg. Chem.* **1996**, *35*, 2100–2112. [CrossRef]
34. Marchivie, M.; Guionneau, P.; Létard, J.F.; Chasseau, D. Photo-induced spin-transition: the role of the iron(II) environment distortion. *Acta Crystallogr. Sect. B Struct. Sci.* **2005**, *61*, 25–28. [CrossRef]
35. Furushou, D.; Hashibe, T.; Fujinami, T.; Nishi, K.; Hagiwara, H.; Matsumoto, N.; Sunatsuki, Y.; Kojima, M.; Iijima, S. Reprint of "facial and meridional geometrical isomers of tris(2-methylimidazol-4-yl-methylideneaminobenzyl)iron(II) with delta- and lambda-configurations and their enantio-discriminative assembly via imidazole···chloride hydrogen bonding and spin crossover. *Polyhedron* **2012**, *44*, 194–203. [CrossRef]

© 2019 by the authors. Licensee MDPI, Basel, Switzerland. This article is an open access article distributed under the terms and conditions of the Creative Commons Attribution (CC BY) license (http://creativecommons.org/licenses/by/4.0/).

Article

# Threefold Spiral Structure Constructed by 1D Chains of [[M(NCS)$_2$(bpa)$_2$]·biphenyl]$_n$ (M = Fe, Co; bpa = 1,2-bis(4-pyridyl)ethane)

Satoshi Tokinobu [1], Haruka Dote [2] and Satoru Nakashima [1,3,*]

[1] Graduate School of Science, Hiroshima University, 1-3-1 Kagamiyama, Higashi-Hiroshima 739-8526, Japan; avogadro-leblanc-8als.t@ezweb.ne.jp
[2] Graduate School of Engineering, Hiroshima University, 1-4-1 Kagamiyama, Higashi-Hiroshima 739-8526, Japan; hdote@hiroshima-u.ac.jp
[3] Natural Science Center for Basic Research and Development, Hiroshima University, 1-4-2 Kagamiyama, Higashi-Hiroshima 739-8526, Japan
* Correspondence: snaka@hiroshima-u.ac.jp

Received: 27 December 2018; Accepted: 12 February 2019; Published: 14 February 2019

**Abstract:** Assembled complexes [[M(NCS)$_2$(bpa)$_2$]·biphenyl]$_n$ (M = Fe, Co; bpa = 1,2-bis(4-pyridyl)ethane) have been synthesized because [Fe(NCBH$_3$)$_2$(bpa)$_2$·biphenyl]$_n$ has a novel threefold spiral structure and shows stepwise spin-crossover phenomenon. We attempted to obtain spiral structures for [[Fe(NCS)$_2$(bpa)$_2$]·biphenyl]$_n$ and [[Co(NCS)$_2$(bpa)$_2$]·biphenyl]$_n$ using a one-step diffusion method, while the reported spiral structure of [[Fe(NCBH$_3$)$_2$(bpa)$_2$]·biphenyl]$_n$ was obtained by diffusion method after synthesizing Fe(II)-pyridine complex. X-ray structural analysis revealed that [[Fe(NCS)$_2$(bpa)$_2$]·biphenyl]$_n$ and [[Co(NCS)$_2$(bpa)$_2$]·biphenyl]$_n$ had a chiral propeller structure of pyridines around the central metal, and they had a novel spiral structure and chiral space group $P3_12_1$ without the presence of chiral auxiliaries. It was shown that the host 1D chain, having a chiral propeller structure of pyridines around the central metal along with its concerted interaction with an atropisomer of biphenyl, made a threefold spiral structure.

**Keywords:** spiral structure; 1,2-bis(4-pyridyl)ethane; supramolecular coordination polymer; chiral propeller structure; atropisomerism

## 1. Introduction

The first transition metal complexes with configurations $d^4$–$d^7$ may exist in either high-spin (HS) or low-spin (LS) state, depending on the ligand field strength. The ground state becomes a HS state in a weak field, while the ground state becomes an LS state in a strong field. The spin state changes between HS and LS states due to external perturbations, such as from changes in temperature, pressure, and light illumination in a medium field [1]. This is called spin-crossover (SCO) phenomenon. Especially in $d^6$ Fe(II) complexes, spin appears (S = 2) and disappears (S = 0) depending on the SCO phenomenon, suggesting it acts as a molecular switch [2]. An interesting application is in solvatochromic spin state switching in SCO compounds [3].

The design and construction of various structures for self-assembled complexes have attracted great interest from many chemists. These self-assembled complexes may have a vacancy, and usually a solvent molecule is enclathrated in the vacancy. The bridging ligand itself is also enclathrated in rare cases [4]. It is known that the structure of self-assembled complexes changes by changing the guest molecule. Therefore, the selection of the guest molecule is an important factor in designing the assembled structure.

The SCO of assembled complexes becomes important because a variety of assembled structures are expected. The spin state is affected by guest molecules and the steepness of the transition is affected by the intermolecular interactions. There are many assembled complexes, such as Hoffman type [5–9], triazole-bridged type [10,11], tetrazole-bridged type [12], and bis(pyridyl) type [13]. Among them, we became interested in the complexes bridged by bis(pyridyl) type ligands, because these complexes easily form vacancies. We have studied SCO phenomenon for the complexes bridged by 1,2-bis(4-pyridyl)ethane [14–17], 1,3-bis(4-pyridyl)propane [18,19], 1,4-bis(4-pyridyl)benzene derivatives [20], and 1,4-bis(4-pyridyl)anthracene [21]. By changing the bridging ligand and guest molecule, the local structure is controlled to propeller, parallel type, or distorted propeller (Scheme 1). Such local structure determines whether SCO occurs or not [22,23]. This shows that the ligands can easily approach iron in the chiral propeller type local structure when the spin state becomes an LS state.

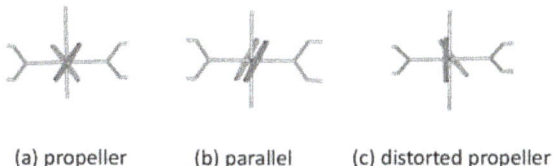

(a) propeller     (b) parallel     (c) distorted propeller

**Scheme 1.** Difference in the Fe-pyridine local structures observed in a variety of the assembled [Fe(NCX)$_2$L$_2$]$_n$ (X = S, Se, and BH$_3$; L = bridging ligand).

It had been reported that [Fe(NCS)$_2$(bpa)$_2$]$_n$ has a 1D chain structure and shows HS state [24]. We became interested in [Fe(NCS)$_2$(bpa)$_2$]$_n$, because the bpa has an *anti-gauche* conformation and NCS$^-$ can be changed with other anionic ligands. We synthesized crystals by diffusion method and obtained several types crystals. We obtained 2D grid structure and 3D interpenetrated structure as well as 1D chain structure [14]. 2D grid structure and 3D interpenetrated structure enclathrated the solvent molecule. The structure changed to a 1D chain structure by desorbing the guest molecule [14]. We thought that the complexes that enclathrated larger guest molecules, such as biphenyl, have a stable structure. Therefore, we have synthesized self-assembled Fe(II) complexes [[Fe(NCX)$_2$(bpa)$_2$]·(guest)]$_n$ (X = S, Se, BH$_3$; bpa = 1,2-bis(4-pyridyl)ethane; guest = biphenyl, 1,4-dichlorobenzene, diphenylmethane, 2-nitrobiphenyl). The *anti-gauche* conformer of bpa contributed greatly to the assembled structure, i.e., *anti* conformer-formed 3D interpenetrated or 2D grid structure, and *gauche* conformer-formed 1D chain structure. Moreover, we revealed that SCO phenomena appeared by having enclathrated a guest molecule. The crystal structures and SCO phenomena are summarized in Table 1 [16]. [[Fe(NCBH$_3$)$_2$(bpa)$_2$]·biphenyl]$_n$ usually had 1D chain structure and showed one-step spin transition [15]. In special cases, the 1D chain self-assembled sheet of [[Fe(NCBH$_3$)$_2$(bpa)$_2$]·biphenyl]$_n$ was stacked spirally, having a threefold axis [25], and this spiral structure showed a stepwise spin transition. Stepwise SCO phenomena play an important role in tuning the spin state precisely. Spiral structure is a key point to showing stepwise transition. Therefore, the mechanism for forming spiral structures becomes an important theme. In the present study, new series of self-assembled complexes [[M(NCS)$_2$(bpa)$_2$]·biphenyl]$_n$ (M = Fe, Co) have been synthesized to obtain other novel spiral structure, and we discuss the formation mechanisms.

**Table 1.** Summary of crystal structures and their spin-crossover (SCO) phenomena for [[Fe(NCX)$_2$(bpa)$_2$]·(guest)]$_n$.

| Anion \ Guest | NCS | NCSe | NCBH$_3$ |
|---|---|---|---|
| biphenyl | 2D grid | Linear | Linear |
| 2-nitrobiphenyl | Interpenetrated | Interpenetrated | 2D grid |
| 1,4-dichlorobenzene | Linear | Linear | Not included |
| diphenylmethane | Interpenetrated | Interpenetrated | 2D grid |

The crystals underlined showed a color change from pale yellow to deep red by cooling with Liq. N$_2$. The color change corresponded well with the SCO phenomenon.

## 2. Materials and Methods

[[Fe(NCS)$_2$(bpa)$_2$]·biphenyl]$_n$ and [[Co(NCS)$_2$(bpa)$_2$]·biphenyl]$_n$ were obtained by diffusion method from FeCl$_2$·4H$_2$O (or FeSO$_4$·7H$_2$O) and CoCl$_2$·6H$_2$O, respectively. FeCl$_2$·4H$_2$O (or FeSO$_4$·7H$_2$O, CoCl$_2$·6H$_2$O) and NaNCS were dissolved to distilled water as bottom layer. Biphenyl was dissolved to a mixed solvent of water and EtOH as intermediate layer. Bpa was dissolved to EtOH as upper layer. From the vessel, block-like crystal and plate-like crystal were obtained for [[Fe(NCS)$_2$(bpa)$_2$]·biphenyl]$_n$ and [[Co(NCS)$_2$(bpa)$_2$]·biphenyl]$_n$, respectively. Anal. found (calcd)%: for [[Fe(NCS)$_2$(bpa)$_2$]·biphenyl]$_n$, C, 65.10 (65.70); H, 4.67 (4.93); N, 12.01 (12.10); S, 9.08 (9.23). Anal. found (calcd)%: for [[Co(NCS)$_2$(bpa)$_2$]·biphenyl]$_n$, C, 65.42 (65.41); H, 4.81 (4.91); N, 12.14 (12.04); S, 8.42 (9.19).

For single crystal X-ray diffraction analysis, all diffraction data were collected by using a Bruker SMART-APEX diffractometer (Bruker, Billerica, MA, USA) equipped with CCD area detector and graphite-monochromated Mo Kα radiation, λ = 0.71073 Å, ω-scan mode (0.3° steps). Semi-empirical absorption corrections on Laue equivalents were applied. The samples were coated with adhesive to avoid desorption of guest molecules. The structures were solved by direct methods and refined by full-matrix least-squares against $F^2$ of all data using SHELXL-2014/6 [26]. The crystal data can be obtained free of charge from the Cambridge Crystallographic Data Centre via www.ccdc.cam.ac.uk/data_request/cif (CCDC: 1892503 and 1892504).

## 3. Results and Discussion

In the synthesis of spiral [[Fe(NCBH$_3$)$_2$(bpa)$_2$]·biphenyl]$_n$, the pyridine complex of Fe(II) was first synthesized and then the diffusion method was used (Scheme 2). In the present study, an easier method was attempted. That is, the diffusion method was used without synthesizing a pyridine complex of Fe(II) (Scheme 2). In the synthesis of iron complex, when FeCl$_2$·4H$_2$O was used, a large crystal having spiral structure was obtained. However, it was easily oxidized in the synthetic process. When FeSO$_4$·7H$_2$O was used, a small crystal having spiral structure was obtained and the oxidation was avoided. The spiral structure was not obtained by direct mixing method.

Figure 1 shows an Oak Ridge Thermal-Ellipsoid Plot Program (ORTEP) drawing of [[Fe(NCS)$_2$(bpa)$_2$]·biphenyl]$_n$. The packing view is shown in Figure 2. The crystal data are shown in Table 2. The structure of the complex showed an octahedral geometry by coordination of the four N atoms of bpa and the two N atoms of NCS$^-$ in the *trans* position. The local structure around iron was chiral propeller type. The biphenyl molecule was enclathrated by Fe(NCS)$_2$(bpa)$_2$ in the ratio of 1:1. The Fe–N$_{CS}$ and Fe–N$_{Py}$ distances were 2.092 and 2.232 Å, respectively, suggesting a HS state. In the crystal, the biphenyl molecule showed atropisomerism. We have analyzed the present complexes in single crystal X-ray structural analysis by using the chiral space group $P3_12_1$ (Flack parameter: 0.022). When analyzing it using $P3_22_1$, the Flack parameter becomes 0.9764, suggesting the space group is $P3_12_1$. From the result of analysis, it was found that the sign of biphenyl's dihedral angle is different between the two chiral space groups. The structure of biphenyl is shown in Figure 3. The sign of biphenyl's dihedral angle is related to the space group. The configuration of biphenyl was $R$ in the present results. This result may suggest that the chiral assembly is controlled by a chiral propeller-type local structure and biphenyl's atropisomerism.

Synthesis of spiral [[Fe(NCBH$_3$)$_2$(bpa)$_2$]·biphenyl)]$_n$

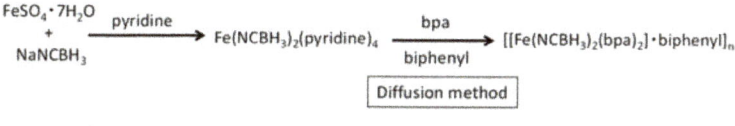

~~~~~~~~~~~~~~~~~~~~~~~~~~~~~~~~~~~~~~~~~~~~~~~~~~~~~~~~~~~~~~~~~~~~~~~~~~~~

Synthesis of spiral [[M(NCS)$_2$(bpa)$_2$]·biphenyl]$_n$ (M = Fe, Co)

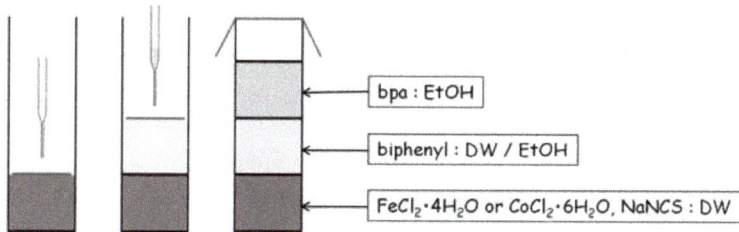

**Scheme 2.** Synthetic scheme to obtain spiral structure.

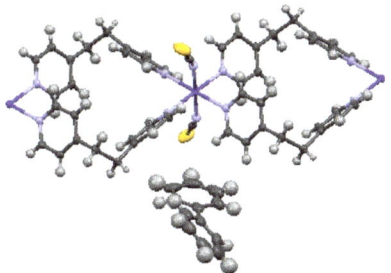

**Figure 1.** ORTEP drawing of [[Fe(NCS)$_2$(bpa)$_2$]·biphenyl]$_n$.

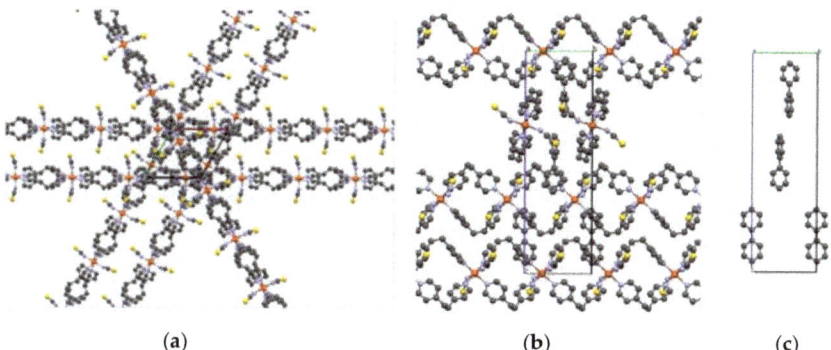

**Figure 2.** The projections of [[Fe(NCS)$_2$(bpa)$_2$]·biphenyl]$_n$ to *ab* plane (**a**), *a* axis (**b**), and the projection of biphenyl to *a* axis (**c**).

Table 2. Crystal data of [[Fe(NCS)$_2$(bpa)$_2$]·biphenyl]$_n$.

| Fe(NCS)$_2$(bpa)$_2$·biphenyl | |
|---|---|
| Temperature | RT |
| Space group | $P3_12_1$ |
| a,b/Å | 10.234(5) |
| c/Å | 30.070(15) |
| α,β/° | 90 |
| γ/° | 120 |
| R1 | 0.0368 |
| wR2 | 0.0909 |
| Goodness of fit | 1.039 |
| Volume/Å$^3$ | 2727(3) |
| Flack parameter | 0.022(26) |
| Flack parameter when using $P3_22_1$ | 0.9764 |

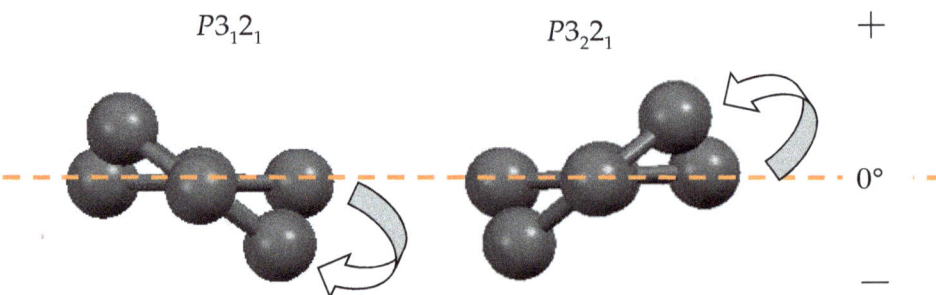

Figure 3. Structure of biphenyl. The dashed line shows a phenyl plane in the rear.

The crystals underlined in Table 1 showed a color change from pale yellow to deep red by cooling with liquid (Liq.) N$_2$. The color change corresponded well with the SCO phenomenon. The color of the present spiral [[Fe(NCS)$_2$(bpa)$_2$]·biphenyl]$_n$ was pale yellow and did not change by cooling with Liq. N$_2$. It was judged that the present spiral [[Fe(NCS)$_2$(bpa)$_2$]·biphenyl]$_n$ does not show SCO, although we could not measure the magnetic susceptibility and $^{57}$Fe Mössbauer spectrum because of too little quantities of crystal.

It was revealed that [[Co(NCS)$_2$(bpa)$_2$]·biphenyl]$_n$ has the same structure with [[Fe(NCS)$_2$(bpa)$_2$]·biphenyl]$_n$. The crystal data are shown in Table 3. There is not much difference in structure between the Fe complex and Co complex. The local structure around cobalt was chiral propeller type. The biphenyl molecule showed an atropisomerism in this crystal. We have analyzed the present complex in single crystal X-ray structural analysis by using chiral space group $P3_12_1$ (Flack parameter: 0.015). When analyzing it using $P3_22_1$, the Flack parameter becomes 0.9822, suggesting the space group is $P3_12_1$.

The projection of [[Fe(NCS)$_2$(bpa)$_2$]·biphenyl]$_n$ to *ab* plane is shown in Figure 2a. In this figure, divalent metal ions were bridged by bpa to form self-assembled 1D chain complex. Several 1D chains gathered together to form 1D chain sheet. The 1D chain sheet was stacked spirally to form novel spiral assembly with threefold axis. Biphenyl was stacked with threefold axis, and it was arranged between upper and lower 1D chain sheets (Figure 2b,c).

Biphenyl in the crystal is shown in Figure 3. Biphenyl molecules were stacked along a threefold spiral structure and the biphenyl was situated between the upper and lower sheet of M(NCS)$_2$(bpa)$_2$, which linked the two sheets. The dihedral angle in biphenyl was −35.72 and −36.55 for [[Fe(NCS)$_2$(bpa)$_2$]·biphenyl]$_n$ and [[Co(NCS)$_2$(bpa)$_2$]·biphenyl]$_n$, respectively. The biphenyl molecule showed atropisomerism in the solid state.

Table 3. Crystal data of [[Co(NCS)$_2$(bpa)$_2$]·biphenyl]$_n$.

| Co(NCS)$_2$(bpa)$_2$·biphenyl | |
|---|---|
| Temperature | 173K |
| Space group | $P3_12_1$ |
| a,b/Å | 10.1607(4) |
| c/Å | 30.0262(13) |
| α,β/° | 90 |
| γ/° | 120 |
| R1 | 0.0232 |
| wR2 | 0.0572 |
| Goodness of fit | 1.113 |
| Volume/Å$^3$ | 2684.6(2) |
| Flack parameter | 0.015(15) |
| Flack parameter when using $P3_22_1$ | 0.9822 |

It is known that self-assembled complexes enclathrate guest molecules in order to fill their vacancies. We investigated the relationship between 1D chain M(NCS)$_2$(bpa)$_2$ (M = Fe, Co) and biphenyl molecule. A Space-filling view of [[Fe(NCS)$_2$(bpa)$_2$]·biphenyl]$_n$ is shown in Figure 4. Figure 4a shows upper and lower 1D chain sheets and guest biphenyl. Figure 4b shows middle 1D chain. The size of 1D chain of M(NCS)$_2$(bpa)$_2$ (M = Fe, Co) and the chiral propeller type local structure around metal center fit in the space made by upper and lower biphenyl molecules. In spite of the good fit between 1D chain of M(NCS)$_2$(bpa)$_2$ (M = Fe, Co) and biphenyl molecule, intermolecular interactions, such as π–π stacking and CH/π interaction, were not observed in space-filling view and short-contact analysis. These results may suggest that biphenyl enclathrated by the host framework causes the biphenyl's dihedral angle to be fixed by a weak interaction. Therefore, the crystal shows a chirality, reflecting the chiral propeller type local structure around the metal center and biphenyl's atropisomerism.

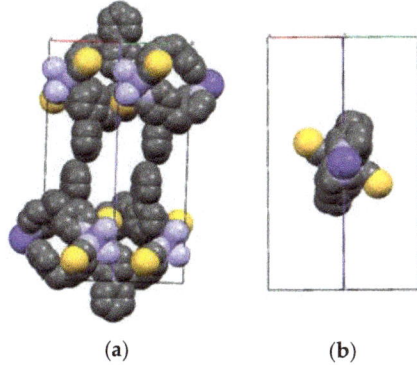

(a)   (b)

**Figure 4.** Space-filling view of [[Fe(NCS)$_2$(bpa)$_2$]·biphenyl]$_n$. Upper and lower 1D chain sheets and guest biphenyl are shown in (a) and middle 1D chain is shown in (b).

The schematic packing mechanism is shown in Figure 5. Threefold spiral structure is explained as shown below. For simplicity, we set the dihedral angle between the two phenyls in the biphenyl molecule at 30°, and we set the dihedral angle between the two phenyls in the intermolecular two biphenyls as 30°. The biphenyls are stacked with these angles, showing a threefold axis. The 1D chain of M(NCS)$_2$(bpa)$_2$ (M = Fe, Co) grows along the space that is formed by the phenyls of the top and bottom of the biphenyls. It can easily be seen that the upper and lower sheets have a 60° torsion angle. Therefore, the 1D chain sheet also stacks spirally with a threefold axis.

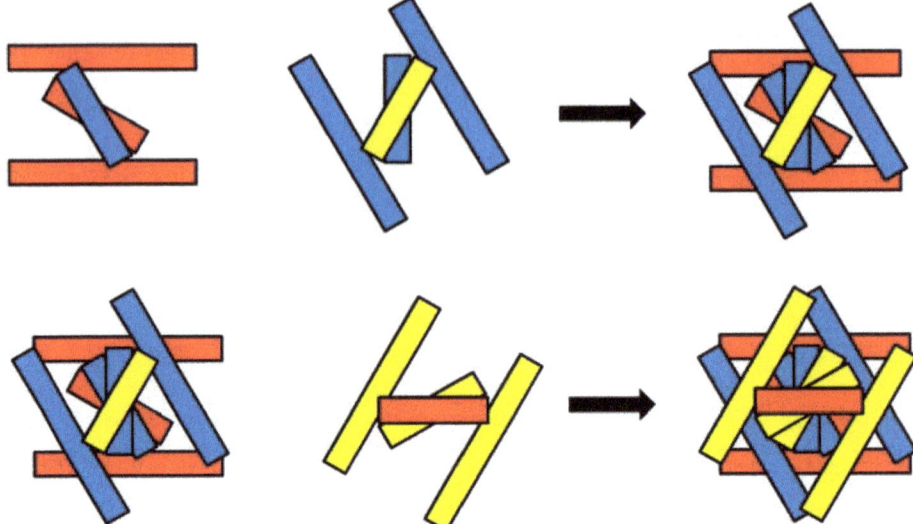

**Figure 5.** Schematic view of packing of spiral [[M(NCS)$_2$(bpa)$_2$]·biphenyl]$_n$ (M = Fe, Co). The long stick shows M(NCS)$_2$(bpa)$_2$ (M = Fe, Co) chain, while the short stick shows biphenyl. Elements with the same color have the same height.

In general, many enclathrated complexes are formed to fit the vacancies constructed by host structure with a guest molecule. In the present spiral complexes, both 1D chains having chiral propeller type structure of pyridines around the central metal and atropisomer of biphenyl constructed a spiral structure. That is, the chiral propeller type local structure determines the chiral structure and atropisomer of the biphenyl determines the "threefold" spiral structure. Moreover, it is expected that the 1D chain of M(NCS)$_2$(bpa)$_2$ (M = Fe, Co), having opposite direction of chiral propeller structure around the central metal and atropisomer of biphenyl, constructs a spiral structure having $P3_22_1$ space group.

## 4. Conclusions

We have synthesized self-assembled complexes [[M(NCS)$_2$(bpa)$_2$]·biphenyl]$_n$ (M = Fe, Co; bpa = 1,2-bis(4-pyridyl)ethane), which have a novel spiral structure, by one-step diffusion method. The present spiral complexes showed a chirality related with chiral propeller structure of pyridines around the central metal and biphenyl's atropisomerism. From the point of view of space-filling, 1D chain and biphenyl molecule formed a crystal, leading to the threefold spiral structure. Such structure conducts stepwise SCO if the anionic ligand is changed from NCS$^-$ to NCBH$_3^-$, as reported by the present author et al. [25]. Although the propeller structure of pyridines around the iron atom is a key point to showing SCO, the slight difference in ligand field between NCS$^-$ to NCBH$_3^-$ also affects the SCO phenomenon.

**Author Contributions:** S.T. was involved in all stages of the work, including conducting experiments and analyzing the data; H.D. was involved in the X-ray structural analysis; S.N. acted as the supervisor and helped with the data analysis and work planning; and S.T. and S.N wrote the paper.

**Conflicts of Interest:** The authors declare no conflict of interest.

## References

1. Gütlich, P.; Bill, E.; Trautwein, A.X. *Mössbauer Spectroscopy and Transition Metal Chemistry*; Springer-Verlag: Berlin/Heidelberg, Germany, 2011; pp. 391–476, ISBN 978-3-540-88427-9.

2. Kahn, O.; Kröber, J.; Jay, C. Spin transition molecular materials for displays and data recording. *Adv. Mater.* **1992**, *4*, 718–728. [CrossRef]
3. Gentili, D.; Demitri, N.; Schäfer, B.; Liscio, F.; Bergenti, I.; Ruani, G.; Ruben, M.; Cavallini, M. Multi-modal sensing in spin crossover compounds. *J. Mater. Chem. C* **2015**, *3*, 7836–7844. [CrossRef]
4. Nakashima, S.; Yamamoto, Y.; Asada, Y.; Koga, N.; Okuda, T. New assembled Fe-*trans*-1,2-Bis(4-pyridyl)ethylene-NCS(NCSe) complexes—Hydrogen bonded and π-π interacted structure and grid structure enclathrating ligand. *Inorg. Chim. Acta* **2005**, *358*, 257–264. [CrossRef]
5. Kitazawa, T.; Gomi, Y.; Takahashi, M.; Takeda, M.; Enomoto, M.; Miyazaki, A.; Enoki, T. Spin-crossover behaviour of the coordination polymer $Fe^{II}(C_5H_5N)_2Ni^{II}(CN)_4$. *J. Mater. Chem.* **1996**, *6*, 119–121. [CrossRef]
6. Niel, V.; Martinez-Agudo, J.M.; Muñoz, M.C.; Gaspar, A.B.; Real, J.A. Cooperative Spin Crossover Behavior in Cyanide-Bridged Fe(II)−M(II) Bimetallic 3D Hofmann-like Networks (M = Ni, Pd, and Pt). *Inorg. Chem.* **2001**, *40*, 3838–3839. [CrossRef]
7. Muñoz, M.C.; Gaspar, A.B.; Galet, A.; Real, J.A. Spin-Crossover Behavior in Cyanide-Bridged Iron(II)–Silver(I) Bimetallic 2D Hofmann-like Metal-Organic Frameworks. *Inorg. Chem.* **2007**, *46*, 8182–8192. [CrossRef]
8. Agustí, G.; Muñoz, M.C.; Gaspar, A.B.; Real, J.A. Spin-Crossover Behavior in Cyanide-bridged Iron(II)−Gold(I) Bimetallic 2D Hofmann-like Metal-Organic Frameworks. *Inorg. Chem.* **2008**, *47*, 2552–2561. [CrossRef]
9. Rodriguez-Velamazan, J.A.; Carbonera, C.; Castro, M.; Palacios, E.; Kitazawa, T.; Letard, J.-F.; Burriel, R. Two-step thermal spin transition and LIESST relaxation of the polymeric spin-crossover compounds Fe(X-py)2[Ag(CN)2]2 (X = H, 3-Methyl, 4-Methyl, 3,4-Dimethyl, 3-Cl). *Chem. Eur. J.* **2010**, *16*, 8785–8796. [CrossRef]
10. Dîrtu, M.M.; Naik, A.D.; Rotaru, A.; Spinu, L.; Poelman, D.; Garcia, Y. $Fe^{II}$ Spin Transition Materials Including an Amino–Ester 1,2,4-Triazole Derivative, Operating at, below, and above Room Temperature. *Inorg. Chem.* **2016**, *55*, 4278–4295. [CrossRef]
11. Roubeau, O. Triazole-Based One-Dimensional Spin-Crossover Coordination Polymers. *Chem. Eur. J.* **2012**, *18*, 15230–15244. [CrossRef]
12. Quesada, M.; Koojiman, H.; Gomez, P.; Costa, J.S.; Koningsbruggen, P.J.; Weinberger, P.; Reissner, M.; Spek, A.L.; Haasnoot, J.G.; Reedjik, J. [Fe(μ-btzmp)2(btzmp)2](ClO4)2: A doubly-bridged 1D spin-transition bistetrazole-based polymer showing thermal hysteresis behavior. *Dalton. Trans.* **2007**, 5434–5440. [CrossRef] [PubMed]
13. Moliner, N.; Muñoz, M.C.; Létard, S.; Salmon, L.; Tuchgues, J.P.; Boussksou, A.; Real, J.A. Mass Effect on the Equienergetic High-Spin/Low-Spin States of Spin-Crossover in 4,4′-Bipyridine-Bridged Iron(II) Polymeric Compounds: Synthesis, Structure, and Magnetic, Mössbauer, and Theoretical Studies. *Inorg. Chem.* **2002**, *41*, 6997–7005. [CrossRef] [PubMed]
14. Morita, T.; Asada, Y.; Okuda, T.; Nakashima, S. Isomerism of assembled iron complex bridged by 1,2-di(4-pyrisyl)ethane and its solid-to-solid transformation accompanied by a change of electronic state. *Bull Chem. Soc. Jpn.* **2006**, *79*, 738–744. [CrossRef]
15. Morita, T.; Nakashima, S.; Yamada, K.; Inoue, K. Occurrence of the spin-crossover phenomenon of assembled complexes, Fe(NCX)2(bpa)2 (X= S, BH3; bpa= 1,2-bis(4-pyridyl)ethane) by enclathrating organic guest molecule. *Chem. Lett.* **2006**, *35*, 1042–1043. [CrossRef]
16. Nakashima, S.; Morita, T.; Inoue, K. Spin-crossover phenomenon of the assembled iron complexes with 1,2-bis(4-pyridyl)ethane as bridging ligand studied by Mössbauer spectroscopy. *Hyperfine Interact.* **2009**, *188*, 107–111. [CrossRef]
17. Dote, H.; Kaneko, M.; Inoue, K.; Nakashima, S. Synthesis of Anion-Mixed Crystals of the Assembled Complexes Bridged by 1,2-Bis(4-pyridyl)ethane and Ligand Field of Fe(NCS)(NCBH3) Unit. *Bull. Chem. Soc. Jpn.* **2018**, *91*, 71–81. [CrossRef]
18. Atsuchi, M.; Higashikawa, H.; Yoshida, Y.; Nakashima, S.; Inoue, K. Novel 2D Interpenetrated Structure and Occurrence of the Spin-crossover Phenomena of Assembled Complexes, Fe(NCX)2(bpp)2 (X = S, Se, BH3; bpp = 1,3-Bis(4-pyridyl)propane. *Chem. Lett.* **2006**, *36*, 1064–1065. [CrossRef]
19. Atsuchi, M.; Inoue, K.; Nakashima, S. Reversible structural change of host framework triggered by desorption and adsorption of guest benzene molecules in Fe(NCS)2(bpp)2·2(benzene) (bpp = 1,3-bis(4-pyridyl)propane). *Inorg. Chim. Acta* **2011**, *370*, 82–88. [CrossRef]

20. Yoshinami, K.; Kaneko, M.; Yasuhara, H.; Nakashima, S. Effect of methyl substituent on the spin state of iron(II) assembled complex using 1,4-bis(4-pyridyl)benzene. *Radioisotopes* **2017**, *66*, 625–632. [CrossRef]
21. Iwai, S.; Yoshinami, K.; Nakashima, S. Structure and Spin State of Iron(II) Assembled Complexes using 9,10-Bis(4-pyridyl)anthracene as Bridging Ligand. *Inorganics* **2017**, *5*, 61. [CrossRef]
22. Kaneko, M.; Tokinobu, S.; Nakashima, N. Density Functional Study on Spin-crossover Phenomena of Assembled Complexes, [Fe(NCX)$_2$(bpa)$_2$]$_n$ (X = S, Se, BH$_3$; bpa = 1,2-bis(4-pyridyl)ethane). *Chem. Lett.* **2013**, *42*, 1432–1434. [CrossRef]
23. Kaneko, M.; Nakashima, S. Computational Study on Thermal Spin-Crossover Behavior for Coordination Polymers Possessing trans-Fe(NCS)$_2$(pyridine)$_4$ Unit. *Bull. Chen. Soc. Jpn.* **2015**, *88*, 1164–1170. [CrossRef]
24. Hernández, M.L.; Barandika, M.G.; Urtiaga, M.K.; Cortés, R.; Lezama, L.; Arriortuac, M.L.; Rojo, T. Structural analysis and magnetic properties of the 1-D compounds [M(NCS)$_2$bpa$_2$] [M = Fe, Co, Ni and bpa = 1,2-bis(4-pyridyl)ethane]. *J. Chem. Soc. Dalton Trans.* **1999**, 1401–1406. [CrossRef]
25. Nakashima, S.; Morita, T.; Inoue, K.; Hayami, S. Spiral assembly of the 1D chain sheet of Fe(NCBH$_3$)$_2$(bpa)$_2$·(biphenyl) (bpa = 1,2-bis(4-pyridyl)ethane) and its stepwise spin-crossover phenomenon. *Polymers* **2011**, *4*, 880–888. [CrossRef]
26. The SHELX Homepage. Available online: http://shelxl.uni-ac.gwdg.de/ (accessed on 14 February 2019).

© 2019 by the authors. Licensee MDPI, Basel, Switzerland. This article is an open access article distributed under the terms and conditions of the Creative Commons Attribution (CC BY) license (http://creativecommons.org/licenses/by/4.0/).

Article

# High-Temperature Cooperative Spin Crossover Transitions and Single-Crystal Reflection Spectra of [Fe$^{III}$(qsal)$_2$](CH$_3$OSO$_3$) and Related Compounds

Kazuyuki Takahashi [1,*], Kaoru Yamamoto [2], Takashi Yamamoto [3], Yasuaki Einaga [3], Yoshihito Shiota [4], Kazunari Yoshizawa [4] and Hatsumi Mori [5]

1. Department of Chemistry, Graduate School of Science, Kobe University, 1-1 Rokkodai, Nada-ku, Kobe, Hyogo 657-8501, Japan
2. Department of Applied Physics, Okayama University of Science, 1-1 Ridaicho, Kita-ku, Okayama, Okayama 700-0005, Japan; yamamoto@dap.ous.ac.jp
3. Department of Chemistry, Graduate School of Science and Technology, Keio University, 3-14-1 Hiyoshi, Kohoku-ku, Yokohama, Kanagawa 223-8522, Japan; takyama@chem.keio.ac.jp (T.Y.); einaga@chem.keio.ac.jp (Y.E.)
4. Institute for Materials Chemistry and Engineering, Kyushu University, 744 Motooka, Nishi-ku, Fukuoka 819-0395, Japan; shiota@ms.ifoc.kyushu-u.ac.jp (Y.S.); kazunari@ms.ifoc.kyushu-u.ac.jp (K.Y.)
5. Institute for Solid State Physics, The University of Tokyo, 5-1-5 Kashiwanoha, Kashiwa, Chiba 277-8581, Japan; hmori@issp.u-tokyo.ac.jp
* Correspondence: ktaka@crystal.kobe-u.ac.jp; Tel.: +81-78-803-5691

Received: 22 January 2019; Accepted: 30 January 2019; Published: 2 February 2019

**Abstract:** New Fe(III) compounds from qsal ligand, [Fe(qsal)$_2$](CH$_3$OSO$_3$) (**1**) and [Fe(qsal)$_2$] (CH$_3$SO$_3$)·CH$_3$OH (**3**), along with known compound, [Fe(qsal)$_2$](CF$_3$SO$_3$) (**2**), were obtained as large well-shaped crystals (Hqsal = $N$-(8-quinolyl)salicylaldimine). The compounds **1** and **2** were in the low-spin (LS) state at 300 K and exhibited a cooperative spin crossover (SCO) transition with a thermal hysteresis loop at higher temperatures, whereas **3** was in the high-spin (HS) state below 300 K. The optical conductivity spectra for **1** and **3** were calculated from the single-crystal reflection spectra, which were, to the best of our knowledge, the first optical conductivity spectra of SCO compounds. The absorption bands for the LS and HS [Fe(qsal)$_2$] cations were assigned by time-dependent density functional theory calculations. The crystal structures of **1** and **2** consisted of a common one-dimensional (1D) array of the [Fe(qsal)$_2$] cation, whereas that of **3** had an unusual 1D arrangement by $\pi$-stacking interactions which has never been reported. The crystal structures in the high-temperature phases for **1** and **2** indicate that large structural changes were triggered by the motion of counter anions. The comparison of the crystal structures of the known [Fe(qsal)$_2$] compounds suggests the significant role of a large non-spherical counter-anion or solvate molecule for the total lattice energy gain in the crystal of a charged complex.

**Keywords:** spin crossover; Fe(III) complex; qsal ligand; thermal hysteresis; structure phase transition; counter-anion; solvate; lattice energy; optical conductivity spectrum

## 1. Introduction

Spin crossover (SCO) between a high-spin (HS) and low-spin (LS) state in a transition metal coordination compound is one of the molecular switching phenomena responsive to various external stimuli such as temperature, pressure, light, magnetic field, and chemicals. Significant attention has been attracted to SCO phenomena in the wide range of fields of chemical and physical sciences [1–5]. The SCO switches not only a spin-state but also color and coordination structure in a transition metal complex. Thus, the utilization of electronic and structural transformation accompanying SCO can lead to potential applications of display, memory, sensing, electronic, and mechanical devices.

A family of coordination compounds from qsal (Hqsal = N-(8-quinoyl)salicylaldimine), which is a π-extended tridentate Schiff base ligand, is known as one of well-studied SCO compounds. Dahl et al. first reported the synthesis, magnetic, and spectral properties of the qsal ligand and its transition metal coordination compounds in 1969 [6]. Dickinson et al. reported on the anomalous magnetic conversion of [Fe(qsal)$_2$](NCS) in 1977 [7]. Successively the [Fe(qsal)$_2$] derivatives with various counter-anions and solvate-molecules have been studied so far [8–16], leading to the appearance of a cooperative SCO transition with a very wide thermal hysteresis in [Fe(qsal)$_2$](NCSe) [9] and giving a rare example of Fe(III) compounds exhibiting the light-induced excited spin trapping (LIESST) effect [11]. Moreover, substitution effects on the qsal ligand have been reported in recent years [17–21]. We focused on the [Fe(qsal)$_2$] compounds showing a cooperative SCO transition probably due to strong π-stacking interactions and developed SCO conductors and magnets by combining the [Fe(qsal)$_2$] cation with redox active functional anions [22,23]. A number of multifunctional materials from the [Fe(qsal)$_2$] cation have been reported to date [24–33]. Recently, heteroleptic Fe(III) compounds [34–37] as well as Fe(II) compounds [38–44] from the qsal and its derivatives, were also reported.

As described above, the SCO compounds from qsal have attracted great deal attention, and some systems have been reported to undergo SCO phase transitions in a critical manner. However, the fundamental mechanism, namely how the spin-state changes in the individual iron sites interact with each other, has been virtually unknown. This is partly because it is often difficult to determine both the HS and LS structures due to deterioration in crystal through the SCO transition or the desorption of solvate molecules. In particular, the existence of solvate molecules in an SCO compound affords various undistinguishable effects on its SCO behavior as well as instability of the crystal. Therefore, non-solvate [Fe(qsal)$_2$] compounds seem to be favorable for the elucidation of the SCO mechanism. Meanwhile, structure-characterized non-solvate [Fe(qsal)$_2$] compounds were rare [11,13,15,24,33].

The optical spectra are useful to interpret the photoresponsive property of SCO compounds. Most absorption spectra for SCO compounds have been obtained by using a solution, KBr pellet, oil-dispersed powder, and cast film. Some diffuse reflectance spectra for powdered samples have also been reported. However, the sample preparations for these measurements often vary the spin-state and SCO behavior of the compound measured. To obtain an accurate optical spectrum of an SCO compound whose structure and spin-state is confirmed, a single-crystal optical spectrum may be most suitable. Although a single-crystal absorption spectrum is the most suitable for this purpose, it is not easy to record it because most Fe(III) compounds and LS Fe(II) compounds are intensely colored. A single-crystal reflection spectrum is the second choice. By the Kramers–Kronig analysis, it can be converted to its optical conductivity spectrum corresponding to the absorption spectrum. Meanwhile, the compound measured has a large area of a flat crystal surface without the occurrence of interference in the wide wavelength range from visible to near-infrared. Thus, it is also difficult to obtain a single-crystal reflection spectrum of an SCO compound.

Recently we found a new non-solvate compound [Fe(qsal)$_2$](CH$_3$OSO$_3$) (**1**) as very large rhombic platelets with a flat crystal surface (Figure 1). As we tried to prepare the related compounds with similar size and shape anions, we obtained relatively large crystals of the known non-solvate compound [Fe(qsal)$_2$](CF$_3$SO$_3$) (**2**) [12] and a new methanol-solvate compound [Fe(qsal)$_2$](CH$_3$SO$_3$)·CH$_3$OH (**3**). The compounds **1** and **2** were in the LS state, whereas compound **3** was in the HS state at room temperature. The compounds **1** and **2** showed a cooperative SCO transition with a thermal hysteresis above 350 K. Fortunately, we were successful to determine the crystal structures of the HS phases for **1** and **2** at 425 and 400 K, respectively. The structural comparison among the [Fe(qsal)$_2$] compounds suggests the role of counter-anion or solvate molecule for the crystal of a charged complex. Moreover, single-crystal reflection spectra could be recorded for **1** and **3** in the wide wavelength range. The optical conductivity spectra from the single-crystal reflection spectra are, to the best our knowledge, first examples for SCO compounds. We report herein the preparation, crystal structures, and spectral characterizations for **1–3** and also discuss the mechanism of the cooperative SCO transitions in the [Fe(qsal)$_2$] compounds.

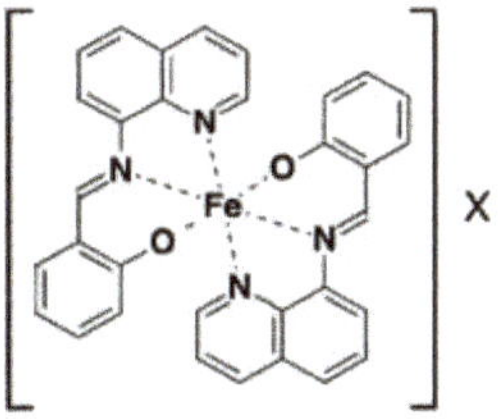

**Figure 1.** Molecular structures of compounds **1–3**.

1: X = CH$_3$OSO$_3$
2: X = CF$_3$SO$_3$
3: X = CH$_3$SO$_3$·CH$_3$OH

## 2. Materials and Methods

*2.1. Synthesis of Compounds 1–3*

All chemicals were purchased and used without further purification. [Fe(qsal)$_2$]Cl·1.5H$_2$O was prepared according to the literature [7].

[Fe(qsal)$_2$](CH$_3$OSO$_3$) (**1**): A saturated methanol solution of [Fe(qsal)$_2$]Cl·1.5H$_2$O was filtered by an Advantec 5B filter paper. Diffusion of an excess amount of (Bu$_3$MeN)(CH$_3$OSO$_3$) to the filtered solution at room temperature gave **1** as large black rhombic platelets. Anal. Calcd. For C$_{33}$H$_{25}$N$_4$O$_6$FeS: C, 59.92; H, 3.81; N, 8.47%. Found: C, 59.85; H, 3.95; N, 8.18%.

[Fe(qsal)$_2$](CF$_3$SO$_3$) (**2**): A saturated methanol solution of [Fe(qsal)$_2$]Cl·1.5H$_2$O was filtered by an Advantec 5B filter paper. Diffusion of an excess amount of Li(CF$_3$SO$_3$) to the filtered solution at room temperature gave **2** as black parallelogrammatic platelets. Anal. Calcd. For C$_{33}$H$_{22}$N$_4$O$_5$F$_3$FeS: C, 56.67; H, 3.17; N, 8.01%. Found: C, 56.54; H, 3.21; N, 7.94%.

[Fe(qsal)$_2$](CH$_3$SO$_3$)·CH$_3$OH (**3**): A saturated methanol solution of [Fe(qsal)$_2$]Cl·1.5H$_2$O was filtered by an Advantec 5B filter paper. Diffusion of an excess amount of (Bu$_4$N)(CH$_3$SO$_3$) followed by cooling to 3 °C gave **3** as black rhombic platelets. Anal. Calcd. for C$_{34}$H$_{29}$N$_4$O$_6$FeS: C, 60.27; H, 4.31; N, 8.27%. Found: C, 60.00; H, 4.46; N, 8.13%.

*2.2. Physical Measurements*

Variable temperature direct current magnetic susceptibilities of polycrystalline samples were measured on a Quantum Design MPMS-XL magnetometer under a field of 0.5 T in the temperature range of 2–320 K. The oven option was used for the measurement in the temperature range of 300–450 K. The magnetic susceptibilities were corrected for diamagnetic contributions estimated by Pascal constants [45].

The Mössbauer spectra were recorded on a constant acceleration spectrometer with a source of $^{57}$Co/Rh in the transmission mode. The measurements at low temperature were performed with a closed-cycle helium refrigerator (Iwatani Co., Ltd., Japan). Velocity was calibrated by using an α-Fe standard. The obtained Mössbauer spectra were fitted with asymmetric Lorentzian doublets by the least squares fitting program (MossWinn).

The polarized reflection spectrum was recorded using an infrared microscope Spectratech IR-Plan combined with an FT-IR spectrometer Thermo Nicolet NEXUS 870 for 5000–12000 cm$^{-1}$, and a multichannel visible spectrograph Atago Macs 320 for 11000–33000 cm$^{-1}$. The spectrum was obtained from the developed plane of the thin plate crystal. The crystal orientation was adjusted so that the infrared reflectivity was maximized for the plane polarized light. The optical conductivity spectrum was calculated from the reflection spectrum by the Kramers–Kronig analysis.

## 2.3. Crystal Structure Determinations of 1–3

A crystal was mounted on a roll of 15 μm thick polyimide film by using the Araldite™ adhesive. A Nihon Thermal Engineering nitrogen gas flow temperature controller was used for the temperature variable measurements. All data were collected on a Bruker APEX II CCD area detector with monochromated Mo-Kα radiation generated by a Bruker Turbo X-ray Source coupled with Helios multilayer optics. All data collections were performed using the APEX2 crystallographic software package (Bruker AXS). The data were collected to a maximum 2θ value of 55.0°. A total of 720 oscillation images were collected. The APEX3 crystallographic software package (Bruker AXS) was used to determine the unit cell parameters. Data were integrated by using SAINT. Numerical absorption correction was applied by using SADABS. The structures at all temperatures were solved by direct methods and refined by full-matrix least-squares methods based on $F^2$ by using the SHELXTL program. All non-hydrogen atoms were refined anisotropically. Hydrogen atoms were generated by calculation and refined using the riding model. CCDC 1891471-1891477 contains the supplementary crystallographic data for this paper. These data can be obtained free of charge via http://www.ccdc.cam.ac.uk/conts/retrieving.html (or from the CCDC, 12 Union Road, Cambridge CB2 1EZ, UK; Fax: +44 1223 336033; E-mail: deposit@ccdc.cam.ac.uk).

## 2.4. Density Functional Theory (DFT) Calculations

All theoretical calculations were performed using the Gaussian 09 program package [46]. The atomic coordinates for the LS and HS states of the [Fe(qsal)$_2$] cation were taken from the single crystal structural data for **1** and **3**, respectively. All geometry optimization and frequency calculations of the compounds were carried out at the B3LYP functional [47,48]. The Wachters-Hay basis set [49,50] for Fe atoms and the 6-31+G(d) basis set [51] for H, C, O, and N atoms were used. No imaginary frequencies were found in the optimized structures. Cartesian coordinates of the LS [Fe(qsal)$_2$] and HS [Fe(qsal)$_2$] cations calculated by the B3LYP level of theory are listed in Tables S1 and S2 in the supplementary materials. The transition energies of all electron transitions of the LS [Fe(qsal)$_2$] and HS [Fe(qsal)$_2$] cations were calculated by using the time-dependent DFT (TD-DFT) method [52] at the CAM-B3LYP level [53].

# 3. Results and Discussion

## 3.1. Synthesis of Compounds 1–3

Compounds **1–3** were prepared by the metathesis reaction between [Fe(qsal)$_2$]Cl·1.5H$_2$O and corresponding anion salts using the diffusion methods. Compound **1** was obtained as very large rhombic platelets (Figure 2a), whereas compounds **2** and **3** gave relatively large parallelogrammatic and rhombic platelets, respectively (Figure 2b,c). The compositions of **1** to **3** were confirmed by microanalyses and crystal analyses described below.

**Figure 2.** Photographs of crystals for **1** (**a**), **2** (**b**), and **3** (**c**). The grid unit is 1 mm long.

## 3.2. Magnetic Susceptibility for 1–3

The temperature variations of magnetic susceptibility for compounds **1–3** are shown in Figure 3. The $\chi_M T$ value for compound **1** at 300 K was 0.56 cm$^3$ K mol$^{-1}$, suggesting **1** was almost in the LS state. Below 300 K, the $\chi_M T$ values were temperature-independent. The $\chi_M T$ value was 0.43 cm$^3$ K mol$^{-1}$ at 10 K. Meanwhile, on heating compound **1** above 300 K, the $\chi_M T$ values smoothly increased up to 410 K and then an abrupt change in $\chi_M T$ was observed ($T_{1/2}\uparrow$ = 414 K). After the steep transition, the $\chi_M T$ values still increased gradually again and reached 3.38 cm$^3$ K mol$^{-1}$ at 450 K, suggesting compound **1** exhibited an incomplete SCO transition. On successive lowering temperatures, a steep decrease in $\chi_M T$ was observed at 406 K ($T_{1/2}\downarrow$ = 402 K), and then the $\chi_M T$ curve traced that measured in the heating scan below 396 K. Thus, the compound **1** showed the cooperative SCO transition with a thermal hysteresis width of 12 K.

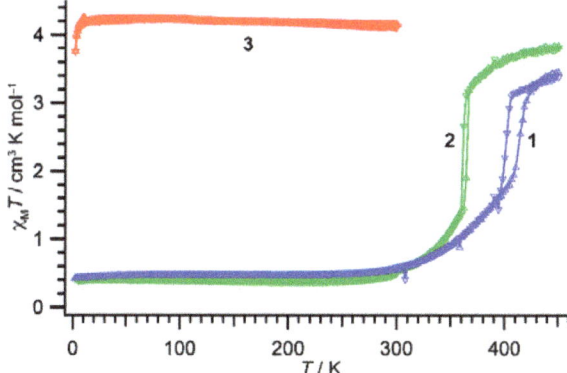

**Figure 3.** The $\chi_M T$ vs. *T* products for **1** (blue triangles), **2** (green triangles), and **3** (red triangles). Triangles up and down indicate heating and cooling processes, respectively. Lines are used to guiding the eye.

The $\chi_M T$ value for compound **2** at 300 K was 0.54 cm$^3$ K mol$^{-1}$, suggesting **2** was also almost in the LS state. On heating the sample, the $\chi_M T$ values gradually rose, and an abrupt increase in $\chi_M T$ was observed at 362 K ($T_{1/2}\uparrow$ = 365 K). The $\chi_M T$ value reached to 3.84 cm$^3$ K mol$^{-1}$ at 450 K. On lowering temperatures, the $\chi_M T$ values traced those of the heating scan and then abruptly decreased at 364 K ($T_{1/2}\downarrow$ = 361 K), resulting in a cooperative transition with a thermal hysteresis width of 4 K.

The $\chi_M T$ value for compound **3** at 300 K was 4.15 cm$^3$ K mol$^{-1}$, suggesting **3** was almost in the HS state. On decreasing temperatures, the $\chi_M T$ values were almost constant in the temperature range of 2–300 K.

## 3.3. Mössbauer Spectroscopy for 1 and 3

To confirm the spin-states of the Fe ion for compound **1** and **3** at 293 K, the Mössbauer spectra for **1** and **3** were recorded (Figure 4). The spectrum of **1** at 293 K consisted of only a sharp asymmetric doublet. The asymmetry of the spectrum may originate from the preferred orientation of the crystals. As compared with the Mössbauer parameters reported in the literature (Table 1), the spectrum of **1** can be ascribed to the LS spectrum, which is consistent with the $\chi_M T$ value of **1** at 293 K. On the other hand, the spectrum of **3** at 293 K consisted of a very broad asymmetric doublet. The isomer shift of **3** is similar to those of the simulated HS spectra reported in the literature. Note that the quadrupole splitting of **3** is the largest among those of the HS spectra in the related compounds and corresponds to that of the [Ni(nmt)$_2$] compound having the significantly distorted coordination octahedron (mnt = maleonitriledithiolate) [33]. This suggests that the compound **3** might have a distorted coordination sphere due to some crystal packing effect.

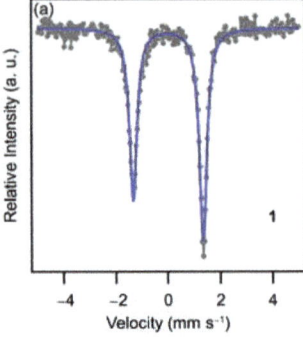
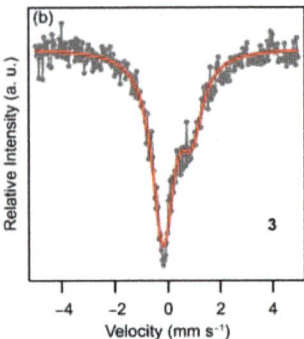

**Figure 4.** Mössbauer spectra for **1** (**a**) and **3** (**b**) at 293 K. Gray circles are recorded data points. The blue and red curves are fitted LS and HS spectra, respectively.

**Table 1.** Mössbauer parameters (mm s$^{-1}$) for [Fe(qsal)$_2$](X)·solv.

| Compound X | Solv | T/K | Spin-State [1] | IS [2] | QS [3] | LW [4] | Ref. |
|---|---|---|---|---|---|---|---|
| CH$_3$OSO$_3$ (1) | – | 293 | LS (100%) | 0.0040(18) | 2.689(4) | 0.342(5) | this work |
| CH$_3$SO$_3$ (3) | CH$_3$OH | 293 | HS (100%) | 0.376(14) | 1.11(2) | 0.98(3) | this work |
| Cl | 1.5H$_2$O | 298 | HS (100%) | 0.28 | 0.68 | – | [7] |
|  |  | 77 | HS | 0.36 | 0.70 | – |  |
|  |  |  | LS | 0.04 | 2.44 | – |  |
|  |  | 4.3 | HS | 0.40 | 0.66 | – |  |
|  |  |  | LS | 0.14 | 2.63 | – |  |
| NCS | – | 288 | HS (100%) | 0.253 | 0 | 1.737 | [8] |
|  |  | 78 | HS (6.5%) | 0.360 | 0 | 1.075 |  |
|  |  |  | LS (93.5%) | 0.191 | 2.660 | 0.288 0.326 |  |
| NCSe | CH$_3$OH | 293 | LS (100%) | 0.07 | 2.52 | – | [9] |
| NCSe | CH$_2$Cl$_2$ | 293 | LS (100%) | 0.11 | 2.47 | – | [9] |
| I$_3$ | – | 293 | HS (100%) | 0.232 | 0.616 | – | [14] |
|  |  | 15 | LS (100%) | 0.075 | 2.878 | – |  |
| [Ni(dmit)$_2$] | 2CH$_3$CN | 293 | HS (100%) | 0.276 | 0.745 | – | [22] |
|  |  | 9 | HS (13%) | 0.32 | 0.68 | – |  |
|  |  |  | LS (87%) | 0.07 | 2.74 | – |  |
| [Ni(mnt)$_2$] | – | 293 | HS (100%) | 0.307(16) | 0.91(2) | 0.79(4) | [33] |
|  |  | 100 | HS (100%) | 0.48(3) | 1.07(3) | 1.54(5) |  |

[1] HS: high-spin, LS: low-spin. The percentage of each spin-state is shown in the parenthesis. [2] Isomer shift. [3] Quadrupole splitting. [4] Linewidth.

## 3.4. Optical Properties of the [Fe(qsal)$_2$] Compounds

### 3.4.1. Reflection and Optical Conductivity Spectra of Single Crystals for 1 and 3

Since **1** and **3** have relatively large flat crystal surfaces, we tried to record reflection spectra at room temperature. The obtained reflection spectrum for **1** and **3** and their optical conductivity spectrum calculated by the Kramers–Kronig analysis are shown in Figure 5a,b, respectively.

As is shown in the panel (**a**), the spectrum of the LS state in **1** displays the two clear absorption maxima: the strong visible absorption at 428 nm and the near-infrared medium band at 904 nm. On the other hand, the spectrum of the HS state in **3** does not show any clear absorption in the near-infrared region, whereas in the visible region there are two absorption bands at 467 and 366 nm.

Some examples of the light-induced excited spin-state trapping (LIESST) effect on the [Fe(qsal)$_2$] compounds have ever been reported to date. The excitation wavelengths used for the LIESST effect were 808 [11] and 830 nm [22,23]. These wavelengths are in good agreement with the absorption band observed for **1**, namely the LS [Fe(qsal)$_2$] compound. Therefore, the LIESST scheme from the LS to HS states in the [Fe(qsal)$_2$] compounds is evidenced.

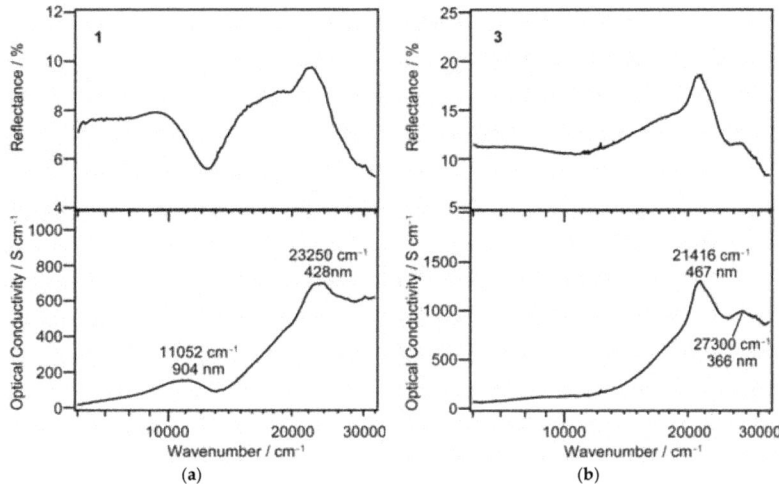

**Figure 5.** Single-crystal reflectance (upper) and optical conductivity (lower) spectra at room temperature for **1** (**a**) and **3** (**b**).

3.4.2. Time-Dependent Density Functional Theory (TD-DFT) Calculations for the LS and HS [Fe(qsal)$_2$] Cations

To provide an insight into the absorption bands for **1** and **3**, the transition energies of all electron transitions in the LS [Fe(qsal)$_2$] and HS [Fe(qsal)$_2$] cations were calculated by using the time-dependent density functional theory (TD-DFT) method. The transition energies were strongly dependent on both the density functionals and [Fe(qsal)$_2$] structures, and the most reasonable energies could be obtained by calculating the B3LYP-optimized structures at the CAM-B3LYP level. The transition wavelengths and assignments are summarized in Table 2.

**Table 2.** Excitation wavelengths, oscillator strengths (*f*), and absorption assignments for the LS [Fe(qsal)$_2$] and HS [Fe(qsal)$_2$] cations.

| | LS | | | HS | | |
|---|---|---|---|---|---|---|
| Wavelength/nm | *f* | Assignment | Wavelength/nm | *f* | Assignment |
| 2138.01 | 0.0000 | d-d, LMCT | 656.61 | 0.0138 | LMCT |
| 1837.95 | 0.0001 | d-d, LMCT | 628.25 | 0.0000 | LMCT |
| 842.64 | 0.0000 | d-d, LMCT | 579.97 | 0.0099 | LMCT |
| 802.03 | 0.0000 | d-d, LMCT | 569.29 | 0.0020 | LMCT |
| 653.38 | 0.0001 | d-d, LMCT | 559.22 | 0.0004 | LMCT |
| 636.23 | 0.0000 | d-d, LMCT | 546.71 | 0.0056 | LMCT |
| 619.84 | 0.0000 | π-π* | 525.34 | 0.0304 | LMCT |
| 615.97 | 0.0000 | π-π*, d-d | 488.80 | 0.0007 | LMCT |
| 515.63 | 0.0005 | d-d, LMCT | 453.77 | 0.0053 | π-π* |
| 500.30 | 0.0196 | LMCT, π-π* | 452.76 | 0.0101 | π-π* |
| 490.33 | 0.0007 | π-π*, d-d | 425.61 | 0.0144 | LMCT |
| 487.20 | 0.0059 | LMCT, π-π* | 416.98 | 0.0001 | LMCT |
| 481.57 | 0.0018 | LMCT, d-d | 385.54 | 0.0580 | LMCT, d-d |
| 479.60 | 0.0001 | d-d, LMCT | 382.47 | 0.0805 | π-π* |
| 464.66 | 0.0021 | LMCT, π-π* | 375.07 | 0.1551 | LMCT, π-π* |
| 413.29 | 0.0474 | LMCT, π-π* | 363.99 | 0.0556 | LMCT |
| 411.68 | 0.0559 | π-π*, LMCT | 363.21 | 0.0017 | LMCT |
| 403.64 | 0.1539 | π-π*, LMCT | 357.28 | 0.0305 | LMCT |
| 400.94 | 0.0967 | π-π* | 353.41 | 0.0003 | LMCT, d-d |
| 383.76 | 0.0044 | LMCT, d-d | 350.25 | 0.0033 | LMCT |

The wavelengths calculated for the HS [Fe(qsal)$_2$] cation were ascribed mainly to ligand-to-metal charge transfer (LMCT) transition and intra-ligand π-π* transition and were shorter than 656 nm, which was consistent with no remarkable absorption band observed in the near-infrared region for **3**. The absorption bands observed at 467 and 366 nm for **3** were ascribed to π-π* and LMCT transitions, respectively. On the other hand, the wavelengths calculated for the LS [Fe(qsal)$_2$] cation were ascribed mainly to d-d and LMCT transitions. Very weak absorption bands for the LS [Fe(qsal)$_2$] cation were calculated in the near-infrared region. This is probably because these absorption bands would originate mainly from forbidden d-d transitions. The observed absorption bands at 904 and 428 nm for **1** can be ascribed to d-d and LMCT transitions, respectively.

### 3.5. Crystal Structures of 1–3

The variable temperature single crystal X-ray structural analyses for **1**, **2**, and **3** were performed using a Bruker AXS APEXII Ultra diffractometer. Crystallographic data are listed in Tables 3 and 4. The crystal structures for **1** and **3** at 293 K belonged to a triclinic system with $P\bar{1}$, whereas the crystal structure of **2** at 293 K belonged to monoclinic $P2_1/n$. Each asymmetrical unit consisted of one [Fe(qsal)$_2$] cation and one corresponding anion, and additionally one methanol molecule for **3**. Fortunately, it was successful to determine the high-temperature phase structures of **1** and **2** at 400 and 425 K, respectively. The crystal space group for **2** maintained at 400 K, whereas that for **1** was changed into monoclinic $P2/n$. As associated with the crystal transformation, the crystallographically independent molecules in **1** at 425 K increased to three and a half of the [Fe(qsal)$_2$] cations and CH$_3$OSO$_3$ anions. The independent [Fe(qsal)$_2$] cations in **1** at 425 K are hereafter designated as A, B, C, and D (a half independent cation). The atomic numbers of the [Fe(qsal)$_2$] cations A to D also add the notations of A to D to the corresponding atomic numbers.

Table 3. Crystallographic data for **1**.

| Formula | \multicolumn{4}{c}{C$_{33}$H$_{25}$FeN$_4$O$_6$S} | | | |
|---|---|---|---|---|
| Formula Weight | | 661.48 | | |
| Color | | black | | |
| Dimension/mm | | 0.30 × 0.30 × 0.15 | | |
| T/K | 293 | 360 | 400 | 425 |
| Crystal System | triclinic | triclinic | triclinic | monoclinic |
| Space Group | $P\bar{1}$ | $P\bar{1}$ | $P\bar{1}$ | $P2/n$ |
| a/Å | 9.7672(6) | 9.7725(7) | 9.743(2) | 33.110(8) |
| b/Å | 11.8138(7) | 11.9003(8) | 12.031(3) | 9.886(3) |
| c/Å | 12.9116(8) | 12.9660(9) | 13.023(3) | 34.385(9) |
| α/° | 70.6370(10) | 70.5486(8) | 70.377(2) | 90 |
| β/° | 85.4390(10) | 85.6245(9) | 85.801(3) | 115.807(3) |
| γ/° | 85.5930(10) | 85.5470(8) | 85.532(2) | 90 |
| V/Å$^3$ | 1399.10(15) | 1415.49(17) | 1431.7(5) | 10133(4) |
| Z | 2 | 2 | 2 | 14 |
| $\rho_{calcd}$/g cm$^{-3}$ | 1.570 | 1.552 | 1.534 | 1.518 |
| μ (Mo-Kα) | 0.671 | 0.663 | 0.655 | 0.648 |
| 2θ$_{max}$/° | 51.36 | 50.06 | 50.04 | 52.74 |
| No. Reflections | 7116 | 6907 | 6956 | 52039 |
| ($R_{int}$) | (0.0145) | (0.0158) | (0.0184) | (0.0361) |
| No. Observations | 5158 | 4922 | 4961 | 20655 |
| ($I > 2.00\sigma(I)$) | (4887) | (4577) | (4483) | (14398) |
| No. Variables | 407 | 407 | 407 | 1423 |
| R1 ($I > 2.00\sigma(I)$) | 0.0282 | 0.0305 | 0.0338 | 0.0747 |
| R (all data) | 0.0298 | 0.0329 | 0.0374 | 0.1033 |
| wR2 (all data) | 0.0782 | 0.0867 | 0.0982 | 0.2109 |
| Residual Electron Density / e Å$^{-3}$ | 0.270 −0.413 | 0.297 −0.362 | 0.282 −0.280 | 1.263 −0.597 |
| Goodness of Fit | 1.067 | 1.061 | 1.047 | 1.009 |

Table 4. Crystallographic data for **2** and **3**.

| Compound | 2 | | 3 |
|---|---|---|---|
| Formula | $C_{33}H_{22}F_3FeN_4O_5S$ | | $C_{34}H_{29}FeN_4O_6S$ |
| Formula Weight | 699.45 | | 677.52 |
| Color | black | | black |
| Dimension/mm | $0.30 \times 0.125 \times 0.03$ | | $0.40 \times 0.15 \times 0.065$ |
| T/K | 293 | 400 | 293 |
| Crystal System | monoclinic | monoclinic | triclinic |
| Space Group | $P2_1/n$ | $P2_1/n$ | $P\bar{1}$ |
| a/Å | 9.8302(15) | 9.934(4) | 8.6110(8) |
| b/Å | 26.985(4) | 26.659(11) | 12.6981(12) |
| c/Å | 11.6733(17) | 12.366(5) | 15.1480(14) |
| $\alpha/°$ | 90 | 90 | 114.7000(10) |
| $\beta/°$ | 110.038(2) | 111.436(4) | 94.1660(10) |
| $\gamma/°$ | 90 | 90 | 92.3320(10) |
| V/Å$^3$ | 2909.1(8) | 3048(2) | 1496.2(2) |
| Z | 4 | 4 | 2 |
| $\rho_{calcd}$/g cm$^{-3}$ | 1.597 | 1.524 | 1.504 |
| $\mu$ (Mo-K$\alpha$) | 0.661 | 0.631 | 0.629 |
| $2\theta_{max}/°$ | 54.96 | 53.19 | 50.04 |
| No. Reflections | 16774 | 16916 | 7231 |
| ($R_{int}$) | (0.0188) | (0.0202) | (0.0152) |
| No. Observations | 6634 | 6710 | 5166 |
| ($I > 2.00\sigma(I)$) | (5595) | (4931) | (4942) |
| No. Variables | 424 | 424 | 417 |
| R1 ($I > 2.00\sigma(I)$) | 0.0358 | 0.0620 | 0.0347 |
| R (all data) | 0.0445 | 0.0839 | 0.0361 |
| wR2 (all data) | 0.0966 | 0.1976 | 0.0991 |
| Residual Electron Density/e Å$^{-3}$ | 0.396 / −0.251 | 0.762 / −0.554 | 0.962 / −0.388 |
| Goodness of Fit | 1.022 | 1.044 | 1.049 |

### 3.5.1. Molecular Structure Description of the [Fe(qsal)$_2$] Cation in **1–3**

The π-ligand qsal anion was coordinated to a central Fe atom as a tridentate chelate ligand and thus two coordinated ligand molecules were arranged in an almost perpendicular manner (Figure 6a–c). The coordination bond lengths and distortion parameters ($\Sigma$, $\Theta$, $\Phi$, see Figure 6d) for **1**, **2**, and **3** along with those of the HS and LS [Fe(qsal)$_2$] compounds confirmed by Mössbauer spectra are listed in Table 5. The distortion parameter $\Sigma$ is the sum of the absolute differences in 12 coordination bite angles from 90°. The distortion parameter $\Theta$ is the sum of the absolute differences in 24 angles from 60° on 8 surface triangles of a coordination octahedron. Both $\Sigma$ and $\Theta$ are zero if the coordination sphere is a regular octahedron. The distortion parameter $\Phi$ is the deviation of the angle between two coordination bonds from Fe to N atoms of two imine groups from 180°. All the distortion parameters increase their values on distorting the coordination octahedron.

As compared with the coordination bond lengths and distortion parameters, the Fe–O and Fe–N bond lengths and distortion parameters for **1** and **2** at 293 K were quite similar to those of the LS NCSe compound, indicating that the [Fe(qsal)$_2$] cations in **1** and **2** were in the LS state at 293 K. On the other hand, the coordination bond lengths and distortion parameters for **3** at 293 K were much larger than those in **1** and **2**, whereas they were similar to those in the HS [Ni(dmit)$_2$] compound (dmit = 4,5-dithiolato-1,3-dithiole-2-thione). Thus, the [Fe(qsal)$_2$] cation in **3** was in the HS state at 293 K. These observations are in good agreement of the spin-states of **1–3** from the magnetic susceptibilities and Mössbauer spectra at 293 K. On increasing temperatures, the coordination bond lengths of **1** were gradually lengthened, which was consistent with the gradual increase in the $\chi_M T$ in **1**. Note that the distortion parameters $\Sigma$ in **1** decreased on increasing temperature to 400 K. This suggests that the distortion parameter $\Sigma$ may not closely correlate with the spin-state of the [Fe(qsal)$_2$] cation.

At 425 K all the coordination bond lengths of three and a half independent [Fe(qsal)$_2$] cations except the Fe-O bond lengths in cation A were longer than those in **1** at 400 K. Although the distortion parameters Σ of the three and a half cations were varied, all the distortion parameters Θ and Φ were larger than those at 400 K. As compared with the LS NCSe and HS [Ni(dmit)$_2$] compounds, the cations A and D may contain certain LS fractions. Therefore, the SCO in **1** proved to be an incomplete cooperative transition accompanying a crystal structure phase transition. On the other hand, the coordination bond lengths and distortion parameters Θ and Φ in **2** at 400 K were a little smaller than those in the HS [Ni(dmit)$_2$] compound. This indicates that the SCO transition in **2** was also an incomplete one.

The Fe-O, Fe-N(imine), and Fe-N(quinoline) bond lengths in the [Fe(qsal)$_2$] compounds showing SCO conversions were varied from 1.869 to 1.914 Å, from 1.943 to 2.143 Å, and from 1.978 to 2.164 Å, respectively (Table 5). The Fe1-N2(quinoline) bond length of 2.172 Å in HS compound **3**, the Fe1-N1(imine) bond length of 1.936 Å in the LS NCSe compound, and the Fe1-O1 and Fe1-N2 bond lengths of 1.9278 and 2.195 Å in the HS [Ni(mnt)$_2$] compound were deviated from the above bond length range. Since the degrees of deviations were small, the occurrence of SCO cannot be judged only from the coordination bond lengths. As mentioned above, the parameters Σ may not be related to the spin-states of the [Fe(qsal)$_2$] cation. On the other hand, the parameters Θ and Φ are probably reflected in their spin-states. In particular, the parameters Φ may be useful to judge the occurrence of SCO because the largest Φ values are observed in the HS [Ni(mnt)$_2$] compound and HS compound **3**. Halcrow reported that the similar deviations of Φ are related to the occurrence of SCO in the Fe(II) complexes from bpp ligands (bpp = 2,6-di(pyrazol-1-yl)pyridine) [54]. These findings indicate that the central donor atoms of tridentate ligands can greatly impact the ligand field splitting energies in their homoleptic complexes, which is in good agreement with the role of the azo-functional group in new anionic SCO Fe(III) complexes from azobisphenolate ligands [55].

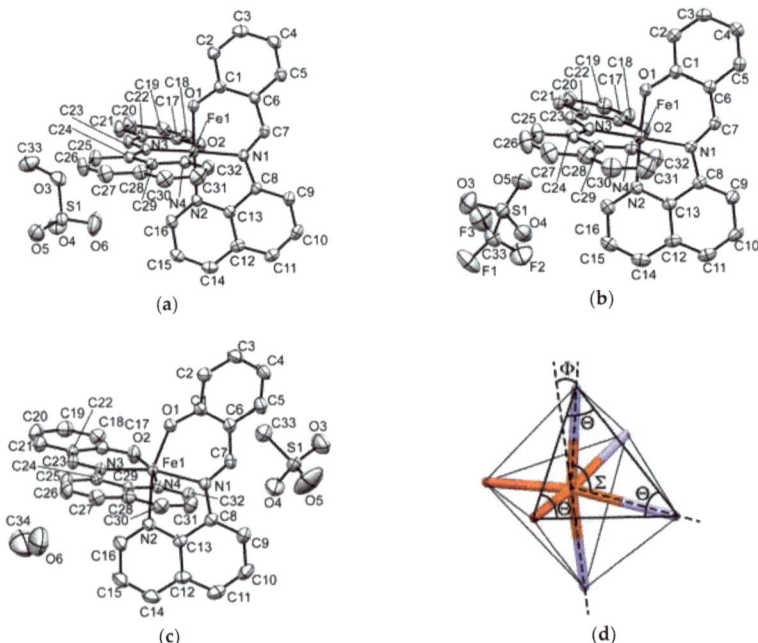

**Figure 6.** ORTEP drawings of 50% probability with atomic numberings for the [Fe(qsal)$_2$] cation, counter anion, and solvate molecule at 293 K for **1** (**a**); for **2** (**b**); for **3** (**c**). Hydrogen atoms are omitted for clarity; (**d**) The angles of a coordination octahedron concerning the distortion parameters.

**Table 5.** Coordination bond lengths and distortion parameters for **1**, **2**, **3**, and related compounds.

| Compound | 1 | 1 | 1 | 1A | 1B | 1C | 1D | 2 | 2 | 3 | [Ni(dmit)$_2$] [4] | NCSe [5] | [Ni(mnt)$_2$] [6] |
|---|---|---|---|---|---|---|---|---|---|---|---|---|---|
| T/K | 293 | 360 | 400 | 425 | 425 | 425 | 425 | 293 | 400 | 293 | 273 | 230 | 293 |
| Fe1-O1/Å | 1.8741(11) | 1.8692(13) | 1.8715(16) | 1.880(3) | 1.881(4) | 1.903(3) | 1.892(4) | 1.8707(12) | 1.900(3) | 1.9158(15) | 1.913(2) | 1.868(2) | 1.9278(19) |
| Fe1-O2/Å | 1.8779(12) | 1.8830(13) | 1.8880(15) | 1.877(4) | 1.900(3) | 1.911(3) | 1.892(4) | 1.8693(12) | 1.894(2) | 1.9085(15) | 1.914(3) | 1.874(2) | 1.908(2) |
| Fe1-N1/Å | 1.9434(13) | 1.9674(14) | 2.0091(16) | 2.019(3) | 2.096(4) | 2.141(4) | 2.062(4) | 1.9500(14) | 2.107(3) | 2.1250(17) | 2.125(3) | 1.936(3) | 2.097(2) |
| Fe1-N2/Å | 1.9804(13) | 2.0083(14) | 2.0497(16) | 2.066(4) | 2.156(4) | 2.151(4) | 2.091(4) | 1.9802(15) | 2.148(3) | 2.1722(18) | 2.150(2) | 1.969(3) | 2.195(2) |
| Fe1-N3/Å | 1.9438(13) | 1.9681(14) | 2.0105(17) | 2.032(3) | 2.101(3) | 2.143(4) | 2.062(4) | 1.9460(15) | 2.103(3) | 2.1291(17) | 2.138(3) | 1.945(3) | 2.131(2) |
| Fe1-N4/Å | 1.9782(14) | 2.0049(16) | 2.0483(18) | 2.068(4) | 2.133(4) | 2.164(4) | 2.091(4) | 1.9842(16) | 2.148(3) | 2.1595(17) | 2.151(3) | 1.986(3) | 2.136(2) |
| $\Sigma$ [1] /° | 51.13(17) | 49.6(2) | 50.6(2) | 52.1(5) | 65.9(5) | 74.5(5) | 52.7(6) | 52.4(2) | 73.6(4) | 77.7(2) | 72.5(3) | 52.1(4) | 83.1(3) |
| $\Theta$ [2] /° | 61.8(2) | 61.9(2) | 66.2(3) | 71.3(6) | 104.1(6) | 143.4(6) | 84.6(7) | 71.0(3) | 114.9(5) | 126.4(3) | 124.2(4) | 69.4(5) | 149.1(4) |
| $\Phi$ [3] /° | 2.50(5) | 3.74(6) | 5.77(6) | 7.54(15) | 12.20(15) | 15.33(14) | 10.8(2) | 1.86(6) | 9.42(11) | 15.80(7) | 13.79(9) | 1.50(14) | 17.83(8) |
| CCDC No. | 1891471 | 1891472 | 1891473 | 1891474 | 1891474 | 1891474 | 1891474 | 1891475 | 1891476 | 1891477 | 289952 | 194657 | 1559560 |

[1] The sum of the absolute differences of bite angles from 90°. [2] The sum of the absolute differences of all the angles of triangle surfaces of a coordination octahedron from 60°. [3] The deviation angle of N1-Fe1-N3 from 180°. [4] [Fe(qsal)$_2$][Ni(dmit)$_2$]·2CH$_3$CN from reference [22]. [5] [Fe(qsal)$_2$](NCSe)·CH$_2$Cl$_2$ from reference [9]. [6] [Fe(qsal)$_2$][Ni(mnt)$_2$] from reference [33].

### 3.5.2. Molecular Arrangement of 1

The molecular arrangement of the [Fe(qsal)$_2$] cations and CH$_3$OSO$_3$ anions in **1** along the $b-c$ direction is shown in Figure 7a. Selected intermolecular distances in **1** are listed in Table 6. The phenyl ring (C1–C6) in the [Fe(qsal)$_2$] cation was stacked with the quinolyl ring (C8–C16,N2) in the nearest neighboring [Fe(qsal)$_2$] cation in a parallel-displaced manner, to form a π-stacking [Fe(qsal)$_2$] dimer (p in Figure 7b). The π-stacking dimers were arranged along the $b-c$ direction through short C···C contacts between the imine and quinolyl moieties (q in Figure 7b) and thus afforded an alternate one-dimensional (1D) [Fe(qsal)$_2$] array. This type of 1D [Fe(qsal)$_2$] arrangement was found in [Fe(qsal)$_2$](NCX)·solv (X = S, Se) [9–11] and [Fe(qsal)$_2$](I$_3$) [13]. Moreover, a parallel-displaced π-stacking between the phenyl rings (C17–C22) (r in Figure 7c) and edge-shared π-stacking between the quinolyl rings (C8–C16,N2) (s in Figure 7d) were observed along the $b+c$ and $a$ directions, respectively. Therefore, the formation of a three-dimensional (3D) π-stacking interaction network is elucidated in **1**. The CH$_3$OSO$_3$ anions were located in the cavity of the 3D π-stacking network of the [Fe(qsal)$_2$] cation. All the edge oxygen atoms in the CH$_3$OSO$_3$ anion were contacted with the hydrogen atoms of four neighboring [Fe(qsal)$_2$] cations.

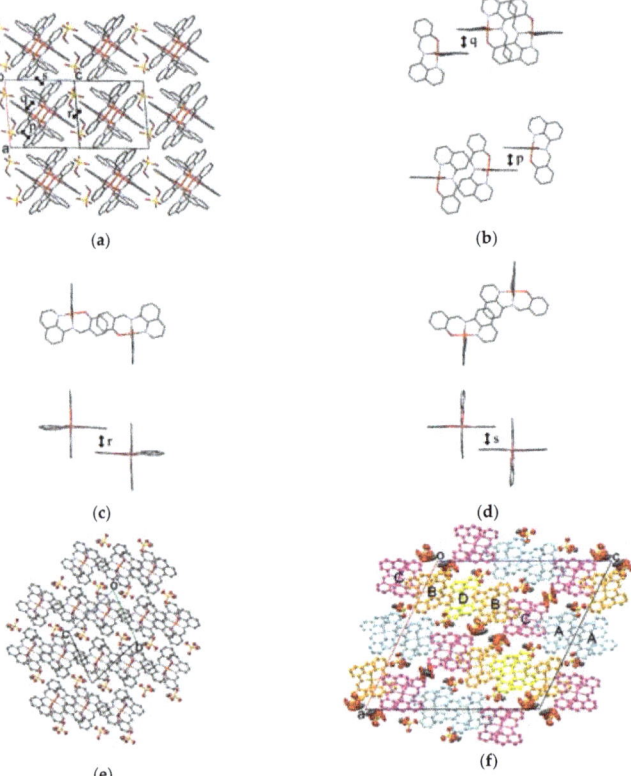

**Figure 7.** (a) Crystal structure viewed along the $b-c$ direction of **1** at 293 K; Top (upper) and side (bottom) views of π-stacking structures of the [Fe(qsal)$_2$] cations for p (**b**), r (**c**), and s (**d**) at 293 K; ORTEP drawings of **1** with 50% probability at 293 K (**e**) and 425 K (**f**). The light blue, orange, magenta, and yellow [Fe(qsal)$_2$] cations correspond to cations A, B, C, and D, respectively. Hydrogen atoms are omitted for clarity.

**Table 6.** Selected intermolecular short distances (Å) at 293, 360, and 400 K in **1**.

| Position [a] | | 293 K | 360 K | 400 K |
|---|---|---|---|---|
| p | Plane(C1−C6)···C10 [b] | 3.243 | 3.252 | 3.261 |
|   | Plane(C8−C16,N2)···C5 [b] | 3.264 | 3.288 | 3.320 |
|   | C4···C11 [b] | 3.267(3) | 3.279(3) | 3.293(4) |
|   | C6···C9 [b] | 3.245(3) | 3.270(3) | 3.297(3) |
| q | C23···C26 [c] | 3.375(3) | 3.403(3) | 3.436(4) |
| r | Plane(C17−C22)···Plane(C17−C22) [d] | 3.535 | 3.561 | 3.581 |
|   | C18···C20 [d] | 3.553(3) | 3.576(3) | 3.594(4) |
| s | Plane(C8−C16,N2)···Plane(C8−C16,N2) [e] | 3.266 | 3.276 | 3.281 |
|   | C10···C11 [e] | 3.377(3) | 3.381(3) | 3.388(3) |
|   | C10···C12 [e] | 3.370(3) | 3.383(3) | 3.384(4) |

[a] The positions corresponding to letters are shown in Figure 7. [b] (1−x, 2−y, 1−z). [c] (1−x, 1−y, 2−z). [d] (1−x, 1−y, 1−z). [e] (2−x, 2−y, 1−z).

On increasing temperatures from 293 to 400 K, the intermolecular distances for **1** were gradually increased (Table 6), but the overlapping modes were not changed. On the other hand, the molecular arrangement of the [Fe(qsal)$_2$] cations and CH$_3$OSO$_3$ anions at 425 K were dramatically changed, which can be easily recognized by comparison between Figure 7e,f. The periodic units of the 1D arrays of the [Fe(qsal)$_2$] cation and CH$_3$OSO$_3$ anion were two and seven at 293 and 425 K, respectively. The incommensurate periodic units suggest that a large structural rearrangement would take place through the structural phase transition accompanying SCO. Within the 1D [Fe(qsal)$_2$] array at 425 K, the π-overlaps similar to p shown in Figure 7b were found both between cations A and between cations B and D. Although the π-overlaps were deformed and twisted, several C···C contacts shorter than the sum of van der Waals (vdW) radii (C: 1.70 Å) [56] were observed between cations A and C, whereas there is only one short C···C contact between cations B and C. Thus, the 1D [Fe(qsal)$_2$] array seemed to consist of C···A···A···C tetramer and D··B···D trimer at 425 K. Between the 1D [Fe(qsal)$_2$] arrays, most π-overlaps similar to s shown in Figure 7d were lengthened in the range of 3.407–3.610 Å along the *b* axis, whereas one half of π-overlaps similar to r shown in Figure 7c disappeared and the remained π-overlaps were observed only between cations A and B along the *a* axis. Thus, the 3D π-stacking interaction network was rearranged and weakened by the deformation of the [Fe(qsal)$_2$] array. Since the orientations of the CH$_3$OSO$_3$ anions were partly different from those below 400 K and their thermal ellipsoids were also much larger than those of the [Fe(qsal)$_2$] cation at 425 K, the crystal structure phase transition in **1** seems to arise from the thermal motion of the CH$_3$OSO$_3$ anions.

3.5.3. Molecular Arrangement of **2**

The molecular arrangement of the [Fe(qsal)$_2$] cations and CF$_3$SO$_3$ anions in **2** along the *a+c* direction at 293 and 400 K are shown in Figure 8a,b. Selected intermolecular distances are listed in Table 7. Similar to the π-stacking [Fe(qsal)$_2$] dimer in **1**, the qsal ligand of the [Fe(qsal)$_2$] cation was stacked with that of the neighboring [Fe(qsal)$_2$] cation in a head-to-tail manner (t in Figure 8c). On the other hand, the π-stacked [Fe(qsal)$_2$] cations were related each other by a symmetry operation of *n* glide plane and thus gave a uniform 1D molecular array along the *a+c* direction. This type of the uniform 1D [Fe(qsal)$_2$] arrangement was found in [Fe(qsal)$_2$](NCS) [11] and [Fe(qsal)$_2$](I) [15]. The π-overlaps between the 1D [Fe(qsal)$_2$] arrays were found between the phenyl moieties along the *a−c* direction (u in Figure 8a,c). Thus, the [Fe(qsal)$_2$] cations formed a two-dimensional (2D) regular network along the *ac* plane. Meanwhile, there was no remarkable short contact between the 2D [Fe(qsal)$_2$] networks.

The CF$_3$SO$_3$ anions were located between the 2D [Fe(qsal)$_2$] networks. The S-O bond lengths imply that the negative charge in the CF$_3$SO$_3$ anion may be delocalized at all oxygen atoms. The O5···C23(imine) and O4···C14(quinolyl) distances were found to be 3.078(3) and 3.179(3) Å, respectively. These suggest the existence of effective Coulomb interactions between the [Fe(qsal)$_2$] cation and CF$_3$SO$_3$ anion.

The molecular arrangement of **2** at 400 K was very similar to that at 293 K. The difference in the 1D π-stacking arrays (Figure 8a,b) and 2D π-stacking network structure (Figure 8c,d) were hardly observed, but the 2D π-stacking layers glided alternately along the *c* axis (Figure 8e,f). Since the short O5···C23(imine) distance was shortened to be 3.039(6) Å at 400 K, this structure transition would be involved in the Coulomb interaction. The CF$_3$SO$_3$ anions in a 2D π-stacking layer were located near the quinolyl rings in the neighboring 2D layer at 293 K, whereas the CF$_3$SO$_3$ anions were shifted from the quinolyl rings in the neighboring 2D layer at 400 K. Moreover, the thermal ellipsoids of the CF$_3$SO$_3$ anion and quinolyl moiety were larger than those of the other molecular components at 400 K. These observations indicate that this structure transition in **2** may result from the thermal motion of the CF$_3$SO$_3$ anion, which is reminiscent of the structural transition in **1**.

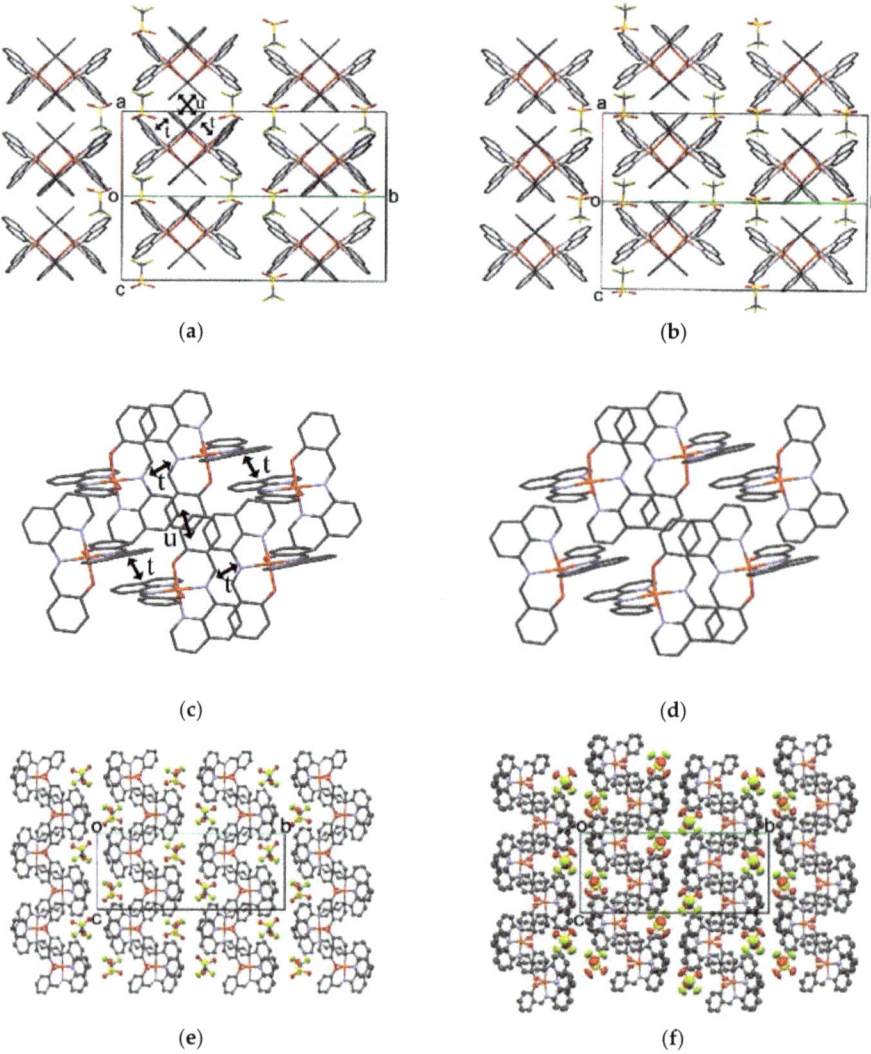

**Figure 8.** Crystal structure viewed along the *a+c* direction of **2** at 293 K (**a**) and 400 K (**b**). Hydrogen atoms are omitted for clarity; π-stacking structures of the [Fe(qsal)$_2$] cations along the *ac* plane at 293 K (**c**) and 400 K (**d**); ORTEP drawings of **2** with 50% probability at 293 K (**e**) and 400 K (**f**).

**Table 7.** Selected intermolecular short distances (Å) in **2**.

| Position [a] | | 293 K | 400 K |
|---|---|---|---|
| t | Plane(C1−C6)⋯C25 [b] | 3.503 | 3.617 |
| | Plane(C8−C16,N2)⋯C20 [b] | 3.282 | 3.428 |
| | Plane(C8−C16,N2)⋯C21 [b] | 3.263 | 3.415 |
| | Plane(C17−C22)⋯C9 [c] | 3.327 | 3.465 |
| | Plane(C17−C22)⋯C10 [c] | 3.328 | 3.457 |
| | Plane(C24−C32,N4)⋯C4 [c] | 3.580 | 3.578 |
| | Plane(C24−C32,N4)⋯C5 [c] | 3.302 | 3.359 |
| | C5⋯C24 [b] | 3.331(3) | 3.394(6) |
| | C5⋯C25 [b] | 3.515(3) | 3.687(7) |
| | C7⋯C23 [b] | 3.255(3) | 3.375(5) |
| | C8⋯C21 [b] | 3.289(3) | 3.440(6) |
| | C9⋯C21 [b] | 3.442(3) | 3.709(7) |
| | C9⋯C22 [b] | 3.476(3) | 3.630(6) |
| | C10⋯C19 [b] | 3.368(3) | 3.485(8) |
| | C10⋯C20 [b] | 3.470(3) | 3.747(8) |
| | C11⋯C20 [b] | 3.430(3) | 3.650(7) |
| u | Plane(C1−C6)⋯C19 [d] | 3.529 | 3.638 |
| | Plane(C17−C22)⋯C2 [d] | 3.421 | 3.546 |
| | Plane(C17−C22)⋯C3 [d] | 3.511 | 3.690 |
| | C1⋯C19 [d] | 3.592(3) | 3.686(7) |
| | C2⋯C20 [d] | 3.463(3) | 3.585(7) |

[a] The positions corresponding to letters are shown in Figure 8. [b] (−0.5+x, 1.5−y, −0.5+z). [c] (0.5+x, 1.5−y, 0.5+z). [d] (0.5+x, 1.5−y, −0.5+z).

### 3.5.4. Molecular Arrangement of 3 at 293 K

The molecular arrangement of the [Fe(qsal)$_2$] cations in **3** was quite different from those found in common [Fe(qsal)$_2$] compounds (Figure 9a). Selected intermolecular distances are listed in Table 8. Relatively strong π-stacking interactions between the [Fe(qsal)$_2$] cations afforded a 1D array along the *a* axis (v in Figure 9b,c). There were two kinds of π-overlaps between the 1D [Fe(qsal)$_2$] arrays. One overlaps were between the quinolyl planes (w in Figure 9b,d), the other ones were between the quinolyl and imine moieties (x in Figure 9b,e). These overlaps were alternately arranged along the *b+c* direction. Meanwhile, π-overlaps y between the quinolyl planes shown in Figure 9b,f seem to be weak, because the shortest distance is 3.679(4) Å. The above-mentioned three π-stacking interactions gave a 2D network parallel to the *b+c* direction (Figure 9b). Since the intermolecular C3⋯C4 and C15⋯C20 distances between the 2D [Fe(qsal)$_2$] networks were 3.620(4) and 3.627(5) Å, respectively, the [Fe(qsal)$_2$] cations in **3** formed the 2D interaction network.

**Table 8.** Selected intermolecular distances (Å) in **3**.

| Position [a] | | 293 K |
|---|---|---|
| v | Plane(C17−C22)⋯C27 [b] | 3.427 |
| | Plane(C24−C32,N4)⋯C19 [c] | 3.353 |
| | C18⋯C28 [b] | 3.379(4) |
| | C20⋯C26 [b] | 3.427(4) |
| | C2⋯C30 [b] (T-shape) | 3.440(4) |
| | C1⋯C30 [b] (T-shape) | 3.451(3) |
| w | Plane(C24−C32,N4)⋯Plane(C24−C32,N4) [d] | 3.493 |
| | C25⋯C30 [d] | 3.525(4) |
| | C27⋯C29 [d] | 3.464(4) |
| x | Plane(C8−C16,N2)⋯C7 [e] | 3.439 |
| | C7⋯C10 [e] | 3.475(3) |
| y | Plane(C8−C16,N2)⋯Plane(C8−C16,N2) [f] | 3.652 |
| | C10⋯C11 [f] | 3.679(4) |

[a] The positions corresponding to letters are shown in Figure 9. [b] (1+x, y, z). [c] (−1+x, y, z). [d] (−x, −y, −z). [e] (1−x, 1−y, 1−z). [f] (−x, 1−y, 1−z).

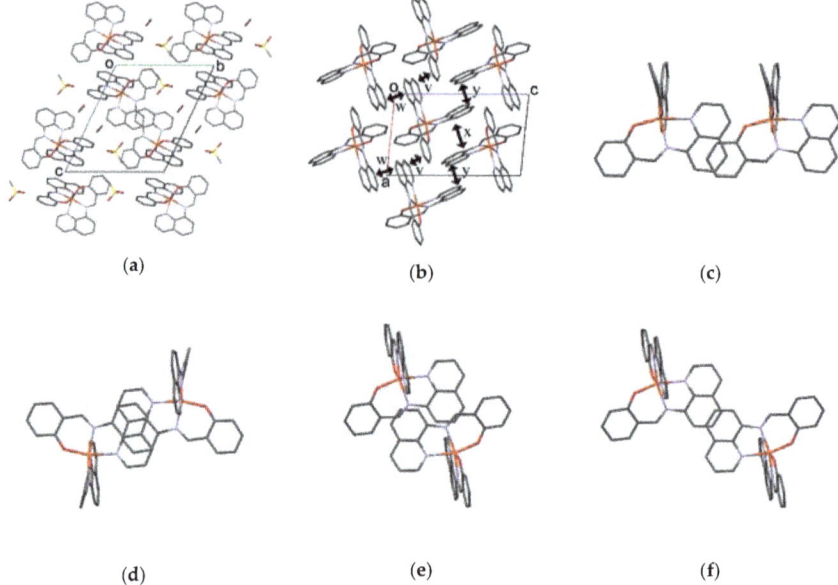

**Figure 9.** (**a**) Crystal structure viewed along the *a* axis of **3** at 293 K; (**b**) Side view of the π-stacking structure of **3**; Top views of the π-stacking structure for v (**c**), w (**d**), x (**e**), and y (**f**). Hydrogen atoms are omitted for clarity.

The $CH_3SO_3$ anions and methanol molecules were located between the 2D $[Fe(qsal)_2]$ networks. The S-O bond lengths imply that the negative charge in the $CH_3SO_3$ anion may be localized mainly at O3 and O4 atoms. The short O4···C7 and O3···O6 distances were found to be 3.070(3) and 2.814(4) Å, respectively. The former suggests the existence of effective Coulomb interactions between the $[Fe(qsal)_2]$ cation and $CH_3SO_3$ anion, the latter indicates that of hydrogen bonding interactions between the $CH_3SO_3$ anion and methanol molecule. We can assume that both strong Coulomb interaction and hydrogen bonding interaction may induce the distortion of a coordination octahedron, leading to the HS state in the whole temperature range.

3.5.5. Correlation between the Crystal Structures and Magnetic Behaviors for the $[Fe(qsal)_2]$ Compounds

The magnetic behaviors and short cation···cation distances for the $[Fe(qsal)_2]$ compounds with nonplanar counter-anions whose crystal structures were deposited to date are summarized in Table 9. Interestingly, the $[Fe(qsal)_2]$ cation arrangements except compound **3** are quite similar to each other. All the $[Fe(qsal)_2]$ compounds except **3** consist of a 1D π-stacking $[Fe(qsal)_2]$ cation array shown in Figures 7b and 8c. On the other hand, the arrangements between the 1D $[Fe(qsal)_2]$ cation arrays are slightly different, for example, between Figures 7a and 8a. Basically, the counter-anions or solvate molecules may determine the molecular arrangement of the $[Fe(qsal)_2]$ cations.

**Table 9.** Magnetic behaviors and selected intermolecular short C···C distances between the [Fe(qsal)$_2$] cations for [Fe(qsal)$_2$](X)·solv.

| Compound | X Solv | CH$_3$OSO$_3$ (1) — | | CF$_3$SO$_3$ (2) — | | NCS — | NCSe — |
|---|---|---|---|---|---|---|---|
| Magnetic behavior [1] | Heating Process | gSCO 290 → 410 K aSCO $T_{1/2}\uparrow$ = 414 K gSCO 420 → 450 K | | gSCO 290 → 360 K aSCO $T_{1/2}\uparrow$ = 365 K gSCO 370 → 450 K | | gSCO 200 → 270 K aSCO $T_{1/2}\uparrow$ = 289 K | aSCO $T_{1/2}\uparrow$ = 215 K gSCO 230 → 270 K aSCO $T_{1/2}\uparrow$ = 282 K |
| | Cooling Process | gSCO 450 → 405 K aSCO $T_{1/2}\downarrow$ = 402 K gSCO 400 → 290 K | | gSCO 450 → 370 K aSCO $T_{1/2}\downarrow$ = 361 K gSCO 355 → 290 K | | gSCO 350 → 220 K aSCO $T_{1/2}\downarrow$ = 205 K | aSCO $T_{1/2}\downarrow$ = 212 K |
| Short distances/Å [2] | Temp. | 293 K | 425 K | 293 K | 400 K | 293 K | 296 K |
| | Spin-State | LS | mHS | LS | mHS | mHS | HS |
| | 1D | 3.245 3.267 3.375 | 3.276 3.309 3.352 | 3.255 3.289 3.331 | 3.375 3.440 3.394 | 3.466 3.527 3.593 | 3.491 3.533 3.545 |
| | B1D(1) | 3.370 | 3.407 | 3.463 | 3.585 | 3.710 | 3.538 |
| | B1D(2) | 3.553 | 3.638 | 3.886 | 3.943 | 4.218 | 3.680 |
| CCDC No. | | 1891471 | 1891474 | 1891475 | 1891476 | 717842 | 717844 |
| Ref. | | this work | | this work | | [11] | [11] |
| Compound | X Solv | I — | NCS CH$_2$Cl$_2$ | NCSe CH$_2$Cl$_2$ | I$_3$ | | CH$_3$SO$_3$ (3) CH$_3$OH |
| Magnetic behavior [1] | | LS 2 – 300 K | LS 113 K | LS 150 – 360 K | gSCO 200 – 260 K | | HS 2 – 300 K |
| Short Distances/Å [2] | Temp. | 296 K | 113 K | 230 K | 50 K | 293 K | 293 K |
| | Spin-State | LS | LS | LS | mLS | mHS | HS |
| | 1D | 3.375 3.409 3.424 | 3.229 3.384 3.388 | 3.317 3.360 3.383 | 3.226 3.266 3.269 | 3.371 3.376 3.395 | 3.379 |
| | B1D(1) | 3.487 | 3.494 | 3.455 | 3.456 | 3.579 | 3.464 3.475 |
| | B1D(2) | 3.910 | 3.865 | 3.956 | 3.527 (CH···π) | 3.688 (CH···π) | 3.620 |
| CCDC No. | | 902864 | 717843 | 194657 | 749309 | 749310 | 1891477 |
| Ref. | | [15] | [11] | [9] | [13] | | this work |

[1] gSCO: Gradual SCO. aSCO: Abrupt SCO. LS: low-spin. HS: high-spin. [2] The temperature of the crystal structure analysis and the spin-state of the [Fe(qsal)$_2$] cation are shown. mLS: mainly low-spin. mHS: mainly high-spin. 1D indicates short distances within the 1D [Fe(qsal)$_2$] cation array. B1D(1) indicates shortest distance between the 1D [Fe(qsal)$_2$] cation arrays in one direction. B1D(2) indicates shortest distance between the 1D [Fe(qsal)$_2$] cation arrays in the other direction.

Let us consider the comparison with the most intriguing compounds [Fe(qsal)$_2$](NCS) and [Fe(qsal)$_2$](NCSe), both of which exhibited a cooperative SCO transition with a large thermal hysteresis loop. It is very difficult to quantitatively evaluate the strength of π-stacking interactions in various stacking manners. However, the number and shrinkage of intermolecular short distances between the atoms can estimate the strength of π-stacking interactions qualitatively. It should be noted that neither C···C nor C···N distance between the [Fe(qsal)$_2$] cations is shorter than the sum of vdW radii (C: 1.70, N: 1.55 Å) [56] for [Fe(qsal)$_2$](NCS) and [Fe(qsal)$_2$](NCSe). The short C···C distances in [Fe(qsal)$_2$](NCS) are 3.466, 3.527, and 3.593 Å along the 1D [Fe(qsal)$_2$] array, whereas those in [Fe(qsal)$_2$](NCSe) are 3.491, 3.533, and 3.545 Å along the 1D [Fe(qsal)$_2$] array, and 3.538 Å between the 1D [Fe(qsal)$_2$] arrays. On the contrary, as shown in Table 9, several C···C distances shorter than the sum of vdW radii (< 3.40 Å) along the 1D [Fe(qsal)$_2$] array are found in the LS [Fe(qsal)$_2$] compounds, and moreover, between the 1D [Fe(qsal)$_2$] arrays for **1**. As compared between [Fe(qsal)$_2$](NCX) and [Fe(qsal)$_2$](NCX)·solv (X = S, Se), the existence of solvate molecules should result in the expansion of intermolecular distances. However, the intermolecular [Fe(qsal)$_2$] cation distances involved in π-stacking interactions in the solvate compounds are much shorter than those in the non-solvate compounds.

The literature [9] disclosed that [Fe(qsal)$_2$](NCSe)·solv were transformed into [Fe(qsal)$_2$](NCSe) by desolvation. Moreover, the time-dependence of the spin-states in [Fe(qsal)$_2$](NCS) was reported in the literature [8], suggesting that the desolvation from [Fe(qsal)$_2$](NCS)·solv to [Fe(qsal)$_2$](NCS) took place at 286 K. Very recently, similar desolvation-induced crystal structure transformations in the charged Fe(II) and Fe(III) compounds were reported [57,58], which seemed to be driven by Coulomb interactions. In general, the Coulomb interaction is much stronger than other intermolecular interactions. Thus, the structure transformation through desolvation in charged complexes may be driven by the Coulomb energy gain, namely shrinking the distances between the cations and anions. Several C···N and C···S distances shorter than the sum of vdW radii are found between the [Fe(qsal)$_2$] cation and NCS anion for [Fe(qsal)$_2$](NCS), indicating the Coulomb interactions may operate the crystal structure transformation for [Fe(qsal)$_2$](NCX)·solv.

One may ask why solvate molecules are included in a crystal lattice although they will reduce Coulomb energy gain. If the energy gain from other intermolecular interactions in a solvate compound is larger than the loss of Coulomb energy gain, the total energy gain can be larger than that in a non-solvate compound. The shorter intermolecular distances involved in π-stacking interactions in [Fe(qsal)$_2$](NCX)·solv suggest that the strong π-stacking interactions may result from the [Fe(qsal)$_2$] cation arrangement given by the inclusion of the solvate molecules. Therefore, we can assume that the solvate molecule in a solvate complex seems to play a significant role in the enhancement of the total lattice energy gain by various intermolecular interactions between its molecular components.

This idea is also applicable to **1**, **2**, and the I$_3$ compound. Although the crystal structure from a charged non-solvate complex should be a typical ionic crystal which has a larger coordination number and shorter cation···anion distances, the crystal structures for **1**, **2**, and the I$_3$ compound were not those of a typical ionic crystal but those similar to [Fe(qsal)$_2$](NCS)·CH$_2$Cl$_2$ and [Fe(qsal)$_2$](NCSe)·CH$_2$Cl$_2$. This suggests that the total lattice energy gain from the strong π-stacking interactions found in **1**, **2**, and the I$_3$ compound may exceed the loss of Coulomb energy gain in the crystal structure of a typical ionic crystal. Consequently, we can discuss the role of large non-spherical counter-anions or solvate molecules for a crystal packing in a charged complex by means of the competition between Coulomb and intermolecular interactions. Further investigations on other charged complexes are needed for verification of the present finding.

Next, we will discuss the large thermal hysteresis loops of the magnetic susceptibility found in [Fe(qsal)$_2$](NCS) and [Fe(qsal)$_2$](NCSe). As shown in Table 9, more than two-step variations in the cooperative SCO transition are one of the characteristic points for **1**, **2**, [Fe(qsal)$_2$](NCS) and [Fe(qsal)$_2$](NCSe). Moreover, large thermal ellipsoids or disorder of counter-anions were observed in the crystal structures of their HS phases. This implies that the thermal motion of the counter-anion may play an important role in their HS crystal structures. The thermal variations in the crystal structures of **1** from 293 to 400 K (Table 6) revealed that strong intermolecular π-stacking interactions were retained despite the gradual SCO conversion. Recently, we found similar gradual SCO conversion in the charged Fe(II) compound having strong intermolecular interactions, in which the large difference in lattice enthalpy between the LS and HS states leads to a gradual SCO conversion despite the existence of strong intermolecular interactions [58]. Therefore, to undergo an abrupt SCO transition with a large hysteresis loop, it may be important to realize a small difference in lattice enthalpy between the LS and HS states by the choice of a counter-anion or solvate molecule.

## 4. Conclusions

New [Fe(qsal)$_2$] compounds, [Fe(qsal)$_2$](CH$_3$OSO$_3$) **1** and [Fe(qsal)$_2$](CH$_3$SO$_3$)·CH$_3$OH **3**, along with the reported compound, [Fe(qsal)$_2$](CF$_3$SO$_3$) **2**, were prepared and characterized. The compounds **1** and **2** exhibited a cooperative SCO transition at higher temperatures. The optical conductivity spectrum from the single-crystal reflection spectrum of **1** revealed that the photo-excitation band for the LIESST effect on the LS [Fe(qsal)$_2$] compounds is attributed to d-d transition. The successful crystal structure determinations of **1** and **2** in the high-temperature phase reveal that large structural changes were triggered by the motion of counter anions. The structural comparison among the [Fe(qsal)$_2$] compounds determined to date suggests that the counter-anions and solvate molecules play a significant role in the total lattice energy gain in a charged complex. The present finding may lead to elucidation of the role of counter-anions and solvate molecules for controlling SCO transition behaviors. To do this end, the quantitative evaluation of lattice enthalpy for each spin-state in SCO compounds may be required. Since the [Fe(qsal)$_2$] derivatives are suitable candidates for this purpose, we are now investigating a family of the [Fe(qsal)$_2$] derivatives.

**Supplementary Materials:** The following are available online at http://www.mdpi.com/2073-4352/9/2/81/s1, Table S1: Cartesian coordinates of the LS [Fe(qsal)$_2$] cation, Table S2: Cartesian coordinates of the HS [Fe(qsal)$_2$] cation.

**Author Contributions:** Conceptualization, K.T.; Methodology, K.T.; Validation, K.T.; Formal analysis, K.T., K.Y., T.Y., and Y.S.; Investigation, K.T., K.Y., T.Y, Y.E., Y.S., K.Y., and H.M.; Resources, K.T. and H.M.; Data curation, K.T., K.Y., T.Y., and Y.S.; Writing—original draft preparation, K.T.; Writing—review and editing, K.Y., T.Y., Y.E., Y.S., K.Y., and H.M.; Visualization, K.T.; Supervision, K.T.; Project administration, K.T.; Funding acquisition, K.T. and H.M.

**Funding:** This research was funded partly by a Grant-in-Aid for Scientific Research on the Priority Area of Molecular Conductors (no. 18028024) and on Innovative Areas of Molecular Degrees of Freedom (no. 20110007), and a Grant-in-Aid for Young Scientists (B) (no. 19750107) and for Scientific Research (C) (no. 25410068) from the Ministry of Education, Culture, Sports, Science, and Technology of Japan.

**Acknowledgments:** K.T. is grateful to T. Ishikawa at Tokyo Institute of Technology for discussing the optical properties of the Fe complexes. K.T. also thanks to A. Sakamoto at the University of Tokyo, and Y. Furuie at Kobe University for performing the elemental analysis.

**Conflicts of Interest:** The authors declare no conflict of interest.

## References

1. Gütlich, P.; Goodwin, H.A. (Eds.) *Spin Crossover in Transition Metal Compounds I−III*; Springer: Berlin/Heidelberg, Germany, 2004; ISBN 3-540-40394-9, 3-540-40396-5 and 3-540-40395-7.
2. Halcrow, M.A. (Ed.) *Spin-Crossover Materials*; John Wiley & Sons, Ltd.: Oxford, UK, 2013; ISBN 978-1-119-99867-9.
3. Takahashi, K. (Ed.) *Spin-Crossover Complexes*; MDPI: Basel, Switzerland, 2018; ISBN 978-3-03842-825-1.
4. Bousseksou, A.; Molnár, G.; Salmon, L.; Nicolazzi, W. Molecular spin crossover phenomenon: Recent achievements and prospects. *Chem. Soc. Rev.* **2011**, *40*, 3313–3335. [CrossRef] [PubMed]
5. Gütlich, P.; Gaspar, A.B.; Garcia, Y. Spin state switching in iron coordination compounds. *Beilstein J. Org. Chem.* **2013**, *9*, 342–391. [CrossRef] [PubMed]
6. Dahl, B.M.; Dahl, O. Studies of Chelates with Heterocyclic Ligands IV. Transition Metal Complexes with N-(8-quinolyl)-salicylaldimine. *Acta Chem. Scand.* **1969**, *23*, 1503–1513. [CrossRef]
7. Dickinson, R.C.; Baker, W.A.; Collins, R.L. The Magnetic Properties of bis[N-(8-quinolyl)-salicylaldimine] halogenoiron(III)·X hydrate, Fe(8-QS)$_2$X·$x$H$_2$O: A Reexamination. *J. Inorg. Nucl. Chem.* **1977**, *39*, 1531–1533. [CrossRef]
8. Oshio, H.; Kitazaki, K.; Mishiro, J.; Kato, N.; Maeda, Y.; Takashima, Y. New Spin-crossover Iron(III) Complexes with Large Hysteresis Effects and Time Dependence of their Magnetism. *J. Chem. Soc. Dalton Trans.* **1987**, 1341–1347. [CrossRef]
9. Hayami, S.; Gu, Z.; Yoshiki, H.; Fujishima, A.; Sato, O. Iron(III) Spin-Crossover Compounds with a Wide Apparent Thermal Hysteresis around Room Temperature. *J. Am. Chem. Soc.* **2001**, *123*, 11644–11650. [CrossRef] [PubMed]
10. Hayami, S.; Kawahara, T.; Juhasz, G.; Kawamura, K.; Uehashi, K.; Sato, O.; Maeda, Y. Iron(III) Spin Transition Compound with a Large Thermal Hysteresis. *J. Radioanal. Nucl. Chem.* **2003**, *255*, 443–447. [CrossRef]
11. Hayami, S.; Hiki, K.; Kawahara, T.; Maeda, Y.; Urakami, D.; Inoue, K.; Ohama, M.; Kawata, S.; Sato, O. Photo-Induced Spin Transition of Iron(III) Compounds with π-π Intermolecular Interactions. *Chem. A Eur. J.* **2009**, *15*, 3497–3508. [CrossRef] [PubMed]
12. Ivanova, T.A.; Ovchinnikov, I.V.; Garipov, R.R.; Ivanova, G.I. Spin Crossover [Fe(qsal)$_2$]X (X = Cl, SCN, CF$_3$SO$_3$) Complexes: EPR and DFT Study. *Appl. Magn. Reson.* **2011**, *40*, 1–10. [CrossRef]
13. Takahashi, K.; Sato, T.; Mori, H.; Tajima, H.; Sato, O. Correlation between the Magnetic Behaviors and Dimensionality of Intermolecular Interactions in Fe(III) Spin Crossover Compounds. *Phys. B Condens. Matter* **2010**, *405*, S65–S68. [CrossRef]
14. Takahashi, K.; Sato, T.; Mori, H.; Tajima, H.; Einaga, Y.; Sato, O. Cooperative Spin Transition and Thermally Quenched High-Spin State in New Polymorph of [Fe(qsal)$_2$]I$_3$. *Hyperfine Interact.* **2012**, *206*, 1–5. [CrossRef]

15. Djukic, B.; Jenkins, H.A.; Seda, T.; Lemaire, M.T. Structural and Magnetic Properties of Homoleptic Iron(III) Complexes Containing N-(8-Quinolyl)-Salicylaldimine [Fe(Qsal)$_2$]$^+$X$^-$ {X = I or (Qsal)FeCl$_3$}. *Transit. Metal Chem.* **2013**, *38*, 207–212. [CrossRef]
16. Tsukiashi, A.; Nakaya, M.; Kobayashi, F.; Ohtani, R.; Nakamura, M.; Harrowfield, J.M.; Kim, Y.; Hayami, S. Intermolecular Interaction Tuning of Spin-Crossover Iron(III) Complexes with Aromatic Counteranions. *Inorg. Chem.* **2018**, *57*, 2834–2842. [CrossRef] [PubMed]
17. Harding, D.J.; Phonsri, W.; Harding, P.; Gass, I.A.; Murray, K.S.; Moubaraki, B.; Cashion, J.D.; Liu, L.; Telfer, S.G. Abrupt Spin Crossover in an Iron(III) Quinolylsalicylaldimine Complex: Structural Insights and Solvent Effects. *Chem. Commun.* **2013**, *49*, 6340–6342. [CrossRef] [PubMed]
18. Harding, D.J.; Sertphon, D.; Harding, P.; Murray, K.S.; Moubaraki, B.; Cashion, J.D.; Adams, H. Fe$^{III}$ Quinolylsalicylaldimine Complexes: A Rare Mixed-Spin-State Complex and Abrupt Spin Crossover. *Chem. A Eur. J.* **2013**, *19*, 1082–1090. [CrossRef]
19. Harding, D.J.; Phonsri, W.; Harding, P.; Murray, K.S.; Moubaraki, B.; Jameson, G.N.L. Abrupt Two-Step and Symmetry Breaking Spin Crossover in an Iron(III) Complex: An Exceptionally Wide [LS-HS] Plateau. *Dalton Trans.* **2014**, *44*, 15079–15082. [CrossRef] [PubMed]
20. Phonsri, W.; Harding, D.J.; Harding, P.; Murray, K.S.; Moubaraki, B.; Gass, I.A.; Cashion, J.D.; Jameson, G.N.L.; Adams, H. Stepped Spin Crossover in Fe(III) Halogen Substituted Quinolylsalicylaldimine Complexes. *Dalton Trans.* **2014**, *43*, 17509–17518. [CrossRef] [PubMed]
21. Phonsri, W.; Harding, P.; Liu, L.; Telfer, S.G.; Murray, K.S.; Moubaraki, B.; Ross, T.M.; Jameson, G.N.L.; Harding, D.J. Solvent Modified Spin Crossover in an Iron(III) Complex: Phase Changes and an Exceptionally Wide Hysteresis. *Chem. Sci.* **2017**, *8*, 3949–3959. [CrossRef] [PubMed]
22. Takahashi, K.; Cui, H.; Kobayashi, H.; Einaga, Y.; Sato, O. The Light-Induced Excited Spin State Trapping Effect on Ni(dmit)$_2$ Salt with an Fe(III) Spin-Crossover Cation: [Fe(qsal)$_2$][Ni(dmit)$_2$]·2CH$_3$CN. *Chem. Lett.* **2005**, *34*, 1240–1241. [CrossRef]
23. Takahashi, K.; Cui, H.-B.; Okano, Y.; Kobayashi, H.; Einaga, Y.; Sato, O. Electrical Conductivity Modulation Coupled to a High-Spin-Low-Spin Conversion in the Molecular System [Fe$^{III}$(qsal)$_2$][Ni(dmit)$_2$]$_3$·CH$_3$CN·H$_2$O. *Inorg. Chem.* **2006**, *45*, 5739–5741. [CrossRef]
24. Faulmann, C.; Dorbes, S.; Lampert, S.; Jacob, K.; Garreau de Bonneval, B.; Molnár, G.; Bousseksou, A.; Real, J.A.; Valade, L. Crystal Structure, Magnetic Properties and Mössbauer Studies of [Fe(qsal)$_2$][Ni(dmit)$_2$]. *Inorg. Chim. Acta* **2007**, *360*, 3870–3878. [CrossRef]
25. Takahashi, K.; Cui, H.-B.; Okano, Y.; Kobayashi, H.; Mori, H.; Tajima, H.; Einaga, Y.; Sato, O. Evidence of the Chemical Uniaxial Strain Effect on Electrical Conductivity in the Spin-Crossover Conducting Molecular System: [Fe$^{III}$(qnal)$_2$][Pd(dmit)$_2$]$_5$·Acetone. *J. Am. Chem. Soc.* **2008**, *130*, 6688–6689. [CrossRef] [PubMed]
26. Takahashi, K.; Mori, H.; Kobayashi, H.; Sato, O. Mechanism of Reversible Spin Transition with a Thermal Hysteresis Loop in [Fe$^{III}$(qsal)$_2$][Ni(dmise)$_2$]·2CH$_3$CN: Selenium Analogue of the Precursor of an Fe(III) Spin-Crossover Molecular Conducting System. *Polyhedron* **2009**, *28*, 1776–1781. [CrossRef]
27. Neves, A.I.S.; Dias, J.C.; Vieira, B.J.C.; Santos, I.C.; Castelo Branco, M.B.; Pereira, L.C.J.; Waerenborgh, J.C.; Almeida, M.; Belo, D.; da Gama, V.A. New Hybrid Material Exhibiting Room Temperature Spin-Crossover and Ferromagnetic Cluster-Glass Behavior. *CrystEngComm* **2009**, *11*, 2160–2168. [CrossRef]
28. Faulmann, C.; Chahine, J.; Valade, L.; Chastanet, G.; Létard, J.-F.; De Caro, D. Photomagnetic Studies of Spin-Crossover-and Photochromic-Based Complexes. *Eur. J. Inorg. Chem.* **2013**, *5*, 1058–1067. [CrossRef]
29. Togo, T.; Amolegbe, S.A.; Yamaguchi, R.; Kuroda-Sowa, T.; Nakaya, M.; Shimayama, K.; Nakamura, M.; Hayami, S. Crystal Structure and Spin-Crossover Behavior of Iron(III) Complex with Nitroprusside. *Chem. Lett.* **2013**, *42*, 1542–1544. [CrossRef]
30. Fukuroi, K.; Takahashi, K.; Mochida, T.; Sakurai, T.; Ohta, H.; Yamamoto, T.; Einaga, Y.; Mori, H. Synergistic Spin Transition between Spin Crossover and Spin-Peierls-like Singlet Formation in the Halogen-Bonded Molecular Hybrid System: [Fe(Iqsal)$_2$][Ni(dmit)$_2$]·CH$_3$CN·H$_2$O. *Angew. Chem. Int. Ed.* **2014**, *53*, 1983–1986. [CrossRef]
31. Vieira, B.J.C.; Dias, J.C.; Santos, I.C.; Pereira, L.C.J.; da Gama, V.; Waerenborgh, J.C. Thermal Hysteresis in a Spin-Crossover Fe$^{III}$ Quinolylsalicylaldimine Complex, Fe$^{III}$(5-Br-qsal)$_2$Ni(dmit)$_2$·solv: Solvent Effects. *Inorg. Chem.* **2015**, *54*, 1354–1362. [CrossRef]

32. Vieira, B.J.C.; Coutinho, J.T.; Dias, J.C.; Nunes, J.C.; Santos, I.C.; Pereira, L.C.J.; Da Gama, V.; Waerenborgh, J.C. Crystal Structure and Spin Crossover Behavior of the [Fe(5-Cl-Qsal)2][Ni(Dmit)2]·2CH3CN Complex. *Polyhedron* **2015**, *85*, 643–651. [CrossRef]
33. Takahashi, K.; Sakurai, T.; Zhang, W.-M.; Okubo, S.; Ohta, H.; Yamamoto, T.; Einaga, Y.; Mori, H. Spin-Singlet Transition in the Magnetic Hybrid Compound from a Spin-Crossover Fe(III) Cation and π-Radical Anion. *Inorganics* **2017**, *5*, 54. [CrossRef]
34. Phonsri, W.; Davies, C.G.; Jameson, G.N.L.; Moubaraki, B.; Murray, K.S. Spin Crossover, Polymorphism and Porosity to Liquid Solvent in Heteroleptic Iron(III) {Quinolylsalicylaldimine/Thiosemicarbazone-Salicylaldimine} Complexes. *Chem. A Eur. J.* **2016**, *22*, 1322–1333. [CrossRef] [PubMed]
35. Phonsri, W.; Macedo, D.; Moubaraki, B.; Cashion, J.; Murray, K. Heteroleptic Iron(III) Spin Crossover Complexes; Ligand Substitution Effects. *Magnetochemistry* **2016**, *2*, 3. [CrossRef]
36. Murata, S.; Takahashi, K.; Mochida, T.; Sakurai, T.; Ohta, H.; Yamamoto, T.; Einaga, Y. Cooperative Spin-Crossover Transition from Three-Dimensional Purely π-Stacking Interactions in a Neutral Heteroleptic Azobisphenolate Fe$^{III}$ Complex with a N$_3$O$_3$ Coordination Sphere. *Dalton Trans.* **2017**, *46*, 5786–5789. [CrossRef] [PubMed]
37. Phonsri, W.; Macedo, D.S.; Davies, C.G.; Jameson, G.N.L.; Moubaraki, B.; Murray, K.S. Heteroleptic Iron(III) Schiff Base Spin Crossover Complexes: Halogen Substitution, Solvent Loss and Crystallite Size Effects. *Dalton Trans.* **2017**, *46*, 7020–7029. [CrossRef] [PubMed]
38. Kuroda-Sowa, T.; Yu, Z.; Senzaki, Y.; Sugimoto, K.; Maekawa, M.; Munakata, M.; Hayami, S.; Maeda, Y. Abrupt Spin Transitions and LIESST Effects Observed in Fe$^{II}$ Spin-Crossover Complexes with Extended π-Conjugated Schiff-Base Ligands Having N$_4$O$_2$ Donor Sets. *Chem. Lett.* **2008**, *37*, 1216–1217. [CrossRef]
39. Yu, Z.; Kuroda-Sowa, T.; Kume, H.; Okubo, T.; Maekawa, M.; Munakata, M. Effects of Metal Doping on the Spin-Crossover Properties of an Iron(II) Complex with Extended π-Conjugated Schiff-Base Ligand Having an N$_4$O$_2$ Donor Set. *Bull. Chem. Soc. Jpn.* **2009**, *82*, 333–337. [CrossRef]
40. Kuroda-Sowa, T.; Kimura, K.; Kawasaki, J.; Okubo, T.; Maekawa, M. Effects of Weak Interactions on Spin Crossover Properties of Iron(II) Complexes with Extended π-Conjugated Schiff-Base Ligands. *Polyhedron* **2011**, *30*, 3189–3192. [CrossRef]
41. Iasco, O.; Rivière, E.; Guillot, R.; Buron-Le Cointe, M.; Meunier, J.-F.; Bousseksou, A.; Boillot, M.-L. Fe$^{II}$(pap-5NO$_2$)$_2$ and Fe$^{II}$(qsal-5NO$_2$)$_2$ Schiff-Base Spin-Crossover Complexes: A Rare Example with Photomagnetism and Room-Temperature Bistability. *Inorg. Chem.* **2015**, *54*, 1791–1799. [CrossRef] [PubMed]
42. Phonsri, W.; Macedo, D.S.; Vignesh, K.R.; Rajaraman, G.; Davies, C.G.; Jameson, G.N.L.; Moubaraki, B.; Ward, J.S.; Kruger, P.E.; Chastanet, G.; et al. Halogen Substitution Effects on N$_2$O Schiff Base Ligands in Unprecedented Abrupt Fe$^{II}$ Spin Crossover Complexes. *Chem. A Eur. J.* **2017**, *23*, 7052–7065. [CrossRef] [PubMed]
43. Kuroda-Sowa, T.; Isobe, R.; Yamao, N.; Fukumasu, T.; Okubo, T.; Maekawa, M. Variety of Spin Transition Temperatures of Iron(II) Spin Crossover Complexes with Halogen Substituted Schiff-Base Ligands, HqsalX(X = F, Cl, Br, and I). *Polyhedron* **2017**, *136*, 74–78. [CrossRef]
44. Atzori, M.; Poggini, L.; Squillantini, L.; Cortigiani, B.; Gonidec, M.; Bencok, P.; Sessoli, R.; Mannini, M. Thermal and Light-Induced Spin Transition in a Nanometric Film of a New High-Vacuum Processable Spin Crossover Complex. *J. Mater. Chem. C* **2018**, *6*, 8885–8889. [CrossRef]
45. König, E. *Landolt-Börnstein Neue Serie Gruppe II, Vol. 2*; Hellwege, K.-H., Hellwege, A.M., Eds.; Springer: Berlin/Heidelberg, Germany; New York, NY, USA, 1966; pp. 16–18.
46. Frisch, M.J.; Trucks, G.W.; Schlegel, H.B.; Scuseria, G.E.; Robb, M.A.; Cheeseman, J.R.; Scalmani, G.; Barone, V.; Mennucci, B.; Petersson, G.A.; et al. *Gaussian 09, Revision D.01*; Gaussian, Inc.: Wallingford, CT, USA, 2009.
47. Becke, A.D. A New Mixing of Hartree-Fock and Local Density-Functional Theories. *J. Chem. Phys.* **1993**, *98*, 1372–1377. [CrossRef]
48. Lee, C.; Yang, W.; Parr, R.G. Development of the Colle-Salvetti Correlation-Energy Formula into a Functional of the Electron Density. *Phys. Rev. B* **1988**, *37*, 785–789. [CrossRef]
49. Wachters, A.J.H. Gaussian Basis Set for Molecular Wavefunctions Containing Third-Row Atoms. *J. Chem. Phys.* **1970**, *52*, 1033–1036. [CrossRef]
50. Hay, P.J. Gaussian Basis Sets for Molecular Calculations. The Representation of 3d Orbitals in Transition-Metal Atoms. *J. Chem. Phys.* **1977**, *66*, 4377–4384. [CrossRef]

51. Ditchfield, R.; Hehre, W.J.; Pople, J.A. Self-Consistent Molecular-Orbital Methods. IX. An Extended Gaussian-Type Basis for Molecular-Orbital Studies of Organic Molecules. *J. Chem. Phys.* **1971**, *54*, 724–728. [CrossRef]
52. Casida, M.E.; Jamorski, C.; Casida, K.C.; Salahub, D.R. Molecular excitation energies to high-lying bound states from time-dependent density-functional response theory: Characterization and correction of the time-dependent local density approximation ionization threshold. *J. Chem. Phys.* **1998**, *108*, 4439–4449. [CrossRef]
53. Yanai, T.; Tew, D.P.; Handy, N.C. A new hybrid exchange-correlation functional using the Coulomb-attenuating method (CAM-B3LYP). *Chem. Phys. Lett.* **2004**, *393*, 51–57. [CrossRef]
54. Halcrow, M.A. Iron(II) Complexes of 2,6-di(pyrazol-1-yl)pyridines—A Versatile System for Spin-Crossover Research. *Coord. Chem. Rev.* **2009**, *253*, 2493–2514. [CrossRef]
55. Takahashi, K.; Kawamukai, K.; Okai, M.; Mochida, T.; Sakurai, T.; Ohta, H.; Yamamoto, T.; Einaga, Y.; Shiota, Y.; Yoshizawa, K. A new family of anionic Fe$^{III}$ spin crossover complexes featuring a weak-field N$_2$O$_4$ coordination octahedron. *Chem. Eur. J.* **2016**, *22*, 1253–1257. [CrossRef]
56. Bondi, A. Van der Waals Volumes and Radii. *J. Phys. Chem.* **1964**, *68*, 441–451. [CrossRef]
57. Murata, S.; Takahashi, K.; Sakurai, T.; Ohta, H.; Yamamoto, T.; Einaga, Y.; Shiota, Y.; Yoshizawa, K. The Role of Coulomb Interactions for Spin Crossover Behaviors and Crystal Structural Transformation in Novel Anionic Fe(III) Complexes from a π-Extended ONO Ligand. *Crystals* **2016**, *6*, 49. [CrossRef]
58. Takahashi, K.; Okai, M.; Mochida, T.; Sakurai, T.; Ohta, H.; Yamamoto, T.; Einaga, Y.; Shiota, Y.; Yoshizawa, K.; Konaka, H.; et al. Contribution of Coulomb Interactions to a Two-Step Crystal Structure Phase Transformation Coupled with a Significant Change in Spin Crossover Behavior for a Series of Charged Fe$^{II}$ Complexes from 2,6-bis(2-methylthiazol-4-yl)pyridine. *Inorg. Chem.* **2018**, *57*, 1277–1287. [CrossRef] [PubMed]

© 2019 by the authors. Licensee MDPI, Basel, Switzerland. This article is an open access article distributed under the terms and conditions of the Creative Commons Attribution (CC BY) license (http://creativecommons.org/licenses/by/4.0/).

Article

# Spatio-temporal Investigations of the Incomplete Spin Transition in a Single Crystal of [Fe(2-pytrz)$_2$\{Pt(CN)$_4$\}]·3H$_2$O: Experiment and Theory

Houcem Fourati [1], Guillaume Bouchez [1], Miguel Paez-Espejo [1], Smail Triki [2] and Kamel Boukheddaden [1,*]

1. Groupe d'Etudes de la Matière Condensée, UMR 8635, CNRS-Université de Versailles, Université Paris Saclay, 45 Avenue des Etats Unis, 78035 Versailles, France; houcem.fourati@uvsq.fr (H.F.); guillaume.bouchez@uvsq.fr (G.B.); miguelangel.paezespejo@gmail.com (M.P.-E.)
2. Univ Brest, CNRS, CEMCA, 6 Avenue Le Gorgeu, C.S. 93837 - 29238 Brest Cedex 3, France; smail.triki@univ-brest.fr
* Correspondence: kamel.boukheddaden@uvsq.fr

Received: 13 December 2018; Accepted: 11 January 2019; Published: 16 January 2019

**Abstract:** Optical microscopy technique is used to investigate the thermal and the spatio-temporal properties of the spin-crossover single crystal [Fe(2-pytrz)$_2$\{Pt(CN)$_4$\}]·3H$_2$O, which exhibits a first-order spin transition from a full high-spin (HS) state at high temperature to an intermediate, high-spin low-spin (HS-LS) state, below 153 K, where only one of the two crystallographic Fe(II) centers switches from the HS to HS-LS state. In comparison with crystals undergoing a complete spin transition, the present transformation involves smaller volume changes at the transition, which helps to preserving the crystal's integrity. By analyzing the spatio-temporal properties of this spin transition, we evidenced a direct correlation between the orientation and shape of HS/HS-LS domain wall with the crystal's shape. Thanks to the small volume change accompanying this spin transition, the analysis of the experimental data by an anisotropic reaction-diffusion model becomes very relevant and leads to an excellent agreement with the experimental observations.

**Keywords:** spin-crossover; optical microscopy; reaction diffusion

## 1. Introduction

Because of the growing societal requirement of information processing and big data storage, the design of devices with reversible and high density storage capacities as well as fast responses becomes mandatory. Multi-functional molecular materials with intrinsic physical properties at the molecular scale constitute serious candidates for their use as nano-memories, nano-switches or nano-probes due to many advantages, like their easy processability and their low cost. In this context, spin crossover (SCO) materials, based on iron (II) complexes, have been recognized as prime candidates, in particular due to their variety of nano-structuration possibilities [1–5] allowing their integration into devices for a various set of currently considered applications, e.g., as display and memory devices [6–9], sensing of temperature [10–12], probes of contact pressure or shocks [13], as well as actuators [14–16]. From the thermodynamic point of view, the SCO phenomenon is an entropic-driven mechanism involving the switching between two different electronic states, namely a diamagnetic low-spin (LS) state and a paramagnetic high-spin (HS) state, as a response to the application of an external stimulus. In the case of d$^6$ Fe(II) based SCO compounds, the electronic configuration of the metal ion in the HS state, which generally emerges at high temperature, is (t$_{2g}^4$e$_g^2$), where the valence electrons fill all electronic orbitals according to the Hund rule leading to the total spin momentum $S = 2$, which makes the SCO systems paramagnetic in this state. In contrast, in the

LS state, which appears at low-temperature, the electronic configuration $t_{2g}^6 e_g^0$ is stabilized, for which corresponds a spin momentum, $S = 0$, leading to a diamagnetic LS state.

On the other hand, due to the antibonding nature of the $e_g$ orbital, the Fe-ligand distances increase by $\sim 10\%$ in the HS state in comparison with those of LS state, which induces an expansion of the unit cell volume in the HS state [17]. This molecular volume expansion and contraction in the HS and LS states, respectively, is delocalized in the lattice over several unit cells and is at the origin of the cooperative effects observed in these molecular solids. Indeed, as a consequence of these long-range interactions between molecules, insured by the lattice phonons, SCO solids can display first-order phase transitions accompanied with thermal hysteresis loops [6,18,19], allowing them to show macroscopic bistability, and then to switch collectively between the LS and the HS states. In addition, the spin transition can be triggered by various external parameters, such as pressure, light, temperature, electric and magnetic fields, [6,20,21], thus increasing their potential application in many areas.

From the microscopic point of view, the spin transition phenomenon has been considered as a vibronic problem [22,23] in which the electronic and vibrational structures of the molecule are strongly coupled, to the extent that in some cases, the Born-Oppenheimer approximation is not valid [24,25]. The microscopic changes of the magnetoelastic properties of the SCO solids at the transition, accompanied with large volume changes resulting from the constructive interferences of the molecular volume changes which deploy at long-range through elastic interactions, cause important variations in the physical properties of the SCO materials, which show thermo-chromic features at the transition as well as significant rigidity changes. As a direct consequence, an important panel of experimental techniques is utilized to study the SCO phenomenon, among them one can quote, the differential scanning calorimetry [26], magnetometry [27], X-ray diffraction [28], Mössbauer spectroscopy [29], diffuse reflectivity [30] as well as optical microscopy (OM) [31–38], which has been found very useful in the understanding of the non-equilibrium properties of the SCO materials. As a matter of fact, OM is now recognized as one of the major techniques allowing a direct spatiotemporal imaging of the thermally-induced spin crossover transition, with the respective spatial and temporal resolutions of 0.3 µm and 10 ms.

OM allows the study of the first-order phase transitions of SCO solids on a unique single crystal, which constitutes an undeniable progress towards a deep understanding of the fundamental aspects of this phenomenon. In addition, this technique provides valuable information not only on the electronic properties of the studied single crystals along the transition but also on the modification of their elastic properties, particularly when the spin transition is accompanied with a significant volume change. Among the large number of SCO compounds already investigated by OM [31–38], it is interesting to recall that the first quantitative experiments [35] were conducted on the fragile SCO single crystals of [Fe(btr)$_2$(NCS)$_2$]H$_2$O (btr = 4,4'-bis-1,2,4-triazole), which easily lose watter molecules at the transition, causing their deterioration. To avoid such a disastrous effect, the single crystals were embedded in oil and the OM measurements have been performed by preventing any dehydration. Despite this, the first cooling process from HS to LS revealed the appearance of irreversible defects in the crystal caused by the mechanical stresses generated by the volume change at the spin transition, which hindered the complete analysis of the spatio-temporal properties along the two branches of the thermal hysteresis, although it was possible to evaluate the HS-LS interface velocity on cooling to $\sim 2.3$ µm·s$^{-1}$. Later on, Chong et al. [36] investigated the spin transition in the hexagonally-shaped single crystals of [Fe(bbtr)$_3$](ClO$_4$)$_2$. Similarly to the previous study, the SCO transition was found to start at the corner of the hexagon, with a clear HS-LS interface which adapts its length and shape along its propagation process. Unfortunately, due to the large lattice parameter misfit between the LS and HS structures, the system does not succeed in finding a stress free interface, which causes irreversible crystal damages after the first cooling. It is worth mentioning that this enhanced brittleness of the crystals around the interface region which stores the excess of elastic energy, is quite well explained in the theoretical electro-elastic models [37,39–42] describing the elastic properties of the SCO solids. Finally, more recent

OM studies, performed on the resilient SCO single crystal [Fe(NCSe)(py)$_2$(μ-bpypz)] (py = pyridine and bpypz = 3,5-bis(2-pyridyl)-pyrazolate) [37,42–44], allowed for the first-time to monitor the motion of the HS-LS interface on both cooling and heating processes, whose corresponding velocities were estimated in the range 4–6 μm·s$^{-1}$. In the latter case, due to the particularly regular shape of the crystals, made of rectangular platelets along the direction of the interface propagation, the shape of the HS-LS interface remained unchanged during its motion. Very recently [38], the SCO compound, [Fe(2-pytrz)$_2$Pd(CN)$_4$]·3H$_2$O (2-pytrz = 4-(2-pyridyl)-1,2,4,4H-triazole), has been investigated by OM and showed a well defined interface during the spin transition. Thanks to the robust character of the investigated single crystals, the dynamical properties of the interface propagation were deeply studied, and their behavior were well reproduced by a reaction diffusion model. In the present work, we report about new OM investigations, realized on a high quality SCO complex of formula [Fe(2-pytrz)$_2$Pt(CN)$_4$]·3H$_2$O (1) for which we selected a reliable single crystal characterized by an irregular shape, in order to investigate the interplay between the mode of interface propagation and the crystal's shape. The expected synergetic effects are considered here as an evidence of the action of the long-range elastic interactions between the SCO molecules within the crystal, caused by the volume change at the transition. One of the consequences will be that mean-field based models become suitable for the theoretical description of the spatiotemporal properties of SCO materials, as will be demonstrated below.

The present manuscript is organized as follows: Section 2 summarizes the experimental findings on the thermo-induced spin transition of the [Fe(2-pytrz)$_2${Pt(CN)$_4$}]·3H$_2$O and presents the data of OM investigations which revealed the spatio-temporal character of the SCO transition. In Section 3, we introduce the theoretical spatio-temporal model, based on reaction diffusion description, allowing to faithfully describe the experimental data. Section 4 concludes and outlines some possible developments of this work.

## 2. Results and Discussion

The present work contains experimental investigations of optical microscopy performed at several temperatures in the thermal hysteresis region in order to observe the spatio-temporal features of the first-order spin transition of the title compound. To understand the physical mechanism of the interface propagation in this particular system, in which the spin transition is accompanied with a small volume change, we developed an appropriate modeling, based on a reaction diffusion description, allowing to reproduce the experimental spatiotemporal features of the current first-order SCO transition.

## 3. Experimental Investigations

### 3.1. Synthesis

Single crystals of [Fe(2-pytrz)$_2${Pt(CN)$_4$}]·3H$_2$O were obtained by slow diffusion, in a fine glass tube (3.0 *mm* diameter) of two different aqueous solutions: the first one was prepared by dissolving K$_2$[Pt(CN)$_4$]·xH$_2$O (37.7 mg, 0.1 mmol) in 10 mL; the second one was obtained by dissolving iron(II) perchlorate salt (25.5 mg, 0.1 mmol) in a solution (10 mL) of 4-(2-pyridyl)-1,2,4,4H-triazole (2-pytrz) (29.2 mg, 0.2 mmol), which became light yellow after a night. A volume of 1 mL of the K$_2$[Pt(CN)$_4$]·xH$_2$O solution was put in the fine glass tube and then similar volume (1 mL) was layered meticulously onto the yellow solution. After two days of slow diffusion, colorless fine square crystals of [Fe(2-pytrz)$_2${Pt(CN)$_4$}]·3H$_2$O were formed [45].

### 3.2. Magnetic and Structural Characterizations

Magnetic measurements of [Fe(2-pytrz)$_2${Pt(CN)$_4$}]·3H$_2$O performed in Ref. [45] and reported in Figure 1, show the cooperative character of this SCO material which leads to a thermally-induced first-order spin transition occurring around ∼152 K on cooling and ∼154 K on heating, accompanied with a thermal hysteresis of ca. 2 K width. As already discussed in Ref. [45], Figure 1a, shows $\chi_m T$

values of 3.51 cm$^3$·K·mol$^{-1}$ compatible with a HS ($S = 2$) state of Fe(II) at high temperature, while below 150 K, this value drops abruptly to ca. 2.0 cm$^3$·K·mol$^{-1}$, indicating an incomplete spin transition, with the presence of ~56% residual HS fraction at low temperature. The slight decrease of this fraction at very low temperature (i.e., below 30 K) is attributed to zero field splitting effects, induced by the magnetic anisotropy of the current hexacoordinated Fe(II) ions. In agreement with X-rays diffraction data measurements [45], it was demonstrated that in the low-temperature regime, the spin transition involves one out of two atoms, as shown in Figure 1b. Indeed, the distortion experienced by one of the iron sites (see Figure 1b), originating from an elastic frustration occurring inside the lattice, induces the trapping of Fe1 in the HS state, thus precluding its transition at low-temperature.

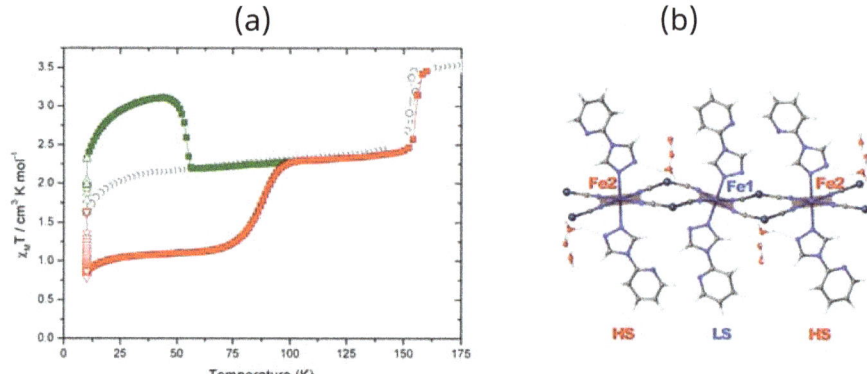

**Figure 1.** (a) Thermal behavior of the magnetic moment of the SCO compound [Fe(2-pytrz)$_2${Pt(CN)$_4$}]·3H$_2$O, adapted from Ref. [45]. Black open circles represent the heating and cooling process without light showing the presence of a first order transition with thermal hysteresis around $T \sim 153$ K, between the HS and the intermediate HS-LS states. The green curve is the heating process after the green light ($\lambda = 510$ nm) irradiation at the lowest temperature, and the red one corresponds to the heating process from the LS to the HS state through HS-LS state after the red light ($\lambda = 830$ nm) irradiation at the lowest temperature. The relaxation temperatures, after green and red light excitations, are $T_{LIESST} \simeq 52$ K and $\simeq 90$ K, respectively. Temperature sweep rate is 0.4 K·min$^{-1}$ for all experiments. (b) Structure of [Fe(2-pytrz)$_2${Pt(CN)$_4$}]·3H$_2$O in the HS-LS phase.

In addition to the thermal hysteresis and the incomplete HS to HS-LS spin transition, Figure 1 reports the photomagnetic response of this compound under green light ($\lambda = 510$ nm) showing a clear evidence of a LIESST (Light-Excited-Spin-State-Trapping) [45–48] effect allowing to access metastable photo-induced HS state at low-temperature. The resultant irradiated HS state relaxes back to the intermediate HS-LS state around ~52 K. On the other hand, the excitation of the intermediate HS-LS low-temperature phase using a red wavelength ($\lambda = 830$ nm) produced the reverse-LIESST effect [45,47–49], leading to the fully "stable" LS state, whose thermal relaxation to the intermediate HS-LS phase takes place at $T \simeq 90$ K. The investigations of the dynamical properties of this LS state allowed in a previous study [45] to demonstrate its stability at very low-temperature. It results that the intermediate HS-LS state is stable on cooling until some temperature located in the region 40–70 K and then becomes metastable due to its thermal freezing at lower temperatures.

This bistability between HS and HS-LS states was confirmed by DSC studies for which the thermal variation of the heat flow shows exo- and endothermic transitions at ~151.8 K and ~154.4 K, respectively (see Figure 2). The phase transition occurs with an enthalpy and entropy changes of $\Delta H = 4.8$ kJ·mol$^{-1}$ and $\Delta S = 30$ J·K$^{-1}$·mol$^{-1}$, respectively, in agreement with the values reported in literature of SCO compounds [38,50]

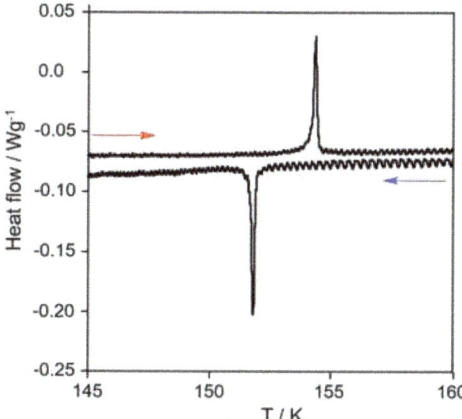

**Figure 2.** DSC study for [Fe(2-pytrz)$_2${Pt(CN)$_4$}]·3H$_2$O (scan rate = 0.3 K·min$^{-1}$) revealing the exo- and the endo-thermic transitions with maxima at 151.8 and 154.4 K, respectively. These values, derived from the maximum of the DSC peaks, are in good agreement with the transition temperatures obtained in magnetic data.

The change of the unit cell parameters as well as $\beta$ angle misfit between HS and HS-LS states is depicted in Table 1. The latter is compared to two other SCO compounds that have been already studied by optical microscopy, namely [{Fe(NCSe)(py)$_2$}$_2$(m-bpypz)] [37,42–44] and [Fe(btr)$_2$(NCS)$_2$]·H$_2$O [35]. One can see that the change of $\beta$ angle is much smaller for the title compound, which then exhibits reduced distortion at the transition.

**Table 1.** Relative variation of lattice parameters between the high- and low-temperature phases of several SCO compounds. The isotropic/anisotropic character of the volume change at the transition affects the HS-LS interface velocity and the resilience of the single crystal during the transition. HS, LS and HS-LS phases denote high spin, low-spin and intermediate HS-LS states, respectively.

| Compound | Phases | $\frac{\Delta a}{a}$ (%) | $\frac{\Delta b}{b}$ (%) | $\frac{\Delta c}{c}$ (%) | $\Delta\beta(°)$ |
|---|---|---|---|---|---|
| [{Fe(NCSe)(py)$_2$}$_2$(m-bpypz)] | HS ↔ LS | 0.6 | 1.58 | 0.3 | −2.329 |
| [Fe(2-pytrz)$_2${Pt(CN)$_4$}]·3H$_2$O | HS ↔ HS-LS | 1.82 | 2.17 | 0.69 | −0.036 |
| [Fe(btr)$_2$(NCS)$_2$](H$_2$O) | HS ↔ LS | −2.04 | 4.12 | 3.1 | −1.22 |

To avoid the strong elastic effects and volume changes which usually degrade the materials, and to compare the crystal shape effects on the interface behavior (shape and velocity), we investigate the spatio-temporal features of the compound [Fe(2-pytrz)$_2${Pt(CN)$_4$}]·3H$_2$O for which the unit cell distortion between the low-temperature and the high-temperature phases is less significant than those of compounds [{Fe(NCSe)(py)$_2$}$_2$(m-bpypz)] and [Fe(btr)$_2$(NCS)$_2$](H$_2$O). Indeed in [Fe(2-pytrz)$_2${Pt(CN)$_4$}]·3H$_2$O, the transition takes place between a fully HS phase at high temperature and an intermediate HS-LS phase at low-temperature, which then involves less volume change.

### 3.3. Optical Microscopy Measurements

Optical microscopy studies have been performed on a single crystal of the SCO compound [Fe(2-pytrz)$_2${Pt(CN)$_4$}]·3H$_2$O [45] of size ∼ 400 μm × 300 μm which exhibits an incomplete transition between a fully HS state at high temperature and a well organized HS-LS phase at low temperature [45].

### 3.3.1. Thermal Hysteresis of One Single Crystal

OM investigations evidencing the spatio-temporal properties of the present single crystal along its thermal spin transition are reported in Figure 3a, which displays a set of selected snapshots on cooling and heating processes, exhibiting different colors in the HS and HS-LS phases. Furthermore, a single domain nucleation and growth mechanism upon spin transition accompanied with a well-defined interface between the two phases, is observed. We exploited the thermochromic character of this transformation, which leads to which leads to different optical densities (OD) $[OD = \log_{10}(\frac{I_0}{I})$, where $I_0$ is the incident bright field intensity and $I$ is the transmitted intensity] for the HS and HS-LS states. The transition temperatures and the thermal hysteresis loop of the system, were evaluated using an image processing treatment, whose procedure was already explained in [38,51], the spatially averaged total OD emerging from the single crystal. The obtained thermal evolution of the OD is given in Figure 3b. It is worth mentioning that the local HS fraction, $n(x, y, T)$, of the system connects linearly to the normalized local OD, defined as $n(x,y,T) = \frac{OD(x,y,T)-OD(HS-LS)}{OD(HS)-OD(HS-LS)}$ (where OD(HS) and OD(HS-LS) are the OD values in the HS and intermediate HS-LS states, respectively).

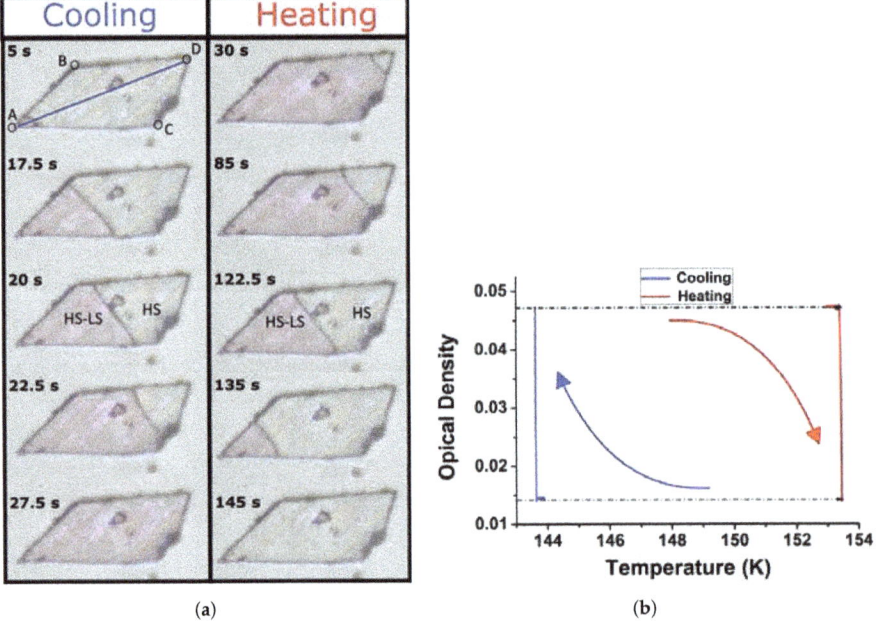

**Figure 3.** (a) Snapshots of the single crystal during the thermo-induced spin transition on cooling and heating. (b) Temperature dependence of the average OD showing the occurrence of a thermal hysteresis loop. The temperature sweep rate was 0.5 K.min$^{-1}$, for both cooling and heating processes.

The transition temperatures deduced from Figure 3b were found to be $T^- \sim 143$ K on cooling and $T^+ \sim 153$ K on heating. Compared to the magnetic data which led to $T^- \sim 152$ K and $T^+ \sim 154$, one may conclude to the existence of a net disagreement. However, one has to consider that the thermal hysteresis of OM data relates to the transition of a unique single crystal, while that of magnetic measurements (Figure 1a) resulted from the average response of a large number of crystalline powder (or micro-crystals) sample with distributed sizes, shapes, transition temperatures and thermal hysteresis.

OM images of Figure 3a, recorded during the spin transition indicate that the first-order transition, on cooling between the HS and the intermediate HS-LS phases, starts from the bottom left corner (point A) and proceeds as a unique domain that propagates until reaching the first upper right corner (point B). During this first stage, the interface has more or less, a circular shape with contact angles perpendicular to the crystal edges. From point B the two crystal borders become parallel, which affects the interface shape becomes sigmoidal (S-shaped). This special shape is imposed by: (i) the border conditions between the interface and the crystal borders (i.e., the contact angles) which remain perpendicular to each others and (ii) the shift existing along the crystal length between the downer and upper interface points. When the interface reaches point C, a similar behavior as that of the first stage (A → B) occurs again, since its shape becomes again curved with contact angles equal to $\sim \frac{\pi}{2}$.

3.3.2. Interface Velocities

As we use the OM data to evaluate the thermal hysteresis loop of Figure 3b, we use it also to monitor the interface position along the diagonal line AD (see Figure 3a) as a function of time, the behavior of which is given in Figure 4. There, one can easily remark that the average velocity of the interface on cooling is higher than that on heating. We can also observe the existence of different propagation regimes with different velocities for both processes, attributed to the irregular shape of the single crystal, in good agreement with a similar study, recently realized on the analogous compound [Fe(2-pytrz)$_2$Pd(CN)$_4$]·3H$_2$O [38]. Figure 4 also evidences the occurrence of an acceleration regime when the interface gets close to the extremity of the single crystal. It is interesting to notice that this result is in good agreement with the predictions of recent theoretical investigations [40,52]. The quantitative evaluation of the average propagation velocities in the first regime before the acceleration are 17 µm·s$^{-1}$ and 2.3 µm·s$^{-1}$ on cooling and heating processes, respectively.

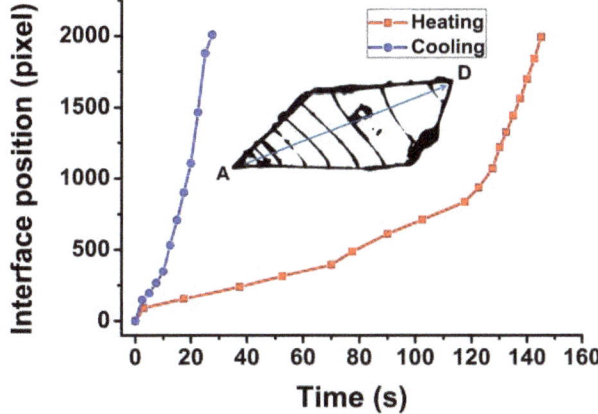

**Figure 4.** The interface position as a function of time, estimated along the diagonal of the crystal, from point A to point D and D to A on cooling and heating, respectively, indicated with blue and red symbols. Inset shows the interface positions on the crystal along the heating process.

In Figure 5, we present the time evolution of the interface position along the crystal edges upon cooling and heating. In Figure 5a, the interface velocity is evaluated along DB and DC directions during the heating process. The obtained curves display a very similar behavior with the presence of a regime change (an acceleration) starting around $t \simeq 80$ s for both of them. In the first regime on DB and DC directions, the measured interface velocities are $v_1 = 1.6$ µm·s$^{-1}$ and $v_1 = 1.4$ µm·s$^{-1}$, respectively, while for the second regime, we found the respective values, $v_2 = 3.7$ µm·s$^{-1}$ and $v_2 = 3.1$ µm·s$^{-1}$. The set of obtained results is summarized in Table 2. The existence of the above two regimes is here attributed to the observable crystal defect, visible in the inset of Figure 5a along DC direction, which perturbs the interface propagation process on heating. Figure 5b presents the time-dependence of the interface position on cooling along AB and AC directions. There also, we remark the existence of two regimes. The first one, in AB and AC directions is characterized by the interface velocities, $v_1 = 10$ μm·s$^{-1}$ and $v_1 = 12$ μm·s$^{-1}$, respectively, while the second regime has higher respective velocities, $v_2 = 22$ μm·s$^{-1}$ and $v_2 = 32$ μm·s$^{-1}$. Interestingly, both curves show a regime change at the same time, $t \simeq 8$ s, which was identified as a particular time at which OM images show the spontaneous emergence of an additional HS-LS phase ahead the front interface. To emphasize this point, we represent in Figure 6 the spatial dependence of the transmitted light intensity across the interface along the blue line reported in the snapshots of Figure 6. This investigation is performed for different positions of the interface during the cooling regime. The intensity profile across the interface (Figure 6) is characterized by the presence of a significant reduction of the light intensity signal in the interface region compared to those of HS and HS-LS phases. This decrease is due to the darker character of the front which results from the light diffusion caused by the HS/(HS-LS) interface. In the right panels of Figure 6, we represented the intensity profile along the blue lines associated to the images of the left panel. For the two regimes showed in (b) and (c) images, the intensity profile shows a single peak corresponding to a unique interface, although that of image (b) appears as little bit broader. In contrast, the intensity profile corresponding to image (a), associated to the particular time value, $t = 8$ s, for which the velocity regime change appeared in Figure 5b, shows the existence of two peaks. This indicates the presence at this particular time of a second interface emerging ahead the main front. The origin of this second interface is attributed to a nucleation starting from the bottom of the crystal and initiated by the presence of defects which accelerate the process of relaxation of the elastic strain accompanying the volume change at the transition.

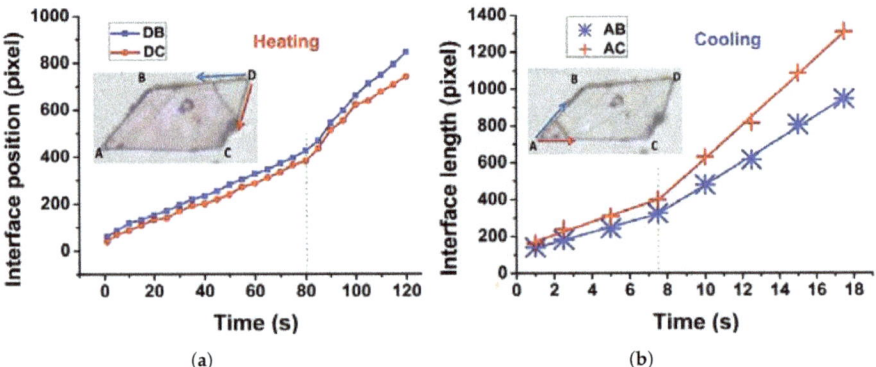

**Figure 5.** Interface velocity evaluated along the crystal edges as indicated in the two insets: (a) on heating along DB (blue) and DC (red) directions and (b) on cooling along AB (blue) and AC (red) directions. The vertical dashed lines indicate the times at which a regime change takes place in the propagation process.

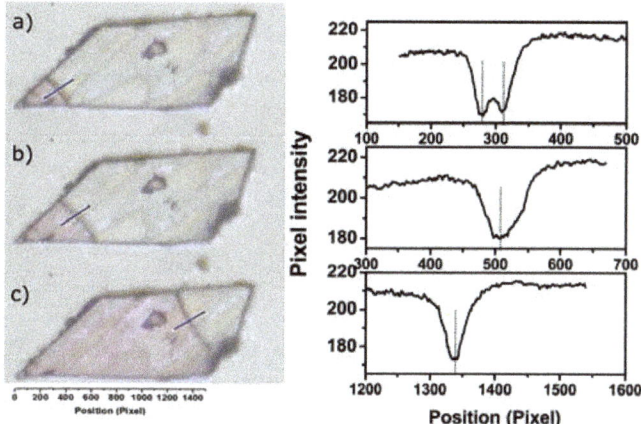

**Figure 6.** Intensity profile across the interface (separating HS and HS-LS phases) along the blue line during the cooling process. The pixel size is 0.3 µm × 0.3 µm.

**Table 2.** The various interface velocities evaluated on heating and cooling along AB, AC, DB and DC directions for the two different regimes identified in Figure 5.

| Direction of Propagation | | Velocity (µm·s$^{-1}$) | |
|---|---|---|---|
| | | 1st Regime | 2nd Regime |
| Heating | DB | 1.6 | 3.7 |
| | DC | 1.4 | 3.1 |
| Cooling | AB | 10 | 22 |
| | AC | 12 | 32 |

OM allows also to derive relevant information about the change of the crystal dimensions during the spin transition process. Using our own image processing software developed on Matlab, we could measure the distances AB, AC and AD in the low-temperature and high temperature phases, from which we deduced their relative variations. The obtained results are summarized in Table 3. Thanks to the good agreement with the relative variation of the lattice parameters extracted from X-rays studies [38,45] and summarized in the same Table 3, one can easily identify that the front propagation takes place in the *a-b* plane, where the most important changes are observed at the transition.

**Table 3.** Relative variation of the crystal sizes derived from OM data and X-rays measurements [38,45] in the HS and HS-LS phases. The pixel size is ~0.3 × 0.3 µm$^2$.

| | X-rays Data | | |
|---|---|---|---|
| Lattice Parameter (Å) | HS | HS-LS | Relative Variation (%) |
| a | 25.248 | 24.795 | 1.82 |
| b | 7.4044 | 7.247 | 2.17 |
| c | 27.293 | 27.105 | 0.69 |
| | Optical microscopy data | | |
| Distances (pixel) | HS | HS-LS | Relative variation (%) |
| $d_{B-D}$ | 1209 | 1197 | 1 |
| $d_{A-B}$ | 950 | 929 | 2.2 |
| $d_{A-C}$ | 1542 | 1522 | 1.2 |

## 4. Theoretical Section

### 4.1. The Hamiltonian

The present experimental results are modeled using the kinetic version of an Ising model [53] in which the low- and the high-temperature phases are assumed to be LS and HS, respectively. We demonstrated in a previous work, that the spatiotemporal version of such a description leads to a reaction diffusion equation [38,54] that governs the time and space dependence of the HS fraction during the spin transition. The Hamiltonian of the Ising-like model is expressed as follows,

$$\mathcal{H} = -J \sum_{ij} S_i S_j + \Delta_{eff} \sum_i S_i. \tag{1}$$

In Equation (1), $S_i$ is a fictitious operator whose eigenvalues $+1$ and $-1$ are respectively associated with the HS and LS states and $J > 0$ stands for the ferro-elastic interactions between the spin states, while $\Delta_{eff} = \Delta - \frac{k_B T}{2} \ln g$ is the effective energy gap, which contains the contributions of the ligand-field energy, $\Delta$, and the degeneracy ratio, $g = \frac{g_{HS}}{g_{LS}}$, between the HS and LS states. The latter contribution enters under the form of an entropic term in the effective ligand-field energy. The mean-field analysis of Hamiltonian (1) leads quite easily to the following expression of the free energy,

$$\mathcal{F}^{hom} = \frac{1}{2} J m^2 - k_B T \ln [2g \cosh(\frac{Jm - \Delta_{eff}}{k_B T})], \tag{2}$$

where $m = <s>$, is the average fictitious magnetization per site and $T$ is the temperature. This net magnetization connects to the HS fraction, $n_{HS}$, through the simple relation

$$n_{HS} = \frac{1+m}{2} \tag{3}$$

The used parameters for this simulation are $\Delta = 380$ K, $\ln g = 5$ and $J = 300$ K. The transition temperature obtained with the model is here $T_{\frac{1}{2}} = \frac{2\Delta}{k_B \ln(g)} \simeq 150$ K. The current model does not account for the elastic effects between the SCO sites nor for the volume change at the transition between LS and HS states. The equilibrium properties of the mean field version of Hamiltonian (1) have been carefully studied in the literature; they are obtained by solving the equation, $\frac{\partial \mathcal{F}^{hom}}{\partial m} = 0$, which gives the state equation,

$$m = \tanh \beta (Jm - \Delta_{eff}), \tag{4}$$

whose resolution leads to specify the conditions of obtaining first-order and gradual spin transition [38,54]. This model is then quite well adapted to mimic the spatio-temporal behavior of SCO materials showing first-order transitions with low volume change. Here, we are mainly interested in the correlation between the dynamics of the HS-LS interface (as well as its shape along the propagation process) and its relation with the global macroscopic crystal's shape. Therefore, to be as realistic as possible, we designed in the simulation a lattice which has a shape resembling that of the current crystal, shown in Figure 3.

### 4.2. Spatio-Temporal Aspects of the HS Fraction

According to the experimental results of OM, summarized in Figure 3, the interface propagates along the diagonal direction of the lattice, with different velocities along the lattice edges (Figure 4). This behavior supports the idea of the existence of strong anisotropic effects in the propagation process. To include this important property, we extended our previous isotropic reaction diffusion model [54], for which we give the main developments below.

Derivation of the Anisotropic Reaction Diffusion Approach

First, we start with the homogeneous mean-field free energy of Equation (2), that we expanded as function of $n_{HS}(\vec{r},t)$ and $\vec{\nabla}n_{HS}$, where $\vec{r}$ is the position and $t$, the time. The density of free energy at position $\vec{r}$ is then written as $f(\vec{r}) = f(n_{HS}, \vec{\nabla}n_{HS})$ for which corresponds the density of the homogeneous free energy $f^{hom}(n_{HS}) = f(n_{HS}, \vec{\nabla}n_{HS} = 0)$ which relates to $\mathcal{F}^{hom}$ through $\mathcal{F}^{hom} = V \times f^{hom}(n_{HS})$, where $V$ denotes the system's volume (here, the surface in 2D). Following a standard procedure, $f(n_{HS}, \vec{\nabla}n_{HS})$ is expanded as a function of successive powers of the gradient of the HS fraction, giving

$$f(n_{HS}, \vec{\nabla}n_{HS}) = f^{hom}(n_{HS}, 0) + \vec{L} \cdot \vec{\nabla}n_{HS} + \frac{1}{2}\vec{\nabla}^T \cdot D_{ij} \cdot \vec{\nabla}n_{HS}, \tag{5}$$

where,

$$\vec{L} = [L_{x_1}, L_{x_2}, L_{x_3}] = \frac{\partial f}{\partial (\partial n_{HS}/\partial x_i)}, \tag{6}$$

is a vector evaluated at zero gradient, and $D_{ij}$ is a tensor property known as the gradient-energy coefficient with components,

$$D_{ij} = \frac{1}{2}\frac{\partial^2 f}{\partial (\partial n_{HS}/\partial x_i)\, \partial (\partial n_{HS}/\partial x_j)}, \tag{7}$$

which should be a symmetric tensor. Since the free energy density does not depend on the direction of the gradients, we take $L = 0$. Furthermore, if the homogeneous system has an inversion center, $\vec{L}$ is automatically zero, and if in addition the system is isotropic or cubic, $D_{ij}$ will be a diagonal tensor with equal components. In Equation (5), $\vec{\nabla}^T$ is the transpose of the gradient operator. So, in the general case of an anisotropic material, the density of free energy is given by,

$$f(n_{HS}, \vec{\nabla}n_{HS}) = f^{hom}(n_{HS}, 0) + \frac{1}{2}\vec{\nabla}^T \cdot D_{ij} \cdot \vec{\nabla}n_{HS}. \tag{8}$$

and the equation of motion of the HS fraction, including the spatial variations, is expressed as,

$$\frac{\partial n_{HS}}{\partial t} = -\frac{\partial f}{\partial n_{HS}}. \tag{9}$$

The local diffusion potential $f(\vec{r})$ at instant $t = t_0$ can be determined from the variation of the rate of the total free energy, $\mathcal{F}$, as follows

$$\mathcal{F}(t_0) = \int_V \left[ f^{hom}(n_{HS}(\vec{r},t_0)) + \vec{L} \cdot \vec{\nabla}n_{HS} + \frac{1}{2}\vec{\nabla}^T n_{HS} D_{ij} \cdot \vec{\nabla}n_{HS} \right] dV, \tag{10}$$

where, we have kept the general expression of $f(\vec{r})$, given in Equation (5). Considering that the order parameter varies with the local velocity, as $n_{HS}(\vec{r},t) = n_{HS}(\vec{r},t_0) + \dot{n}_{HS} \times t$, we can estimate the rate of variation of $\mathcal{F}$ at $t_0$, as follows

$$\left.\frac{d\mathcal{F}}{dt}\right|_{t_0} = \int_V \left[ \frac{\partial f^{hom}(n_{HS}(\vec{r},t_0))}{\partial n_{HS}}\dot{n}_{HS} + \vec{L} \cdot \vec{\nabla}\dot{n}_{HS} + \vec{\nabla}^T \dot{n}_{HS} D_{ij} \cdot \vec{\nabla}n_{HS} \right] dV. \tag{11}$$

Using the general identity

$$\vec{\nabla}^T \left( \dot{p} D_{ij} \vec{\nabla} p \right) = \dot{p}\vec{\nabla}^T D_{ij} \vec{\nabla} p + \vec{\nabla}^T \dot{p} D_{ij} \cdot \vec{\nabla} p, \tag{12}$$

which is combined to the divergence theorem, and applying Newman border conditions which zeros the gradient of the HS fraction at the lattice borders one easily finds the following expression for the rate of variation of $\mathcal{F}$,

$$\left.\frac{d\mathcal{F}}{dt}\right|_{t_0} = \int_V \left[\frac{\partial f^{hom}(n_{HS}(\vec{r},t_0))}{\partial n_{HS}} - \vec{\nabla}^T D_{ij}.\vec{\nabla} n_{HS}\right] \dot{n}_{HS} dV. \tag{13}$$

Since the lattice parameters varies with a quantity $\delta n_{HS} = \dot{n}_{HS} \times \delta t$, the change in the total potential, $\delta \mathcal{F}$ is then the sum of local variations and can be written as,

$$\delta \mathcal{F} = \int_V \left[\frac{\partial f^{hom}(n_{HS}(\vec{r},t_0))}{\partial n_{HS}} - \vec{\nabla}^T D_{ij}.\vec{\nabla} n_{HS}\right] \delta n_{HS}(\vec{r}) dV. \tag{14}$$

The quantity between brackets in Equation (14) is the change of the density of local free energy resulting from the change of HS fraction. The equation of motion is then simply given by

$$\frac{\partial n_{HS}}{\partial t} = -\frac{\delta\left(\frac{d\mathcal{F}}{dV}\right)}{\delta n_{HS}}. \tag{15}$$

### 4.3. The Spatio-Temporal Equation of Motion of the HS Fraction

The general spatio-temporal equation of motion of the HS fraction for an anisotropic diffusion of the spin states, is deduced from Equation (15) as follows,

$$\frac{\partial n_{HS}}{\partial t} = -\Gamma \frac{\partial \mathcal{F}^{hom}}{\partial n_{HS}} + \vec{\nabla}^T D_{ij}.\vec{\nabla} n_{HS}, \tag{16}$$

where $\Gamma$ is a time scale factor. The anisotropy of the diffusion constant may have several origins, among which we quote (i) the anisotropic deformation of the unit cell at the transition between the LS and HS states, as already demonstrated for the SCO compound [{Fe(NCSe)(py)$_2$}$_2$(m-bpypz)] [37], or (ii) the existence of direction-dependent elastic constants acting between the SCO units, as well as (iii) the presence of a crystalline misorientation (a jump in angle across the boundary) which leads to an inclination angle of the boundary plane between the LS and the HS phases. On the other hand, ideally the gradient $\vec{\nabla} n_{HS}$ and the diffusion tensor $D_{ij}$ (given in Equation (7)) should be calculated in the crystalline frame, which is not the same in the intermediate HS-LS and HS states of the title compound.

Because of the change of the unit cell structure at the transition, the direct relation between the tensor $D_{ij}$ and crystalline frame is far from obvious. However, if one considers that the crystal structure remains unchanged, and the misorientation angle at the interface is zero, then the diffusion tensor can be written in the general form as $D_{ij} = R_\theta^T D_{ij}^{diag} R_\theta$, where $D_{ij}^{diag}$ is the diagonal diffusion constant, and $R_\theta$ is the transfer matrix where the superscript "$T$" stands here for the transpose. The matrices $D_{ij}^{diag}$ and $R_\theta$ have the following expressions,

$$D_{ij}^{diag} = \begin{pmatrix} D_1 & 0 \\ 0 & D_2 \end{pmatrix} \quad \text{and} \quad R_\theta = \begin{pmatrix} \cos\theta & \sin\theta \\ -\sin\theta & \cos\theta \end{pmatrix}. \tag{17}$$

By developing $D_{ij}$ as function of $D_1$, $D_2$ and $\theta$, whose expressions are introduced in Equation (16), the following general equation of motion is obtained:

$$\frac{\partial n_{HS}}{\partial t} = -\Gamma_n \frac{\partial \mathcal{F}^{hom}}{\partial n_{HS}} + \left(D_{xx}\frac{\partial^2}{\partial x^2} + D_{yy}\frac{\partial^2}{\partial y^2} + D_{xy}\frac{\partial^2}{\partial x \partial y}\right) n_{HS}, \tag{18}$$

where, the expressions of the three components of the diffusion tensor, connect to $D_1$, $D_2$ and $\theta$, as follows,

$$\begin{aligned} D_{xx} &= D_1 \cos^2\theta + D_2 \sin^2\theta, \\ D_{yy} &= D_1 \sin^2\theta + D_2 \cos^2\theta, \\ D_{xy} &= (D_1 - D_2) \cos\theta \sin\theta. \end{aligned} \quad (19)$$

According to Equation (7), the elements of the diffusion tensor, $D_{ij}$, have the following expressions,

$$\begin{aligned} D_{xx} &= \frac{1}{2} \frac{\partial^2 f}{\partial^2 (\partial n_{HS}/\partial x)}, \\ D_{yy} &= \frac{1}{2} \frac{\partial^2 f}{\partial^2 (\partial n_{HS}/\partial y)} \\ D_{xy} &= \frac{1}{2} \frac{\partial^2 f}{\partial (\partial n_{HS}/\partial x) \partial (\partial n_{HS}/\partial y)}. \end{aligned} \quad (20)$$

Equation (18) expresses the motion of a front interface in the case of an anisotropic diffusion of spin states. In these qualitative simulations, we used $D_1 = 2.0\ \mu\text{m}^2 \cdot \text{s}^{-1}$, $D_2 = 15.0\ \mu\text{m}^2 \cdot \text{s}^{-1}$ and $\theta = 30°$, which were found to well reproduce the experimental shapes of the interface. However, it should be noted that other triplets of solutions may also exist.

It is interesting to discuss briefly some limiting cases. First of all, in the stationary and uniform state, i.e., $\frac{\partial n_{HS}}{\partial t} = 0$ and and $\nabla^2 n_{HS} = 0$, Equation (18) gives the self-consistent mean-field Equation (4), which describes the equilibrium properties of the SCO material. Furthermore, one can easily see that in the case of an isotropic diffusion of the spin states, i.e., $D_1 = D_2 = D$, Equation (18) leads to $D_{xy} = 0$ and $D_{xx} = D_{xy} = D$, which allows to recover the previous equation of motion,

$$\frac{\partial n_{HS}}{\partial t} = -\Gamma_n \frac{\partial \mathcal{F}^{hom}}{\partial n_{HS}} + D \Delta n_{HS}, \quad (21)$$

that we have established in Ref. [54] to describe the front propagation observed in the SCO compound [{Fe(NCSe)(py)$_2$}$_2$(m-bpypz)]. The solutions of Equation (18) have the form (in 1D) $n_{HS}(x,t) = a + b \tanh(\frac{x-vt}{\delta})$ where $v$ is the interface velocity and $\delta$ is its width. Here $v$ is proportional to $\sqrt{D}$, and the propagation direction takes place along the crystal length with a right interface as we demonstrated in a previous report [54], which does no meet the experimental results of Figure 3.

Unfortunately, the analytical resolution of the anisotropic reaction diffusion Equation (18) is out of reach, although one may expect that the existence of several components in the Laplacian of the HS fraction will drive the front propagation along a direction depending on the ratio $\frac{D_2}{D_1}$.

Equation (18) is then solved numerically, after discretization of time and space using finite difference method, where the used space $(dx, dy)$ and time $(dt)$ steps values are, $dx = dy = 0.2\ \mu\text{m}$ and $dt = 0.001$ s.

As clearly shown by Figure 7, the calculated snapshots are in excellent agreement with the experimental results proving the reliability of the anisotropic reaction diffusion model in the description of the spatio-temporal features of the front propagation accompanying the first-order transitions of [Fe(2-pytrz)$_2${Pt(CN)$_4$}]·3H$_2$O SCO single crystal. In particular, it is found that the anisotropy of the spin states diffusion tensor, $D_{ij}$, allows here to control the direction of propagation of the interface, which depends on the angle $\theta$ and the ratio $\frac{D_2}{D_1}$. Furthermore, the shape of the simulated HS/LS interface, its variation along the transformation process and the obtained contact angles with the "crystal" edges are very similar to the experimental ones (see Figure 7).

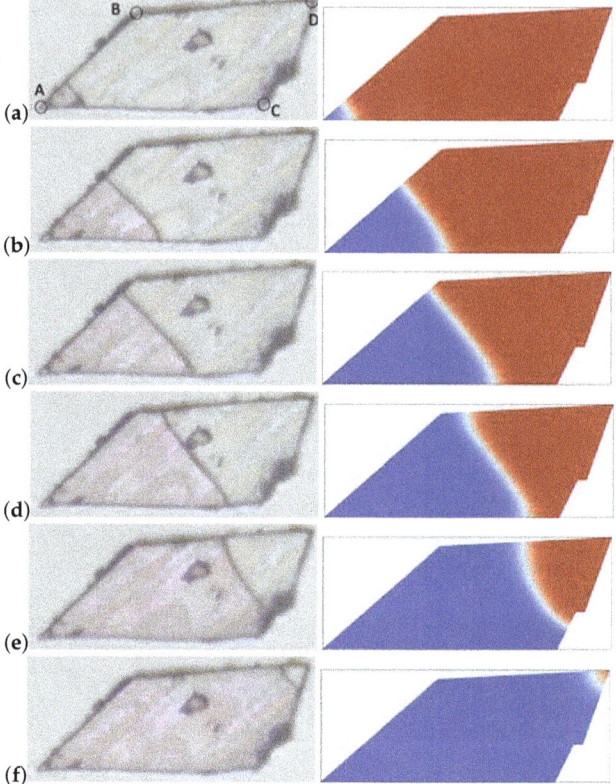

**Figure 7.** (a–f) Left panels present the experimental optical microscopy images of the crystal transformation of the compound [Fe(2-pytrz)$_2${Pt(CN)$_4$}]·3H$_2$O on cooling along the thermal hysteresis of Figure 3b. Right panels present the simulated snapshots obtained from the resolution of the anisotropic Equation (18), showing the spatiotemporal configurations of the lattice along the nucleation and growth of the LS phase (in blue) inside the HS (in red) phase on cooling. The parameter values are $D_1 = 2.0$ μm$^2$·s$^{-1}$, $D_2 = 15.0$ μm$^2$·s$^{-1}$ and $\theta = 30°$ and the temperature is 150 K.

Furthermore, it is an interesting extension to consider the direct connection between the tensor elements $D_1$, $D_2$ and $\theta$ or $D_{xx}$, $D_{xy}$, $D_{yy}$ and the crystallographic data. For that the gradients of HS fraction, should be expressed in the crystallographic basis of the material. Let us denote by $\vec{a}$ and $\vec{b}$ the two vectors of unit cell attached to the material in which a point $M$ has the coordinates $(x', y')$ and by $\vec{i}, \vec{j}$ the orthonormal system attached to the laboratory framework, where the same point has the coordinates $(x, y)$, so that we have $\vec{OM} = x\vec{i} + y\vec{j} = x'\vec{a} + y'\vec{b}$. The transfer matrix between the two basis, denoted $P$, is made of scalar products between the two vectors of the sets of basis. It is easy to demonstrate that gradient operator in the crystal frame is expressed,

$$\vec{\nabla}' = A\frac{\partial}{\partial x}\vec{a} + B\frac{\partial}{\partial y}\vec{b}, \tag{22}$$

where $A = \dfrac{1}{\frac{\partial x'}{\partial x} + \frac{\partial y'}{\partial x}}$ and $B = \dfrac{1}{\frac{\partial x'}{\partial y} + \frac{\partial y'}{\partial y}}$ depend on the scalar products between the crystal and laboratory set of basis defined above.

Since the objective of this part is to re-write the equation of motion (18) of the HS fraction in the crystal frame, it is enough to focus on the diffusion part, i.e., $\vec{\nabla}'.D'.\vec{\nabla}'$, whose developed expression can be written,

$$\vec{\nabla}'.D'_{ij}.\vec{\nabla}' = A^2 a^2 D'_{x'x'} \frac{\partial^2}{\partial x^2} + B^2 b^2 D'_{y'y'} \frac{\partial^2}{\partial y^2} + A\vec{a}.\vec{b}\left(D'_{x'y'} + D'_{y'x'}\right) \frac{\partial^2}{\partial x \partial y} \qquad (23)$$

The comparison between Equations (18) and (23) in which the diffusion parts should be equal, allows to connect the tensor elements of the diffusion tensors $D'$ and $D$, as follows:

$$\begin{aligned} D_{xx} &= A^2 a^2 D'_{x'x'}, \\ D_{yy} &= B^2 b^2 D'_{y'y'}, \\ D_{xy} + D_{yx} &= A \times B \vec{a}.\vec{b} \left(D'_{x'y'} + D'_{y'x'}\right). \end{aligned} \qquad (24)$$

These equations help to better understand how the lattice structure affects the diffusion tensor and so the direction of propagation of the HS/LS interface. These interesting features will be examined in a further work.

## 5. Conclusions

We have presented the spatio-temporal aspects of the spin transition derived from OM experiments, of a new robust single crystal of the compound [Fe(2-pytrz)$_2${Pt(CN)$_4$}]·3H$_2$O showing a clear correlation between the HS/HS-LS interface properties at the thermal transition and the macroscopic crystal shape. In particular, the shape, the length and the velocity of the interface adapts to the change of the macroscopic crystal shape during the propagation process. This interplay between the electro-elastic HS/HS-LS interface and the macroscopic characteristics of the crystal validates our belief that the origin of the elastic interactions in SCO materials are of long-range nature, as predicted in related elastic models [37,39–42]. In addition, the analysis of the total crystal length change derived from OM data enabled to find an excellent agreement with X-rays data [45], thus enhancing the relevance of this technique. Finally, the present experimental spatio-temporal findings are adequately described by a reaction diffusion formalism which allowed to reproduce the main experimental features, such as (i) the diagonal character of the front propagation and (ii) the change of the interface shape according to that of the crystal during the transition. To reproduce these aspects, we extended our previous spatio-temporal mean-field dynamics to include anisotropic effects of the spin state "diffusion", resulting in a more general description, which can be studied for its own. This work highlights the importance of synthesis of cooperative SCO materials showing incomplete transitions, which present less volume change, thus enhancing their resilience at the transition. Experiments on light driven interface control are in progress to explore the dynamical response of the interface to such excitations.

**Author Contributions:** H.F. did the optical microscopy measurements and made the image processing of the data. G.B. helped in the technical developments of the experimental setup of cryogenic optical microscopy. M.P.-E. worked on the experimental modeling of the spatiotemporal data and did the simulations of the reaction diffusion model. S.T., synthesized the sample, did and analyzed the crystallographic and the calorimetric measurements. K.B. supervised the optical microscopy and theoretical works and wrote the paper with the contribution of all authors.

**Funding:** This research was funded by the CNRS, Université de Versailles Saint-Quentin-en-Yvelines, Université Paris Saclay, Université de Brest, and by "Agence Nationale de la Recherche" (ANR project BISTA-MAT: ANR-12-BS07-0030-01).

**Acknowledgments:** The authors are indebted to CNRS and the Université de Versailles St-Quentin, member of the Université Paris-Saclay, for their financial support.

**Conflicts of Interest:** There are no conflicts to declare.

## References

1. Cavallini, M. Status and perspectives in thin films and patterning of spin crossover compounds. *Phys. Chem. Chem. Phys.* **2012**, *14*, 11867–11876. [CrossRef] [PubMed]
2. Shepherd, H.J.; Molnár, G.; Nicolazzi, W.; Salmon, L.; Bousseksou, A. Spin Crossover at the Nanometre Scale. *Eur. J. Inorg. Chem.* **2013**, *2013*, 653–661. [CrossRef]
3. Cavallini, M.; Bergenti, I.; Milita, S.; Kengne, J.C.; Gentili, D.; Ruani, G.; Salitros, I.; Meded, V.; Ruben, M. Thin deposits and patterning of room-temperature-switchable one-dimensional spin-crossover compounds. *Langmuir* **2011**, *27*, 4076–81. [CrossRef] [PubMed]
4. Naik, A.D.; Stappers, L.; Snauwaert, J.; Fransaer, J.; Garcia, Y. A Biomembrane Stencil for Crystal Growth and Soft Lithography of a Thermochromic Molecular Sensor. *Small* **2010**, *6*, 2842–2846. [CrossRef] [PubMed]
5. Basak, S.; Hui, P.; Chandrasekar, R. Flexible and Optically Transparent Polymer Embedded Nano/Micro Scale Spin Crossover Fe(II) Complex Patterns/Arrays. *Chem. Mater.* **2013**, *25*, 3408–3413. [CrossRef]
6. Gütlich, P.; Goodwin, H.A. Spin crossover—An overall perspective. In *Spin Crossover in Transition Metal Compounds I; Topics in Current Chemistry*; Springer: Berlin/Heidelberg, Germany, 2004; Volume 233, pp. 1–47.
7. Linares, J.; Codjovi, E.; Garcia, Y. Pressure and temperature spin crossover sensors with optical detection. *Sensors* **2012**, *12*, 4479–4492. [CrossRef]
8. Bousseksou, A.; Molnár, G.; Salmon, L.; Nicolazzi, W. Molecular spin crossover phenomenon: Recent achievements and prospects. *Chem. Soc. Rev.* **2011**, *40*, 3313–3335. [CrossRef]
9. Gütlich, P.; Gaspar, A.B.; Garcia, Y. Spin state switching in iron coordination compounds. *Beilstein J. Org. Chem.* **2013**, *9*, 342–391. [CrossRef]
10. Cavallini, M.; Melucci, M. Organic Materials for Time–Temperature Integrator Devices. *ACS Appl. Mater. Interfaces* **2015**, *7*, 16897–16906. [CrossRef]
11. Gentili, D.; Demitri, N.; Schäfer, B.; Liscio, F.; Bergenti, I.; Ruani, G.; Ruben, M.; Cavallini, M. Multi-modal sensing in spin crossover compounds. *J. Mater. Chem. C* **2015**, *3*, 7836–7844. [CrossRef]
12. Naik, A.D.; Robeyns, K.; Meunier, C.F.; Léonard, A.F.; Rotaru, A.; Tinant, B.; Filinchuk, Y.; Su, B.L.; Garcia, Y. Selective and Reusable Iron(II)-Based Molecular Sensor for the Vapor-Phase Detection of Alcohols. *Inorg. Chem.* **2014**, *53*, 1263–1265. [CrossRef] [PubMed]
13. Boukheddaden, K.; Ritti, M.H.; Bouchez, G.; Sy, M.; Dîrtu, M.M.; Parlier, M.; Linares, J.; Garcia, Y. Quantitative Contact Pressure Sensor Based on Spin Crossover Mechanism for Civil Security Applications. *J. Phys. Chem. C* **2018**, *122*, 7597–7604. [CrossRef]
14. Shepherd, H.J.; Gural'skiy, I.A.; Quintero, C.M.; Tricard, S.; Salmon, L.; Molnar, G.; Bousseksou, A. Molecular actuators driven by cooperative spin-state switching. *Nat. Commun.* **2013**, *4*, 2607. [CrossRef] [PubMed]
15. Sy, M.; Garrot, D.; Slimani, A.; Paez-Espejo, M.; Varret, F.; Boukheddaden, K. Reversible Control by Light of the High-Spin Low-Spin Elastic Interface inside the Bistable Region of a Robust Spin-Transition Single Crystal. *Angew. Chem. Int. Ed.* **2016**, *55*, 1755–1759. [CrossRef]
16. Paez-Espejo, M.; Sy, M.; Boukheddaden, K. Unprecedented Bistability in Spin-Crossover Solids Based on the Retroaction of the High Spin Low-Spin Interface with the Crystal Bending. *J. Am. Chem. Soc.* **2018**, *140*, 11954–11964. [CrossRef] [PubMed]
17. König, E. Nature and dynamics of the spin-state interconversion in metal complexes. In *Complex Chemistry*; Springer: Berlin/Heidelberg, Germany, 1991; pp. 51–152.
18. Gütlich, P.; Hauser, A.; Spiering, H. Thermal and optical switching of iron(II) complexes. *Angew. Chem. Int. Ed.* **1994**, *33*, 2024–2054. [CrossRef]
19. Köhler, C.P.; Jakobi, R.; Meissner, E.; Wiehl, L.; Spiering, H.; Gütlich, P. Nature of the phase transition in spin crossover compounds. *J. Phys. Chem. Solids* **1990**, *51*, 239–247. [CrossRef]
20. Bousseksou, A.; Negre, N.; Goiran, M.; Salmon, L.; Tuchagues, J.P.; Boillot, M.L.; Boukheddaden, K.; Varret, F. Dynamic triggering of a spin-transition by a pulsed magnetic field. *Eur. Phys. J. B* **2000**, *13*, 451–456.
21. Hauser, A.; Jeftic, J.; Romstedt, H.; Hinek, R.; Spiering, H. Cooperative phenomena and light-induced bistability in iron(II) spin-crossover compounds. *Coord. Chem. Rev.* **1999**, *190*, 471–491. [CrossRef]
22. Kambara, T. The Effect of Iron Concentration on the High-Spin Low-Spin Transitions in Iron Compounds. *J. Phys. Soc. Jpn.* **1980**, *49*, 1806–1811. [CrossRef]

23. Sasaki, N. Theory of cooperative high-spin low-spin transitions in iron (III) compounds induced by the molecular distortions. *J. Chem. Phys.* **1981**, *74*, 3472. [CrossRef]
24. D'Avino, G.; Painelli, A.; Boukheddaden, K. Vibronic model for spin crossover complexes. *Phys. Rev. B* **2011**, *84*. [CrossRef]
25. Klinduhov, N.; Boukheddaden, K. Vibronic Theory of Ultrafast Intersystem Crossing Dynamics in a Single Spin-Crossover Molecule at Finite Temperature beyond the Born-Oppenheimer Approximation. *J. Phys. Chem. Lett.* **2016**, *7*, 722–727. [CrossRef] [PubMed]
26. Castro, M.; Roubeau, O.; Pineiro-Lopez, L.; Real, J.A.; Rodriguez-Velamazan, J.A. Pulsed-Laser Switching in the Bistability Domain of a Cooperative Spin Crossover Compound: A Critical Study through Calorimetry. *J. Phys. Chem. C* **2015**, *119*, 17334–17343. [CrossRef]
27. De Gaetano, Y.; Jeanneau, E.; Verat, A.Y.; Rechignat, L.; Bousseksou, A.; Matouzenko, G.S. Ligand Induced Distortions and Magneto Structural Correlations in a Family of Dinuclear Spin Crossover Compounds with Bipyridyl Like Bridging Ligands. *Eur. J. Inorg. Chem.* **2013**, *2013*, 1015–1023. [CrossRef]
28. Pillet, S.; Hubsch, J.; Lecomte, C. Single crystal diffraction analysis of the thermal spin conversion in [Fe(btr)$_2$(NCS)$_2$](H$_2$O): Evidence for spin-like domain formation. *Eur. Phys. J. B* **2004**, *38*, 541–552. [CrossRef]
29. Gawali-Salunke, S.; Varret, F.; Maurin, I.; Enachescu, C.; Malarova, M.; Boukheddaden, K.; Codjovi, E.; Tokoro, H.; Ohkoshi, S.; Hashimoto, K. Magnetic and Mössbauer Investigation of the Photomagnetic Prussian Blue Analogue Na$_{0.32}$Co[Fe(CN)$_6$]$_{0.74}$3.4H2O: Cooperative Relaxation of the Thermally Quenched State. *J. Phys. Chem. B* **2005**, *109*, 8251–8256. [CrossRef]
30. Mishra, V.; Mukherjee, R.; Linares, J.; Balde, C.; Desplanches, C.; Letard, J.F.; Collet, E.; Toupet, L.; Castro, M.; Varret, F. Temperature-dependent interactions and disorder in the spin-transition compound [Fe$^{II}$(L)$_2$][ClO$_4$]$_2$.C$_7$H$_8$ through structural, calorimetric, magnetic, photomagnetic, and diffuse reflectance investigations. *Inorg. Chem.* **2008**, *47*, 7577–7587. [CrossRef]
31. Varret, F.; Chong, C.; Goujon, A.; Boukheddaden, K. Light-induced phase separation (LIPS) in [Fe(ptz)$_6$](BF$_4$)$_2$ spin-crossover single crystals: Experimental data revisited through optical microscope investigation. *J. Phys. Conf. Ser.* **2009**, *148*, 012036. [CrossRef]
32. Goujon, A.; Varret, F.; Boukheddaden, K.; Chong, C.; Jeftic, J.; Garcia, Y.; Naik, A.D.; Ameline, J.C.; Collet, E. An optical microscope study of photo-switching and relaxation in single crystals of the spin transition solid [Fe(ptz)$_6$](BF$_4$)$_2$, with image processing. *Inorg. Chim. Acta* **2008**, *361*, 4055–4064. [CrossRef]
33. Chong, C.; Mishra, H.; Boukheddaden, K.; Denise, S.; Bouchez, G.; Collet, E.; Ameline, J.C.; Naik, A.D.; Garcia, Y.; Varret, F. Electronic and Structural Aspects of Spin Transitions Observed by Optical Microscopy. The Case of [Fe(ptz)$_6$](BF$_4$)$_2$. *J. Phys. Chem. B* **2010**, *114*, 1975–1984. [CrossRef]
34. Slimani, A.; Varret, F.; Boukheddaden, K.; Chong, C.; Mishra, H.; Haasnoot, J.; Pillet, S. Visualization and quantitative analysis of spatiotemporal behavior in a first-order thermal spin transition: A stress-driven multiscale process. *Phys. Rev. B* **2011**, *84*, 094442. [CrossRef]
35. Varret, F.; Slimani, A.; Boukheddaden, K.; Chong, C.; Mishra, H.; Collet, E.; Haasnoot, J.; Pillet, S. The Propagation of the Thermal Spin Transition of [Fe(btr)$_2$(NCS)$_2$](H$_2$O) Single Crystals, Observed by Optical Microscopy. *New J. Chem.* **2011**, *35*, 2333. [CrossRef]
36. Chong, C.; Slimani, A.; Varret, F.; Boukheddaden, K.; Collet, E.; Ameline, J.C.; Bronisz, R.; Hauser, A. The kinetics features of a thermal spin transition characterized by optical microscopy on the example of [Fe(bbtr)$_3$](ClO$_4$)$_2$ single crystals: Size effect and mechanical instability. *Chem. Phys. Lett.* **2011**, *504*, 29–33. [CrossRef]
37. Sy, M.; Varret, F.; Boukheddaden, K.; Bouchez, G.; Marrot, J.; Kawata, S.; Kaizaki, S. Structure-Driven Orientation of the High-Spin–Low-Spin Interface in a Spin-Crossover Single Crystal. *Angew. Chem.* **2014**, *126*, 7669–7672. [CrossRef]
38. Fourati, H.; Milin, E.; Slimani, A.; Chastanet, G.; Abid, Y.; Triki, S.; Boukheddaden, K. Interplay between a crystal's shape and spatiotemporal dynamics in a spin transition material. *Phys. Chem. Chem. Phys.* **2018**, *20*, 10142–10154. [CrossRef]
39. Nishino, M.; Enachescu, C.; Miyashita, S.; Rikvold, P.A.; Boukheddaden, K.; Varret, F. Macroscopic nucleation phenomena in continuum media with long-range interactions. *Sci. Rep.* **2011**, *1*, 162. [CrossRef]

40. Slimani, A.; Boukheddaden, K.; Varret, F.; Nishino, M.; Miyashita, S. Properties of the low-spin high-spin interface during the relaxation of spin-crossover materials, investigated through an electro-elastic model. *J. Chem. Phys.* **2013**, *139*, 194706. [CrossRef]
41. Slimani, A.; Boukheddaden, K.; Yamashita, K. Effect of intermolecular interactions on the nucleation, growth, and propagation of like-spin domains in spin-crossover materials. *Phys. Rev. B* **2015**, *92*, 014111. [CrossRef]
42. Traiche, R.; Oubouchou, H.; Zergoug, M.; Boukheddaden, K. Spatio-temporal aspects of the domain propagation in a spin-crossover lattice with defect. *Phys. B Condens. Matter* **2017**, *516*, 77–84. [CrossRef]
43. Slimani, A.; Varret, F.; Boukheddaden, K.; Garrot, D.; Oubouchou, H.; Kaizaki, S. Velocity of the high-spin low-spin interface inside the thermal hysteresis loop of a spin-crossover crystal, via photothermal control of the interface motion. *Phys. Rev. Lett.* **2013**, *110*, 087208–087213. [CrossRef] [PubMed]
44. Sy, M.; Traiche, R.; Fourati, H.; Singh, Y.; Varret, F.; Boukheddaden, K. Spatiotemporal Investigations on Light-Driven High-Spin– Low-Spin Interface Dynamics in the Thermal Hysteresis Region of a Spin-Crossover Single Crystal. *J. Phys. Chem. C* **2018**, *122*, 20952–20962. [CrossRef]
45. Milin, E.; Patinec, V.; Triki, S.; Bendeif, E.E.; Pillet, S.; Marchivie, M.; Chastanet, G.; Boukheddaden, K. Elastic Frustration Triggering Photoinduced Hidden Hysteresis and Multistability in a Two-Dimensional Photoswitchable Hofmann-Like Spin-Crossover Metal Organic Framework. *Inorg. Chem.* **2016**, *55*, 11652–11661. [CrossRef] [PubMed]
46. Decurtins, S.; Gütlich, P.; Köhler, C.P.; Spiering, H.; Hauser, A. Light-induced excited spin state trapping in a transition-metal complex: The hexa-1-propyltetrazole-iron (II) tetrafluoroborate spin-crossover system. *Chem. Phys. Lett.* **1984**, *105*, 1–4. [CrossRef]
47. Boukheddaden, K.; Sy, M. Direct Optical Microscopy Observation of Photo-Induced Effects and Thermal Relaxation in a Spin Crossover Single Crystal. *Curr. Inorg. Chem.* **2016**, *6*, 40–48. [CrossRef]
48. Chastanet, G.; Desplanches, C.; Baldé, C.; Rosa, P.; Marchivie, M.; Guionneau, P. A critical review of the T(LIESST) temperature in spin-crossover materials What it is and what it is not. *Chem. Sq.* **2018**, *2*. [CrossRef]
49. Hauser, A. Reversibility of light-induced excited spin state trapping in the $Fe(ptz)_6(BF_4)_2$, and the $Zn_{1-x}Fe_x(ptz)_6(BF_4)_2$ spin-crossover systems. *Chem. Phys. Lett.* **1986**, *124*, 543–548. [CrossRef]
50. Sorai, M.; Seki, S. Phonon coupled cooperative low-spin $^1A_1 \rightleftharpoons$ high-spin $^5T_2$ transition in $[Fe(phen)_2(NCS)_2]$ and $[Fe(phen)_2(NCSe)_2]$ crystals. *J. Phys. Chem. Solids* **1974**, *35*, 555–570. [CrossRef]
51. Varret, F.; Chong, C.; Slimani, A.; Garrot, D.; Garcia, Y.; Naik, A.D. Real-Time Observation of Spin-Transitions by Optical Microscopy. In *Spin-Crossover Materials*; John Wiley & Sons Ltd.: New York, NY, USA, 2013; pp. 425–441.
52. Nishino, M.; Enachescu, C.; Miyashita, S.; Boukheddaden, K.; Varret, F. Intrinsic Effects of the Boundary Condition on Switching Processes in Effective Long-Range Interactions Originating from Local Structural Change. *Phys. Rev. B Condens. Matter Mater. Phys.* **2010**, *82*, 020409. [CrossRef]
53. Boukheddaden, K.; Shteto, I.; Hoo, B.; Varret, F. Dynamical model for spin-crossover solids. I. Relaxation effects in the mean-field approach. *Phys. Rev. B* **2000**, *62*, 14796–14805. [CrossRef]
54. Paez-Espejo, M.; Sy, M.; Varret, F.; Boukheddaden, K. Quantitative macroscopic treatment of the spatiotemporal properties of spin crossover solids based on a reaction diffusion equation. *Phys. Rev. B* **2014**, *89*, 024306. [CrossRef]

© 2019 by the authors. Licensee MDPI, Basel, Switzerland. This article is an open access article distributed under the terms and conditions of the Creative Commons Attribution (CC BY) license (http://creativecommons.org/licenses/by/4.0/).

Article

# Anion Influence on Spin State in Two Novel Fe(III) Compounds: [Fe(5F-sal$_2$333)]X

Sriram Sundaresan [1], Irina A. Kühne [1,2], Conor T. Kelly [1], Andrew Barker [1], Daniel Salley [1], Helge Müller-Bunz [1], Annie K. Powell [2,3] and Grace G. Morgan [1,*]

[1] Centre for Synthesis and Chemical Biology, School of Chemistry and Chemical Biology, University College Dublin (UCD), Belfield, Dublin 4, Ireland; sriram.sundaresan@ucdconnect.ie (S.S.); irina.kuhne@ucd.ie (I.A.K.); conor.kelly@ucdconnect.ie (C.T.K.); andrew.barker@ucdconnect.ie (A.B.); danielsalley1@gmail.com (D.S.); helge.muellerbunz@ucd.ie (H.M.-B.)
[2] Institute for Inorganic Chemistry (AOC), KIT (Karlsruhe Institute of Technology), Engessterstr. 15, 76131 Karlsruhe, Germany; annie.powell@kit.edu
[3] Institute for Nanotechnology (INT), KIT (Karlsruhe Institute of Technology), Hermann-von-Helmholtz-Platz 1, 76344 Eggenstein-Leopoldshafen, Germany
* Correspondence: grace.morgan@ucd.ie; Tel.: +353-1-716-2295

Received: 12 December 2018; Accepted: 22 December 2018; Published: 29 December 2018

**Abstract:** Structural and magnetic data on two iron (III) complexes with a hexadentate Schiff base chelating ligand and Cl$^-$ or BPh$_4{}^-$ counterions are reported. In the solid state, the Cl$^-$ complex [Fe(5F-sal$_2$333)]Cl, **1**, is high spin between 5–300 K while the BPh$_4{}^-$ analogue [Fe(5F-sal$_2$333)]BPh$_4$, **2**, is low spin between 5–250 K, with onset of a gradual and incomplete spin crossover on warming to room temperature. Structural investigation reveals different orientations of the hydrogen atoms on the secondary amine donors in the two salts of the [Fe(5F-sal$_2$333)]$^+$ cation: high spin complex [Fe(5F-sal$_2$333)]Cl, **1**, crystallizes with non-*meso* orientations while the spin crossover complex [Fe(5F-sal$_2$333)]BPh$_4$, **2**, crystallizes with a combination of *meso* and non-*meso* orientations disordered over one crystallographic site. Variable temperature electronic absorption spectroscopy of methanolic solutions of **1** and **2** suggests that both are capable of spin state switching in the solution.

**Keywords:** Fe(III) coordination complexes; hexadentate ligand; Schiff base; spin crossover; UV-Vis spectroscopy; SQUID; EPR spectroscopy

---

## 1. Introduction

Spin crossover complexes (SCO) constitute an interesting class of materials exhibiting interconversion between different electronic states by varying temperature or pressure or by light. Many potential applications have been suggested for their use, including their utilization in data storage, sensors, and display technologies. Fe(II) SCO complexes are more studied in the literature in comparison to Fe(III), and spin crossover is observed only very rarely in Mn(III). We have studied the effect of ligand flexibility on spin state choices in both Fe(III) and Mn(III) using some of the families of hexadentate Schiff base ligands of the type shown in Figure 1 [1–6].

Such chelates are formed by a condensation reaction of linear tetra-amines and substituted salicylaldehydes, and our studies to date have focused mostly on complexes from the "222", "323", and "232" series, where the numbers indicate the number of methylene groups connecting adjacent nitrogen atoms. Here, we show the results of our studies into the effect of the longer chain ligand formed from the "333" polyamine on spin state choices in the resultant Fe(III) complexes. The results of our studies into spin state choices with various metal–ligand combinations, including those reported here, are summarized in Table 1.

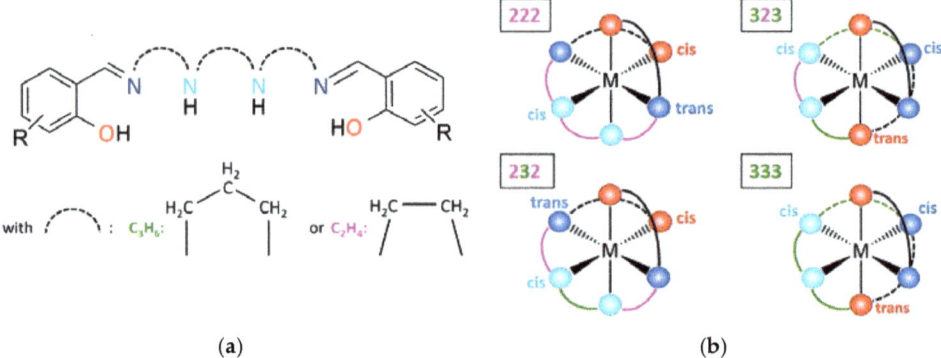

**Figure 1.** (a) Depiction of hexadentate ligands with different chain lengths and (b) resulting coordination geometry around a trivalent 3-D metal ion.

**Table 1.** Donor orientations and spin states for Fe(III) and Mn(III) complexes with selected hexadentate Schiff base ligands. (LS = low spin; HS = high spin; SCO = spin crossover).

| Bond Type | R-Sal$_2$-222 | R-Sal$_2$-232 | R-Sal$_2$-323 | R-Sal$_2$-333 |
|---|---|---|---|---|
| O | cis | cis | trans | trans |
| N$_{amine}$ | cis | cis | cis | cis |
| N$_{imine}$ | trans | trans | cis | cis |
| Mn(III) | HS [3,7] | HS [a] | LS, HS, SCO [1–3,7–13] | HS [14] |
| Fe(III) | LS, HS, SCO [15–24] | LS, HS, SCO [4,24] | LS [24–26] | HS, SCO [24,25] |

[a] G.G. Morgan unpublished results.

In both Fe(III) and Mn(III) complexes, three possible spin arrangements are possible in mononuclear complexes. These comprise the fully spin paired or low spin (LS) arrangement, the fully unpaired or high spin (HS) combination, and a mixture of paired and unpaired, typically termed the intermediate spin (IS) choice. In Fe(III) complexes, these descriptions (LS, IS, and HS) are regularly and accurately used. However, due to the prevalence of the $S = 2$ HS state in Mn(III), any other spin state was historically considered an oddity and the use of "LS" to describe the $S = 1$ state became common usage. In Table 1, "LS" is therefore used to describe the $S = 1$ state as is common in the literature. The true low spin $S = 0$ state has not yet been observed in any manganese (III) complex. Table 1 summarizes the results of our investigations on spin state choices for Fe(III) and Mn(III) in a range of coordination geometries engineered by binding to the type of $N_4(O_2)^-$ chelating ligands depicted in Figure 1. The first point to note is that the $S = 2$ state is dominant for Mn(III) across the four ligand types. However in 2006, our group discovered that the R-Sal$_2$-323 ligand family promoted SCO in Mn(III) [3] and several crystal engineering studies on such complexes followed [1,2,9,11,13]. Although most of these SCO transitions occur below room temperature, some Mn(III) complexes of the R-Sal$_2$-323 ligand family also persist in the $S = 1$ state up to room temperature, and these are defined in Table 1 as LS.

In contrast to manganese, iron (III) shows a range of observed spin states with the four ligand types highlighted in Table 1. It has long been known that the R-Sal$_2$-222 ligand family promotes SCO in iron(III) in addition to stabilizing both HS and LS complexes across a temperature range [15–24]. The Fe(III) R-Sal$_2$-222 complexes have, in recent years, been extensively developed as new switchable materials due to the ease of derivatization which has led to the synthesis of ionic liquids [27], liquid crystals [28], Langmuir-Blodgett film formation with amphiphilic complexes [6,29], and preparation of templated nanowires [21]. We have also observed SCO in Fe(III) complexes from the R-Sal$_2$-232 ligand type [4], but the majority of this class of compound remains HS from room temperature down to 5 K. In contrast, the R-Sal$_2$-323 ligand type has been shown by Reedjik [26] to promote the LS state in iron (III).

Less is known, however, about the iron (III) spin state preferences that would be conferred by coordination to the R-Sal$_2$-333 ligand type. An early work by Ito and co-workers [25] reported the solid-state and methanolic solution-state properties of the nitrate salts of the iron (III) complexes with R-Sal$_2$-323 and R-Sal$_2$-333 ligands, where R was hydrogen [25]. In the solid state, the iron (III) complex with H-Sal$_2$-323 was LS over the measured range, while that with H-Sal$_2$-323 was HS. Ito also used the Evans' NMR technique and variable temperature electronic absorption spectroscopy to monitor the spin state of both complexes in methanol solution. The LS complex with H-Sal$_2$-323 showed no change in solution, i.e., remained LS with only weak thermochromism in methanol. The HS complex with H-Sal$_2$-333, however, demonstrated a strong temperature dependence as shown by both NMR and UV-Vis absorption and a clear isosbestic point is apparent in the electronic absorption spectra recorded between 268–322 K.

At the outset of this work, the nitrate salt of the Fe(III) complex with H-Sal$_2$-333 reported by Ito in 1983 was the only ferric complex with this ligand type in the literature. Given that SCO for this complex was detected in solution, the R-Sal$_2$-333 ligand type was deemed to constitute a good basis for further investigations into the choice of Fe(III) spin state when coordinated to this ligand type. Here, we report two new Fe(III) complexes with 5-Fluoro-Sal$_2$-333 in two crystalline lattices [Fe(5F-sal$_2$333)]Cl, **1**, and [Fe(5F-sal$_2$333)]BPh$_4$, **2**. Both compounds were examined by single crystal diffraction, and an important and new result to emerge from this study was the variation in orientation of the hydrogen atoms on the two secondary amine nitrogen atoms in the 9-carbon length ligand backbone between non-*meso* (Type A) and a disordered combination of *meso* and non-*meso* co-crystallized on the same site (Type B).

To recall the definition of *meso* and non-*meso* in stereochemistry, there is the special case where a molecule exhibits two stereo centres but is achiral, since one conformation shows an intramolecular C$_s$-symmetry which is then defined as the *meso* form. By using a hexadentate Schiff base ligand formed from condensation of a substituted salicylaldehyde and N,N'-bis(3-aminopropyl)-propylenediamine (333) and by forming the iron (III) complex, the binding amine nitrogen atoms, N2 and N3 (shown in turquoise in Figure 2), can have their attached hydrogen atoms either both pointing in the same direction leading to the *meso* form or in opposite directions, leading to the non-*meso* form. The results of our structural studies into the iron (III) complexes **1** and **2** reveal two structural types (Figure 2): (i) pure non-*meso* (Type A) and (ii) a disordered combination of *meso* and non-*meso*, which co-crystallize on the same Fe site (Type B).

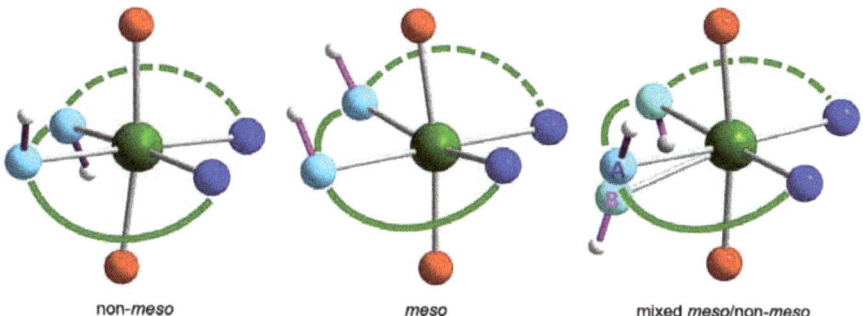

**Figure 2.** Hydrogen atoms at the amine nitrogen of the R-Sal$_2$333 backbone, forming the non-*meso* Type A form (**Left**) and *meso* form (**Middle**) of the ligand. Co-crystallization of both modes (**Right**) was observed in the tetraphenylborate complex **2** and this crystallization mode was termed Type B.

## 2. Materials and Methods

### 2.1. Materials and Instrumentation

All chemicals used were supplied as described: N,N'-Bis-(3-aminopropyl)-1,3-propandiamine (Sigma-Aldrich, 90%), 5-fluorosalicylaldehyde (Fluorochem, 98%), iron (III) chloride hexahydrate (Sigma-Aldrich, 97%), sodium tetraphenylborate (Sigma-Aldrich, $\geq$ 99.5%), acetonitrile (Fisher Scientific, 99.8%), and ethanol (Fisher Scientific, $\geq$ 99.5%). Elemental analysis was recorded on an Exeter Analytical CE-440 CHN analyzer, for Carbon, Hydrogen, and Nitrogen. Infrared spectra were recorded using a Bruker Alpha Platinum attenuated total reflection (ATR) spectrometer. Mass spectrometry was recorded using a Waters 2695 separations module electrospray spectrometer on acetonitrile solutions of **1** and **2**.

### 2.2. Synthesis

The synthesis of compounds **1** and **2** is straightforward, and the complexes were recovered in varying yields. Solid 5-fluorosalicylaldehyde (1.0 mmol) was added to a solution of N,N'-bis(3-aminopropyl)-1,3-propandiamine (0.5 mmol) in 50:50 ethanol/acetonitrile (15 mL), causing a yellow color to form. After stirring for 15 min, solid iron (III) chloride hexahydrate (0.5 mmol) was added whereupon the solution turned dark black. For synthesis of complex **2**, solid sodium tetraphenylborate (0.5 mmol) was then added. For both complexes **1** and **2**, the dark solution was stirred at room temperature for 30 min, then gravity filtered and allowed to stand for slow evaporation for 3–4 days. Dark purple crystals were collected for both compounds **1** (ca. 15%) and **2** (ca. 20%) which were suitable for single crystal X-ray structural analysis.

Complex **1**, [Fe(5F-sal$_2$333)]Cl: Elemental analysis calculated for $C_{23}H_{28}N_4O_2F_2ClFe$. Calculated: C 52.94, H 5.41, and N 10.74. Found: C 52.86, H 5.37, and N 10.72. Mass Spec: 486.33 ES$^+$.

Complex **2**, [Fe(5F-sal$_2$333)]BPh$_4$: Elemental analysis calculated for $C_{47}H_{48}BN_4O_2F_2Fe$. Calculated: C 67.33, H 5.77, and N 6.68. Found: C 67.10, H 5.75, and N 6.57. Mass Spec: 519.24 ES$^+$.

### 2.3. Single-Crystal X-Ray Structure Determinations

X-ray crystallography was carried out on suitable single crystals using an Oxford Supernova diffractometer (Oxford Instruments, Oxford, United Kingdom). Datasets were measured using monochromatic Cu-K$\alpha$ and Mo-K$\alpha$ radiation for **1** and **2** respectively and corrected for absorption. The temperature was controlled with an Oxford Cryosystem instrument. A complete dataset was collected, assuming that the Friedel pairs are not equivalent. An analytical absorption correction based on the shape of the crystal was performed [30]. All structures were solved by dual-space direct methods (SHELXT) [31] and refined by full matrix least-squares on $F^2$ for all data using SHELXL-2016 [31]. The hydrogen atoms attached to nitrogen were located in the difference Fourier map and allowed to refine freely. All other hydrogen atoms were added at calculated positions and refined using a riding model. Their isotropic displacement parameters were fixed to 1.2 times the equivalent one of the parent atom. Anisotropic displacement parameters were used for all non-hydrogen atoms. Crystallographic details for both compounds are summarized in Table A1 (Appendix A) and crystallographic data for the structures reported in this paper have been deposited with the Cambridge Crystallographic Data Centre as supplementary publication numbers CCDC-1884365 (**1**, 100 K), CCDC-1884366 (**2**, 100 K), and CCDC-1884367 (**2**, 293 K).

### 2.4. Magnetic Measurements

The magnetic susceptibility measurements were obtained using a Quantum Design Magnetic Property Measurement System, the MPMS-XL SQUID Magnetometer (Quantum Design, San Diego, CA, USA) operating between 5 and 300 K. Direct current (DC) measurements were performed on a polycrystalline sample of 11.1 mg of complex **1** and of 11.9 mg of complex **2**. Each sample was wrapped in a polyethylene membrane, and susceptibility data were collected at 0.1 T between 5–300 K in cooling and warming mode. The magnetization data was collected at 100 K in order to check for ferromagnetic impurities, which were found to be absent in the samples. Diamagnetic corrections were applied to correct for contribution from the sample holder, and the inherent diamagnetism of the sample was estimated with the use of Pascal's constants.

*2.5. UV-Vis and Electron Paramagnetic Resonance (EPR) Spectroscopy*

UV-Vis solution spectra of [Fe(5F-sal$_2$333)]Cl, **1**, and [Fe(5F-sal$_2$333)]BPh$_4$, **2**, were recorded on an Agilent UV-Vis spectrometer fitted with an Oxford Instruments cryostat insert. Solid state variable temperature EPR spectra were recorded on a Magnettech X-band EPR spectrometer (Freiberg Instruments, Freiberg, Germany) (9.430 GHz) at variable temperatures. A modulation amplitude of 0.7 mT was used in conjunction with a microwave power of 0.1 mW and a gain of 10.

## 3. Results

*3.1. Synthetic Route*

Synthesis of [Fe(5F-sal$_2$333)]Cl, **1**, and [Fe(5F-sal$_2$333)]BPh$_4$, **2**, was achieved in a facile reaction by condensation of N,N'-bis(3-aminopropyl)-1,3-propanediamine (333) with two equivalents of 5-fluorosalicylaldehyde followed by addition of hydrated iron (III) chloride (complex **1**) with further addition of sodium tetraphenylborate in the case of complex **2**, Scheme 1. The filtered reaction mixture on slow evaporation yielded complexes **1** and **2** as dark purple crystals.

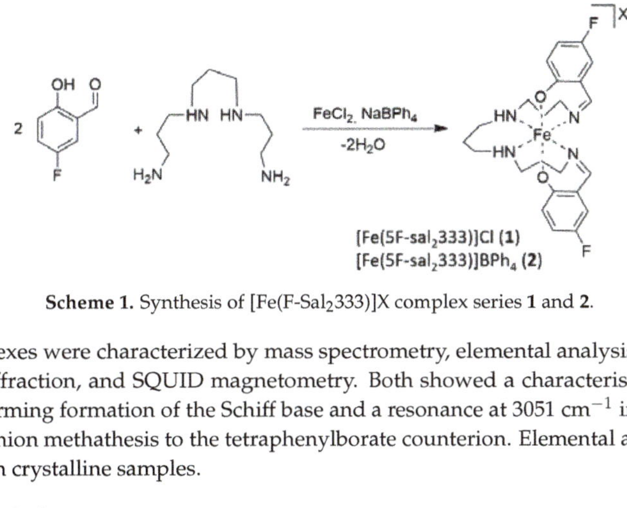

**Scheme 1.** Synthesis of [Fe(F-Sal$_2$333)]X complex series **1** and **2**.

Both complexes were characterized by mass spectrometry, elemental analysis, IR spectroscopy, single crystal diffraction, and SQUID magnetometry. Both showed a characteristic C=N stretch at 1611 cm$^{-1}$, confirming formation of the Schiff base and a resonance at 3051 cm$^{-1}$ in the case of **2** and confirming the anion methathesis to the tetraphenylborate counterion. Elemental analysis confirmed the purity of both crystalline samples.

*3.2. Structural Analysis*

For crystallographic details for complexes **1** and **2**, see Table A1 (Appendix A). Complex **1** crystallized in orthorhombic and non-centrosymmetric space group *Pccn* where the asymmetric unit comprises half of one complex [FeL]$^+$ cation and half of one chloride counterion. The two halves of the complex cation are related by a C2 axis which passes through the central carbon of the middle propylene chain on the ligand backbone. The well-ordered non-*meso* arrangement of the amine on N2 is illustrated in Figure 3.

The chloride counterion within the crystal lattice exhibits short contacts to the hydrogen atoms of the amine nitrogen atoms of the ligand backbone, which leads to the formation of a 1D-chain, Figure 4. Bond length data at 100 K are in line with an S = 5/2 spin state assignment at this temperature as the three bond types, Fe-O$_{phenolate}$ (1.9426(9) Å), Fe-N$_{imine}$ (2.1398(11) Å), and Fe-N$_{amine}$ (2.1837(11) Å) are all typical for the HS state with these types of donor [21].

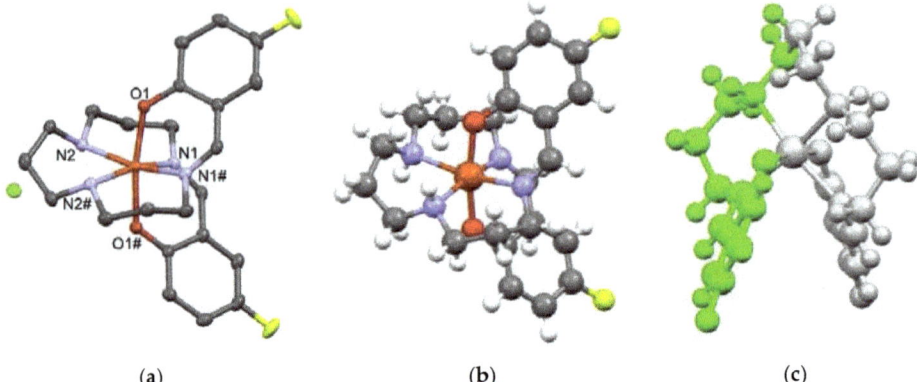

**Figure 3.** (a) View of [Fe(5F-sal$_2$333)]Cl, complex **1**, showing symmetry equivalence of donor atoms related by the central C2 axis, (b) view of the complex cation showing the orientation of hydrogen atoms, including the non-*meso* arrangement of amine hydrogens, and (c) depiction of the symmetry relationship of the two halves of the complex.

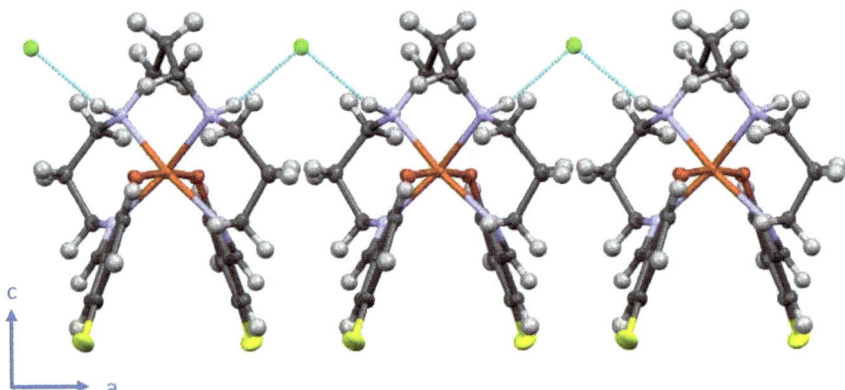

**Figure 4.** View of 1-D hydrogen bonding chain between adjacent complexes, mediated by chloride counterions in [Fe(5F-sal$_2$333)]Cl, **1**.

Complex **2**, [Fe(5F-sal$_2$333)]BPh$_4$, crystallizes in triclinic space group P-1 where the asymmetric unit at 100 K comprises one full occupancy disordered [FeL]$^+$ cation and one full occupancy well-ordered BPh$_4^-$ anion, Figure 5. Both the *meso* and non-*meso* orientations of the amine hydrogens co-crystallize on a single site leading to the disorder of most of the nine propylene carbons and one of the amine nitrogen atoms over two positions. Each component of the disorder was modelled separately within the crystal structure, and the *meso* and non-*meso* orientations of the hydrogen atoms on both amine nitrogen positions co-crystallized on single site are clear, Figure 5.

Structural data for complex **2** was collected initially at 100 K and some months later at 293 K on a different crystal, after analysis of the SQUID data which revealed the change in spin state between the two temperatures. Bond length data, Table 2, indicate a LS state at 100 K with markedly shorter bond lengths for the three bond types, Fe-O$_{phenolate}$ (*ca.* 1.86 Å), Fe-N$_{imine}$ (*ca.* 1.95 Å), and Fe-N$_{amine}$ (with disorder component *ca.* 2–2.10 Å) than those for HS complex **1** at the same temperature. The bond lengths for **1** at 100 K are in line with other LS complexes with comparable donors [21]. At 293 K SQUID data for **2** indicate a small HS fraction, and this is reflected in the small increase in bond lengths compared with those at 100 K: Fe-O$_{phenolate}$ (*ca.* 1.87 Å), Fe-N$_{imine}$ (*ca.* 1.97 Å), and Fe-N$_{amine}$ (with disorder component *ca.* 2–2.15 Å), suggesting that the majority of sites remain LS. The absence of hydrogen bond donors or acceptors on the BPh$_4^-$ counterions and the large distance between the complex cations means no hydrogen bond network emerges to tether the complexes together as was the case with the fully HS analogue, complex **1**. The absence of hydrogen bonding in complex **2** may contribute to the different spin state choices in the two complexes.

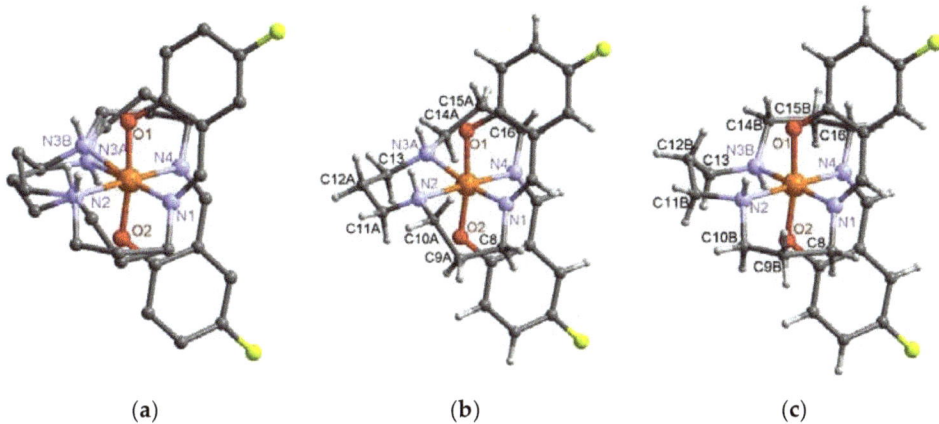

**Figure 5.** View of asymmetric unit of [Fe(5F-sal$_2$333)]BPh$_4$, **2**, showing (**a**) disorder of three propylene groups in ligand backbone due to co-crystallization of *meso* and non-*meso* forms; view of (**b**) the *meso* form and (**c**) the pure non-*meso* orientation.

**Table 2.** Bond length data for complexes **1** and **2**.

|  | [Fe(5F-sal$_2$333)]Cl (**1**) 100 K | Fe-Donor | [Fe(5F-sal$_2$333)]BPh$_4$ (**2**) 100 K | [Fe(5F-sal$_2$333)]BPh$_4$ (**2**) 293 K |
|---|---|---|---|---|
| Fe-O$_{phenolate}$ | 1.9426(9) | Fe-O(2) | 1.8521(14) | 1.8569(15) |
|  |  | Fe-O(1) | 1.8721(15) | 1.8801(16) |
| Fe-N$_{imine}$ | 2.1398(11) | Fe-N(1) | 1.9480(15) | 1.9667(18) |
|  |  | Fe-N(4) | 1.9582(14) | 1.9778(16) |
| Fe-N$_{amine}$ | 2.1837(11) | Fe-N(3A) | 2.014(3) | 2.017(5) |
|  |  | Fe-N(2) | 2.0624(16) | 2.0734(19) |
|  |  | Fe-N(3B) | 2.134(4) | 2.162(7) |

*3.3. Magnetic Characterization*

Magnetic susceptibility of complexes **1** and **2** were recorded on an MPMS-XL magnetometer between 5–300 K in warming and cooling modes. The expected $\chi_M T$ values for S = 5/2 and S = 1/2 are 4.25 and 0.375 cm$^3$ K/mol respectively, and plots of $\chi_M T$ versus *T*, Figure 6, indicate that complex **1** remains HS over the measured temperature range. Complex **2** persists in the predominantly LS state on warming from 5 K to around 250 K above which the $\chi_M T$ value starts to rise, indicating some thermal population of the HS state.

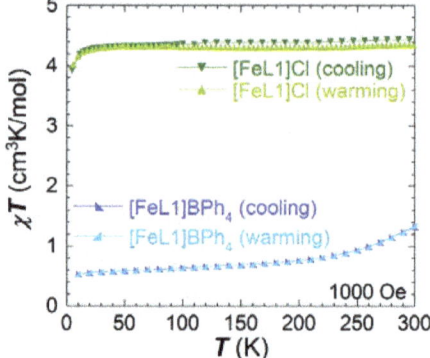

**Figure 6.** Plots of $\chi_M T$ versus $T$ for complexes **1** (green) and **2** (blue) between 5–300 K in cooling and warming mode.

### 3.4. Solid State EPR Spectroscopy

Solid state variable temperature EPR spectra of complexes **1** and **2** were recorded on a Magnettech X-band EPR spectrometer, Figure 7. Complex **1** shows the characteristic broad S = 5/2 with g = 2 over the whole temperature range which fits well with the HS assignment from SQUID magnetometry. The EPR spectra of complex **2**, [Fe(5F-sal$_2$333)]BPh$_4$, are also in line with the SQUID data, showing a gradual thermal SCO in the solid state. A characteristic S = 1/2 signal at g = 2 with differentiation of the $x$, $y$, and $z$ components is apparent at low temperatures for complex **2**, which is diminished on warming to 353 K.

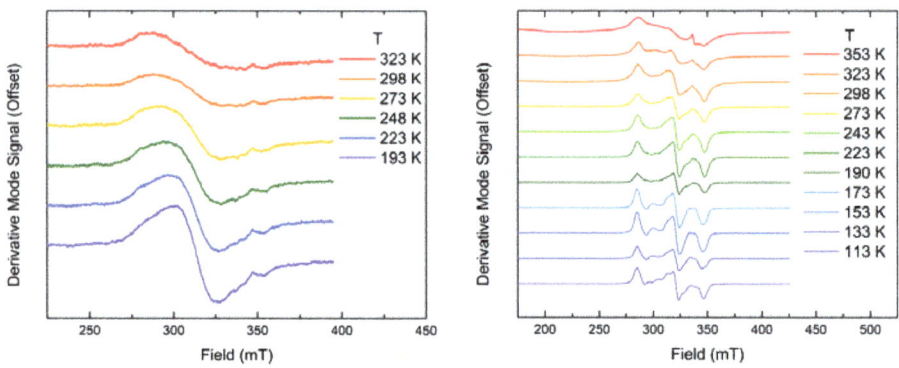

**Figure 7.** Variable temperature EPR spectra of complex **1** (**left**) and **2** (**right**) at variable temperature.

### 3.5. UV-Vis Solution Studies

It was also possible to collect solution state variable temperature electronic absorption for complexes **1** and **2** in methanol, Figure 8, using an Oxford Instruments cryostat insert for a benchtop UV-Vis spectrometer using a $1.0 \times 10^{-4}$ mol/L methanolic solution of **1** and $1.46 \times 10^{-4}$ mol/L methanolic solution of **2**. The spectra suggest that thermal SCO could be achieved in both complexes in this medium despite the fixed HS moment observed between 5–300 K in the crystalline form of complex **1**. The spectra of both complexes show two broad absorptions at around 380 and 620 nm; the latter of which were attributed to charge transfer absorptions rather than d-d transitions. The higher energy band is likely due to ligand only transitions. A strong similarity between the electronic spectra of **1** and **2** in solution is to be expected given that the cation is identical in each. It is also to be expected that the fixed *meso*/non-*meso* differences arise in the solid state due to significant differences in packing between the two variously sized anions. In solution, it is most likely that the complex cation may be in a dynamic exchange between the two forms, and therefore, the spectra of **1** and **2** in methanol should be similar. The higher energy band at 380 nm grows on cooling for both compounds while that at 620 nm narrows on cooling, suggesting population of the LS state which has a narrower vibrational energy well.

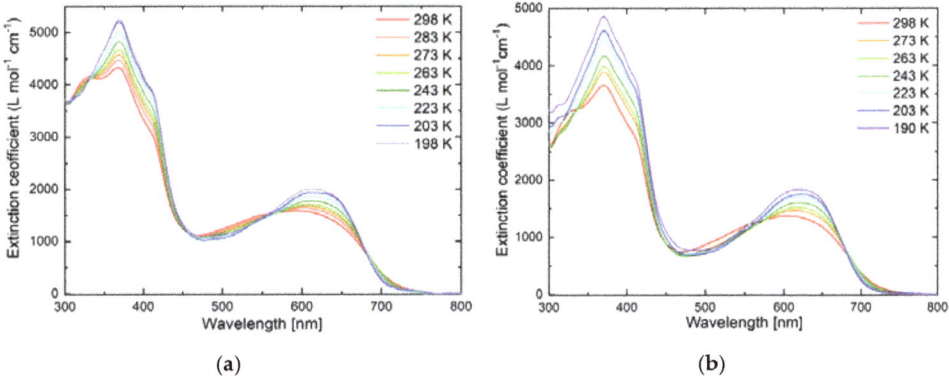

**Figure 8.** Variable temperature UV-Vis characterization of complex **1** (**a**) and **2** (**b**).

## 4. Discussion

In this work, we have investigated the influence of geometry and counterion effects in determining the spin state choices in an iron (III) complex with the 5-Fluoro-sal$_2$333 ligand in a Cl$^-$ or BPh$_4^-$ lattice. In the solid state the [Fe(5F-sal$_2$333)]$^+$ complex adopts the $S = 5/2$ spin state in a chloride lattice and the $S = \frac{1}{2}$ spin state in the tetraphenylborate lattice. The BPh$_4^-$ complex shows onset of a gradual spin crossover on warming from 5 K, but it is still mainly in the LS state by room temperature. An interesting result to emerge from the investigation was the observation that there were two possible orientations (*meso* and non-*meso*) for the hydrogen atoms on the amine nitrogen donors and that a magnetostructural correlation may be present. Complex **1**, which adopted the pure non-*meso* form, showed a preference for the HS state while complex **2**, which crystallized as a mixture of *meso* and non-*meso*, showed preference for the LS state with onset of a gradual SCO only above 250 K. In conclusion, we have established that R-sal$_2$333 ligands can promote SCO in Fe(III) both in the solid state and in solution. Moreover, we have shown that a new type of ligand distortion (*meso*/non-*meso*) exists that may affect the spin state choice and profile of the spin crossover thermal evolution. Future work will include detailed Mössbauer spectroscopy on existing samples reported here and preparation and characterization of further members of the series.

**Author Contributions:** Conceptualization, G.G.M.; Data curation, G.G.M. and S.S.; formal analysis, G.G.M., S.S., I.A.K., C.T.K., A.B., and H.M.-B.; funding acquisition, G.G.M.; investigation, S.S., A.B., D.S., and H.M.-B.; methodology, G.G.M. and S.S.; project administration, G.G.M.; resources, G.G.M. and A.K.P.; supervision, G.G.M.; visualization, S.S. and I.A.K.; writing—original draft, G.G.M. and I.A.K.; writing—review and editing, G.G.M. and I.A.K.

**Funding:** We thank Science Foundation Ireland (SFI) for generous support *via* an Investigator Project Award (12/IP/1703 to G.G.M) and a Walton Fellowship (11/W.1/I1954 to A.K.P). This research was also supported by the Irish Research Council GOIPD/2016/503 fellowship (I.A.K.) and GOIPG/2018/2510 (C.T.K.), EU COST Actions CA15128 Molecular Spintronics (MOLSPIN) and CM1305, Explicit Control over Spin-states in Technology and Biochemistry, (ECOSTBio) and by the Deutsche Forschungsgesellschaft (DFG) through SFB/TRR 88 "3MET" and POF-STN (A.K.P.)

**Conflicts of Interest:** The authors declare no conflict of interest.

## Appendix A

Table A1. Crystallographic details for complexes **1** and **2**.

| Compound | [Fe(5F-sal$_2$333)]Cl (**1**) | [Fe(5F-sal$_2$333)]BPh$_4$ (**2**) | [Fe(5F-sal$_2$333)]BPh$_4$ (**2**) |
|---|---|---|---|
| Empirical formula | C$_{23}$H$_{28}$N$_4$O$_2$F$_2$ClFe | C$_{47}$H$_{48}$BN$_4$O$_2$F$_2$Fe | C$_{47}$H$_{48}$BN$_4$O$_2$F$_2$Fe |
| Formula weight | 521.79 | 805.55 | 805.55 |
| Crystal system | orthorhombic | triclinic | triclinic |
| Space group | *Pccn* | *P*−1 | *P*−1 |
| Crystal size (nm) | 0.195 × 0.114 × 0.033 | 0.3335 × 0.2632 × 0.2017 | 0.405 × 0.350 × 0.337 |
| $a$ (Å) | 7.39810(6) | 10.7385(2) | 10.80774(8) |
| $b$ (Å) | 16.3083(2) | 13.9687(3) | 14.0136(1) |
| $c$ (Å) | 18.6359(2) | 14.7851(2) | 14.9232(2) |
| $\alpha$ (°) | 90 | 102.984(2) | 103.0383(7) |
| $\beta$ (°) | 90 | 94.039(1) | 94.2739(6) |
| $\gamma$ (°) | 90 | 109.415(2) | 107.6673(7) |
| $V$ (Å$^3$) | 2248.43(4) | 2012.64(6) | 2073.03(4) |
| $Z$ | 4 | 2 | 2 |
| $d_{calc}$ (g cm$^{-3}$) | 1.541 | 1.329 | 1.291 |
| $T$ (K) | 100(2) | 100(2) | 293(2) |
| $\mu$ (mm$^{-1}$) | 6.871 | 0.429 | 0.416 |
| F(000) | 1084 | 846 | 846 |
| Limiting indices | h = ± 9, k = ± 20, l = ± 23 | h = ± 13, k = ± 17, l = ± 8 | h = ± 13, k = ± 17, l = ± 18 |
| Reflections collected/unique | 21831/2378 | 35866/8833 | 79115/7853 |
| R(int) | 0.0288 | 0.0315 | 0.0191 |
| Completeness to $\Theta$ (%) | 99.7 | 99.5 | 99.7 |
| Data/restraints/parameters | 2378/0/151 | 8833/0/578 | 7853/0/578 |
| GooF on F$^2$ | 1.047 | 1.041 | 1.079 |
| Final R indices (I > 2$\sigma$ (I)) | R$_1$ = 0.0274, wR$_2$ = 0.0760 | R$_1$ = 0.0414, wR$_2$ = 0.0928 | R$_1$ = 0.0396, wR$_2$ = 0.1068 |
| R indices (all data) | R$_1$ = 0.0292, wR$_2$ = 0.0777 | R$_1$ = 0.0479, wR$_2$ = 0.0963 | R$_1$ = 0.0420, wR$_2$ = 0.1085 |
| Largest diff. peak/hole (e$^-$Å$^{-3}$) | 0.255 and −0.457 | 0.332 and −0.723 | 0.292 and −0.584 |
| CCDC no. | 1884365 | 1884366 | 1884367 |

## References

1. Martinho, P.N.; Gildea, B.; Harris, M.M.; Lemma, T.; Naik, A.D.; Müller-Bunz, H.; Keyes, T.E.; Garcia, Y.; Morgan, G.G. Cooperative Spin Transition in a Mononuclear Manganese(III) Complex. *Angew. Chem. Int. Ed.* **2012**, *51*, 12597–12601. [CrossRef] [PubMed]
2. Gildea, B.; Harris, M.M.; Gavin, L.C.; Murray, C.A.; Ortin, Y.; Müller-Bunz, H.; Harding, C.J.; Lan, Y.; Powell, A.K.; Morgan, G.G. Substituent Effects on Spin State in a Series of Mononuclear Manganese(III) Complexes with Hexadentate Schiff-Base Ligands. *Inorg. Chem.* **2014**, *53*, 6022–6033. [CrossRef] [PubMed]
3. Morgan, G.G.; Murnaghan, K.D.; Müller-Bunz, H.; McKee, V.; Harding, C.J. A Manganese(III) Complex That Exhibits Spin Crossover Triggered by Geometric Tuning. *Angew. Chem. Int. Ed.* **2006**, *45*, 7192–7195. [CrossRef] [PubMed]
4. Griffin, M.; Shakespeare, S.; Shepherd, H.J.; Harding, C.J.; Létard, J.F.; Desplanches, C.; Goeta, A.E.; Howard, J.A.K.; Powell, A.K.; Mereacre, V.; et al. A Symmetry-Breaking Spin-State Transition in Iron(III). *Angew. Chem. Int. Ed.* **2011**, *50*, 896–900. [CrossRef] [PubMed]
5. Murray, C.; Gildea, B.; Müller-Bunz, H.; Harding, C.J.; Morgan, G.G. Co-Crystallisation of Competing Structural Modes in Geometrically Constrained Jahn–Teller Manganese(III) Complexes. *Dalton Trans.* **2012**, *41*, 14487–14489. [CrossRef] [PubMed]
6. Martinho, P.N.; Kühne, I.A.; Gildea, B.; McKerr, G.; O'Hagan, B.; Keyes, T.E.; Lemma, T.; Gandolfi, C.; Albrecht, M.; Morgan, G.G. Self-Assembly Properties of Amphiphilic Iron(III) Spin Crossover Complexes in Water and at the Air–Water Interface. *Magnetochemistry* **2018**, *4*, 49. [CrossRef]
7. Gandolfi, C.; Cotting, T.; Martinho, P.N.; Sereda, O.; Neels, A.; Morgan, G.G.; Albrecht, M. Synthesis and Self-Assembly of Spin-Labile and Redox-Active Manganese(III) Complexes. *Dalton Trans.* **2011**, *40*, 1855–1865. [CrossRef] [PubMed]
8. Wang, S.; Ferbinteanu, M.; Marinescu, C.; Dobrinescu, A.; Ling, Q.-D.; Huang, W. Case Study on a Rare Effect: The Experimental and Theoretical Analysis of a Manganese(III) Spin-Crossover System. *Inorg. Chem.* **2010**, *49*, 9839–9851. [CrossRef]

9. Gildea, B.; Gavin, L.C.; Murray, C.A.; Müller-Bunz, H.; Harding, C.J.; Morgan, G.G. Supramolecular Modulation of Spin Crossover Profile in Manganese(III). *Supramol. Chem.* **2012**, *24*, 641–653. [CrossRef]
10. Wang, S.; Xu, W.-T.; He, W.-R.; Takaishi, S.; Li, Y.-H.; Yamashita, M.; Huang, W. Structural Insights into the Counterion Effects on the Manganese(III) Spin Crossover System with Hexadentate Schiff-base Ligands. *Dalton Trans.* **2016**, *45*, 5676–5688. [CrossRef]
11. Fitzpatrick, A.J.; Trzop, E.; Müller-Bunz, H.; Dîrtu, M.M.; Garcia, Y.; Collet, E.; Morgan, G.G. Electronic vs. Structural Ordering in a Manganese(III) Spin Crossover Complex. *Chem. Commun.* **2015**, *51*, 17540–17543. [CrossRef] [PubMed]
12. Fitzpatrick, A.J.; Stepanovic, S.; Müller-Bunz, H.; Gruden-Pavlović, M.A.; García-Fernández, P.; Morgan, G.G. Challenges in Assignment of Orbital Populations in a High Spin Manganese(III) Complex. *Dalton Trans.* **2016**, *45*, 6702–6708. [CrossRef] [PubMed]
13. Pandurangan, K.; Gildea, B.; Murray, C.; Harding, C.J.; Müller-Bunz, H.; Morgan, G.G. Lattice Effects on the Spin-Crossover Profile of a Mononuclear Manganese(III) Cation. *Chem. Eur. J.* **2012**, *18*, 2021–2029. [CrossRef] [PubMed]
14. Wang, S.; He, W.-R.; Ferbinteanu, M.; Li, Y.-H.; Huang, W. Tetragonally Compressed High-Spin Mn(III) Schiff Base Complex: Synthesis, Crystal Structure, Magnetic Properties and Theoretical Calculations. *Polyhedron* **2013**, *52*, 1199–1205. [CrossRef]
15. Floquet, S.; Carmen Muñoz, M.; Rivière, E.; Clément, R.; Audière, J.-P.; Boillot, M.-L. Structural Effects on the Magnetic Properties of Ferric Complexes in Molecular Materials or a Lamellar CdPS$_3$ Host Matrix. *New J. Chem.* **2004**, *28*, 535–541. [CrossRef]
16. Pritchard, R.; Barrett, S.A.; Kilner, C.A.; Halcrow, M.A. The Influence of Ligand Conformation on the Thermal Spin Transitions in Iron(III) Saltrien Complexes. *Dalton Trans.* **2008**, 3159–3168. [CrossRef] [PubMed]
17. Nishida, Y.; Kino, K.; Kida, S. X-Ray structural study of [Fe(saltrien)]X (X = Br·2H$_2$O, BPh$_4$ or PF$_6$). Origin of Unusual Magnetic Behaviour of the Spin-Crossover Complex [Fe(saltrien)]PF$_6$. *J. Chem. Soc. Dalton Trans.* **1987**, 1957–1961. [CrossRef]
18. Sinn, E.; Sim, G.; Dose, E.V.; Tweedle, M.F.; Wilson, L.J. Iron(III) Chelates with Hexadentate Ligands from Triethylenetetramine and b-diketones or Salicylaldehyde. Spin State Dependent Crystal and Molecular Structures of [Fe(acac)$_2$trien]PF$_6$ (S = 5/2), [Fe(acacCl)$_2$trien]PF$_6$ (S = 5/2), [Fe(sal)$_2$trien]Cl·2H$_2$O (S = 1/2). *J. Am. Chem. Soc.* **1978**, *100*, 3375–3390. [CrossRef]
19. Martinho, P.N.; Harding, C.J.; Müller-Bunz, H.; Albrecht, M.; Morgan, G.G. Inducing Spin Crossover in Amphiphilic Iron(III) Complexes. *Eur. J. Inorg. Chem.* **2010**, *2010*, 675–679. [CrossRef]
20. Gandolfi, C.; Moitzi, C.; Schurtenberger, P.; Morgan, G.G.; Albrecht, M. Improved Cooperativity of Spin-Labile Iron(III) Centers by Self-Assembly in Solution. *J. Am. Chem. Soc.* **2008**, *130*, 14434–14435. [CrossRef] [PubMed]
21. Martinho, P.N.; Lemma, T.; Gildea, B.; Picardi, G.; Müller-Bunz, H.; Forster, R.J.; Keyes, T.E.; Redmond, G.; Morgan, G.G. Template Assembly of Spin Crossover One-Dimensional Nanowires. *Angew. Chem. Int. Ed.* **2012**, *51*, 11995–11999. [CrossRef] [PubMed]
22. Vieira, B.J.C.; Coutinho, J.T.; Santos, I.C.; Pereira, L.C.J.; Waerenborgh, J.C.; da Gama, V. [Fe(nsal$_2$trien)]SCN, a New Two-Step Iron(III) Spin Crossover Compound, with Symmetry Breaking Spin-State Transition and an Intermediate Ordered State. *Inorg. Chem.* **2013**, *52*, 3845–3850. [CrossRef]
23. Nemec, I.; Herchel, R.; Šalitroš, I.; Trávníček, Z.; Moncoľ, J.; Fuess, H.; Ruben, M.; Linert, W. Anion Driven Modulation of Magnetic Intermolecular Interactions and Spin Crossover Properties in an Isomorphous Series of Mononuclear Iron(III) Complexes with a Hexadentate Schiff Base Ligand. *CrystEngComm* **2012**, *14*, 7015–7024. [CrossRef]
24. Hayami, S.; Matoba, T.; Nomiyama, S.; Kojima, T.; Osaki, S.; Maeda, Y. Structures and Magnetic Properties of Some Fe(III) Complexes with Hexadentate Ligands: In Connection with Spin-Crossover Behavior. *Bull. Chem. Soc. Jpn.* **1997**, *70*, 3001–3009. [CrossRef]
25. Ito, T.; Sugimoto, M.; Ito, H.; Toriumi, K.; Nakayama, H.; Mori, W.; Sekizaki, M. A Chelate Ring Size Effect on Spin States of Iron(III) Complexes with Hexadentate Ligands Derived from Salicylaldehyde and 4,8-Diazaundecane-1,11-diamine(3,3,3-tet) or 4,7-Diazadecane-1,10-diamine(3,2,3-tet), and their X-Ray Structures. *Chem. Lett.* **1983**, *12*, 121–124. [CrossRef]

26. Kannappan, R.; Tanase, S.; Mutikainen, I.; Turpeinen, U.; Reedijk, J. Low-Spin Iron(III) Schiff-Base Complexes with Symmetric Hexadentate Ligands: Synthesis, Crystal Structure, Spectroscopic and Magnetic Properties. *Polyhedron* **2006**, *25*, 1646–1654. [CrossRef]
27. Fitzpatrick, A.J.; O'Connor, H.M.; Morgan, G.G. A Room Temperature Spin Crossover Ionic Liquid. *Dalton Trans.* **2015**, *44*, 20839–20842. [CrossRef]
28. Seredyuk, M.; Gaspar, A.B.; Ksenofontov, V.; Galyametdinov, Y.; Verdaguer, M.; Villain, F.; Gütlich, P. Spin-Crossover and Liquid Crystal Properties in 2D cyanide-bridged $Fe^{II}$-$M^{I/II}$ Metalorganic Frameworks. *Inorg. Chem.* **2010**, *49*, 10022–10031. [CrossRef]
29. Soyer, H.; Mingotaud, C.; Boillot, M.-L.; Delhaes, P. Spin Crossover of a Langmuir−Blodgett Film Based on an Amphiphilic Iron(II) Complex. *Langmuir* **1998**, *14*, 5890–5895. [CrossRef]
30. Clark, R.C.; Reid, J.S. The Analytical Calculation of Absorption in Multifaceted Crystals. *Acta Crystallogr. Sect. A* **1995**, *51*, 887–897. [CrossRef]
31. Sheldrick, G.M. Crystal Structure Refinement with SHELXL. *Acta Crystallogr. Sect. C Struct. Chem.* **2015**, *71*, 3–8. [CrossRef] [PubMed]

© 2018 by the authors. Licensee MDPI, Basel, Switzerland. This article is an open access article distributed under the terms and conditions of the Creative Commons Attribution (CC BY) license (http://creativecommons.org/licenses/by/4.0/).

Article

# Effect of Transition Metal Substitution on the Charge-Transfer Phase Transition and Ferromagnetism of Dithiooxalato-Bridged Hetero Metal Complexes, $(n\text{-}C_3H_7)_4N[Fe^{II}_{1-x}Mn^{II}_xFe^{III}(dto)_3]$

Masaya Enomoto [1,*], Hiromichi Ida [1], Atsushi Okazawa [2] and Norimichi Kojima [3]

[1] Department of Chemistry, Faculty of Science Division I, Tokyo University of Science, Tokyo 162-8601, Japan; h.ida1228@gmail.com
[2] Department of Basic Science, Graduate School of Arts and Sciences, The University of Tokyo, Tokyo 153-8902, Japan; cokazawa@mail.ecc.u-tokyo.ac.jp
[3] Toyota Physical and Chemical Research Institute, Aichi 480-1192, Japan; kojima@toyotariken.jp
* Correspondence: menomoto@rs.kagu.tus.ac.jp; Tel.: +81-3-3260-4272 (ext. 5755)

Received: 2 October 2018; Accepted: 21 November 2018; Published: 28 November 2018

**Abstract:** The dithiooxalato-bridged iron mixed-valence complex $(n\text{-}C_3H_7)_4N[Fe^{II}Fe^{III}(dto)_3]$ (dto = dithiooxalato) undergoes a novel charge-transfer phase transition (CTPT) accompanied by electron transfer between adjacent $Fe^{II}$ and $Fe^{III}$ sites. The CTPT influences the ferromagnetic transition temperature according to the change of spin configuration on the iron sites. To reveal the mechanism of the CTPT, we have synthesized the series of metal-substituted complexes $(n\text{-}C_3H_7)_4N[Fe^{II}_{1-x}Mn^{II}_xFe^{III}(dto)_3]$ ($x = 0\text{--}1$) and investigated their physical properties by means of magnetic susceptibility and dielectric constant measurements. With increasing $Mn^{II}$ concentration, $x$, $Mn^{II}$-substituted complexes show the disappearance of CTPT above $x = 0.04$, while the ferromagnetic phase remains in the whole range of $x$. These results are quite different from the physical properties of the $Zn^{II}$-substituted complex, $(n\text{-}C_3H_7)_4N[Fe^{II}_{1-x}Zn^{II}_xFe^{III}(dto)_3]$, which is attributed to the difference of ion radius as well as the spin states of $Mn^{II}$ and $Zn^{II}$.

**Keywords:** charge-transfer phase transition; iron mixed-valence complex; hetero metal complex; dithiooxalato ligand; substitution of 3d transition metal ion; ferromagnetism; dielectric response; [57]Fe Mössbauer spectroscopy

## 1. Introduction

Oxalate dianion (ox) is one of the most efficient building components in molecule-based magnets. Owing to its versatile bridging modes [1–10] as well as its remarkable ability to mediate a strong magnetic interaction between paramagnetic metal ions [11], a large number of ox-based coordination compounds with wide ranges of structures and magnetic properties have been reported [12,13]. Among these compounds, ox-bridged bimetallic complexes $[M^{II}M^{III}(ox)_3]^-$ have been a fascinating target for materials chemistry since the discovery of ferromagnetism in the layered complexes $(n\text{-}C_4H_9)_4N[M^{II}Cr^{III}(ox)_3]$ (M = Cr, Mn, Fe, Co, Ni, Cu) [14]. These complexes are composed of a molecular building block of trioxalato-coordinated metal anion, $[M^{III}(ox)_3]^{3-}$, and a divalent transition metal ion, exhibiting a two-dimensional (2D) sheet or 3D network structure depending on the size, charge and geometry of the counter cation which acts as a template of the formation of the anionic network [13]. In particular, 2D layered complexes accommodate various functional cations as a cation template, obtaining additional functions such as molecular magnetism [15–17], spin-crossover [18–21], photochromism [22], electrical or proton conduction [23–28], dielectricity [29], and nonlinear optics [30–32] to cooperate with the magnetism of the ox-bridged bimetallic layer.

In the ox-bridged bimetallic network, the ox can be replaced by its bis-sulfur analogue, 1,2-dithiooxalate (dto), and dto-bridged anionic layered complexes [M$^{II}$M$^{III}$(dto)$_3$]$^-$ have actually been developed [33,34]. In this family, the iron mixed-valence complex (n-C$_3$H$_7$)$_4$N[Fe$^{II}$Fe$^{III}$(dto)$_3$] indicates a reversible charge-transfer phase transition (CTPT) with thermal hysteresis at around 120 K, which is induced by electron transfer between adjacent Fe$^{II}$ and Fe$^{III}$ sites as shown in Figure 1 [35–37].

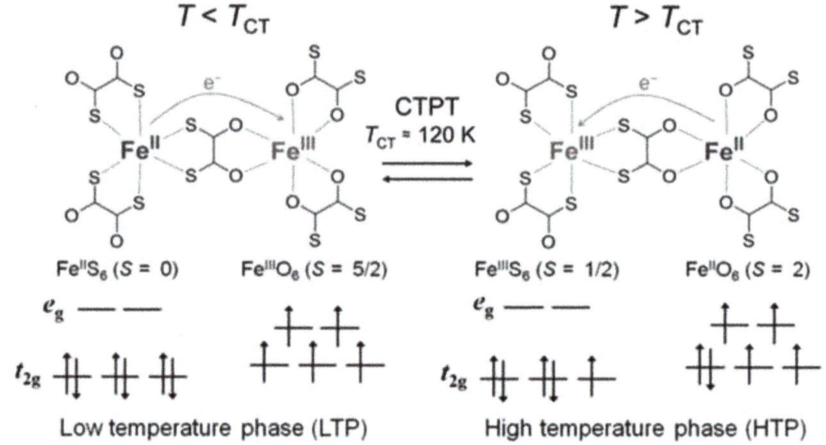

**Figure 1.** Schematic representation of the charge-transfer phase transition (CTPT) in (n-C$_3$H$_7$)$_4$N[Fe$^{II}$Fe$^{III}$(dto)$_3$].

In the high-temperature phase (HTP; $T$ > 120 K), Fe$^{III}$ with a low-spin state ($S$ = 1/2) is coordinated by six S atoms, while Fe$^{II}$ with a high-spin state ($S$ = 2) is coordinated by six O atoms. On the other hand, in the low-temperature phase (LTP; $T$ < 120 K), the FeO$_6$ and FeS$_6$ sites change to high-spin Fe$^{III}$ ($S$ = 5/2) and diamagnetic low-spin Fe$^{II}$ ($S$ = 0) species, respectively, according to the charge transfer. (n-C$_4$H$_9$)$_4$N[Fe$^{II}$Fe$^{III}$(dto)$_3$] also exhibits an incomplete CTPT at around 140 K, which results in the coexistence of the HTP and LTP below the CTPT temperature. Contrary to this, (n-C$_n$H$_{2n+1}$)$_4$N[Fe$^{II}$Fe$^{III}$(dto)$_3$] with n = 5 and 6 do not show the CTPT at ambient pressure, and remain in the spin state corresponding to an HTP down to 2 K [38].

The occurrence or absence of the CTPT in (n-C$_n$H$_{2n+1}$)$_4$N[Fe$^{II}$Fe$^{III}$(dto)$_3$] (n = 3–6) affects the ferromagnetic transition temperature ($T_C$), which strongly correlates with the cation size [38,39]. The ferromagnetic transition for (n-C$_3$H$_7$)$_4$N[Fe$^{II}$Fe$^{III}$(dto)$_3$] occurs at 7 K in the LTP, where the spin state of the Fe$^{II}$ ion is diamagnetic. For (n-C$_n$H$_{2n+1}$)$_4$N[Fe$^{II}$Fe$^{III}$(dto)$_3$] (n = 5–6), the ferromagnetically coupled Fe$^{II}$ ($S$ = 2) and Fe$^{III}$ ($S$ = 1/2) ions contribute a higher $T_C$ of ~20 K, which is thanks to the absence of the CTPT. Furthermore, the coexistence of the LTP and HTP in (n-C$_4$H$_9$)$_4$N[Fe$^{II}$Fe$^{III}$(dto)$_3$] affords respective magnetic ordering at 7 K and 13 K. Thus, the CTPT is a quite important ingredient for the comprehension of magnetic behavior based on the spin states of the metal ions.

Recently, we investigated the magnetic dilution effect on the ferromagnetic transition and the CTPT behavior of (n-C$_3$H$_7$)$_4$N[Fe$^{II}$Fe$^{III}$(dto)$_3$] with the employment of the magnetic diluted system, (n-C$_3$H$_7$)$_4$N[Fe$^{II}_{1-x}$Zn$^{II}_x$Fe$^{III}$(dto)$_3$] ($x$ = 0–1) [40,41]. Judging from the results of magnetic susceptibility and dielectric constant measurements, the CTPT is rapidly suppressed by the substitution of a diamagnetic Zn$^{II}$ for the Fe$^{II}$ ions in the low substituted ratio $x$, and it is absent in $x$ > 0.13. Such a low critical Zn$^{II}$-substituted ratio arises from the high cooperativity of the electron transfer in the CTPT. As a result of the suppression of the CTPT, the $T_C$ was enhanced with the substituted ratio $x$ increased from 0.00 to 0.05. With further increasing of $x$, $T_C$ monotonically decreases and disappears above $x$ = 0.96. The introduction of nonmagnetic Zn$^{II}$ ions into the Fe$^{II}$ sites of the [Fe$^{II}$Fe$^{III}$(dto)$_3$]$^-$ layer causes the disconnection of the ferromagnetic exchange pathway, which causes the decrement of $T_C$.

Thus, nonmagnetic dilution in $(n\text{-}C_3H_7)_4N[Fe^{II}Fe^{III}(dto)_3]$ simultaneously induces both a disconnection of the ferromagnetic exchange pathway and the suppression of the CTPT; therefore, a further experiment should be performed to elucidate metal-ion substitution effects on magnetic properties in $(n\text{-}C_3H_7)_4N[Fe^{II}Fe^{III}(dto)_3]$. Substitution with a paramagnetic metal ion is expected to suppress the CTPT without the disconnection of the ferromagnetic exchange pathway. From this viewpoint, we have synthesized new $Mn^{II}$-substituted complexes $(n\text{-}C_3H_7)_4N[Fe^{II}_{1-x}Mn^{II}_xFe^{III}(dto)_3]$ to investigate the effect of magnetic-ion substitution on the CTPT and the magnetic properties of $(n\text{-}C_3H_7)_4N[Fe^{II}Fe^{III}(dto)_3]$ by means of magnetic and dielectric constant measurements. Furthermore, a magnetic phase diagram of this system is discussed.

## 2. Materials and Methods

### 2.1. Sample Preparation

Potassium dithiooxalate, $K_2(dto)$, was prepared according to reference [42,43]. The precursor $KBa[Fe(dto)_3]\cdot 3H_2O$ was also obtained in accordance with the literature [44]. Commercially available reagents and solvents were used without further purification for raw materials.

$(n\text{-}C_3H_7)_4N[Fe^{II}_{1-x}Mn^{II}_xFe^{III}(dto)_3]$ were prepared by a similar way to the previously reported method [40,41], except for $MnCl_2\cdot 4H_2O$ (KANTO CHAMICAL CO., INC., Tokyo, Japan) being used instead of $ZnCl_2$ (KANTO CHAMICAL CO., INC., Tokyo, Japan). The appropriate amount ($x$ equivalent to the $Fe^{III}$ source) of $MnCl_2\cdot 4H_2O$ was used according to a reduced amount of $FeCl_2\cdot 4H_2O$ (KANTO CHAMICAL CO., INC., Tokyo, Japan) (see Table 1 in the Section 3.1.1).

### 2.2. Characterization

Since the molar fractions of raw materials in the reaction mixture were not directly reflected in the substituted ratio of $Mn^{II}$ for $(n\text{-}C_3H_7)_4N[Fe^{II}_{1-x}Mn^{II}_xFe^{III}(dto)_3]$, the composition of transition-metal ions in $(n\text{-}C_3H_7)_4N[Fe^{II}_{1-x}Mn^{II}_xFe^{III}(dto)_3]$ was determined by energy-dispersive X-ray spectroscopy (EDS; JEOL, EX-37001) in a field-emission scanning electron microscope (SEM; JEOL, JSM-7001F/SHL). Energy spectra of EDS for selected samples are shown in Figure A1. The peaks of Fe $K\alpha$ and Mn $K\alpha$ were used for the determination of the substituted ratio.

The powder X-ray diffraction pattern of all these complexes was measured by Rigaku, RINT2500 using Cu $K\alpha$ radiation at room temperature.

### 2.3. Measurements of Physical Properties

The static magnetic susceptibility was measured by a superconducting quantum interference device (SQUID) susceptometer (Quantum Design Japan, Tokyo, Japan, MPMS-5 or MPMS-XL7AC). The measuring temperature range and static field were set to 2–300 K and 5000 Oe, respectively. The diamagnetic contributions were corrected by the application of Pascal's law. The magnetic moment and the magnetic interaction were estimated by fitting the temperature dependence of the magnetic susceptibility in a high temperature region with the Curie–Weiss law, $\chi = C/(T-\theta)$. $C$ and $\theta$ denote the Curie constant and the Weiss temperature, respectively. The temperature dependence of the zero-field-cooled magnetization (ZFCM) and the field-cooled magnetization (FCM) were measured in the temperature range of 2–30 K under 30 Oe. The remnant magnetization (RM) was measured in the same temperature range under a zero field. The ac magnetic susceptibility measurements were performed in the temperature range of 2–40 K under an ac magnetic field of 3 Oe and frequency range of 10–1000 Hz.

The temperature dependence of dielectric constants was measured by an impedance gain phase analyzer (AMETEK Japan, Tokyo, Japan, Solartron 1260 equipped with a Solartron 1269). The sample was shaped as a pellet and contacted by thin gold wire using the two-probe method. The temperature and the frequency range were selected as 4–300 K and 1 Hz to 1 MHz, respectively.

## 3. Results

### 3.1. Characterization

#### 3.1.1. The Composition of Metallic Ions

The actual ratio of metallic ions was determined by EDS. Table 1 shows the preparation ratio of raw materials and the resultant substitution ratio of $x$ for $(n\text{-}C_3H_7)_4N[Fe^{II}_{1-x}Mn^{II}_xFe^{III}(dto)_3]$. We often found a difference of the molar fractions between the reaction mixtures and resulting powdered samples. The $Mn^{II}$-substitution ratios tend to become lower than the expected value of the corresponding prepared starting materials. The result indicates that the $Mn^{II}$ ion is not efficiently incorporated into the dto layer compared with the $Fe^{II}$ ion, while the $Zn^{II}$-ion uptake into the layer is significantly preferred to the $Fe^{II}$ ion [40,41]. Such a $M^{II}$-substitution tendency reflects the Irving–Williams order ($Mn^{II} < Fe^{II} < Zn^{II}$) of the stability for the ox ligand [45,46].

**Table 1.** The relation between the molar fraction of raw materials in the reaction mixture and resulting values of $x$ for $(n\text{-}C_3H_7)_4N[Fe^{II}_{1-x}Mn^{II}_xFe^{III}(dto)_3]$ determined by EDS. EDS: energy-dispersive X-ray spectroscopy.

| Ratio of raw materials | 0.00 | 0.01 | 0.07 | 0.05 | 0.28 | 0.50 | 0.80 | 1.00 |
|---|---|---|---|---|---|---|---|---|
| Resulting values of $x$ | 0.00 | 0.01 | 0.02 | 0.04 | 0.09 | 0.31 | 0.77 | 1.00 |

#### 3.1.2. Powder X-Ray Analysis

Figure 2 shows the powder X-ray diffraction patterns of $(n\text{-}C_3H_7)_4N[Fe^{II}_{1-x}Mn^{II}_xFe^{III}(dto)_3]$.

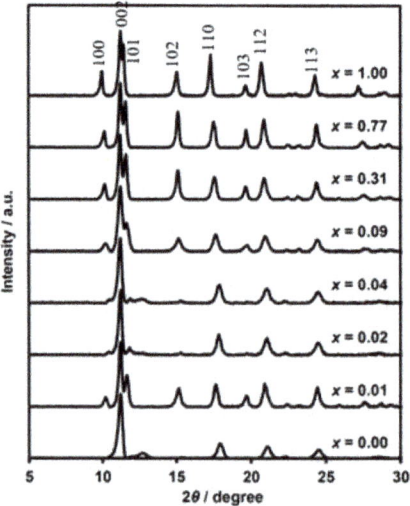

**Figure 2.** Powder X-ray diffraction patterns of $(n\text{-}C_3H_7)_4N[Fe^{II}_{1-x}Mn^{II}_xFe^{III}(dto)_3]$ at 300 K.

The numbers $hkl$ in Figure 2 indicate the indices of Bragg reflections, which are based on the single crystal X-ray diffraction analysis of $(n\text{-}C_3H_7)_4N[Fe^{II}Fe^{III}(dto)_3]$ [37]. Judging from the crystal structure of the parent $(n\text{-}C_3H_7)_4N[Fe^{II}Fe^{III}(dto)_3]$ in the $P6_3$ space group [38], the maximum peaks at around $2\theta$ of $11°$ for all complexes can be assigned to the 002 reflection from interlayer stacking of $[Fe^{II}_{1-x}Mn^{II}_xFe^{III}(dto)_3]^-$ honeycomb sheet along the $c$-axis. The $2\theta$ value of this reflection is almost independent of the substituted ratio $x$, indicating that the layer distance is regulated by the size of the intercalated cation. The smallest $2\theta$ peaks around $10°$ are derived from the 100 reflection correlated with the lattice length along the intralayer direction. The peaks apparently shifted to a lower angle with the increasing of $x$. This finding indicates the elongation of the unit cell length of $a$ (= $b$), which is reasonable considering the order of metal-ion radii ($Mn^{II} > Fe^{II}$). Similarly, the 110 peak reflected in the intralayer direction shows a tendency to shift toward a lower angle with the increasing of $x$.

The reflections of $10l$ are quite weak in some complexes, while the $11l$ reflections remain intense for all measuring complexes. In layered structures, the existence of stacking faults should often be taken into consideration, because the interlayer interaction is weak in general. The difference between hexagonal and cubic close packing is the simple example of periodicity along the stacking layers. In fact, ox-bridged hetero metal complexes show mixed structures between the space groups of $P6_3$ and $R3c$ [47–51]. Indeed, $(n\text{-}C_3H_7)_4N[Mn^{II}Fe^{III}(ox)_3]$ was determined as a biphasic structure of $P6_3$ and $R3c$ with a 20–30% faulting probability of layer stacking, judging from the simulation of the powder X-ray diffraction pattern [52]. A high faulting probability in complexes causes the broadening of some diffraction peaks (e.g., $11l$ indices) related with stacking vectors along the $a + b$ direction.

It should be mentioned that the magnetic properties of the $[Fe^{II}_{1-x}Mn^{II}_xFe^{III}(dto)_3]^-$ layer systematically depend on the change of the substituted ratio $x$ (see below) although the stacking manner is different among the series of $(n\text{-}C_3H_7)_4N[Fe^{II}_{1-x}Mn^{II}_xFe^{III}(dto)_3]$.

*3.2. Physical Properties for $(n\text{-}C_3H_7)_4N[Fe^{II}_{1-x}Mn^{II}_xFe^{III}(dto)_3]$*

3.2.1. Magnetism of $(n\text{-}C_3H_7)_4N[Fe^{II}Fe^{III}(dto)_3]$ and a Series of Nonmagnetic Substituted Complexes

In the case of $(n\text{-}C_3H_7)_4N[Fe^{II}Fe^{III}(dto)_3]$, the characteristic behavior of CTPT in magnetic susceptibility shows the existence of a thermal hysteresis loop, which is induced by the cooperative effect of electron transfer between $Fe^{II}$–$Fe^{III}$ sites. The CTPT also provides a lower $T_C$ for LTP with the spin configuration of $Fe^{II}(S = 0)$-$Fe^{III}(S = 5/2)$ compared with the $T_C$ for HTP with the spin configuration of $Fe^{II}(S = 2)$-$Fe^{III}(S = 1/2)$. The magnetic interaction in the LTP is weaker than that in the HTP because of the diamagnetic nature of $Fe^{II}$ [38].

Moreover, the substitution of diamagnetic $Zn^{II}$ for $Fe^{II}$ indicated the following behavior. (a) The effective magnetic moment ($\mu_{eff}$) decreases with an increased substituted ratio, $x$, in the temperature range of the paramagnetic region. (b) The CTPT is suppressed by the substitution of $Zn^{II}$ for $Fe^{II}$ at a critical substituted ratio between $x = 0.05$ and $0.13$. (c) The $T_C$ is once enhanced in the low substituted ratio in $x < 0.05$ and then decreases with the increasing of $x$. This peculiar substituted ratio dependence of the $T_C$ is explained by the switching between the LTP and HTP with the small amount of substitution with $Zn^{II}$.

Based on the characteristic magnetic properties of $(n\text{-}C_3H_7)_4N[Fe^{II}Fe^{III}(dto)_3]$ and its $Zn^{II}$ substituted system, we have investigated the effect of the substitution of $Mn^{II}$ for $Fe^{II}$ on the CTPT and ferromagnetism.

3.2.2. Magnetism of $(n\text{-}C_3H_7)_4N[Fe^{II}_{1-x}Mn^{II}_xFe^{III}(dto)_3]$

Figure 3a shows the temperature dependence of the molar magnetic susceptibility multiplied by temperature, $\chi T$, in the $Mn^{II}$-substituted system with selected substituted ratios of $x = 0.00, 0.01, 0.02, 0.77$ and $1.00$. The $\chi T$ values at 300 K fall into the range of 4.5–5.1 emu K mol$^{-1}$. Upon cooling below 50 K, the $\chi T$ values for the complexes with $x \leq 0.77$ increase rapidly to reach a maximum and then decrease. This behavior is the signature of the presence of ferromagnetic ordering in these complexes. As for $x = 1.00$, on the other hand, the $\chi T$ value decreases slowly on cooling down to 7 K, then increases toward a small maximum at around 5 K, and then decreases down to 2 K, which is shown in the inset of Figure 3a. As discussed later, the dominant magnetic interaction for $x = 1.00$ is a ferromagnetic one; thus, the small maximum at around 5 K can be attributed to the presence of ferromagnetic ordering. The decrease of $\chi T$ with decreasing temperature, except for the maximum point at around 5 K, presumably arises from the orbital contribution of the $Fe^{III}$ ion. In connection with this, the following should be noted. In the case of the low-spin state ($t_{2g}^5$, $S = 1/2$) of $Fe^{III}$ with a first-order orbital angular momentum in isolated octahedra, $\mu_{\text{eff}}(\text{Fe}^{\text{III}})$ decreases with decreasing temperature [53,54]. In fact, $\mu_{\text{eff}}(\text{Fe}^{\text{III}})$ in KBa[Fe$^{\text{III}}$(dto)$_3$]·3H$_2$O gradually drops with the temperature decreasing below 50 K [55].

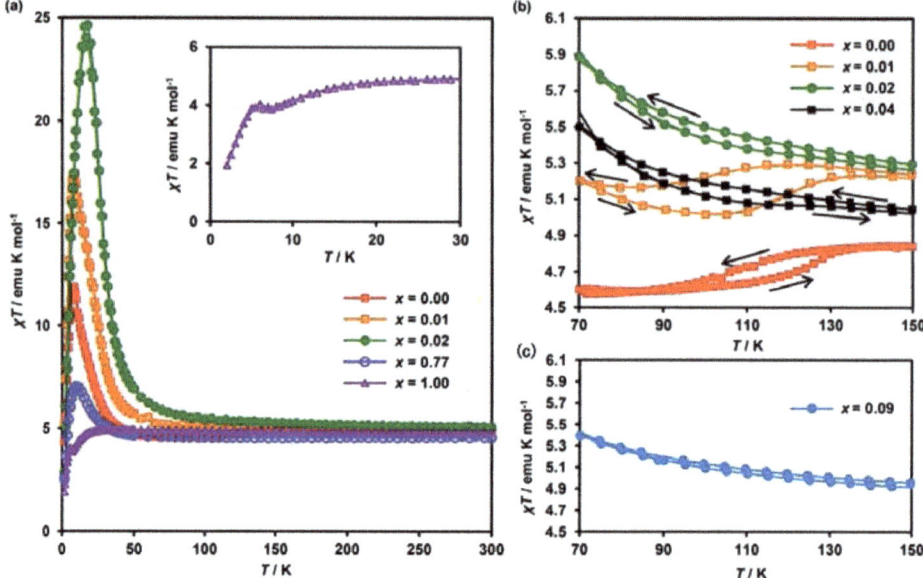

**Figure 3.** Magnetic behavior of $(n\text{-C}_3\text{H}_7)_4\text{N}[\text{Fe}^{\text{II}}_{1-x}\text{Mn}^{\text{II}}_x\text{Fe}^{\text{III}}(\text{dto})_3]$. (**a**) Temperature dependence of the molar susceptibility multiplied by temperature ($\chi T$) for $(n\text{-C}_3\text{H}_7)_4\text{N}[\text{Fe}^{\text{II}}_{1-x}\text{Mn}^{\text{II}}_x\text{Fe}^{\text{III}}(\text{dto})_3]$, (**b**,**c**) are expanded figures for $x$ = 0.00 to 0.04 and 0.09, respectively.

For the complexes with a quite low Mn$^{\text{II}}$ concentration of $0.00 \leq x \leq 0.04$, the $\chi T$ curves exhibit a thermal hysteresis owing to the CTPT at around 110 K (Figure 3b). This thermal hysteresis in $\chi T$ completely disappears for $x \geq 0.09$ (Figure 3c). Furthermore, in the vicinity of the CTPT, a small drop in the $\chi T$ value is observed for $x$ = 0.00 and 0.01, which arises from the difference in the $\mu_{\text{eff}}$ between HTP and LTP. These results suggest that the substitution of Mn$^{\text{II}}$ for Fe$^{\text{II}}$ successively suppresses the CTPT in the low Mn$^{\text{II}}$ concentration region and completely suppresses it for $x \geq 0.09$.

The application of the Curie–Weiss law to these data gives the Curie constant, $C$, and the Weiss temperature, $\theta$, for $(n\text{-C}_3\text{H}_7)_4\text{N}[\text{Fe}^{\text{II}}_{1-x}\text{Mn}^{\text{II}}_x\text{Fe}^{\text{III}}(\text{dto})_3]$. The obtained $C$ value ranges within 4.31–4.86 emu K mol$^{-1}$, which is slightly larger than that expected for the spin-only values of the constituent metal ions, Fe$^{\text{II}}$ ($S$ = 2), Mn$^{\text{II}}$ ($S$ = 5/2) and Fe$^{\text{III}}$ ($S$ = 1/2). This difference can be explained by the anisotropic $g$-value of Fe$^{\text{II}}$ as mentioned in the previous work [38]. Figure 4 shows the Mn$^{\text{II}}$-substituted ratio, with $x$ dependent on the Weiss temperature, $\theta$. Although the $\theta$ value tends to decrease with increasing Mn$^{\text{II}}$ concentration, it remains positive over the whole $x$ range.

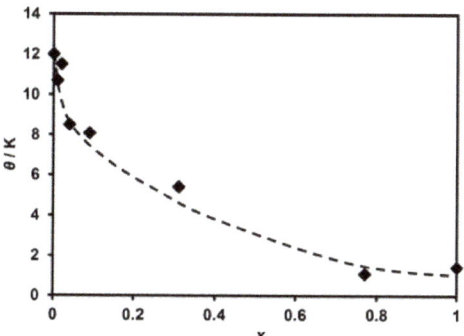

**Figure 4.** Weiss temperature as a function of substituted ratio ($x$) for ($n$-C$_3$H$_7$)$_4$N[Fe$^{II}_{1-x}$Mn$^{II}_x$Fe$^{III}$(dto)$_3$].

In order to elucidate the ferromagnetic phase transition in this system, we investigated the temperature dependences of the FCM, ZFCM and RM. These results are shown in Figure 5. For $x$ = 0.00 (i.e., ($n$-C$_3$H$_7$)$_4$N[Fe$^{II}$Fe$^{III}$(dto)$_3$]), a ferromagnetic phase transition occurred at $T_C$ = 7 K, being estimated from the bifurcation of the FCM and ZFCM curves and the vanishing point of the RM (Figure 5a). For $x$ = 0.01, a slight splitting between FCM and ZFCM curves in addition to non-zero RM was observed below 12 K, while a large deviation of these curves was found below 7 K (Figure 5b). Such a two-step transition behavior in FCM and ZFCM implies the coexistence of two ferromagnetic phases accompanied by an incomplete CTPT. It should be noted that the diamagnetic low-spin state of Fe$^{II}$ in the LTP fragment is responsible for the lower $T_C$ value. A similar two-step transition also progresses in the complexes for $x$ = 0.02 and 0.04 (Figure 5c,d). The $T_C$ values for the HTP fragment in these complexes apparently increased, which are estimated to be 20 and 15 K, respectively. Meanwhile, the lower $T_C$ values derived from the LTP fragment are supposed to be almost invariant. This implies that the Mn$^{II}$ substitution does not affect the magnetism of the LTP. For $x \geq 0.09$, the lower-temperature transition corresponding to the LTP fragment disappears (Figure 5e–h) owing to the absence of the CTPT. The $T_C$ values for these complexes are estimated at 12 K, 10 K, 5 K and 4 K for $x$ = 0.09, 0.31, 0.77 and 1.00, respectively.

To further confirm the coexistence of two ferromagnetic phases and to determine $T_C$, the ac magnetic susceptibility measurements were performed for $x$ = 0.00–0.04. The temperature dependences of the in-phase signal ($\chi'$) and out-of-phase one ($\chi''$) are shown in Figure 6. For $x$ = 0.01, the $\chi'$ peaks were observed as broad maxima at 15 K and 7 K, together with the increased $\chi''$ value foreshowing a maximum or shoulder peak, which corresponds to the development of a ferromagnetic ordered state coming from the HTP component (Figure 6b). Similarly, we can evaluate the $T_C$s of the HTP and LTP for $x$ = 0.02 and 0.04 as 16 K and 7 K, respectively (Figure 6c,d). These data clearly confirm the presence of two ferromagnetic phases for $x$ = 0.01, 0.02 and 0.04. The $T_C$ for the LTP fragment in these complexes are determined to be 7 K, which is identical to the $T_C$ value for $x$ = 0.00 (Figure 6a). Such an independence of the $T_C$ value for the LTP fragment on the Mn$^{II}$-substituted ratio indicates that the vicinity of the dopant Mn$^{II}$ ions is no longer in the LTP spin state because of the suppression of the CTPT, and hence the ferromagnetic phase transition in the LTP component in ($n$-C$_3$H$_7$)$_4$N[Fe$^{II}_{1-x}$Mn$^{II}_x$Fe$^{III}$(dto)$_3$] is unaffected by dopant Mn$^{II}$ ions. In connection with this, it should be noted that similar behavior has already been reported for the magnetic dilution system, ($n$-C$_3$H$_7$)$_4$N[Fe$^{II}_{1-x}$Zn$^{II}_x$Fe$^{III}$(dto)$_3$] [40,41].

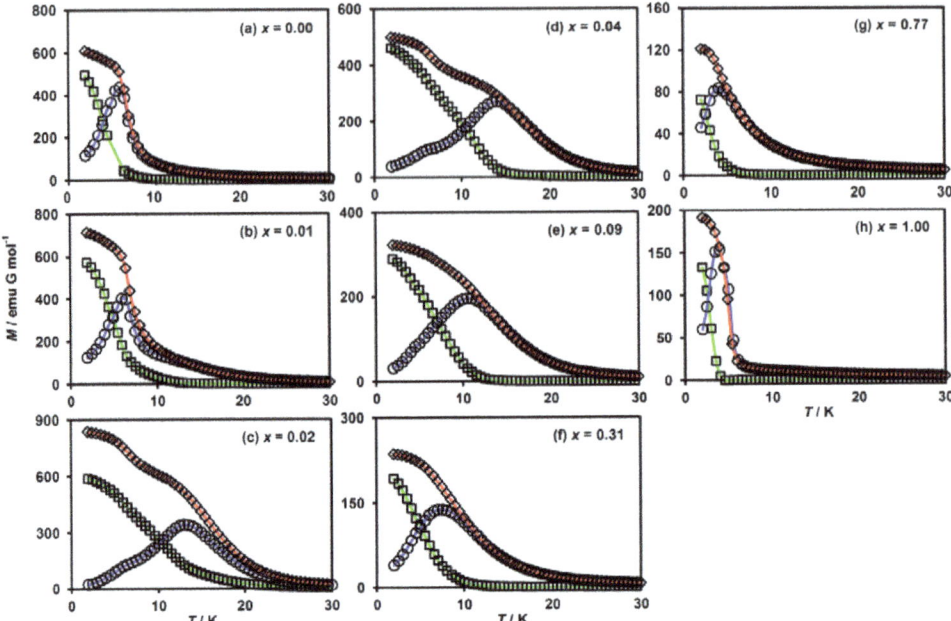

**Figure 5.** Temperature dependence of magnetization for $(n\text{-}C_3H_7)_4N[Fe^{II}_{1-x}Mn^{II}_xFe^{III}(dto)_3]$. FC (rhombus): field-cooled, ZFC (circle): zero-field-cooled and RM (square): remnant magnetization. $H = 30$ Oe. (**a–h**) These figures correspond to the magnetization curves for $x =$ 0.00, 0.01, 0.02, 0.04, 0.09, 0.31, 0.77 and 1.00, respectively.

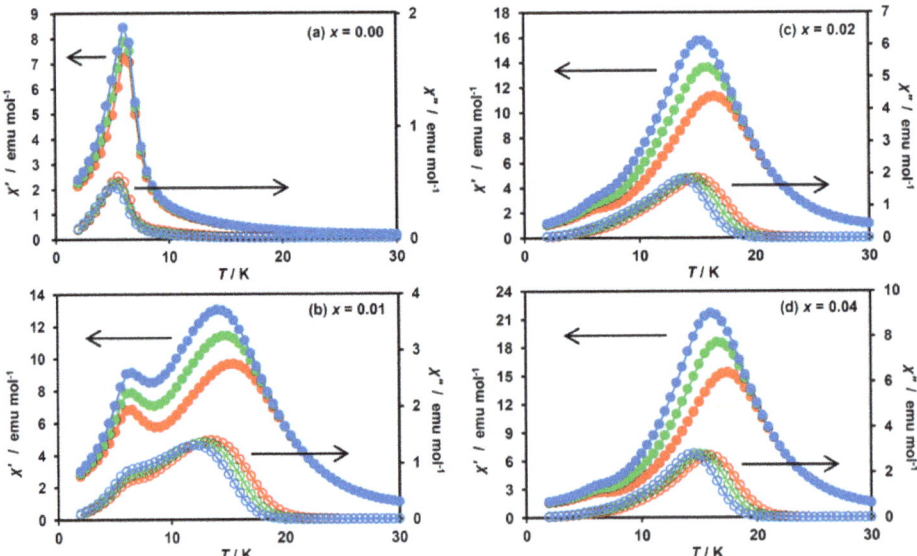

**Figure 6.** Temperature dependences of the in-phase ($\chi'$, filled circle) and out-of-phase ($\chi''$, open circle) ac magnetic susceptibilities for $(n\text{-}C_3H_7)_4N[Fe^{II}_{1-x}Mn^{II}_xFe^{III}(dto)_3]$ at 10 (blue), 100 (green) and 1000 Hz (red). (**a–d**) These figures correspond to the magnetization curves for $x = $ 0.00, 0.01, 0.02 and 0.04, respectively.

### 3.2.3. Dielectric Constant Measurements of $(n\text{-}C_3H_7)_4N[Fe^{II}_{1-x}Mn^{II}_xFe^{III}(dto)_3]$

The dielectric constant measurement is suitable to detect the CTPT in the series of $(n\text{-}C_3H_7)_4N[Fe^{II}_{1-x}Mn^{II}_xFe^{III}(dto)_3]$ as well as the $Zn^{II}$-substituted analogues [40,41]. Figure 7 shows the temperature dependence of the dielectric constants ($\varepsilon'$) for $(n\text{-}C_3H_7)_4N[Fe^{II}_{1-x}Mn^{II}_xFe^{III}(dto)_3]$ measured at 0.1 MHz. For the complexes with a low $Mn^{II}$ concentration of $x \leq 0.04$, an anomalous enhancement of $\varepsilon'$ accompanied by a thermal hysteresis was observed at around 120 K, which corresponds to the occurrence of the CTPT in these complexes (Figure 7a–d). For the complexes with $x \geq 0.09$, such an anomaly is no longer observed (Figure 7e–h). These results confirm that the complexes in the low $Mn^{II}$-substituted region ($0.00 \leq x \leq 0.04$) exhibit CTPT, whereas it is completely suppressed for $x \geq 0.09$. As a result of the dielectric constant and magnetic measurements, we can define the critical substituted ratio for the disappearance of CTPT, estimated at around $0.09 > x > 0.04$.

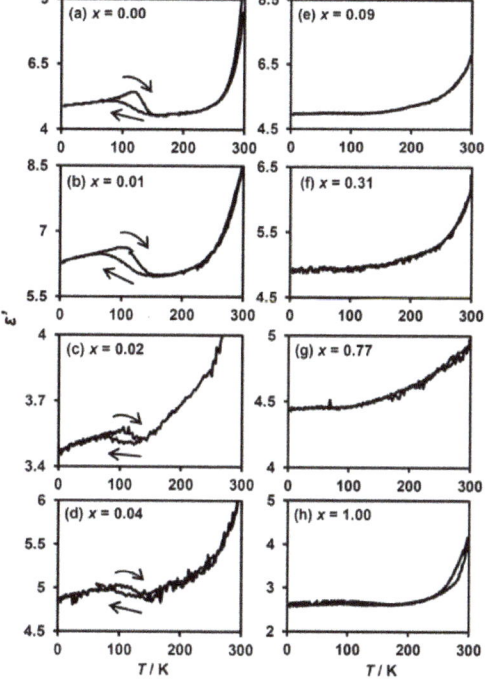

**Figure 7.** Temperature dependence of the dielectric constant, $\varepsilon'$, for $(n\text{-}C_3H_7)_4N[Fe^{II}_{1-x}Mn^{II}_xFe^{III}(dto)_3]$ at an electric field of 0.1 MHz. (**a**–**h**) These figures correspond to the magnetization curves for x = 0.00, 0.01, 0.02, 0.04, 0.09, 0.31, 0.77 and 1.00, respectively.

## 4. Discussion

As shown in Figure 4, the $\theta$ value for $(n\text{-}C_3H_7)_4N[Fe^{II}_{1-x}Mn^{II}_xFe^{III}(dto)_3]$ decreases abruptly in the low $Mn^{II}$ concentration ratio of $x$ and tends to decrease monotonically with increasing $Mn^{II}$ concentration, while $\theta$ remains positive even at $x = 1.00$. This result indicates the ferromagnetic exchange coupling of the nearest neighbor $Mn^{II}$–$Fe^{III}$ pair as well as the $Fe^{II}$–$Fe^{III}$ one.

Figure 8 shows the schematic mechanism of ferromagnetic ordering for the LTP and HTP of $(n\text{-}C_3H_7)_4N[Fe^{II}Fe^{III}(dto)_3]$ and $(n\text{-}C_3H_7)_4N[Mn^{II}Fe^{III}(dto)_3]$. As mentioned in the previous literature [38,56], the LTP contains the low-spin state of $Fe^{II}$ ($t_{2g}^6$: S = 0) and the high-spin state of $Fe^{III}$ ($t_{2g}^4 e_g^2$: S = 5/2), respectively, where the super-exchange interaction via the continuous bridging structure of $Fe^{III}$–dto–$Fe^{II}$–dto–$Fe^{III}$ is presumably very small and antiferromagnetic if 3D electrons are localized. In the case of the LTP of $(n\text{-}C_3H_7)_4N[Fe^{II}Fe^{III}(dto)_3]$, the charge transfer interaction between the $Fe^{II}$ and $Fe^{III}$ sites gives a perturbation to the wave function of the ground state; therefore, the wave function can be described as $\Psi = \sqrt{1-\alpha^2}\{\varphi_i(Fe^{II}(t_{2g}^6))\varphi_j(Fe^{III}(t_{2g}^3 e_g^2))\} + \alpha\{\varphi_i(Fe^{III}(t_{2g}^5))\varphi_j(Fe^{II}(t_{2g}^4 e_g^2))\}$, where $\alpha$ denotes the normalization coefficient for the component of charge transfer interaction. Each $Fe^{III}$ site in LTP accepts a $t_{2g}$ electron with down spin, because both of the $t_{2g}$ and $e_g$ orbitals in the $Fe^{III}$ site are half occupied. Therefore, the perturbed term of $\varphi_i(Fe^{III}(t_{2g}^5))\varphi_j(Fe^{II}(t_{2g}^4 e_g^2))$ achieves the ferromagnetic coupling between the spin configuration of $Fe^{II}$ and $Fe^{III}$. Consequently, the coupling between the ground configuration of $\varphi_i(Fe^{II}(t_{2g}^6))\varphi_j(Fe^{III}(t_{2g}^3 e_g^2))$ and the forward charge transfer configuration of $\varphi_i(Fe^{III}(t_{2g}^5))\varphi_j(Fe^{II}(t_{2g}^4 e_g^2))$ stabilizes the ground state and therefore favors the ferromagnetic ordering. In this way, the charge transfer between the $Fe^{II}$ ($S = 0$) and $Fe^{III}$ ($S = 5/2$) sites induces the ferromagnetic ordering in the LTP.

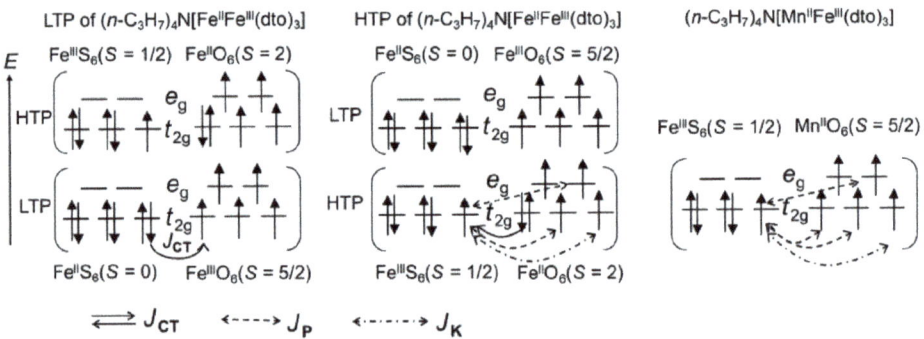

**Figure 8.** Origin of the ferromagnetic ordering for the LTP and HTP of $(n\text{-}C_3H_7)_4N[Fe^{II}Fe^{III}(dto)_3]$ and $(n\text{-}C_3H_7)_4N[Mn^{II}Fe^{III}(dto)_3]$. $J_{CT}$, $J_P$, and $J_K$ indicate the charge transfer interaction, potential exchange interaction, and kinetic exchange interaction, respectively.

In the case of the HTP fragment of $(n\text{-}C_3H_7)_4N[Fe^{II}Fe^{III}(dto)_3]$, each $Fe^{III}$ site with a low-spin state $(t_{2g}^5: S = 1/2)$ accepts a $t_{2g}$ electron with down spin from the adjacent $Fe^{II}$ site with a high-spin state $(t_{2g}^4 e_g^2: S = 2)$. Therefore, the hybridization between the ground state of $\varphi_i(Fe^{III}(t_{2g}^5))\varphi_j(Fe^{II}(t_{2g}^4 e_g^2))$ and the forward charge transfer state of $\varphi_i(Fe^{II}(t_{2g}^6))\varphi_j(Fe^{III}(t_{2g}^3 e_g^2))$ stabilizes the ground state, which favors the ferromagnetic ordering. Furthermore, in addition to the charge transfer interaction ($J_{CT}$), there are also three potential exchange interactions ($J_P$) causing ferromagnetic interaction due to the orbital orthogonality and one kinetic exchange interaction ($J_K$) causing antiferromagnetic interaction due to the orbital overlap between the adjacent $Fe^{III}S_6$ and $Fe^{II}O_6$ sites. The sum of $J_{CT}$ and $J_P$ is considered to be stronger than $J_K$, which is responsible for the ferromagnetic ordering with higher $T_C$ for the HTP fragment of $(n\text{-}C_3H_7)_4N[Fe^{II}Fe^{III}(dto)_3]$.

In the case of $(n\text{-}C_3H_7)_4N[Mn^{II}Fe^{III}(dto)_3]$, there are four $J_P$s and one $J_K$. The sum of the potential exchange interaction is considered to be stronger than the kinetic exchange interaction, which is responsible for the ferromagnetic ordering. Actually, the ferromagnetic ordering of $(n\text{-}C_3H_7)_4N[Mn^{II}Fe^{III}(dto)_3]$ has already been reported by Carling et al. [57], in which both $T_C$ and $\theta$ were estimated at 10 K from the analysis of $1/\chi$ as a function of temperature.

$T_C$ as a function of $x$ for $(n\text{-}C_3H_7)_4[Fe^{II}_{1-x}Mn^{II}_xFe^{III}(dto)_3]$ is shown in Figure 9. An enhancement of $T_C$ from 7 K to 20 K ($x = 0.00$ to $0.02$) is ascribed to the appearance of the HTP fragment exhibiting the higher $T_C$. As with the case of $(n\text{-}C_3H_7)_4[Fe^{II}_{1-x}Zn^{II}_xFe^{III}(dto)_3]$ [40,41], the LTP fragment is unaffected by dopant $Mn^{II}$ ions, and thus the LTP fragment in all these complexes possesses the same $T_C$ value of 7 K. With the further increasing of $x$ above 0.04, $T_C$ decreases monotonically, corresponding to the lowering of the ferromagnetic interaction, to reach a minimum value of 4 K for $x = 1.00$.

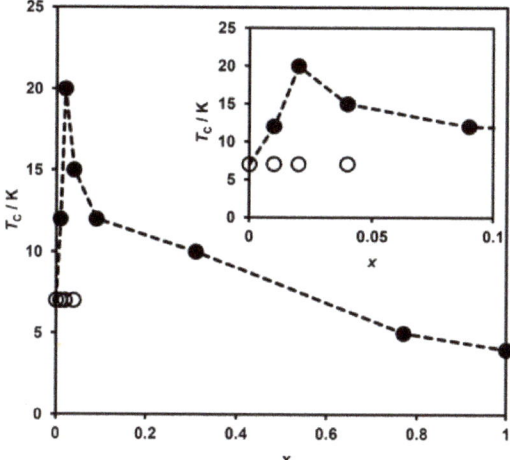

**Figure 9.** Variation of the ferromagnetic transition temperature ($T_C$) with $x$ for ($n$-C$_3$H$_7$)$_4$N[Fe$^{II}_{1-x}$Mn$^{II}_x$Fe$^{III}$(dto)$_3$] in (●) the high-temperature phase and (○) low-temperature phase. Inset shows the extended figure in the low Mn$^{II}$ concentration region. The dashed lines are a guide for the eyes.

The detection of the CTPT can be achieved by both the magnetic susceptibility and dielectric constant measurements. From the results of magnetic susceptibility and dielectric constant measurements, the phase diagram for ($n$-C$_3$H$_7$)$_4$N[Fe$^{II}_{1-x}$Mn$^{II}_x$Fe$^{III}$(dto)$_3$] is determined as shown in Figure 10. The legends in this diagram were assigned as follows. $T_C$(HTP or LTP): the ferromagnetic transition temperature for the HTP or LTP, determined by the FCM, ZFCM, and RM measurements; $T_↑$ or $T_↓$(CT): the upper or lower limit of the thermal hysteresis in the dielectric constant measurement; P$_{HTP}$ or F$_{HTP}$: paramagnetic or ferromagnetic phase with the HTP spin configuration; P$_{mix}$ or F$_{mix}$: paramagnetic or ferromagnetic phase with a mixed state of the HTP and LTP spin configuration, respectively.

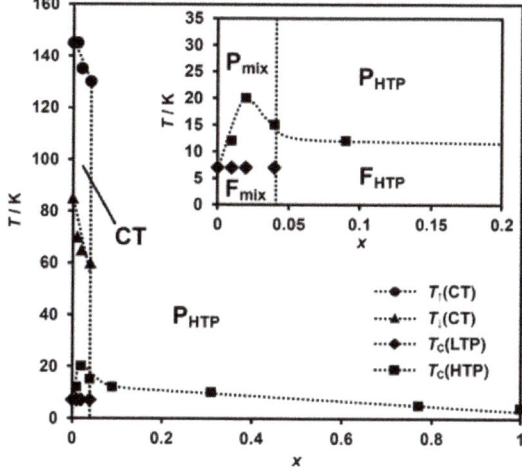

**Figure 10.** Phase diagram of ($n$-C$_3$H$_7$)$_4$N[Fe$^{II}_{1-x}$Mn$^{II}_x$Fe$^{III}$(dto)$_3$].

As shown in Figure 10, the paramagnetic phase corresponding to the HTP ($P_{HTP}$) appears within the whole $Mn^{II}$ concentration region at room temperature. The $Mn^{II}$-substituted complexes with the low $x$ values of $0.00 \leq x \leq 0.04$ exhibit the CT phase in the temperature range between $T_\uparrow$(CT) and $T_\downarrow$(CT). In this region, the CTPT raises the dynamic oscillation between $Fe^{II}$ and $Fe^{III}$, which is found by μSR measurement [58,59]. Below $T_\downarrow$(CT), the paramagnetic phase for the complexes with $0.00 < x \leq 0.04$, which is denoted as $P_{mix}$ in Figure 10, becomes the mixing state between the LTP and HTP as a result of the partial suppression of the CTPT. The complexes in this $x$ region undergo a ferromagnetic phase transition within the HTP domain at $T_C$(HTP), below which the ferromagnetic phase with a spin state of a mixture of the LTP and HTP ($F_{mix}$) appears as already mentioned in Section 3.2.2. The complexes with further high $x$ region (i.e. $x > 0.04$) stay in the $P_{HTP}$ phase down to $T_C$(HTP), reflecting the complete suppression of the CTPT, and undergo a ferromagnetic phase transition within the HTP spin state at this temperature. Below $T_C$(HTP), the complexes in this $x$ region are in the ferromagnetic phase with the spin state of the HTP ($F_{HTP}$).

Although the phase diagram of the magnetically substituted complexes $(n\text{-}C_3H_7)_4N[Fe^{II}_{1-x}Mn^{II}_xFe^{III}(dto)_3]$ is essentially similar to that for $(n\text{-}C_3H_7)_4N[Fe^{II}_{1-x}Zn^{II}_xFe^{III}(dto)_3]$, the major difference between these two systems is the ferromagnetic ordering in the high dopant concentration region; i.e., the ferromagnetic phase appears in the whole $Mn^{II}$ concentration range in the present case, while it was not found for $(n\text{-}C_3H_7)_4N[Fe^{II}_{1-x}Zn^{II}_xFe^{III}(dto)_3]$ [40,41]. The disappearance of the ferromagnetic phase in the latter case is a result of the disconnection of the ferromagnetic exchange pathway by $Zn^{II}$-substitution, since there is no magnetic interaction between $Fe^{III}$ and nonmagnetic $Zn^{II}$. In contrast to this, on the basis of the analysis of the magnetic data, the ferromagnetic interaction between $Mn^{II}$ and $Fe^{III}$ through the dto bridge in $(n\text{-}C_3H_7)_4N[Fe^{II}_{1-x}Mn^{II}_xFe^{III}(dto)_3]$ is operating, though weaker than that for $Fe^{II}$–$Fe^{III}$, and hence the ferromagnetic exchange pathway is maintained over the whole substitution range. This feature is responsible for the existence of the ferromagnetic phase across the whole substitution range in the phase diagram of this system.

Moreover, the critical substituted ratio for the disappearance of the CTPT is unexpectedly low compared with that for the series of $Zn^{II}$-substituted complexes, whose CTPT is completely suppressed in the substituted ratio between $x = 0.13$ and $0.26$ [40,41]. Considering the ion radii between $Mn^{II}$ and $Fe^{II}$, the substitution of $Mn^{II}$ for $Fe^{II}$ tends to expand the honeycomb structure in the magnetic layer of $[Fe^{II}Fe^{III}(dto)_3]^-$. It causes the suppression of HTP in the same manner for the cation-extended complex, $(n\text{-}C_5H_{11})_4N[Fe^{II}Fe^{III}(dto)_3]$ [38]. A cooperative effect of the CTPT in the low dimensional system has been effectively terminated by the substitution of $Mn^{II}$ with remaining the ferromagnetic interaction.

## 5. Conclusions

We investigated the effect of metal substitution on the CTPT and the ferromagnetic phase transition for $(n\text{-}C_3H_7)_4N[Fe^{II}_{1-x}Mn^{II}_xFe^{III}(dto)_3]$ ($x = 0–1$). The existence of the CTPT strongly depends on the $Mn^{II}$-substituted ratio of $x$. The series of $Mn^{II}$-substituted complexes consist of the structures combined in the space group of $P6_3$ and $R3c$ because of the stacking fault between adjacent magnetic layers of $[Fe^{II}_{1-x}Mn^{II}_xFe^{III}(dto)_3]^-$. However, since the magnetic behavior is mainly governed by the intralayer magnetic structure, we can discuss the substituted ratio dependence of their physical properties.

From the results of the magnetic and dielectric measurements, the substitution of $Mn^{II}$ suppressed the CTPT, leading to the disappearance of CTPT above $x = 0.04$. The finding indicates that $Mn^{II}$ substitution is more effective at diminishing the CTPT compared with $Zn^{II}$ substitution $(n\text{-}C_3H_7)_4N[Fe^{II}_{1-x}Zn^{II}_xFe^{III}(dto)_3]$ (critical substituted ratio: $0.13 < x < 0.26$) due to the large $Mn^{II}$-ion radius in addition to the high cooperativity of the charge transfer phenomenon, as discussed in the case of $Zn^{II}$-substituted complexes.

In contrast to such a substitution effect on the CTPT, the ferromagnetic phase was observed in the whole range of $x$ for $(n\text{-}C_3H_7)_4N[Fe^{II}_{1-x}Mn^{II}_xFe^{III}(dto)_3]$, while it disappears above $x = 0.83$ for the $Zn^{II}$-substituted one. In particular, the ferromagnetic transition temperature ($T_C$) was enhanced in a lower region of $x$ = 0.2–0.4, although the magnetic interaction between $Mn^{II}$ and $Fe^{III}$ ions is supposed to be weaker than that of $Fe^{II}$–$Fe^{III}$ ions considering the estimated Weiss temperatures. Such an enhancement of $T_C$ is caused by the increment of the high-temperature phase with the higher $T_C$, which originates in the suppression of the CTPT.

The difference between the $Mn^{II}$ and $Zn^{II}$-substituted complexes is based on the magnetic interaction between $M^{II}$ and $Fe^{III}$. In the case of the $Zn^{II}$-substituted complex, the nonmagnetic $Zn^{II}$ prevents the ferromagnetic interaction between the $Fe^{II}$ and $Fe^{III}$ and induces an antiferromagnetic exchange pathway of $Fe^{III}$-$Zn^{II}$-$Fe^{III}$ through the medium of nonmagnetic $Zn^{II}$. The antiferromagnetic interaction between $Fe^{III}$ and $Fe^{III}$ compensates for the ferromagnetic interaction between $Fe^{II}$ and $Fe^{III}$ at around $x$ = 0.83, and the $\theta$ becomes zero. Above $x$ = 0.83, the absolute value of the negative Weiss temperature rapidly increases with increasing $x$. On the other hand, in the case of the $Mn^{II}$-substituted complex, both the $Fe^{II}$-$Fe^{III}$ and $Mn^{II}$-$Fe^{III}$ magnetic interactions are ferromagnetic, which is responsible for the positive Weiss temperature in the whole range of $x$, in contrast to the $Zn^{II}$-substituted complexes, which is due to the substitution of the magnetic ions.

**Author Contributions:** Conceptualization, M.E.; methodology, M.E.; validation, M.E., H.I., A.O. and N.K.; formal analysis, H.I. and A.O.; investigation, M.E. and N.K.; resources, M.E.; data curation, H.I., A.O. and M.E.; writing—original draft preparation, M.E.; writing—review and editing, A.O. and N.K.; visualization, M.E.; supervision, M.E.; project administration, M.E.; funding acquisition, M.E.

**Funding:** This research was funded by a Grant-in-Aid for Young Scientists (No. 19750105) from the Ministry of Education, Culture, Sports, Science and Technology, Japan.

**Conflicts of Interest:** The authors declare no conflict of interest.

## Appendix A

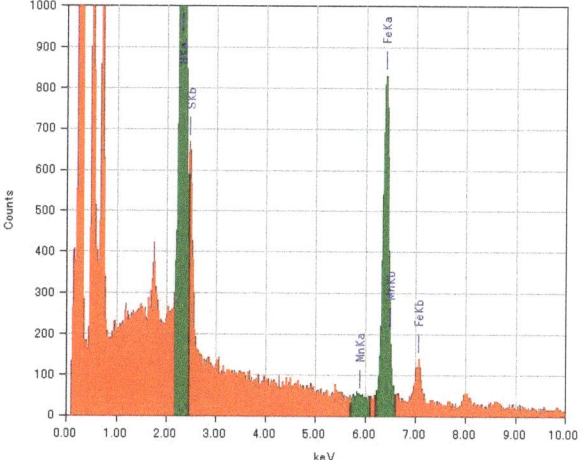

(a)

**Figure A1.** *Cont.*

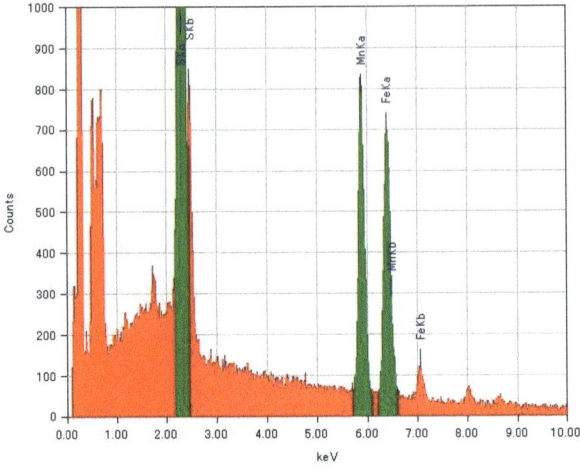

(b)

**Figure A1.** Energy spectra of EDS for $(n\text{-}C_3H_7)_4N[Fe^{II}_{1-x}Mn^{II}_xFe^{III}(dto)_3]$ with (**a**) $x = 0$ and (**b**) $x = 1.00$.

## References

1. Castillo, O.; Luque, A.; Sertucha, J.; Román, P.; Lloret, F. Synthesis, Crystal Structure, and Magnetic Properties of a One-Dimensional Polymeric Copper(II) Complex Containing an Unusual 1,1′-Bicoordinated Oxalato Bridge. *Inorg. Chem.* **2000**, *39*, 6142–6144. [CrossRef] [PubMed]
2. Kim, J.C.; Cho, J.; Lough, A. Syntheses, isolation, and structures of nickel(II) and copper(II) coordination polymers with a tetraaza macrocyclic ligand. *J. Inorg. Chim. Acta* **2001**, *317*, 252–258. [CrossRef]
3. Park, H.; Kim, J.C.; Lough, A.J.; Lee, B.M. One-dimensional macrocyclic zinc(II) coordination polymer containing an unusual bis-monodentate oxalate bridge. *Inorg. Chem. Commun.* **2007**, *10*, 303–306. [CrossRef]
4. Kim, J.; Park, A.H.; Kim, J.C.; Lough, A.J.; Pyun, S.Y.; Roh, J.; Lee, B.M. 1D copper(II) and zinc(II) coordination polymers containing an unusual twisted oxalate bridge. *Inorg. Chim. Acta* **2008**, *361*, 2087–2093. [CrossRef]
5. Pei, Y.; Journaux, Y.; Kahn, O. Ferromagnetic interactions between $t_{2g}^3$ and $e_g^2$ magnetic orbitals in a $Cr^{III}Ni^{II}_3$ tetranuclear compound. *Inorg. Chem.* **1989**, *28*, 100–103. [CrossRef]
6. Ohba, M.; Tamaki, H.; Matsumoto, N.; Ōkawa, H. Oxalate-bridged dinuclear chromium(III)-M(II) (M = copper, nickel, cobalt, iron, manganese) complexes: Synthesis, structure, and magnetism. *Inorg. Chem.* **1993**, *32*, 5385–5390. [CrossRef]
7. Glerup, J.; Goodson, P.A.; Hodgson, D.J.; Michelsen, K. Magnetic Exchange through Oxalate Bridges: Synthesis and Characterization of (μ-Oxalato)dimetal(II) Complexes of Manganese, Iron, Cobalt, Nickel, Copper, and Zinc. *Inorg. Chem.* **1995**, *34*, 6255–6264. [CrossRef]
8. Vivas, C.Y.; Delgado, F.S.; Ruiz-Pérez, C.; Julve, M. Preparation and crystal structure of the oxalato-bridged $Cr^{III}$–$Ag^I$ two-dimensional compound {Ag$_3$(H$_2$O)[Cr(dpa)(ox)$_2$]$_3$}$_n$·2nH$_2$O (dpa = 2,2′-dipyridylamine). *CrystEngComm* **2004**, *6*, 11–18. [CrossRef]
9. Armentano, D.; De Munno, G.; Lloret, F.; Julve, M. Bis and tris(oxalato)ferrate(III) complexes as precursors of polynuclear compounds. *CrystEngComm* **2005**, *7*, 57–66. [CrossRef]
10. Ballester, G.; Coronado, E.; Giménez-Saiz, C.; Romero, F.M. Nitroxide Radicals as Templating Agents in the Synthesis of Magnets Based on Three-Dimensional Oxalato-Bridged Heterodimetallic Networks. *Angew. Chem. Int. Ed.* **2001**, *40*, 792–795. [CrossRef]
11. Kahn, O. Dinuclear Complexes with Predictable Magnetic Properties. *Angew. Chem. Int. Ed.* **1985**, *24*, 834–850. [CrossRef]
12. Rao, C.N.R.; Natarajan, S.; Vaidhyanathan, R. Metal Carboxylates with Open Architectures. *Angew. Chem. Int. Ed.* **2004**, *43*, 1466–1496. [CrossRef] [PubMed]

13. Clemente-León, M.; Coronado, E.; Martí-Gastaldo, C.; Romero, F.M. Multifunctionality in hybrid magnetic materials based on bimetallic oxalate complexes. *Chem. Soc. Rev.* **2011**, *40*, 473–497. [CrossRef] [PubMed]
14. Tamaki, H.; Zhong, Z.J.; Matsumoto, N.; Kida, S.; Koikawa, M.; Achiwa, N.; Hashimoto, Y.; Ōkawa, H. Design of metal-complex magnets. Syntheses and magnetic properties of mixed-metal assemblies {NBu$_4$[MCr(ox)$_3$]}$_x$ (NBu$_4^+$ = tetra(n-butyl)ammonium ion; ox$^{2-}$ = oxalate ion; M = Mn$^{2+}$, Fe$^{2+}$, Co$^{2+}$, Ni$^{2+}$, Cu$^{2+}$, Zn$^{2+}$). *J. Am. Chem. Soc.* **1992**, *114*, 6974–6979. [CrossRef]
15. Clemente-León, M.; Coronado, E.; Galán-Mascarós, J.R.; Gómez-García, C.J. Intercalation of decamethylferrocenium cations in bimetallic oxalate-bridged two-dimensional magnets. *Chem. Commun.* **1997**, *0*, 1727–1728. [CrossRef]
16. Coronado, E.; Galán-Mascarós, J.R.; Gómez-García, C.J.; Ensling, J.; Gütlich, P. Hybrid Molecular Magnets Obtained by Insertion of Decamethylmetallocenium Cations into Layered, Bimetallic Oxalate Complexes: [Z$^{III}$Cp*$_2$][M$^{II}$M$^{III}$(ox)$_3$] (Z$^{III}$=Co, Fe; M$^{III}$=Cr, Fe; M$^{II}$=Mn, Fe, Co, Cu, Zn; ox=oxalate; Cp*=pentamethylcyclopentadienyl). *Chem. Eur. J.* **2000**, *6*, 552–563. [CrossRef]
17. Coronado, E.; Giménez-Saiz, C.; Gómez-García, C.J.; Romero, F.M.; Tarazón, A. A bottom-up approach from molecular nanographenes to unconventional carbon materials. *J. Mater. Chem.* **2008**, *18*, 929–934. [CrossRef]
18. Clemente-León, M.; Coronado, E.; López-Jordà, M.; Waerenborgh, J.C. Multifunctional Magnetic Materials Obtained by Insertion of Spin-Crossover Fe$^{III}$ Complexes into Chiral 3D Bimetallic Oxalate-Based Ferromagnets. *Inorg. Chem.* **2011**, *50*, 9122–9130. [CrossRef] [PubMed]
19. Clemente-León, M.; Coronado, E.; López-Jordà, M.; Mínguez, E.; Soriano-Portillo, A.; Waerenborgh, J.C. Multifunctional Magnetic Materials Obtained by Insertion of a Spin-Crossover Fe$^{III}$ Complex into Bimetallic Oxalate-Based Ferromagnets. *Chem. Eur. J.* **2010**, *16*, 2207–2219. [CrossRef] [PubMed]
20. Clemente-León, M.; Coronado, E.; López-Jordà, M.; Desplanches, C.; Asthana, S.; Wang, H.; Létard, J.-F. A hybrid magnet with coexistence of ferromagnetism and photoinduced Fe(III) spin-crossover. *Chem. Sci.* **2011**, *2*, 1121–1127. [CrossRef]
21. Clemente-León, M.; Coronado, E.; López-Jordà, M. 2D and 3D bimetallic oxalate-based ferromagnets prepared by insertion of different Fe$^{III}$ spin crossover complexes. *Dalton Trans.* **2010**, *39*, 4903–4910. [CrossRef] [PubMed]
22. Bénard, S.; Rivière, E.; Yu, P.; Nakatani, K.; Delouis, J.F. A Photochromic Molecule-Based Magnet. *Chem. Mater.* **2001**, *13*, 159–162. [CrossRef]
23. Coronado, E.; Galán-Mascarós, J.R.; Gómez-García, C.J.; Laukhin, V. Coexistence of ferromagnetism and metallic conductivity in a molecule-based layered compound. *Nature* **2000**, *408*, 447–449. [CrossRef] [PubMed]
24. Alberola, A.; Coronado, E.; Galán-Mascarós, J.R.; Giménez-Saiz, C.; Gómez-García, C.J. A Molecular Metal Ferromagnet from the Organic Donor Bis(ethylenedithio)tetraselenafulvalene and Bimetallic Oxalate Complexes. *J. Am. Chem. Soc.* **2003**, *125*, 10774–10775. [CrossRef] [PubMed]
25. Coronado, E.; Galán-Mascarós, J.R.; Gómez-García, C.J.; Martínez-Ferrero, E.; van Smaalen, S. Incommensurate Nature of the Multilayered Molecular Ferromagnetic Metals Based on Bis(ethylenedithio)tetrathiafulvalene and Bimetallic Oxalate Complexes. *Inorg. Chem.* **2004**, *43*, 4808–4810. [CrossRef] [PubMed]
26. Coronado, E.; Galán-Mascarós, J.R. Hybrid molecular conductors. *J. Mater. Chem.* **2005**, *15*, 66–74. [CrossRef]
27. Coronado, E.; Curreli, S.; Giménez-Saiz, C.; Gómez-García, C.J. The Series of Molecular Conductors and Superconductors ET$_4$[AFe(C$_2$O$_4$)$_3$]·PhX (ET = bis(ethylenedithio)tetrathiafulvalene; (C$_2$O$_4$)$^{2-}$ = oxalate; A$^+$ = H$_3$O$^+$, K$^+$; X = F, Cl, Br, and I): Influence of the Halobenzene Guest Molecules on the Crystal Structure and Superconducting Properties. *Inorg. Chem.* **2012**, *51*, 1111–1126. [CrossRef] [PubMed]
28. Ōkawa, H.; Shigematsu, A.; Sadakiyo, M.; Miyagawa, T.; Yoneda, K.; Ohba, M.; Kitagawa, H. Oxalate-Bridged Bimetallic Complexes {NH(prol)$_3$}[MCr(ox)$_3$] (M = Mn$^{II}$, Fe$^{II}$, Co$^{II}$; NH(prol)$_3^+$ = Tri(3-hydroxypropyl)ammonium) Exhibiting Coexistent Ferromagnetism and Proton Conduction. *J. Am. Chem. Soc.* **2009**, *131*, 13516–13522. [CrossRef] [PubMed]
29. Pardo, E.; Train, C.; Liu, H.; Chamoreau, L.-M.; Dkhil, B.; Boubekeur, K.; Lloret, F.; Nakatani, K.; Tokoro, H.; Ohkoshi, S.; et al. Multiferroics by Rational Design: Implementing Ferroelectricity in Molecule-Based Magnets. *Angew. Chem. Int. Ed.* **2012**, *51*, 8356–8360. [CrossRef] [PubMed]
30. Bénard, S.; Yu, P.; Coradin, T.; Rivière, E.; Nakatani, K.; Clément, R. Design of strongly NLO-active molecularly-based ferromagnets. *Adv. Mater.* **1997**, *9*, 981–984. [CrossRef]

31. Bénard, S.; Yu, P.; Audière, J.P.; Rivière, E.; Clément, R.; Guilhem, J.; Tchertanov, L.; Nakatani, K. Structure and NLO Properties of Layered Bimetallic Oxalato-Bridged Ferromagnetic Networks Containing Stilbazolium-Shaped Chromophores. *J. Am. Chem. Soc.* **2000**, *122*, 9444–9454. [CrossRef]
32. Lacroix, P.G.; Malfant, I.; Bénard, S.; Yu, P.; Rivière, E.; Nakatani, K. Hybrid Molecular-Based Magnets Containing Organic NLO Chromophores: A Search toward an Interplay between Magnetic and NLO Behavior. *Chem. Mater.* **2001**, *13*, 441–449. [CrossRef]
33. Ōkawa, H.; Mitsumi, M.; Ohba, M.; Kodera, M.; Matsumoto, N. Dithiooxalato(dto)-Bridged Bimetallic Assemblies $\{NPr_4[MCr(dto)_3]\}_x$ (M = Fe, Co, Ni, Zn; $NPr_4$ = Tetrapropylammonium Ion): New Complex-Based Ferromagnets. *Bull. Chem. Soc. Jpn.* **1994**, *67*, 2139–2144. [CrossRef]
34. Kojima, N.; Aoki, W.; Seto, M.; Kobayashi, Y.; Maeda, Y. Reversible charge-transfer phase transition in $[(n-C_3H_7)_4N][Fe^{II}Fe^{III}(dto)_3]$ (dto = $C_2O_2S_2$). *Synth. Met.* **2001**, *121*, 1796–1797. [CrossRef]
35. Kojima, N.; Aoki, W.; Itoi, M.; Ono, Y.; Seto, M.; Kobayashi, Y.; Maeda, Y. Charge transfer phase transition and ferromagnetism in a mixed-valence iron complex, $(n-C_3H_7)_4N[Fe^{II}Fe^{III}(dto)_3]$ (dto=$C_2O_2S_2$). *Solid State Commun.* **2001**, *120*, 165–170. [CrossRef]
36. Nakamoto, T.; Miyazaki, Y.; Itoi, M.; Ono, Y.; Kojima, N.; Sorai, M. Heat Capacity of the Mixed-Valence Complex $\{[(n-C_3H_7)_4N][Fe^{II}Fe^{III}(dto)_3]\}_\infty$, Phase Transition because of Electron Transfer, and a Change in Spin-State of the Whole System. *Angew. Chem. Int. Ed.* **2001**, *40*, 4716–4719. [CrossRef]
37. Itoi, M.; Taira, A.; Enomoto, M.; Matsushita, N.; Kojima, N.; Kobayashi, Y.; Asai, K.; Koyama, K.; Nakano, T.; Uwatoko, Y. Crystal structure and structural transition caused by charge-transfer phase transition for iron mixed-valence complex $(n-C_3H_7)_4N[Fe^{II}Fe^{III}(dto)_3]$ (dto = $C_2O_2S_2$). *J. Solid State Commun.* **2004**, *130*, 415–420. [CrossRef]
38. Itoi, M.; Ono, Y.; Kojima, N.; Kato, K.; Osaka, K.; Takata, M. Charge-Transfer Phase Transition and Ferromagnetism of Iron Mixed-Valence Complexes $(n-C_nH_{2n+1})_4N[Fe^{II}Fe^{III}(dto)_3]$ ($n$ = 3–6; dto = $C_2O_2S_2$). *Eur. J. Inorg. Chem.* **2006**, 1198–1207. [CrossRef]
39. Kida, N.; Hikita, M.; Kashima, I.; Okubo, M.; Itoi, M.; Enomoto, M.; Kato, K.; Tanaka, M.; Kojima, N. Control of Charge Transfer Phase Transition and Ferromagnetism by Photoisomerization of Spiropyran for an Organic−Inorganic Hybrid System, $(SP)[Fe^{II}Fe^{III}(dto)_3]$ (SP = spiropyran, dto = $C_2O_2S_2$). *J. Am. Chem. Soc.* **2009**, *131*, 212–220. [CrossRef] [PubMed]
40. Ida, H.; Okazawa, A.; Kojima, N.; Shimizu, R.; Yamada, Y.; Enomoto, M. Effect of Nonmagnetic Substitution on the Magnetic Properties and Charge-Transfer Phase Transition of an Iron Mixed-Valence Complex, $(n-C_3H_7)_4N[Fe^{II}Fe^{III}(dto)_3]$ (dto = $C_2O_2S_2$). *Inorg. Chem.* **2012**, *51*, 8989–8996. [CrossRef] [PubMed]
41. Enomoto, M.; Kojima, N. Magnetic dilution effect on the charge transfer phase transition and the ferromagnetic transition for an iron mixed-valence complex, $(n-C_3H_7)_4N[Fe^{II}_{1-x}Zn^{II}_xFe^{III}(dto)_3]$ (dto = $C_2O_2S_2$). *Polyhedron* **2009**, *28*, 1826–1829. [CrossRef]
42. Jones, H.O.; Tasker, H.S. CCXII.—The action of mercaptans on acid chlorides. Part II. The acid chlorides of phosphorus, sulphur, and nitrogen. *J. Chem. Soc.* **1909**, *95*, 1910–1918. [CrossRef]
43. Robinson, C.S.; Jones, H.O. VII.—Complex thio-oxalates. *J. Chem. Soc.* **1912**, *101*, 62–76. [CrossRef]
44. Dwyer, F.P.; Sargeson, A.M. The Resolution of the Tris-(thio-oxalato)[1] Complexes of Co(III), Cr(III) and Rh(III). *J. Am. Chem. Soc.* **1959**, *81*, 2335–2336. [CrossRef]
45. Irving, H.; Williams, R.J.P. The Stability of Transition-metal Complexes. *J. Phys. Chem.* **1953**, *0*, 3192–3210. [CrossRef]
46. Sigel, H.; McCormick, D.B. Discriminating Behavior of Metal Ions and Ligands with Regard to Their Biological Significance. *Acc. Chem. Res.* **1970**, *3*, 201–208. [CrossRef]
47. Atovymann, O.L.; Shilov, G.V.; Lyubovskaya, R.N.; Zhilyaeva, E.I.; Ovaneseyan, N.S.; Pirumova, S.I.; Gusakovskaya, I.G.; Morozov, Y.G. Crystal structure of the molecular ferromagnet $NBu_4[MnCr(C_2O_4)_3]$ (Bu=$n$-$C_4H_9$). *JETP Lett.* **1993**, *58*, 766–769.
48. Decurtins, S.; Schmale, H.W.; Oswald, H.R.; Linden, A.; Ensling, J.; Gütlich, P.; Hauser, A. A polymeric two-dimensional mixed-metal network. Crystal structure and magnetic properties of $\{[P(Ph)_4][MnCr(ox)_3]\}$. *Inorg. Chim. Acta* **1994**, *216*, 65–73. [CrossRef]
49. Decurtins, S.; Schmalle, H.W.; Pellaux, R.; Schneuwly, P.; Hauser, A. Chiral, Three-Dimensional Supramolecular Compounds: Homo- and Bimetallic Oxalate- and 1,2-Dithiooxalate-Bridged Networks. A Structural and Photophysical Study. *Inorg. Chem.* **1996**, *35*, 1451–1460. [CrossRef] [PubMed]

50. Ovanesyan, N.S.; Shilov, G.V.; Sanina, N.A.; Pyalling, A.A.; Atovmyan, L.O.; Bottyán, L. Structural and Magnetic Properties of Two-Dimensional Oxalate-Bridged Bimetallic Compounds. *Mol. Cryst. Liq. Cryst.* **1999**, *335*, 91–104. [CrossRef]
51. Ovanesyan, N.S.; Makhaev, V.D.; Aldoshin, S.M.; Gredin, P.; Boubekeur, K.; Train, C.; Gruselle, M. Structure, magnetism and optical properties of achiral and chiral two-dimensional oxalate-bridged anionic networks with symmetric and symmetric ammonium cations. *Dalton Trans.* **2005**, *0*, 3101–3107. [CrossRef] [PubMed]
52. Nuttall, C.J.; Day, P. Modeling Stacking Faults in the Layered Molecular-Based Magnets $AM^{II}Fe(C_2O_4)_3$ {$M^{II}$ = Mn, Fe; A = Organic Cation}. *J. Solid State Chem.* **1999**, *147*, 3–10. [CrossRef]
53. Kotani, M. On the Magnetic Moment of Complex Ions. (I). *J. Phys. Soc. Jpn.* **1949**, *4*, 293–297. [CrossRef]
54. Kotani, M. Properties of d-Electrons in Complex Salts. Part I: Paramagnetism of Complex Salts. *Prog. Theor. Phys. Suppl.* **1960**, *14*, 1–16. [CrossRef]
55. Ono, Y.; Okubo, M.; Kojima, N. Crystal Structure and Ferromagnetism of $(n-C_3H_7)_4N[Co^{II}Fe^{III}(dto)_3]$ (dto = $C_2O_2S_2$). *Solid State Commun.* **2003**, *126*, 291–296. [CrossRef]
56. Mayoh, B.; Day, P. Charge transfer in mixed-valence solids. Part VIII. Contribution of valence delocalisation to the ferromagnetism of Prussian Blue. *J. Chem. Soc. Dalton Trans.* **1976**, *0*, 1483–1486. [CrossRef]
57. Carling, S.G.; Bradley, J.M.; Visser, D.; Day, P. Magnetic and structural characterization of the layered materials $AMnFe(C_2S_2O_2)_3$. *Polyhedron* **2003**, *22*, 2317–2324. [CrossRef]
58. Kida, N.; Enomoto, M.; Watanabe, I.; Suzuki, T.; Kojima, N. Spin dynamics of the charge-transfer phase transition of an iron mixed-valence complex observed using muon spin relaxation spectroscopy. *Phys. Rev. B* **2008**, *77*, 144427. [CrossRef]
59. Enomoto, M.; Kida, N.; Watanabe, I.; Suzuki, T.; Kojima, N. Spin dynamics of the ferromagnetic transition in iron mixed-valence complexes, $(n-C_nH_{2n+1})_4N[Fe^{II}Fe^{III}(dto)_3]$ (dto = $C_2O_2S_2$, n = 3-5) by μSR. *Physica B* **2009**, *404*, 642–644. [CrossRef]

© 2018 by the authors. Licensee MDPI, Basel, Switzerland. This article is an open access article distributed under the terms and conditions of the Creative Commons Attribution (CC BY) license (http://creativecommons.org/licenses/by/4.0/).

Article

# Synthesis, Structure, and Photomagnetic Properties of a Hydrogen-Bonded Lattice of [Fe(bpp)$_2$]$^{2+}$ Spin-Crossover Complexes and Nicotinate Anions

Verónica Jornet-Mollá, Carlos Giménez-Saiz and Francisco M. Romero *

Instituto de Ciencia Molecular, Universitat de València, P.O. Box 22085, 46071-Valencia, Spain; Veronica.Jornet@uv.es (V.J.-M.); Carlos.Gimenez@uv.es (C.G.-S.)
* Correspondence: fmrm@uv.es; Tel.: +34-963-54-44-05; Fax: +34-963-54-32-73

Received: 3 November 2018; Accepted: 19 November 2018; Published: 21 November 2018

**Abstract:** In this paper, we report on the synthesis, crystal structure, and photomagnetic properties of the spin-crossover salt of formula [Fe(bpp)$_2$](C$_6$H$_4$NO$_2$)$_2$·4H$_2$O (**1**·4H$_2$O) (bpp = 2,6-bis(pyrazol-3-yl)pyridine; C$_6$H$_4$NO$_2^-$ = nicotinate anion). This compound exhibits a 3D supramolecular architecture built from hydrogen bonds between iron(II) complexes, nicotinate anions, and water molecules. As synthesized, the hydrated material is low-spin and desolvation triggers a low-spin (LS) to high-spin (HS) transformation. Anhydrous phase **1** undergoes a partial spin crossover ($T_{1/2}$= 281 K) and a LS to HS photomagnetic conversion with a $T$(LIESST) value of 56 K.

**Keywords:** spin-crossover; LIESST effect; hydrogen bonding; π-π interactions

## 1. Introduction

Switching magnetic materials represent a prominent avenue for the construction of multifunctional materials with several applications in different fields, such as chemical and pressure sensing [1], data storage [2], and spintronics [3,4]. The most studied spin-crossover (SCO) centres are based on iron(II) complexes that exhibit labile electronic configurations switchable between $^1A_1$ low-spin (LS; $S$ = 0) and $^5T_2$ high-spin (HS; $S$ = 2) states as a consequence of a given external perturbation, such as light irradiation, variation of temperature and/or pressure [5–11]. The response to these external stimuli leads to different changes in magnetism, colour, and structure. In particular, important variations of the Fe−N metal−ligand bond lengths (0.1–0.2 Å) and N−Fe−N angles (0.5–8°) are observed upon spin crossover [12]. Depending on the cooperativity of the system, the spin transition may be abrupt with hysteretic behaviour (memory effect) and with drastic changes in optical and magnetic properties. The appearance of hysteresis and thus cooperativity in a solid system can be achieved when the geometrical distortion is propagated to the whole framework, providing the material with a bistable character.

Therefore, under the same conditions, a bistable system can be localized either in the LS or in the HS state and the possibility of finding a specific state depends on the history of the material. This is a desirable situation for the development of technological applications such as quantum logic operators or components in memory storage devices, mainly when the effects proceed at room temperature [13–17]. In order to achieve high levels of cooperativity, various synthetic strategies have been developed [18]. One way consists in the use of rigid bridging ligands as connectors between the Fe$^{2+}$ cations; this strategy has yielded a wide range of 1-3D SCO coordination polymers [19–22]. Another type of communication between the coordination centres is provided by hydrogen bonding, which is likely to be responsible for the dependence of magnetic properties on the extent of solvation observed in many compounds [23,24].

In this context, the family of bis-chelated iron(II) complexes with the formula [Fe(bpp)$_2$]X$_2$ (bpp being the tridentate ligand 2,6-bis(pyrazol-3-yl)pyridine, Chart 1) is very interesting. In these spin-crossover salts with pseudo $C_{2v}$ symmetry, the imine N atom coordinates to the Fe$^{2+}$ cation and the non-coordinating NH groups allow for interaction through hydrogen bonding between the bpp ligands and anions or solvate molecules present in the lattice. The paramount role of the hydrogen bonds on the SCO features is largely documented in molecular compounds [25,26] and has been extensively studied in these salts [27].

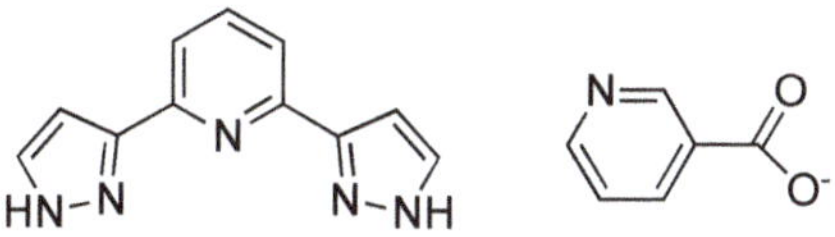

**Chart 1.** Structures of bpp and nicotinate anion.

In addition, these compounds are attractive due to the fact that the desolvated material usually exhibits very abrupt transitions and presents light-induced excited spin-state trapping (LIESST) effects [28], with long-lived lifetimes of the photoinduced metastable phases [29–31].

In a recent report, we introduced a rational design of SCO materials exhibiting ferroelectricity [32]. This is based on the use of hydrogen bonds in the assembly of the [Fe(bpp)$_2$]$^{2+}$ complexes with isonicotinate anions. From the point of view of the connectivity of the lattice, the iron(II) complex acts as a 4-fold pseudotetrahedral H-bond donor, whereas the isonicotinate anion acts as a non-centrosymmetric H-bond acceptor. This necessarily yields a non-centrosymmetric diamondoid lattice. In the present work, the isonicotinate anion has been replaced with the nicotinate anion (nic = 3-pyridinecarboxylate, Chart 1) in order to study the effect of the position of the nitrogen atom on the properties of the resulting material. Herein, we report on the synthesis, crystal structure, and photomagnetic properties of the title compound, [Fe(bpp)$_2$](nic)$_2$·4H$_2$O (1·4H$_2$O).

## 2. Materials and Methods

### 2.1. Physical Measurements

Magnetic susceptibility measurements were performed on polycrystalline samples using a magnetometer (Quantum Design MPMS-XL-5, San Diego, CA, USA) equipped with a SQUID (Superconducting Quantum Interference Device) sensor. Variable temperature measurements were carried out in the temperature range 2-400 K in a magnetic field of 0.1 T. The temperature sweeping rates were as follows: 1 K·min$^{-1}$ (2–20 K), 2.25 K·min$^{-1}$ (20–200 K), and 1.25 K·min$^{-1}$ (200–400 K). A dehydrated sample of **1** was obtained *in situ* by maintaining the sample in the SQUID at 400K for 1 h (until a constant magnetic signal was obtained) and on plastic capsules perforated in order to favour solvent loss. For photomagnetic measurements, the sample was prepared in a thin layer to promote full light penetration. First, the sample was dehydrated in situ in the SQUID device at 400 K for 1 h. After slow cooling to 10 K, the sample was irradiated with green light ($\lambda$ = 532.06 nm) and the magnetization was measured. When the saturation point was reached, the laser was switched off and the temperature increased at a rate of 0.3 K min$^{-1}$ to determine the $T$(LIESST) value given by the minimum of the $\delta(\chi T)/\delta T$ versus $T$ curve for the relaxation process.

Thermogravimetric (TG) measurements of Ag(C$_6$H$_4$NO$_2$) were carried out under O$_2$ atmosphere in a Setsys TGA-ATD16/8 apparatus (Setaram Instrumentation, Caluire, France) the 298–1000 K temperature range at a scan rate of 10 K min$^{-1}$. TG analysis of **1**·4H$_2$O was performed in a TGA/SDTA/851e apparatus (Mettler Toledo, Columbus, OH, USA) under N$_2$ atmosphere in the 298–973 K range at a scan rate of 10 K min$^{-1}$.

Differential scanning calorimetry (DSC) measurements under nitrogen atmosphere were performed in a DSC 821e apparatus (Mettler Toledo, OH, USA) ith warming and cooling rates equal to 10 K·min$^{-1}$. A correction from the sample holder was automatically applied.

Infrared (IR) transmission measurements of potassium bromide pellets were recorded at room temperature with a Nicolet Avatar 320 FT-IR spectrophotometer (Thermo Electron, Waltham, MA, USA) in the range 4000–400 cm$^{-1}$. CHN elemental analyses were carried out in a EA 1110 CHNS analyser (CE Instruments, Wigan, United Kingdom).

Powder X-ray diffraction measurements were collected using Cu K$\alpha$ radiation ($\lambda$ = 1.54056 Å) at room temperature in a $2\theta$ range from 2 to 40°. A polycrystalline sample of **1**·4H$_2$O was lightly ground in an agate mortar and filled into a 0.5 mm borosilicate capillary prior to being mounted and aligned on an Empyrean powder diffractometer (Panalytical, Cambridge, United Kingdom) The simulated diffractogram was obtained from single crystal X-ray data using the CrystalDiffract software (CrystalMaker Software Ltd., Begbroke, United Kingdom).

### 2.2. Synthesis

Ligand bpp was prepared using the previously published procedure [33]. All other reagents and solvents were purchased from commercial sources, with no further purification being undertaken.

Ag(C$_6$H$_4$NO$_2$)

A suspension of nicotinic acid (1.230 g, 10 mmol) in 25 ml of a mixture 3:2 EtOH/H$_2$O was treated with a suspension of silver carbonate (1.371 g, 5 mmol) in 25 ml H$_2$O. Then, the mixture was refluxed for 3 h and 30 min until CO$_2$ ceased to evolve. There was a change in the colour of the solid from yellow-brown to white. The product was collected using filtration and washed with water and acetone to yield 1.988 g (87 %) of the desired compound. We found: C, 30.58; H, 1.80; N, 6.01. C$_6$H$_4$AgNO$_2$ requires C, 31.33; H, 1.75; N, 6.09 (the sample contained some humidity, equivalent to 0.25 H$_2$O molecules per formula). Thermogravimetric analysis (Figure S1) confirmed the anhydrous character of this salt. $\nu_{max}$/cm$^{-1}$: 3400.4, 3042.9, 1593.9, 1549.3, 1387.4, 1196.0, 1088.4, 1022.2, 840.4, 758.4, 704.3, 511.2.

[Fe(bpp)$_2$](C$_6$H$_4$NO$_2$)$_2$·4H$_2$O (**1**·4H$_2$O)

FeCl$_2$·4H$_2$O (0.0994 g, 0.5 mmol) was added as a solid to a degassed solution of bpp (0.212 g, 1.0 mmol) in 20 ml of a mixture 4:1 MeOH/H$_2$O. A deep red colour appeared. After complete dissolution of the iron reagent, a mixture of Ag(C$_6$H$_4$NO$_2$) (230 mg, 1 mmol) in 20 ml H$_2$O was added. After stirring for 1 h at 65 °C, the yellowish precipitate of AgCl was filtered through a low-porosity frit. The filtrate was left undisturbed. Red prisms suitable for X-ray analysis appeared after a few days, yielding 246.3 mg (62%). We found: C, 51.63; H, 4.13; N, 21.83. C$_{34}$H$_{34}$FeN$_{12}$O$_8$ requires C, 51.40; H, 4.31; N, 21.15 (the sample was partially dehydrated, losing 0.55 H$_2$O molecules per formula prior to analysis). $\nu_{max}$/cm$^{-1}$: 3422.8, 1601.0, 1560.3, 1438.4, 1385.0, 1281.1, 1234.5, 1146.1, 1094.1, 1028.3, 897.0, 831.3, 756.8, 697.9, 619.8.

### 2.3. X-ray Crystallography

A suitable crystal of **1**·4H$_2$O was coated with Paratone N oil, suspended on a small fiber loop, and placed in a stream of cold nitrogen (120 K) on an Oxford Diffraction Supernova diffractometer equipped with a graphite-monochromated Enhance Mo X-Ray Source ($\lambda$ = 0.71073 Å). The data collection routines, unit cell refinements, and data processing were carried out using the CrysAlisPro v38.46 software package (Rigaku, The Woodlands, TX, USA). The structure was solved using the SHELX package [34]. The asymmetric unit of **1**·4H$_2$O contains one [Fe(bpp)$_2$]$^{2+}$ cation, two nicotinate anions, and four water molecules, all in general positions. H atoms bonded to carbon atoms were included at calculated positions and refined with a rigid model. H atoms of bpp amino groups were all found in Fourier difference maps and refined positionally with the geometrical restraint N–H = 0.89 Å. Moreover, all H atoms on water molecules were found in difference maps and refined using geometrical restraints (O–H = 0.82 Å and H···H = 1.30 Å). The chemical formula of this compound includes the total number of hydrogen atoms and all non-hydrogen atoms were refined anisotropically. CCDC1832023 contain the supplementary crystallographic data for **1**·4H$_2$O at 120 K. These data were provided free of charge by the Cambridge Crystallographic Data Centre.

## 3. Results

*3.1. Synthesis and Thermal Properties*

In order to avoid the inclusion of undesired counterions in the spin-crossover network, it is necessary to obtain a solution without any other hydrogen-bond acceptors that could be in competition with the nicotinate anions. To achieve this goal, it has been shown that the use of a metathesis reaction gives excellent results [32,35–37]. Based on this, we performed a metathesis reaction of silver nicotinate with [Fe(bpp)$_2$]Cl$_2$ that resulted in the precipitation of AgCl, yielding a solution free from the undesired chloride anions.

TG analysis of **1**·4H$_2$O under nitrogen atmosphere (Figure S2) was performed in order to determine at which temperature the dehydration of the sample starts and the temperature range of the stability of the compound. This will determine the most suitable thermal conditions for magnetism experiments. The first loss of water molecules takes place at 327 K with a weight decrease of 6.3 %, which is related to the loss of about three water molecules per iron cation. The fourth water molecule comes off at temperatures near the thermal decomposition of the sample (2.2%, in the 443–493 K temperature range). These results show a good agreement with the formulation determined from the X-ray crystal structure experiment and elemental analysis. The fact that the loss of water molecules takes place in two separate steps indicates the existence of different types of water molecules in the crystal structure.

The DSC measurement (Figure S3) is in accordance with the thermogravimetric data. The as synthesized material does not show any feature upon cooling down to 133 K (black line). In the first heating process (curve 1), there is a very intense endothermic peak around 334 K that fits in the temperature range where the loss of three water molecules is observed. Therefore, this endothermic peak has been associated with the dehydration of the sample. The subsequent cooling plot (curve 2) shows a broad feature between 373 and 323 K and three very week exothermic peaks at 282, 262, and 218 K. The intensities ascribed to these peaks are very low, meaning that this might be either a result of a partial spin-crossover process and/or evidence of a phase transition. The second heating process (curve 3) presents three endothermic peaks centred at 229, 264, and 289 K, indicating that these processes are reversible and proceed with thermal hysteresis. Values of enthalpy and entropy changes associated with them are gathered in Table S1. Summing up the contributions of the three peaks gives a total enthalpy change $\Delta H$ = 4.8 KJ·mol$^{-1}$ and entropy change $\Delta S$ = 20.6 J·K$^{-1}$·mol$^{-1}$ (average values obtained from the cooling and warming curves). These parameters are weak but lie within the range expected for spin crossover.

*3.2. Structural Description*

**1**·4H$_2$O crystallizes in the non-centrosymmetric and polar *Pc* monoclinic space group (Table 1). The crystal structure contains only one crystallographically independent (Figure 1) Fe(II) centre, Fe1, and two inequivalent anions in the structure. The independent Fe$^{2+}$ site is located in a general position and is bound to two bpp tridentate ligands in mutual perpendicular orientations. Therefore, the iron(II) coordination environment is the well-known octahedral FeN$_6$ with the highest possible $D_{2d}$ local symmetry of the [Fe(bpp)$_2$]$^{2+}$ cations. The Fe-N(pyridine) bond distances (mean value ≈ 1.9264(25)) are slightly shorter than those corresponding to the Fe–N(pyrazole) bonds (mean value ≈ 1.964(8)). In any case, the Fe–N bond distances are lower than 2.0 Å, thus revealing a low-spin ground state for the Fe$^{2+}$ cation, which is the expected stable spin state observed in similar hydrated compounds [37–39]. With respect to the second iron coordination sphere, each [Fe(bpp)$_2$]$^{2+}$ complex is hydrogen-bonded to only two nicotinate anions and two water molecules. The N atoms of the isonicotinate anion do not establish interactions with the bpp ligand and are hydrogen-bonded to crystallization water molecules present in the network.

Table 1. Summary of crystal data.

| Compound | 1·4H$_2$O |
|---|---|
| Formula | C$_{34}$H$_{34}$FeN$_{12}$O$_8$ |
| Formula weight | 794.58 |
| Crystal system | Monoclinic |
| Space group | Pc |
| a/Å | 8.17633(10) |
| b/Å | 8.15854(9) |
| c/Å | 26.9671(3) |
| α/° | - |
| β/° | 92.0138(10) |
| γ/° | - |
| V/Å$^3$ | 1797.78(4) |
| Z | 2 |
| T/K | 120(2) |
| $\rho_{calcd}$/g·cm$^{-3}$ | 1.468 |
| λ/Å | 0.71073 |
| θ-range/° | 2.870–29.963 |
| No. of rflns. collected | 84313 |
| No. of indep. rflns./$R_{int}$ | 9867/0.0635 |
| Restraints/parameters | 18/532 |
| $R_1/wR_2$ ($I > 2\sigma(I)$) [1] | 0.0324/0.0664 |
| $R_1/wR_2$ (all data) [1] | 0.0398/0.0711 |
| $\Delta\rho_{max}$ and $\Delta\rho_{min}$ /e·Å$^{-3}$ | 0.265/−0.395 |
| Absolute structure parameter | −0.016(5) |

[1] $R_1 = \Sigma(F_o - F_c)/\Sigma(F_o)$; $wR_2 = [\Sigma w(F_o^2 - F_c^2)^2/\Sigma w(F_o^2)^2]^{1/2}$.

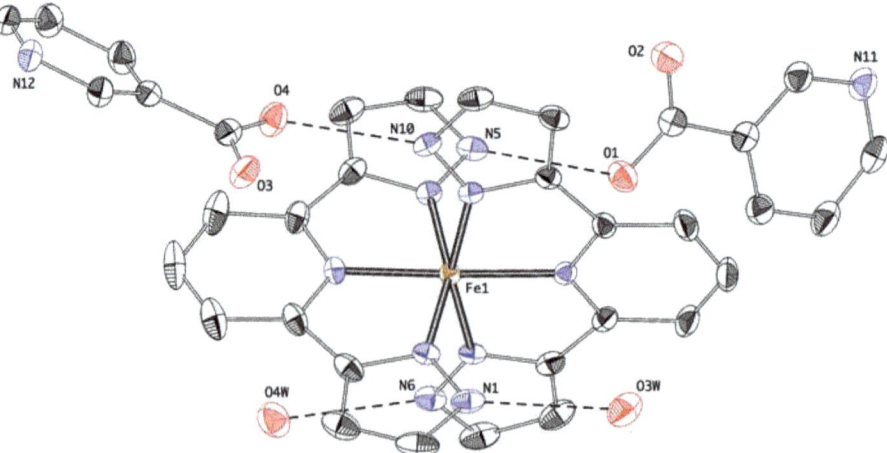

**Figure 1.** Thermal ellipsoid plot of the crystal structure of **1**·4H$_2$O showing the [Fe(bpp)$_2$]$^{2+}$ complex, the two crystallographically independent nicotinate anions, and the two water molecules present in the second coordination sphere of the iron site. Dashed lines refer to H bonds. Thermal ellipsoids are drawn at a 50% probability level. H atoms are not shown.

The crystal structure is best described as a packing of layers of [Fe(bpp)$_2$]$^{2+}$ cations and nicotinate anions that alternate along the c axis (Figure 2). These cationic and anionic layers are held together thanks to the existence of hydrogen bonds, either directly or involving water molecules of crystallization.

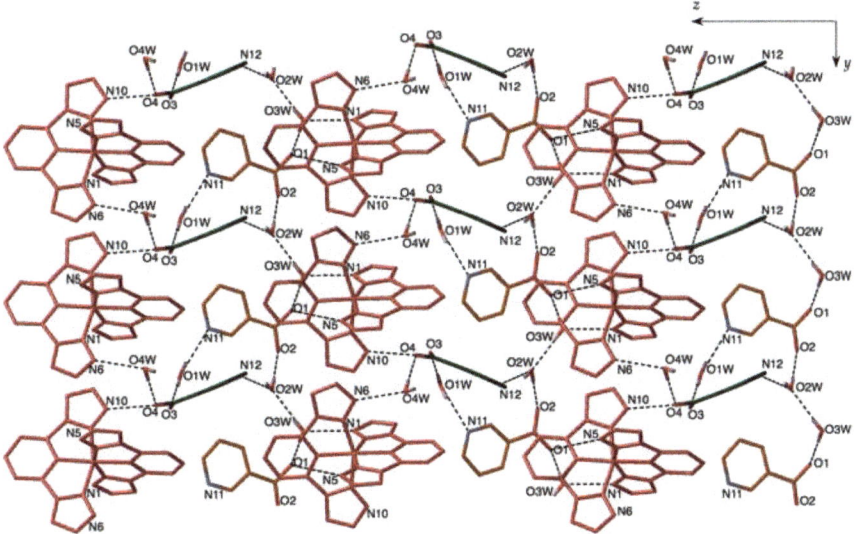

**Figure 2.** A view of the crystal structure of **1**·4H$_2$O along the *a* axis, showing the alternation of [Fe(bpp)$_2$]$^{2+}$ cationic layers and nicotinate anionic layers. The iron complexes are illustrated in red whereas the two independent nicotinate anions are depicted in orange and green. The dashed lines refer to hydrogen bonds. Only the H atoms of water molecules are shown.

The nicotinate anions and water molecules occupy the space between the [Fe(bpp)$_2$]$^{2+}$ cationic layers. A view parallel to the sheets (Figure 2) shows the two independent nicotinate anions depicted in different colours. The first one (illustrated in orange in Figure 2) is hydrogen-bonded to three water molecules (O1W, O2W and O3W) and one [Fe(bpp)$_2$]$^{2+}$ cation (N5). The other nicotinate anion (depicted in green) is also hydrogen-bonded to three water molecules (O1W, O2W and O4W) and one [Fe(bpp)$_2$]$^{2+}$ cation (N10). Therefore, the common feature is that they are connected to only one Fe$^{2+}$ centre using one oxygen atom that is simultaneously hydrogen-bonded to one water molecule, the other oxygen atom and the nitrogen of the pyridine unit being engaged in hydrogen bonding with additional water molecules. Both anions alternate in the interlayer space along the *y* direction. Relevant hydrogen-bonding parameters are gathered in Table S2.

The cationic layers exhibit the typical pseudotetragonal *terpyridine embrace* motif, in which each [Fe(bpp)$_2$]$^{2+}$ cation interacts with four neighbouring iron units via π-π stacking interactions (Figure 3). Equivalent bpp ligands form stacks along the *a* and *b* axes with the shortest distances between adjacent pyrazolyl mean planes equal to 3.256(3) Å (C11···N1) and 3.284(3) Å (C22···N6), respectively. Across the third direction, the iron complexes of consecutive layers are twisted with respect to each other and shifted along the *y* direction, resulting in an alternated packing of the type AA'AA'.

A projection of the crystal packing onto the *ab* plane (Figure 4) shows the presence of two different types of water molecules: the ones that form the layers through direct hydrogen bonds with the bpp ligands (O3W and O4W), and the ones that fill the voids within the layers (O1W and O2W).

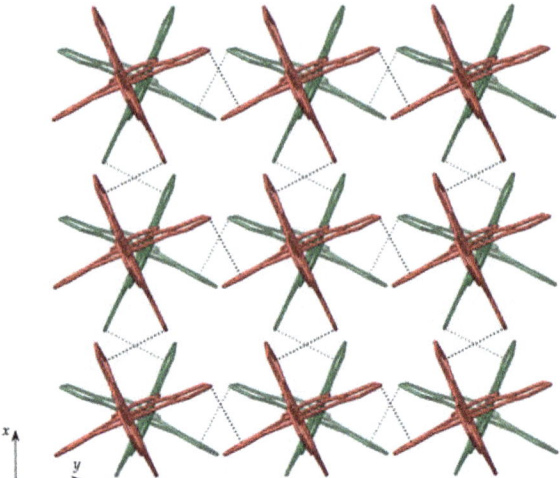

**Figure 3.** A view of the crystal packing of **1**·4H$_2$O parallel to the *c* axis showing two consecutive [Fe(bpp)$_2$]$^{2+}$ cationic layers in different colours (nicotinate anions and water molecules are omitted). The dotted lines refer to π-π stacking interactions. H atoms are not shown.

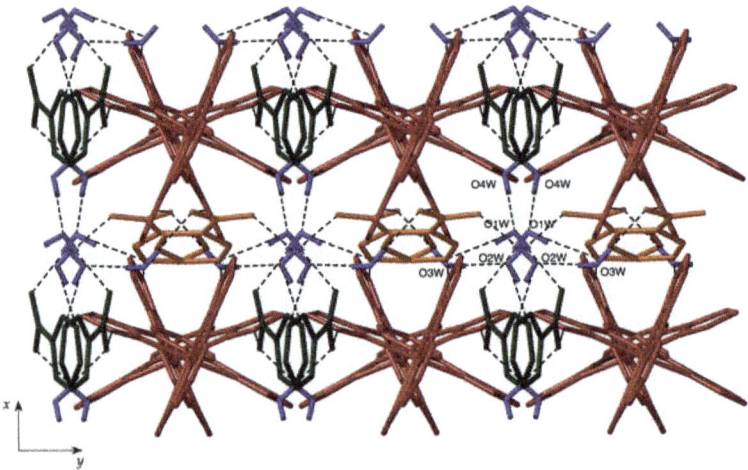

**Figure 4.** A view of the 2D packing of **1**·4H$_2$O along the *c* axis, showing the two different kinds of water molecules that are represented in blue: those filling the voids (O1W and O2W), and those forming the layers (O3W and O4W). Only the H atoms of water molecules are shown.

Powder X-ray diffraction (PXRD) experiments have been performed and Figure 5 shows the comparison between the X-ray diffractogram measured at room temperature of the bulk sample and the simulated diffractogram from the single crystal data recorded at 120 K. It can be seen that there is a good agreement between both patterns, discarding the presence of impurities and excluding the existence of a phase transition between 120 K and room temperature.

Temperature-dependent PXRD data (Figure S4) show that the sample maintained the same diffraction pattern upon heating from 298 K to 353 K, where at least three H$_2$O molecules were lost. Instead, heating up to 400 K resulted in amorphization. We attribute this change to complete dehydration (loss of the fourth water molecule). The discrepancy between this behaviour and the thermogravimetric data comes from the fact that TGA is performed at a high-temperature sweeping rate, whereas PXRD is done in static conditions. We also used PXRD to check the integrity of the sample after dehydration. For that purpose, a sample of $1 \cdot 4H_2O$ was placed in a Schlenk tube and heated under vacuum at 400 K for 2 h. Then, the sample was allowed to rehydrate. An identical diffractogram to the one recorded for the original sample was obtained (Figure S5), even after another dehydration-rehydration cycle, confirming both the integrity of the sample and the reversibility of the dehydration process.

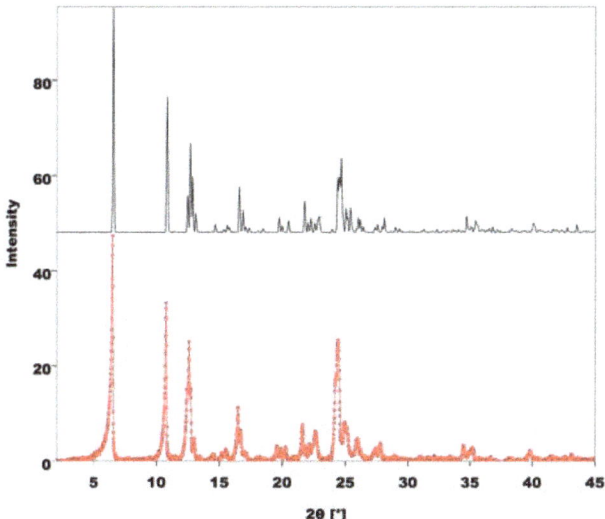

**Figure 5.** Comparison of the powder diffractogram of $1 \cdot 4H_2O$ (in red) with the simulation obtained from single crystal data (in black).

*3.3. Magnetic Properties*

The temperature dependence of the $\chi T$ product ($\chi$ = molar magnetic susceptibility; $T$ = absolute temperature) is shown in Figure 6. At low temperatures, the sample is diamagnetic, as expected for a low-spin ground state for this Fe(II) complex. Above room temperature, $\chi T$ increases continuously on heating up to 400 K, where it equals 3.21 emu·K·mol$^{-1}$. This value is the one expected for a high-spin $Fe^{2+}$ cation per formula ($\geq$3.0 emu·K·mol$^{-1}$), indicating a complete spin crossover triggered by the dehydration of the sample. Upon cooling, $\chi T$ decreases continuously and reaches a critical point at 283 K (corresponding to the first transition seen in DSC, cooling mode), where $\chi T$ = 2.21 emu·K·mol$^{-1}$ (fraction of high-spin centres $\gamma_{HS} \approx 2/3$). Then, it decreases more abruptly until $T$ = 273 K, and then smoothly to reach a plateau below 83 K, where $\chi T$ = 0.41 emu·K·mol$^{-1}$ (fraction of high-spin centres $\gamma_{HS} \approx 1/8$). Below 15 K, $\chi T$ decreases sharply due to zero-field splitting effects.

The second heating curve matches the cooling plot until 245 K, when a hysteresis loop opens ($\Delta T \approx 30$ K). The hysteretic behaviour disappears above a critical point located at 307 K, where the cooling and heating curves merge and both reach practically a saturation value corresponding to a 100% HS material. The width of the thermal hysteresis loop is highly dependent on the thermal treatment at 400 K. If this annealing is avoided, a very similar $\chi T = f(T)$ plot is obtained (showing two discontinuities around 280 K and 290 K, and a limiting low-temperature value corresponding to $\gamma_{HS} \approx 1/8$), but now the hysteresis disappears. It is thus important to ensure the complete dehydration and thermal annealing of the sample to obtain reproducible results.

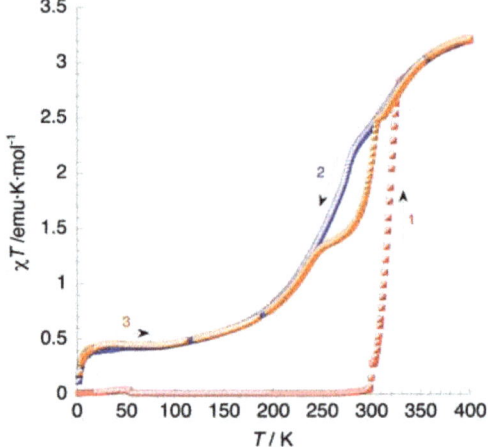

**Figure 6.** Thermal variation of $\chi T$ of **1**·4H$_2$O. Curve 1: first heating process (the bump observed around 50 K is due to the presence of oxygen). Curves 2 and 3: first cooling and second heating processes, respectively.

## 3.4. Photomagnetic Studies

Figure 7 shows the results of the photomagnetic experiments recorded using a SQUID magnetometer. Before irradiation, the sample was dehydrated in situ in the SQUID device at 400 K for 1 h. Then, at 10 K, the LS sample (initially with almost all the iron centres in their LS states) was irradiated using green laser light ($\lambda$ = 532.06 nm, no photomagnetic response was observed under red light). There was an abrupt increase in magnetic susceptibility, which reached saturation at a value of ca. 2 emu·K·mol$^{-1}$, indicating approximately a 2/3 conversion of the iron sites to the metastable HS state. The fraction of iron sites that is reluctant to photoexcitation corresponds to the fraction of HS centres undergoing SCO at high temperatures (higher than room temperature). The light was then turned off and the sample warmed in the dark at a heating rate of 0.3 K/min, according to the standardised $T$(LIESST) procedure [28]. The compound remains in its trapped HS state until $T$(LIESST) = 56 K (calculated as the minimum of the derivative $\delta(\chi T)/\delta T$). At higher temperatures, complete relaxation of the metastable HS content is seen and $\chi T$ reaches a minimum value of 0.5 emu·K·mol$^{-1}$, close to the residual fraction of HS centres (1/8) measured before irradiation.

A database containing more than sixty iron(II) spin-crossover materials with nitrogen-donor ligands [28,40] made it possible to establish empirical correlations between $T$(LIESST) and the thermal spin transition temperature, $T_{1/2}$. Based on these data, it has been shown that the following linear expression:

$$T(\text{LIESST}) = T_0 - 0.3 T_{1/2} \tag{1}$$

governs the photomagnetic properties of most of these compounds, where $T_0$ is an empirical parameter that depends mainly on the distortion of the octahedral coordination sphere (for [Fe(bpp)$_2$]$^{2+}$ complexes: $T_0 = 150$ K) [28].

Considering that $T_{1/2} = 281$ K (the average value between $T_{1/2}^{\uparrow}$ and $T_{1/2}^{\downarrow}$), Equation 1 yields $T$(LIESST) = 66 K, which is in rough agreement with the experimental value.

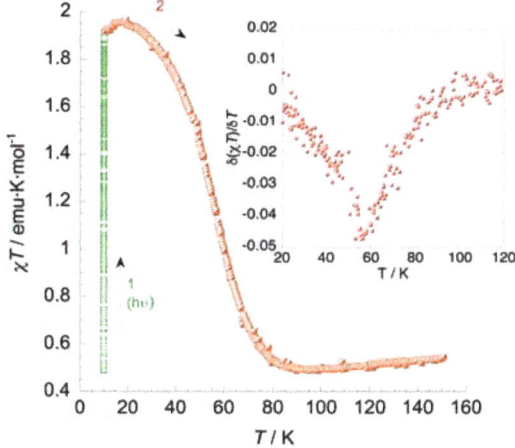

**Figure 7.** Photomagnetic properties of **1**. Curve 1 (green): Irradiation at 10 K. Curve 2 (orange): Thermal variation of $\chi T$ upon heating at a scan rate of 0.3 K·min$^{-1}$. The inset shows the first derivative plot of the $T$(LIESST) curve.

## 4. Conclusions

In this paper, we have described the synthesis, structure, and magnetic properties of a new hydrogen-bonded network that contains [Fe(bpp)$_2$]$^{2+}$ cations and nicotinate anions. It is tempting to compare **1**·4H$_2$O with the previously reported [Fe(bpp)$_2$](isonic)$_2$·2H$_2$O [32]. In this compound, thanks to the linearity of the isonicotinate anion, robust hydrogen bonds can be established between the bpp ligands and the isonicotinate anions occupying the four available positions in the second iron coordination sphere. This provides a strong directionality and allows for the prediction of the connectivity of the resulting structure. Nevertheless, in the case of compound **1**·4H$_2$O, the angular character of the nicotinate anion makes the formation of these interactions more difficult, allowing for the presence of water molecules in the second iron coordination sphere. Another possible reason for the different behaviour of these similar salts could be the slightly different basicities of the nicotinate (p$K$ = 4.77) and isonicotinate (p$K$ = 4.90) anions, the latter being a stronger N-donor [41]. In any case, the final result is that π-π stacking interactions are predominant and the *terpyridine embrace* motif is stabilized.

Differential scanning calorimetry analysis and magnetic characterization of compound **1**·4H$_2$O show a conversion from the LS state to the HS state after the loss of water molecules. The anhydrous material obtained exhibits spin crossover with relatively high cooperativity and with an associated hysteretic behaviour. Furthermore, this compound presents an LIESST effect and can be partially switched from the diamagnetic LS state to the paramagnetic HS state using green light at 10 K. The value of $T$(LIESST) obtained (56 K) is consistent with those previously reported for other iron(II) complexes of bpp ligands [37]. Unfortunately, the compound loses its crystallinity upon dehydration and the lack of a structural characterization of the anhydrous phase precludes the establishment of structure-property correlations.

**Supplementary Materials:** The following are available online at http://www.mdpi.com/2073-4352/8/11/439/s1: Figure S1: Thermogravimetric analysis of Ag(C$_6$H$_4$NO$_2$); Figure S2: Thermogravimetric analysis of **1**·4H$_2$O; Figure S3: Differential scanning calorimetry of **1**·4H$_2$O; Figures S4–S5: Powder X-ray diffractograms of **1**·4H$_2$O after dehydration and rehydration; Table S1: Entalphy and entropy values for peaks observed in the DSC curve; Table S2: Intermolecular hydrogen bonds in the crystal structure of **1**·4H$_2$O; CIF data.

**Author Contributions:** V.J.-M. and F.M.R. jointly designed the experiments, made the synthesis and characterization of the compounds, analysed their thermal, structural, and magnetic properties, and wrote the manuscript. C.G.-S. collected the X-ray data and solved the crystal structure.

**Acknowledgments:** We wish to thank José M. Martínez-Agudo and Gloria Agustí for the magnetic measurements. We acknowledge financial support from the Spanish Ministerio de Ciencia, Innovación y Universidades (MICINN project CTQ2017-87201-P). V.J.-M. also extends thanks to the MICINN for an FPU fellowship (FPU15/02804).

**Conflicts of Interest:** The authors declare no conflict of interest.

## References

1. Bartual-Murgui, C.; Akou, A.; Thibault, C.; Molnár, G.; Vieu, C.; Salmon, L.; Bousseksou, A. Spin-crossover metal-organic frameworks: Promising materials for designing gas sensors. *J. Mater. Chem. C* **2015**, *3*, 1277–1285. [CrossRef]
2. Kahn, O. Spin-transition polymers: From molecular materials toward memory devices. *Science* **1998**, *279*, 44–48. [CrossRef]
3. Senthil Kumar, K.; Ruben, M. Emerging trends in spin crossover (SCO) based functional materials and devices. *Coord. Chem. Rev.* **2017**, *346*, 176–205. [CrossRef]
4. Baadji, N.; Sanvito, S. Giant resistance change across the phase transition in spin-crossover molecules. *Phys. Rev. Lett.* **2012**, *108*, 1–5. [CrossRef] [PubMed]
5. Kahn, O. Spin-crossover molecular materials. *Curr. Opin. Solid State Mater. Sci.* **1996**, *1*, 547–554. [CrossRef]
6. Kahn, O. *Molecular Magnetism*; Wiley-VCH: New York, NY, USA; pp. 10010–14906.
7. Matsumoto, T.; Newton, G.N.; Shiga, T.; Hayami, S.; Matsui, Y.; Okamoto, H.; Kumai, R.; Murakami, Y.; Oshio, H. Programmable spin-state switching in a mixed-valence spin-crossover iron grid. *Nature Comm.* **2014**, *5*, 1–8. [CrossRef] [PubMed]
8. Li, Z.Y.; Ohtsu, H.; Kojima, T.; Dai, J.W.; Yoshida, T.; Breedlove, B.K.; Zhang, W.X.; Iguchi, H.; Sato, O.; Kawano, M.; Yamashita, M. Direct observation of ordered high-spin-low-spin intermediate states of an iron(III) three-step spin-srossover somplex. *Angew. Chem. Int. Ed.* **2016**, *55*, 5184–5189. [CrossRef] [PubMed]
9. Osorio, E.A.; Moth-Poulsen, K.; Van Der Zant, H.S.J.; Paaske, J.; Hedegård, P.; Flensberg, K.; Bendix, J.; Bjørnholm, T. Electrical manipulation of spin states in a single electrostatically gated transition-metal complex. *Nano Lett.* **2010**, *10*, 105–110. [CrossRef] [PubMed]
10. Baadji, N.; Piacenza, M.; Tugsuz, T.; Della Sala, F.; Maruccio, G.; Sanvito, S. Electrostatic spin crossover effect in polar magnetic molecules. *Nature Mater.* **2009**, *8*, 813–817. [CrossRef] [PubMed]
11. Pinkowicz, D.; Rams, M.; Misek, M.; Kamenev, K.V.; Tomkowiak, H.; Katrusiak, A.; Sieklucka, B. Enforcing multifunctionality: A pressure-induced spin-crossover photomagnet. *J. Am. Chem. Soc.* **2015**, *137*, 8795–8802. [CrossRef] [PubMed]
12. Gallois, B.; Real, J.A.; Hauw, C.; Zarembowitch, J. Structural changes associated with the spin transition in bis(isothiocyanato)bis(1,10-phenanthroline)iron: A single-crystal x-ray investigation. *Inorg. Chem.* **1990**, *29*, 1152–1158. [CrossRef]
13. Miyamachi, T.; Gruber, M.; Davesne, V.; Bowen, M.; Boukari, S.; Joly, L.; Scheurer, F.; Rogez, G.; Yamada, T.K.; Ohresser, P.; et al. Robust spin crossover and memristance across a single molecule. *Nature Commun.* **2012**, *3*, 936–938. [CrossRef] [PubMed]
14. Bousseksou, A.; Molnár, G.; Demont, P.; Menegotto, J. Observation of a thermal hysteresis loop in the dielectric constant of spin crossover complexes: Towards molecular memory devices. *J. Mater. Chem.* **2003**, *13*, 2069–2071. [CrossRef]
15. Bousseksou, A.; Molnár, G.; Salmon, L.; Nicolazzi, W. Molecular spin crossover phenomenon: Recent achievements and prospects. *Chem. Soc. Rev.* **2011**, *40*, 3313–3335. [CrossRef] [PubMed]
16. Aromí, G.; Aguilà, D.; Gamez, P.; Luis, F.; Roubeau, O. Design of magnetic coordination complexes for quantum computing. *Chem. Soc. Rev.* **2012**, *41*, 537–546. [CrossRef] [PubMed]
17. Khusniyarov, M.M. How to switch spin-crossover metal complexes at constant room temperature. *Chem. Eur. J.* **2016**, *22*, 15178–15191. [CrossRef] [PubMed]
18. Gütlich, P.; Gaspar, A.B.; Garcia, Y. Spin state switching in iron coordination compounds. *Beilstein J. Org. Chem.* **2013**, *9*, 342–391. [CrossRef] [PubMed]
19. Niel, V.; Thompson, A.L.; Muñoz, M.C.; Galet, A.; Goeta, A.E.; Real, J.A. Crystalline-state reaction with allosteric effect in spin-crossover, interpenetrated networks with magnetic and optical bistability. *Angew. Chem. Int. Ed.* **2003**, *42*, 3760–3763. [CrossRef] [PubMed]

20. Niel, V.; Thompson, A.L.; Goeta, A.E.; Enachescu, C.; Hauser, A.; Galet, A.; Muñoz, M.C.; Real, J.A. Thermal- and photoinduced spin-state switching in an unprecedented three-dimensional bimetallic coordination polymer. *Chem. Eur. J.* **2005**, *11*, 2047–2060. [CrossRef] [PubMed]
21. Piñeiro-López, L.; Valverde-Muñoz, F.J.; Seredyuk, M.; Muñoz, M.C.; Haukka, M.; Real, J.A. Guest induced strong cooperative one- and two-step spin transitions in highly porous iron(ii) hofmann-type metal-organic frameworks. *Inorg. Chem.* **2017**, *56*, 7038–7047. [CrossRef] [PubMed]
22. Murphy, M.J.; Zenere, K.A.; Ragon, F.; Southon, P.D.; Kepert, C.J.; Neville, S.M. Guest programmable multlstep spin crossover in a porous 2-D Hofmann-type material. *J. Am. Chem. Soc.* **2017**, *139*, 1330–1335. [CrossRef] [PubMed]
23. Scudder, M.L.; Craig, D.C.; Goodwin, H.A. Hydrogen bonding influences on the properties of heavily hydrated chloride salts of iron(II) and ruthenium(II) complexes of 2,6-bis(pyrazol-3-yl)pyridine, 2,6-bis(1,2,4-triazol-3-yl)pyridine and 2,2′:6′,2″-terpyridine. *Cryst. Eng. Comm.* **2005**, *7*, 642–649. [CrossRef]
24. Sugiyarto, K.H.; Scudder, M.L.; Craig, D.C.; Goodwin, H.A. Electronic and structural properties of the spin crossover systems bis(2,6-bis(pyrazol-3-yl)pyridine)iron(II)thiocyanate and selenocyanate. *Aust. J. Chem.* **2000**, *53*, 755–765. [CrossRef]
25. Lochenie, C.; Bauer, W.; Railliet, A.P.; Schlamp, S.; Garcia, Y.; Weber, B. Large thermal hysteresis for iron(II) spin crossover complexes with N-(Pyrid-4-yl)isonicotinamide. *Inorg. Chem.* **2014**, *53*, 11563–11572. [CrossRef] [PubMed]
26. Weber, B.; Bauer, W.; Obel, J. An iron(II) spin-crossover complex with a 70 K wide thermal hysteresis loop. *Angew. Chem. Int. Ed.* **2008**, *47*, 10098–10101. [CrossRef] [PubMed]
27. Craig, G.A.; Roubeau, O.; Aromí, G. Spin state switching in 2,6-bis(pyrazol-3-yl)pyridine (3-bpp) based Fe(II) complexes. *Coord. Chem. Rev.* **2014**, *269*, 13–31. [CrossRef]
28. Létard, J.F.; Guionneau, P.; Nguyen, O.; Costa, J.S.; Marcén, S.; Chastanet, G.; Marchivie, M.; Goux-Capes, L. A guideline to the design of molecular-based materials with long-lived photomagnetic lifetimes. *Chem. Eur. J.* **2005**, *11*, 4582–4589. [CrossRef] [PubMed]
29. Marcén, S.; Lecren, L.; Capes, L.; Goodwin, H.A.; Létard, J.F. Critical temperature of the LIESST effect in a series of hydrated and anhydrous complex salts [Fe(bpp)$_2$]X$_2$. *Chem. Phys. Lett.* **2002**, *358*, 87–95. [CrossRef]
30. Buchen, T.; Gütlich, P.; Sugiyarto, K.H.; Goodwin, H.A. High-spin → low-spin relaxation in [Fe(bpp)$_2$](CF$_3$SO$_3$)$_2$·H$_2$O after LIESST and thermal spin-state trapping - Dynamics of spin transition versus dynamics of phase transition. *Chem. Eur. J.* **1996**, *2*, 1134–1138. [CrossRef]
31. Clemente-León, M.; Coronado, E.; Giménez-López, M.C.; Romero, F.M.; Asthana, S.; Desplanches, C.; Létard, J.-F. Structural, thermal and photomagnetic properties of spin crossover [Fe(bpp)$_2$]$^{2+}$ salts bearing [Cr(L)(ox)$_2$]$^-$ anions. *Dalton Trans.* **2009**, *38*, 8087–8095. [CrossRef] [PubMed]
32. Jornet-Mollá, V.; Duan, Y.; Giménez-Saiz, C.; Tang, Y.Y.; Li, P.F.; Romero, F.M.; Xiong, R.G. A ferroelectric iron(ii) spin crossover material. *Angew. Chem. Int. Ed.* **2017**, *56*, 14052–14056. [CrossRef] [PubMed]
33. Lin, Y.; Lang, S.A. Novel two step synthesis of pyrazoles and isoxazoles from aryl methyl ketones. *J. Heterocycl. Chem.* **1977**, *14*, 345–347. [CrossRef]
34. Sheldrick, G.M. SHELX Version 2014/7. Available online: http://shelx.uni-ac.gwdg.de/SHELX/index.php (accessed on 20 November 2018).
35. Clemente-León, M.; Coronado, E.; Giménez-López, M.C.; Romero, F.M. Structural, thermal, and magnetic study of solvation processes in spin-crossover [Fe(bpp)$_2$][Cr(L)(Ox)$_2$]$_2$ nH$_2$O complexes. *Inorg. Chem.* **2007**, *46*, 11266–11276. [CrossRef] [PubMed]
36. Coronado, E.; Giménez-López, M.C.; Giménez-Saiz, C.; Romero, F.M. Spin crossover complexes as building units of hydrogen-bonded nanoporous structures. *Cryst. Eng. Comm.* **2009**, *11*, 2198–2203. [CrossRef]
37. Jornet-Mollá, V.; Duan, Y.; Giménez-Saiz, C.; Waerenborgh, J.C.; Romero, F.M. Hydrogen-bonded networks of [Fe(bpp)$_2$]$^{2+}$ spin crossover complexes and dicarboxylate anions: Structural and photomagnetic properties. *Dalton Trans.* **2016**, *45*, 17918–17928. [CrossRef] [PubMed]
38. Sugiyarto, K.H.; McHale, W.A.; Craig, D.C.; Rae, A.D.; Scudder, M.L.; Goodwin, H.A. Spin transition centres linked by the nitroprusside ion. the cooperative transition in bis(2,6-bis(pyrazol-3-yl)-pyridine)iron(II) nitroprusside. *Dalton Trans.* **2003**, *12*, 2443–2448. [CrossRef]
39. Sugiyarto, K.H.; Goodwin, H.A. Coordination of Pyridine-Substituted Pyrazoles and Their Influence on the Spin State of Iron(II). *Aust. J. Chem.* **1988**, *41*, 1645–1663. [CrossRef]

40. Létard, J.F. Photomagnetism of iron(ii) spin crossover complexes—The T(LIESST) approach. *J. Mater. Chem.* **2006**, *16*, 2550–2559. [CrossRef]
41. Jaffé, H.H.; Doak, G.O. The basicities of substituted pyridines and their 1-oxides. *J. Am. Chem. Soc.* **1955**, *77*, 4441–4444. [CrossRef]

 © 2018 by the authors. Licensee MDPI, Basel, Switzerland. This article is an open access article distributed under the terms and conditions of the Creative Commons Attribution (CC BY) license (http://creativecommons.org/licenses/by/4.0/).

Article

# Soft X-ray Absorption Spectroscopy Study of Spin Crossover Fe-Compounds: Persistent High Spin Configurations under Soft X-ray Irradiation

Ahmed Yousef Mohamed [1,2], Minji Lee [1], Kosuke Kitase [3], Takafumi Kitazawa [3,4], Jae-Young Kim [5] and Deok-Yong Cho [1,*]

1. IPIT & Department of Physics, Chonbuk National University, Jeonju 54896, Korea; yousef@jbnu.ac.kr (A.Y.M.); mj9834@jbnu.ac.kr (M.L.)
2. Department of Physics, South Valley University, Qena 83523, Egypt
3. Department of Chemistry, Toho University, Chiba 274-8510, Japan; 6117004k@st.toho-u.ac.jp (K.K.); kitazawa@chem.sci.toho-u.ac.jp (T.K.)
4. Research Centre for Materials with Integrated Properties, Toho University, Chiba 274-8510, Japan
5. Center for Artificial Low Dimensional Electronic Systems, Institute for Basic Science (IBS), Pohang 37673, Korea; jaeyoung@ibs.re.kr
* Correspondence: zax@jbnu.ac.kr; Tel.: +82-63-270-3444

Received: 19 October 2018; Accepted: 16 November 2018; Published: 19 November 2018

**Abstract:** Metal-organic complex exhibiting spin crossover (SCO) behavior has drawn attention for its functionality as a nanoscale spin switch. The spin states in the metal ions can be tuned by external stimuli such as temperature or light. This article demonstrates a soft X-ray–induced excited spin state trapping (SOXEISST) effect in Hofmann-like SCO coordination polymers of $Fe^{II}$(4-methylpyrimidine)$_2$[Au(CN)$_2$]$_2$ and $Fe^{II}$(pyridine)$_2$[Ni(CN)$_4$]. A soft X-ray absorption spectroscopy (XAS) study on these polymers showed that the high spin configuration (HS; S = 2) was prevalent in $Fe^{2+}$ ions during the measurement even at temperatures much lower than the critical temperatures (>170 K), manifesting HS trapping due to the X-ray irradiation. This is in strong contrast to the normal SCO behavior observed in $Fe^{II}$(1,10-phenanthroline)$_2$(NCS)$_2$, implying that the structure of the ligand chains in the polymers with relatively loose Fe-N coordination might allow a structural adaptation to stabilize the metastable HS state under the soft X-ray irradiation.

**Keywords:** spin crossover; X-ray absorption spectroscopy; soft X-ray induced excited spin state trapping; high spin

## 1. Introduction

Since the first report on spin-crossover (SCO) molecules from Cambi and Szegö in 1931 [1], metal-organic complexes have been the subjects of many studies [2–5] due to the possibility of their use as nanoscale molecular spin switches [6–10]. The SCO phenomenon which involves a change in the spin state of the metal ion, can be caused by many kinds of external stimuli including variation of temperature [11,12] or pressure [13], or the influence of ligand chemistry [14], light or X-ray irradiation [15–22] or electromagnetic field [23,24]. When the structures of the molecular networks change in a cooperative manner associated with the spin crossover, the transition occurs with steepness and/or sometimes is accompanied by a hysteresis loop (the first order transition). Thus, the SCO molecular materials can offer many possible applications in the fields of electronics, information storage, digital display, photonics or photo-magnetism, etc.

In most cases of the SCO metal complexes reported so far, the change in the spin state occurs from a low spin (LS) ground state configuration to a high spin (HS) metastable configuration of the metal ion's $d$ electrons. At low temperatures below a transition temperature, $T_C$, the LS configuration is prevalent, while at temperatures higher than $T_C$, HS configuration becomes dominant. The population ratio between the LS and HS configurations at a given temperature, is determined by the competition of the Hund coupling in the $d$ shell with the strength of the ligand fields from the nearest atoms. According to the ligand field theory, the splitting of the $d$ orbital energies is determined by the coordination symmetry and bond lengths of the metal ions.

Figure 1 illustrates the metal's $d$ orbital exemplary splitting under octahedral ligand field for $Fe^{2+}$ ($d^6$) ions. The $d$ orbital energy splits into a triplet $t_{2g}$ level at a lower energy and a doublet $e_g$ level at a higher energy under the octahedral ligand field, and six electrons fill the levels. When all the electrons occupy the $t_{2g}$ levels only following the order of the ligand field splitting energy, the value of the total spin (S) should be zero, and thereby, the electron configuration is called LS [25]. Meanwhile, when two of the electrons occupy $e_g$ levels following the preference of the spin alignment (Hund's coupling), the value of S should be maximized (S = 2); thus, the configuration is called HS. To attain a HS configuration, each of the two electrons should overcome the ligand field splitting energies (10 Dq), but at the same time, obtain the energies of Hund exchanges with the other four electrons ($-4$ J), so that the energy difference between the HS and LS should be $\Delta E = 20\,Dq - 8\,J$. It has been shown that the ligand field strength (10 Dq) can significantly influence the population ratio [26,27]. Generally, as bond lengths of the metal ions to the ligands are longer, the HS configuration can be favored more because of the lower 10 Dq. Since the bond lengths tend to increase with increasing temperature, HS can be stabilized more at higher temperatures.

The SCO phenomena can be explored by various magnetic and spectroscopic techniques including magnetic susceptibility measurement, Mössbauer spectroscopy [28,29], optical absorption and X-ray absorption spectroscopy (XAS) [4,30,31]. In those experimental techniques, external stimuli are applied to the specimen in order to observe the responses related to the spin states. For instance, XAS at Fe $L_{2,3}$-edge utilizes the electronic excitation from the $2p$ core level to the $3d$ unoccupied state following the X-ray perturbation. Previous Fe $L_{2,3}$-edge XAS studies on a representative SCO molecule, Fe(phen)$_2$(NCS)$_2$ at low and high temperatures, have established a set of spectroscopic fingerprints for the HS and LS states [32–35].

However, one cannot exclude a possibility that the stimulus itself can change the properties of the specimen. Indeed, in the case of soft XAS, it has been reported that the spin state can change under the X-ray irradiation during the measurement. Such soft X-ray–induced excited spin state trapping (SOXIESST) effects can lead to persistent HS configuration even below $T_C$ [21,22]. The SOXIESST effect has been regarded as a result of temporary trapping in the HS due to the cascade of excitations by illumination, similar to the case of light-induced excited spin state trapping (LIESST) [26,27]. For instance, the SOXIESST effect was described in detail on Fe(phen)$_2$(NCS)$_2$ and Fe{[Me$_2$Pyrz]$_3$BH}$_2$ by V. Davesne et al. [21,36].

Earlier studies have suggested that the mechanism of the SOXIESST effect is analogous to that of the LIESST effect [19,21,22,35]; their similarities and differences were discussed by V. Davesne et al. [21,27,36]. Namely, the light or X-ray introduction leads to an excitation of the system from the LS ground state configuration to the excited state configurations and it follows that the excited state configurations decay into a metastable spin triplet HS configuration (intersystem crossing). Then the spin states can remain trapped in this excited HS state, unless the temperature is high enough to overcome the energy barrier between the HS and LS configurations via the thermal fluctuations [19,21,22,35]. However, when X-rays are turned off, the system again favors the LS ground state configuration. In this regard, SOXIESST is a reversible process. Meanwhile, the soft X-ray photochemistry (SOXPC) effect can alter the chemistry of specimens in an irreversible way. D. Collison et al. [22] showed that although the (reversible) SOXIESST effect is predominant in Fe(phen)$_2$(NCS)$_2$, the (irreversible) SOXPC effect also exists and plays in a competitive manner.

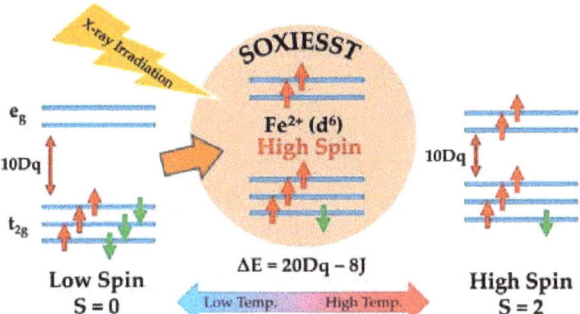

**Figure 1.** Illustration of the $Fe^{2+}$ ($d^6$) $d$ orbital energy splitting under octahedral ligand field and the resultant LS (low spin) and HS (high spin) configurations. The SCO (spin crossover) from LS to HS with increasing temperature can be explained by the weakened ligand field, i.e., lower 10 Dq. If a soft X-ray–induced excited spin state trapping (SOXIESST) effect is activated, HS configuration is persistent even at temperatures below $T_C$.

This article reports a SOXIESST effect in Hofmann-like SCO coordination polymers, $Fe(4\text{-methylpyrimidine})_2[Au(CN)_2]_2$ ($Fe(4\text{-methylpmd})_2[Au(CN)_2]_2$) and $Fe(pyridine)_2[Ni(CN)_4]$ ($Fe(py)_2[Ni(CN)_4]$). Figure 2 illustrates the molecular structures of the two polymers in comparison to a well-known SCO complex, $Fe(1,10\text{-phenanthroline})_2(NCS)_2$ ($Fe(phen)_2(NCS)_2$). Compared to the latter, the two polymers (Figure 2a) have relatively loose Fe-N coordinations, in that all the N ions in the ligands are in separate chains, while many of the N ions in $Fe(phen)_2(NCS)_2$ are bound to benzene-like rings (Figure 2b). The less rigid Fe-N coordination could somehow offer a ground for spin state transition upon external stimuli such as X-ray irradiation. Therefore, soft XAS was performed to examine the possible SOXIESST effects in the loosely coordinated $Fe^{2+}$ ions in the polymeric structures.

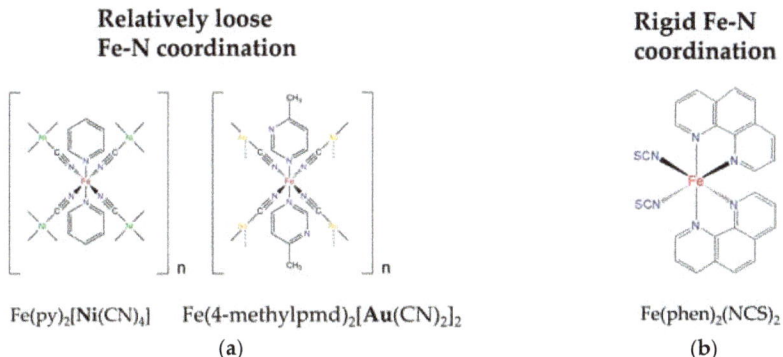

**Figure 2.** Molecular structures of (**a**) Hofmann-like SCO polymers of $Fe(4\text{-methylpmd})_2[Au(CN)_2]_2$ and ($Fe(py)_2[Ni(CN)_4]$), and (**b**) a reference $Fe(phen)_2(NCS)_2$.

## 2. Materials and Methods

$Fe(4\text{-methylpmd})_2[Au(CN)_2]_2$, $Fe(py)_2[Ni(CN)_4]$ and $Fe(phen)_2(NCS)_2$ powders were synthesized by the vapor diffusion methods. Details on the preparation for $Fe(py)_2[Ni(CN)_4]$ and $Fe(phen)_2(NCS)_2$ were the same as reported in References [8,37]. $Fe(4\text{-methylpmd})_2[Au(CN)_2]_2$ was prepared by the same method for the synthesis of $Fe(4\text{-methylpy})_2[Au(CN)_2]_2$ [9].

The compositions and structures of the three powder samples were identified by elemental analysis, powder X-ray diffraction (XRD), and infrared (IR) spectroscopy. The elemental analysis for C, H and N was carried out with CHN CORDER JM10 (Yanaco Corp., Tokyo, Japan). The results of the elemental analysis suggest a Hofmann-type-like formula of Fe(4-methylpmd)$_2$[Au(CN)$_2$]$_2$ (Found: C, 22.57%; H, 1.75%; N, 15.07%; Calculated: C, 22.66%; H, 1.68%; N, 15.10%), the Hofmann-type formula of Fe(py)$_2$[Ni(CN)$_4$] [8] (Found: C, 44.40%; H, 1.75%; N, 22.02%; Calculated: C, 44.62%; H, 1.68%; N, 22.31%), and the mononuclear complex formula of Fe(phen)$_2$(NCS)$_2$ [37] (Found: C, 58.44%; H, 3.32%; N, 15.63%; Calc. C, 58.65%; H, 3.03%; N, 15.791%).

IR spectra were obtained by the Nujol mull method using JASCO FTIR-4100 spectrometer (JASCO Corp., Hachioji, Japan). The spectra for the three samples are shown in Figure S1 in the Supplementary Information (SI). For Fe(4-methylpmd)$_2$[Au(CN)$_2$]$_2$, wavenumbers of the dips, $\nu_{max}$'s were 2170, 1619, 1599, 1556, 1492, 1325, 1014 and 845 in cm$^{-1}$ (Figure S1a). The CN stretching band at 2170 cm$^{-1}$ originates from the linkage between Au$^+$ and Fe$^{2+}$ ions. For Fe(py)$_2$[Ni(CN)$_4$], $\nu_{max}$'s were 2158, 1603, 1573, 1218, 1152, 1038, 1011, 751, 690, 626, 437 and 419 in cm$^{-1}$ (Figure S1b). The CN stretching band (2158 cm$^{-1}$) is due to linkage between Ni$^{2+}$ and Fe$^{2+}$ ions. For Fe(phen)$_2$(NCS)$_2$, $\nu_{max}$'s were 2072 and 2060 in cm$^{-1}$ (Figure S1c).

Powder XRD pattern of Fe(4-methylpmd)$_2$[Au(CN)$_2$]$_2$ was obtained by using a Rigaku RINT2500 diffractometer (Rigaku Corp., Akishima, Japan) with graphite-monochromated Cu K$\alpha$ radiation ($\lambda$ = 1.5406 Å). Figure S2 in SI shows the XRD pattern. It is similar to that from Fe(4-methylpy)$_2$[Au(CN)$_2$]$_2$ [9].

Thermal decomposition of Fe(4-methylpmd)$_2$[Au(CN)$_2$]$_2$ was investigated on TG/DTA6200 (SII Nano Technology, Chiba, Japan) under a dry N$_2$ gas flow by recording the thermogravimetric (TG) curve. Figure S3 in SI shows the TG curve. The 25.4% weight loss at the plateau between 577 K and 617 K corresponds to the thermal decomposition of Fe(4-methylpmd)$_2$[Au(CN)$_2$]$_2$ into Fe[Au(CN)$_2$]$_2$.

Magnetic susceptibility was measured in the temperature range of 4–300 K with a cooling and heating rate of 2 K/min in a 0.1 Tesla field using a MPMS-XL Quantum Design SQUID magnetometer.

For the XAS, the three powder samples were attached gently on carbon tapes. Fe L$_{2,3}$-edge XAS was performed at 2 A beamline in Pohang Light Source. Absorption coefficients were collected with increasing photon energy in total electron yield mode at various temperatures. The intensity of the incident X-rays was estimated to be approximately 3 × 10$^{11}$ photons/s/mm$^2$. The beam flux was sufficiently low so that a SOXIESST did not occur in the reference Fe(phen)$_2$(NCS)$_2$ powder as is evident in the XAS spectra. The samples were not intentionally exposed to X-rays before the measurement, and the measurement time for each data was approximately 10 min.

## 3. Results and Discussion

### 3.1. Magnetic Susceptibility

Figure 3 shows the magnetic susceptibility of the specimens. Theoretically, magnetic susceptibility ($\chi_m$) of a paramagnet is proportional to $J(J + 1)/T$, where $J$ is the total angular momentum quantum number and $T$ is the temperature. All the data from the specimens show clear step-like jumps of $\chi_m T$ with increasing temperature at $T_C$'s ranging from 175 K to 205 K [38,39], suggesting an increase of the $J$ value at high temperature. Generally, such an increase is caused by the increase in the spin angular momentum, while the angular momentum in the metal ion under the octahedral coordination is generally much smaller than the nominal value (two for $d$-orbitals) due to the orbital momentum quenching effect. Therefore, we can tell that for instance, at $T$ = 110 K, the three powder samples were all in the LS (S = 0) state, while at room temperature, those were all in the HS (S = 2) state. Thus, it is clearly shown in the magnetization data that all the samples show strong SCO behaviors (when no soft X-ray beams were applied). It might be interesting to note that in the case of Fe(4-methylpy)$_2$[Au(CN)$_2$]$_2$, two steps were observed without a hysteresis, similar to the case of Fe(3-F-4-methylpy)$_2$[Au(CN)$_2$]$_2$ [10]. Details in the multistep SCO will be discussed elsewhere. The irregular noises in the data of Fe(phen)$_2$(NCS)$_2$ at temperatures below 170 K, presumably originate from nonuniform sample packing due to a relatively small amount of the specimen.

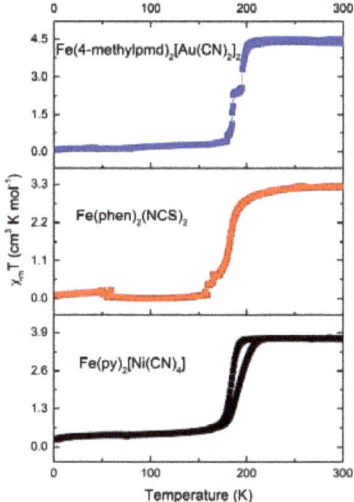

**Figure 3.** Magnetic susceptibility multiplied by temperature ($\chi_m T$) showing the LS-to-HS SCO with increasing temperature across $T_C$'s ranging from 175 K to 205 K.

*3.2. Soft X-ray Absorption Spectroscopy*

Figure 4 shows the Fe $L_{2,3}$-edge XAS spectra of Fe(phen)$_2$(NCS)$_2$ complex recorded at 300 and 110 K. The spectra show that the peak positions and the lineshapes evolved with temperature remarkably. Compared to the spectrum taken at 300 K, the spectrum taken at 110 K shifted to higher energies overall; 1.7 eV for the $L_3$ (features around 710 eV) and 0.6 eV for the $L_2$-edge maxima (features around 723 eV). Also, the $L_2$-edge features become consolidated at the lower temperature.

For comparison, the spectra for the same composition from a literature [34] are appended as the dotted curves in the figure. Each of the spectra (taken at 300 K and 110 K) resemble the respectively dotted curves. In many literatures, the lineshapes of the spectra have been interpreted as the fingerprints of the HS Fe$^{2+}$ (for 300 K) and LS Fe$^{2+}$ (for 110 K) under octahedral ligand fields [22,24,32,34,40]. The clear spectral evolution upon the temperature increase suggests that a certain SOXIESST effect has not occurred in the Fe(phen)$_2$(NCS)$_2$ complex. Therefore, we can conclude that the intensity of the soft X-rays (~3 × 10$^{11}$ photons/sec/mm$^2$) was not high enough to evoke a SOXIESST effect in the reference Fe(phen)$_2$(NCS)$_2$ powder. Compared to the case of thin film samples [32–35], the powder specimen appears more resistant to the SOXIESST effect upon X-ray exposure. Probably, this is because the powders are distributed sparsely so that the actual intensity of the beam shed on the SCO complex should be much weaker than the nominal value. As can be seen in Figure 2, the Fe-N coordination in Fe(phen)$_2$(NCS)$_2$ complex is rigid, being surrounded by benzene-like rings, so the Fe-N bond lengths, which are closely related to the preference of HS/LS population, would hardly change under the external stimuli like X-ray irradiation with a low flux.

On the other hand, the spectra of the two polymers clearly shows the SOXIESST effect. Figure 5 shows the Fe $L_{2,3}$-edge XAS spectra of the Fe$^{II}$ complexes with polymeric structures. For both compositions of (a) Fe(4-methylpmd)$_2$[Au(CN)$_2$]$_2$ and (b) Fe(py)$_2$[Ni(CN)$_4$], the lineshapes and the peak positions in the spectra were mostly similar to that of a HS Fe$^{2+}$, manifesting a prevalence of HS even at 110 K, far below the $T_C$'s (>170 K). Note that, however, the persistent HS is not consistent with the observation in the magnetic susceptibility measurement (Figure 3), which showed clear LS-to-HS SCO behaviors for all the samples. Then the persistent HS observed in soft XAS can be understood only by considering certain excitation effects of the soft X-ray introduction;

under the X-ray irradiation, the system might be excited to possess a metastable HS configuration even at low temperatures.

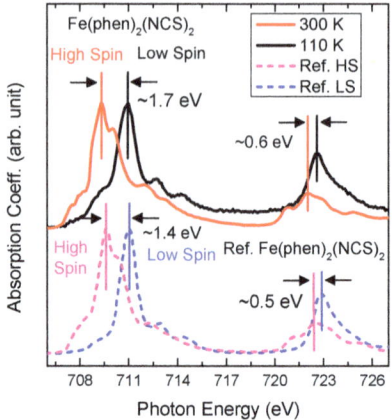

**Figure 4.** Fe $L_{2,3}$-edge XAS spectra of Fe(phen)$_2$(NCS)$_2$ complex recorded at 300 K (above $T_C$) and 110 K (below $T_C$), with reference spectra from the same composition [34], clearly showing a LS-HS SCO without any SOXIESST effect.

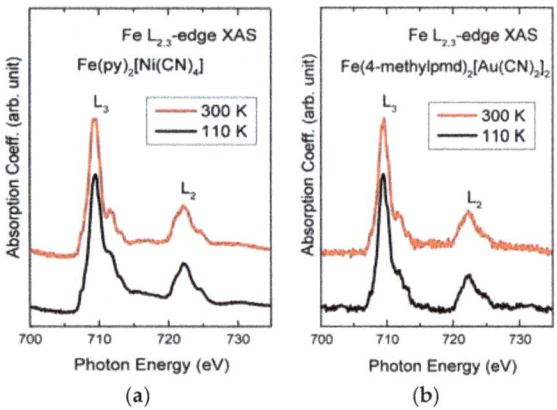

**Figure 5.** Fe $L_{2,3}$-edge XAS spectra of (**a**) Fe(4-methylpmd)$_2$[Au(CN)$_2$]$_2$ and (**b**) Fe(py)$_2$[Ni(CN)$_4$] polymers. HS configuration was dominant even at 110 K (below $T_C$), suggesting SOXIESST effects for the loosely coordinated polymers.

This is in clear contrast to the case of reference, Fe(phen)$_2$(NCS)$_2$ (Figure 4), which showed no SOXIESST effect. Although the three spectra were taken with the same beam flux, the two polymer samples appear to possess HS state while Fe(phen)$_2$(NCS)$_2$ was relatively resistant to the weak X-rays. The distinction might originate from the different structures of the bonds. Since the bond chains in the two polymers are formed only in radial directions with respect to the Fe$^{2+}$ ions at the center (see Figure 2a), the Fe-N coordination would be relatively free to rotate or to be elongated. Such structural freedom in the coordination polymers can somehow act to facilitate the stabilization of the metastable HS state.

## 4. Conclusions

In conclusion, results of soft XAS on Hofmann-like $Fe^{II}$(4-methylpyrimidine)$_2$[Au(CN)$_2$]$_2$ and $Fe^{II}$(pyridine)$_2$[Ni(CN)$_4$] polymers showed that the high spin configuration prevailed even at temperatures far below the $T_C$'s (despite the normal SCO behaviors in magnetic susceptibility), suggesting that the SOXIESST effects were activated to suppress the SCO effects. This was in contrast to the reference Fe(phen)$_2$(NCS)$_2$ complex, in which a clear SCO was observed by soft XAS; thereby, we suspect that the relatively loose Fe-N coordination in the Hofmann-like coordination polymers might be responsible for the SOXIESST effects.

**Supplementary Materials:** The following are available online at http://www.mdpi.com/2073-4352/8/11/433/s1, Figure S1: FT-IR spectroscopy analysis of (a) Fe(4-methylpmd)2[Au(CN)2]2, (b) Fe(py)2[Ni(CN)4], and (c) Fe(phen)2(NCS)2, Figure S2: Powder XRD pattern of Fe(4-methylpmd)2[Au(CN)2]2, Figure S3: Thermogravimetric analysis (TGA) of Fe(4-methylpmd)2[Au(CN)2]2.

**Author Contributions:** Conceptualization, T.K. and D.-Y.C.; Soft XAS, A.Y.M., M.L., J.-Y.K. and D.-Y.C.; Samples, Magnetization and preliminary measurements, K.K. and T.K.

**Funding:** This research was funded by National Research Foundation of Korea, grant number NRF-2018R1D1A1B07043427 and Japan Society for the Promotion Science (JSPS) KAKENHI Grant Number 15K05485.

**Acknowledgments:** The authors thank Daisuke Akahoshi and Toshiaki Saito in Department of Physics, Toho University for the SQUID measurements.

**Conflicts of Interest:** The authors declare no conflict of interest. The funders had no role in the design of the study; in the collection, analyses, or interpretation of data; in the writing of the manuscript, or in the decision to publish the results.

## References

1. Cambi, L.; SzegzI, L. Über die magnetische Suszeptibilität der komplexen Verbindungen. *Berichte Dtsch. Chem. Gesellachaft B* **1931**, *64*, 2591–2598. [CrossRef]
2. Gütlich, P.; Goodwin, H.A. *Spin Crossover in Transition Metal I*; Gütlich, P., Goodwin, H.A., Eds.; Springer: Berlin/Heidelberg, Germany, 2004. [CrossRef]
3. Gütlich, P.; Garcia, Y.; Goodwin, H.A. Spin crossover phenomena in Fe(ii) complexes. *Chem. Soc. Rev.* **2000**, *29*, 419–427. [CrossRef]
4. Real, J.A.; Gaspar, A.B.; Munoz, M.C. Thermal, pressure and light switchable spin-crossover materials. *Dalton. Trans.* **2005**, 2062–2079. [CrossRef] [PubMed]
5. Murray, K.S. *Spin-Crossover Materials*; Halcrow, M.A., Ed.; John Wiley & Sons, Ltd.: London, UK, 2013.
6. Létard, J.-F.; Guionneau, P.; Goux-Capes, L. Towards Spin Crossover Applications. *Top. Curr. Chem.* **2004**, *235*, 221–249. [CrossRef]
7. Marchivie, M.; Guionneau, P.; Howard, J.A.K.; Chastanet, G.; Létard, J.-F.; Goeta, A.E.; Chasseau, D. Structural Characterization of a Photoinduced Molecular Switch. *J. Am. Chem. Soc.* **2002**, *124*, 194–195. [CrossRef] [PubMed]
8. Kitazawa, T.; Gomi, Y.; Takahashi, M.; Takeda, M.; Enomoto, M.; Miyazakib, A.; Enoki, T. Spin-crossover behaviour of the coordination polymer $Fe^{II}$(C$_5$H$_5$N)$_2$Ni$^{II}$(CN)$_4$. *J. Mater. Chem.* **1996**, *6*, 119–121. [CrossRef]
9. Kosone, T.; Tomori, I.; Kanadani, C.; Saito, T.; Mochida, T.; Kitazawa, T. Unprecedented three-step spin-crossover transition in new 2-dimensional coordination polymer {$Fe^{II}$(4-methylpyridine)$_2$[Au$^I$(CN)$_2$]$_2$}. *Dalton Trans.* **2010**, *39*, 1719–1721. [CrossRef] [PubMed]
10. Kosone, T.; Kawasaki, T.; Tomori, I.; Okabayashi, J.; Kitazawa, T. Modification of Cooperativity and Critical Temperatures on a Hofmann-Like Template Structure by Modular Substituent. *Inorganics* **2017**, *5*, 55. [CrossRef]
11. Wu, Z.; Justo, J.F.; da Silva, C.R.S.; de Gironcoli, S.; Wentzcovitch, R.M. Anomalous thermodynamic properties in ferropericlase throughout its spin crossover. *Phys. Rev. B* **2009**, *80*. [CrossRef]
12. Wentzcovitch, R.M.; Justo, J.F.; Wu, Z.; da Silva, C.R.; Yuen, D.A.; Kohlstedt, D. Anomalous compressibility of ferropericlase throughout the iron spin cross-over. *Proc. Natl. Acad. Sci. USA* **2009**, *106*, 8447–8452. [CrossRef] [PubMed]

13. Ksenofontov, V.; Gaspar, A.B.; Gütlich, P. Pressure Effect Studies on Spin Crossover and Valence Tautomeric Systems. In *Spin Crossover in Transition Metal III*; Gütlich, P., Goodwin, H.A., Eds.; Springer: Berlin/Heidelberg, Germany, 2004. [CrossRef]
14. Boillot, M.-L.; Zarembowitch, J.; Sour, A. Ligand-Driven Light-Induced Spin Change (LD-LISC): A Promising Photomagnetic Effect. In *Spin Crossover in Transition Metal II*; Gütlich, P., Goodwin, H.A., Eds.; Springer: Berlin/Heidelberg, Germany, 2004. [CrossRef]
15. Naggert, H.; Bannwarth, A.; Chemnitz, S.; von Hofe, T.; Quandt, E.; Tuczek, F. First observation of light-induced spin change in vacuum deposited thin films of iron spin crossover complexes. *Dalton Trans.* **2011**, *40*, 6364–6366. [CrossRef] [PubMed]
16. Cannizzo, A.; Milne, C.J.; Consani, C.; Gawelda, W.; Bressler, C.; van Mourik, F.; Chergui, M. Light-induced spin crossover in Fe(II)-based complexes: The full photocycle unraveled by ultrafast optical and X-ray spectroscopies. *Coord. Chem. Rev.* **2010**, *254*, 2677–2686. [CrossRef]
17. Herber, R.; Casson, L.M. Light-Induced Excited-Spin-State Trapping: Evidence from VTFTIR Measurements. *Inorg. Chem.* **1986**, *25*, 847–852. [CrossRef]
18. Hauser, A. Light-Induced Spin Crossover and the High-Spin → Low-Spin Relaxation. In *Spin Crossover in Transition Metal II*; Gütlich, P., Goodwin, H.A., Eds.; Springer: Berlin/Heidelberg, Germany, 2004. [CrossRef]
19. Decurtins, S.; Gutlich, P.; Hasselbach, K.M.; Hauser, A.; Spieringt, H. Light-Induced Excited-Spin-State Trapping in Iron(II) Spin-Crossover Systems. Optical Spectroscopic and Magnetic Susceptibility Study. *Inorg. Chem.* **1985**, *24*, 2174–2178. [CrossRef]
20. Hauser, A. Reversibility of light-induced excited spin state trapping in the $Fe(ptz)_6(BF_4)_2$ and the $Zn_{1-x}Fe_x(ptz)_6(BF_4)_2$ spin-crossover systems. *Chem. Phys. Lett.* **1986**, *124*, 543–548. [CrossRef]
21. Davesne, V.; Gruber, M.; Miyamachi, T.; Da Costa, V.; Boukari, S.; Scheurer, F.; Joly, L.; Ohresser, P.; Otero, E.; Choueikani, F.; et al. First glimpse of the soft X-ray induced excited spin-state trapping effect dynamics on spin cross-over molecules. *J. Chem. Phys.* **2013**, *139*, 074708. [CrossRef] [PubMed]
22. Collison, D.; Garner, C.D.; McGrath, C.M.; Mosselmans, J.F.W.; Roper, M.D.; Seddon, J.M.W.; Sinn, E.; Young, N.A. Soft X-ray induced excited spin state trapping and soft X-ray photochemistry at the iron $L_{2,3}$ edge in $[Fe(phen)_2(NCSe)_2]$ and $[Fe(phen)_2(NCS)_2]$ (phen = 1,10-phenanthroline). *J. Chem. Soc. Dalton Trans.* **1997**, *22*, 4371–4376. [CrossRef]
23. Bousseksou, A.; Varret, F.; Goiran, M.; Boukheddaden, K.; Tuchagues, J.P. The Spin Crossover Phenomenon Under High Magnetic Field. In *Spin Crossover in Transition Metal III*; Gütlich, P., Goodwin, H.A., Eds.; Springer: Berlin/Heidelberg, Germany, 2004. [CrossRef]
24. Miyamachi, T.; Gruber, M.; Davesne, V.; Bowen, M.; Boukari, S.; Joly, L.; Scheurer, F.; Rogez, G.; Yamada, T.K.; Ohresser, P.; et al. Robust spin crossover and memristance across a single molecule. *Nat. Commun.* **2012**, *3*, 938. [CrossRef] [PubMed]
25. Cotton, F.; Wilkinson, G.; Gaus, P. *Basic Inorganic Chemistry*, 3rd ed.; Cotton, F., Wilkinson, G., Gaus, P., Eds.; Wiley: New York, NY, USA, 1995. [CrossRef]
26. Wäckerlin, C.; Donati, F.; Singha, A.; Baltic, R.; Decurtins, S.; Liu, S.-X.; Rusponi, S.; Dreiser, J. Excited Spin-State Trapping in Spin Crossover Complexes on Ferroelectric Substrates. *J. Phys. Chem. C* **2018**, *122*, 8202–8208. [CrossRef]
27. Davesne, V. Organic Spintronics: An Investigation on Spin-Crossover Complexes from Isolated Molecules to the Device. Ph.D. Thesis, Physique et Chimie Physique Université de strasbourg, Strasbourg, France, 2013.
28. Kitazawa, T.; Kawasaki, T.; Shiina, H.; Takahashi, M. Mössbauer Spectroscopic Study on Hofmann-like Coordination Polymer $Fe(4-Clpy)_2[Ni(CN)_4]$. *Croat. Chem. Acta* **2016**, *89*, 111–115. [CrossRef]
29. Kitazawa, T.; Kishida, T.; Kawasaki, T.; Takahashi, M. Spin crossover behaviour in Hofmann-like coordination polymer $Fe(py)_2[Pd(CN)_4]$ with 57Fe Mössbauer spectra. *Hyperfine Interact.* **2017**, *238*. [CrossRef]
30. Gütlich, P.; Goodwin, H.A. Spin Crossover—An Overall Perspective. *Top. Curr. Chem.* **2004**, *233*, 1–47. [CrossRef]
31. Ueki, Y.; Okabayashi, J.; Kitazawa, T. Guest Molecule Inserted Spin Crossover Complexes: Fe[4-(3-Pentyl) pyridine]$_2$[Au(CN)$_2$]$_2$·Guest. *Chem. Lett.* **2017**, *46*, 747–749. [CrossRef]
32. Moulin, C.C.D.; Rudolf, P.; Flank, A.M.; Chen, C.T. Spin Transition Evidenced by Soft X-ray Absorption Spectroscopy. *J. Phys. Chem.* **1992**, *96*, 6196–6198. [CrossRef]

33. Moulin, C.C.D.; Flank, A.M.; Rudolf, P.; Chen, C.T. Electronic Structure from Iron L-edge Spectroscopy: An Example of Spin Transition Evidenced by Soft X-ray Absorption Spectroscopy. *Jpn. J. Appl. Phys.* **1993**, *32* (Suppl. 2), 308–310. [CrossRef]
34. Briois, V.; Moulin, C.C.D.; Sainctavit, P.; Brouder, C.; Flank, A.-M. Full Multiple Scattering and Crystal Field Multiplet Calculations Performed on the Spin Transition Fe$^{II}$(phen)$_2$(NCS)$_2$ Complex at the Iron K and L$_{2,3}$ X-ray Absorption Edges. *J. Am. Chem. Soc.* **1995**, *117*, 1019–1026. [CrossRef]
35. Kipgen, L.; Bernien, M.; Ossinger, S.; Nickel, F.; Britton, A.J.; Arruda, L.M.; Naggert, H.; Luo, C.; Lotze, C.; Ryll, H.; et al. Evolution of cooperativity in the spin transition of an iron(II) complex on a graphite surface. *Nat. Commun.* **2018**, *9*, 2984. [CrossRef] [PubMed]
36. Davesne, V.; Gruber, M.; Studniarek, M.; Doh, W.H.; Zafeiratos, S.; Joly, L.; Sirotti, F.; Silly, M.G.; Gaspar, A.B.; Real, J.A.; et al. Hysteresis and change of transition temperature in thin films of Fe{[Me$_2$Pyrz]$_3$BH}$_2$, a new sublimable spin-crossover molecule. *J. Chem. Phys.* **2015**, *142*, 194702. [CrossRef] [PubMed]
37. Madeja, K. Darstellung und magnetisches Verhalten von [Fe(phen)$_2$X$_2$]-Komplexen. *Chemické Zvesti* **1965**, *19*, 186–191.
38. Baker, W.A., Jr.; Bobonich, H.M. Magnetic Properties of Some High-Spin Complexes of Iron(II). *Inorg. Chem.* **1964**, *3*, 1184–1188. [CrossRef]
39. Blundell, S. *Magnetism in Condensed Matter*; Oxford University Press Inc.: New York, NY, USA, 2001.
40. Lee, J.-J.; Sheu, H.; Lee, C.-R.; Chen, J.-M.; Lee, J.-F.; Wang, C.-C.; Huang, C.-H.; Wang, Y. X-ray Absorption Spectroscopic Studies on Light-Induced Excited Spin State Trapping of an Fe(II) Complex. *J. Am. Chem. Soc.* **2000**, *122*, 5742–5747. [CrossRef]

© 2018 by the authors. Licensee MDPI, Basel, Switzerland. This article is an open access article distributed under the terms and conditions of the Creative Commons Attribution (CC BY) license (http://creativecommons.org/licenses/by/4.0/).

*Article*

# New Iron(II) Spin Crossover Complexes with Unique Supramolecular Networks Assembled by Hydrogen Bonding and Intermetallic Bonding

**Takashi Kosone [1,\*], Itaru Tomori [2], Daisuke Akahoshi [3], Toshiaki Saito [3] and Takafumi Kitazawa [3]**

1. Department of Creative Technology Engineering Course of Chemical Engineering, Anan College, 265 Aoki, Minobayashi, Anan, Tokushima 774-0017, Japan
2. Department of Chemistry, Faculty of Science, Toho University, 2-2-1 Miyama, Funabashi, Chiba 274-8510, Japan; synapse_yf@yahoo.co.jp
3. Department of Physics, Faculty of Science, Toho University, 2-2-1, Miyama 274-8510, Japan; daisuke.akahoshi@sci.toho-u.ac.jp (D.A.); saito@ph.sci.toho-u.ac.jp (T.S.); kitazawa@chem.sci.toho-u.ac.jp (T.K.)
* Correspondence: kosone@anan-nct.ac.jp; Tel.: +81-884-23-7195

Received: 25 September 2018; Accepted: 20 October 2018; Published: 5 November 2018

**Abstract:** Two spin crossover (SCO) coordination polymers assembled by combining $Fe^{II}$ octahedral ion, 4-cyanopyridine (4-CNpy) and $[Au(CN)_2]^-$ liner unit are described. These compounds, $Fe(4-CNpy)_2[Au(CN)_2]_2 \cdot 1/2(4-CNpy)$ (**1a**) and $\{Fe(4-CNpy)_2[Au(CN)_2]_2\}-\{Fe(H_2O)_2[Au(CN)_2]_2\}$ (**1b**), present quite different supramolecular networks that show different magnetic behaviors. Compound **1a** crystallizes in the centrosymmetric space group *Pbcn*. The asymmetric unit contains two 4-CNpy, one type of $Fe^{2+}$, and two types of crystallographically distinct $[Au(CN)_2]^-$ units which form Hofmann-like two dimensional layer structures with guest spaces. The layers are combined with another layer by strong gold-gold intermetalic interactions. Compound **1b** crystallizes in the centrosymmetric space group *Pnma*. The bent bismonodentate $[Au^I(CN)_2]$ units and $Fe^{II}$ ions form a complicated interpenetrated three dimensional structure. In addition, **1b** exhibits ferromagnetic interaction.

**Keywords:** coordination polymer; supramolecular isomerism; spin crossover; crystal engineering

## 1. Introduction

The designing of supramolecular networks is essential for practical spin crossover (SCO) materials [1–4]. The networks enhance the cooperativity in the entire crystal structure. Strong cooperativity leads to steep spin transition with a wide hysteresis loop [5,6]. From the viewpoint of constructing supramolecular networks, coordination polymers are useful material. However, systematic designing of networks is still hard because of the unexpected occurrence of supramolecular isomerism in the process of self-assembling. On the other hand, this structural diversity can result in unanticipated and interesting materials. Therefore, control of structural diversity represents fundamental research in crystal engineering. Since we reported the first Hofmann like two-dimensional (2-D) SCO coordination polymer $\{Fe(py)_2[Ni(CN)_4]\}_n$ (py = pyridine) [7], many 2-D layers of $\{Fe^{II}(L)_2[M^I(CN)_2]_2\}_n$ [8–16] ($M^I$ = Ag, or Au, L = monodentate pyridine derivatives) have been developed. These compounds show an almost similar bilayer structure because of their strongly determinate self-assembly process in which they link octahedral metal centers through the N atoms of the bidentate $[Au(CN)_2]^-$ unit with strong aurophilic interaction between layers. This structural constancy enables us to precisely modify its crystal structure and properties. However, the applicable ligands for this system are still determinative. For instance, 3-cyano pyridine (3-CNpy) displays three different polymorphs [8]. A strong polarity of cyano substituent must cause variations of the supramolecular networks, which strongly affects SCO properties. Therefore, more applicable ligands for this structural system must be investigated. Vice versa, cyano substituent offers new interesting networks and properties in cyano-bridged coordination polymers. Here, we report new supramolecular isomers of the general formula $Fe(4\text{-CNpy})_2[Au(CN)_2]_2 \cdot 1/2(4\text{-CNpy})$ (**1a**) and $\{Fe(4\text{-CNpy})_2[Au(CN)_2]_2\}\text{-}\{Fe(H_2O)_2[Au(CN)_2]_2\}$ (**1b**).

## 2. Materials and Methods

### 2.1. Materials

All the chemicals were purchased from commercial sources and used without any further purification.

### 2.2. Synthesis

#### 2.2.1. Preparation of Compound **1a**

$FeSO_4 \cdot (NH_4)_2SO_4 \cdot 6H_2O$ (0.0397 g, $1.01 \times 10^{-4}$ mol) ascorbic acid (0.0208 g, $1.18 \times 10^{-4}$ mol) and $K[Au(CN)_2]$ (0.0582 g, $2.02 \times 10^{-4}$ mol) were dissolved in 2 mL of water. The other solution contained 4-CNpy (0.0204 g, $1.96 \times 10^{-4}$ mol) in 5 mL water. The two solutions were mixed together. Yellow single crystals suitable for single crystal X-ray diffraction were formed over a day. The powder sample for superconducting quantum interference device (SQUID), X-ray powder diffraction (XRPD), thermos gravimetry/differential thermal analysis (TG/DTA) and elemental analysis was also prepared. One of these contained a mixture of $FeSO_4 \cdot (NH_4)_2SO_4 \cdot 6H_2O$ (0.0400 g, $1.02 \times 10^{-4}$ mol), ascorbic acid (0.0204 g, $1.36 \times 10^{-4}$ mol) and $K[Au^I(CN)_2]$ (0.0570 g, $1.98 \times 10^{-4}$ mol) in 1 mL water. The other contained a 1 mL ethanol–water (1:1) solution of 4-CNpy (0.0206 g, $1.98 \times 10^{-4}$ mol). Yellow powder sample of **1a** was formed immediately. The powder sample was checked by XRPD data (see Figure S1). Impurity and isomers were observed as almost absent. Elem. Anal. Calcd for $C_{19}H_8Au_2FeN_9$: C, 28.03; H, 1.24; N, 15.48. Found: C, 27.92; H, 1.38; N, 15.35. IR(cm$^{-1}$): 2237 (νCN (4-CNpy)), 2157, 2169 (νCN).

#### 2.2.2. Preparation of Compound **1b**

Complex **1b** was prepared by the same procedure as **1a**. The reaction mixture was allowed to stand undisturbed for 2 days. After forming yellow single crystals (**1a**), orange crystals (**1b**) slowly grew. The crystalline sample for SQUID measurement was picked up using a binocular lens. The samples were checked by XRPD data (Figure S1). Impurity of the samples was observed as almost absent. Due to the small amount of sample picked, the background of the diffraction data were very high. Elem. Anal. Calcd for $C_{20}H_{11}Au_4Fe_2N_{12}O_2$: C, 17.77; H, 0.89; N, 12.43. Found: C, 17.75; H, 1.13; N, 12.15. IR (cm$^{-1}$): 2250 (νCN (4-CNpy)), 2169 (νCN)

### 2.3. X-ray Crystallography

Data collection was performed on a BRUKER APEX SMART CCD area-detector diffractometer for **1a** and **1b** with Monochromed Mo–Kα radiation (λ = 0.71073 Å) (Bruker, Billerica, MA, USA). A selected single crystal was carefully mounted on a thin glass capillary and immediately placed under liquid $N_2$ cooled $N_2$ stream in each case. The diffraction data were treated using SMART and SAINT, and absorption correction was performed using SADABS [17]. The structures were solved by using direct methods with *SHELXTL* [18]. All non-hydrogen atoms were refined anisotropically, and the hydrogen atoms were generated geometrically. Pertinent crystallographic parameters and selected metric parameters for **1a** and **1b** are displayed in Tables 1–3. Diffraction data of **1a** in high spin (HS) state was measured at 150 K in order to suppress the thermal motion of the guest molecules. When the sample was sufficiently cooled, both compounds showed a drastic and reversible change of color from yellow (**1a**) or orange (**1b**) to purple. The crystal structure of **1a** in low spin

(LS) state could not be determined. The low-quality data at 90 K was likely due to the occurrence of a sharp phase transition that provoked a notable increase of the mosaicity of the whole crystal structure. We described here the only HS state at 150 K. Crystallographic data have been deposited with Cambridge Crystallographic Data Centre: Deposition numbers CCDC-1869343 for compound **1a** (150 K), CCDC-1869342 for **1b** (298 K), and CCDC-1869341 for **1b** (90 K). These data can be obtained free of charge via http://www.ccdc.cam.ac.uk/conts/retrieving.html.

*2.4. Magnetic Measurements*

Measurements of the temperature dependence of the magnetic susceptibility of the complexes **1a** and **1b** of the powdered samples in the temperature range 2–300 K with a cooling and heating rate of 2 K·min$^{-1}$ in a 1 kOe field were measured on a MPMS-XL Quantum Design SQUID magnetometer. The diamagnetism of the samples and sample holders were taken into account.

Table 1. Crystal data and structure refinement for compounds **1a** and **1b**.

|  | 1a (150 K) | 1b (298 K) | 1b (90 K) |
| --- | --- | --- | --- |
| Empirical formula | $C_{19}H_8Au_2FeN_9$ | $C_{20}H_{11}Au_4Fe_2N_{12}O_2$ | $C_{20}H_{11}Au_4Fe_2N_{12}O_2$ |
| Formula weight | 812.13 | 1350.97 | 1350.97 |
| Crystal size/mm$^3$ | 0.55 × 0.28 × 0.22 | 0.15 × 0.13 × 0.05 | 0.38 × 0.15 × 0.05 |
| Crystal system | Orthorhombic | Orthorhombic | Orthorhombic |
| $a$/Å | 21.658(2) | 7.6273(7) | 7.6243(4) |
| $b$/Å | 13.7647(15) | 30.007(3) | 29.1960(16) |
| $c$/Å | 15.5895(17) | 13.7694(13) | 13.5000(7) |
| $V$/Å$^3$ | 4647.4(9) | 3151.5(5) | 3005.1(3) |
| Space group | *Pbcn* | *Pnma* | *Pnma* |
| Z value | 8 | 4 | 4 |
| $D_{calc}$ | 2.321 | 2.847 | 2.986 |
| F(000) | 2952 | 2396 | 2396 |
| No. of reflections | 28,466 | 19,996 | 20,977 |
| No. of observations | 5443 | 3976 | 3808 |
| Parameters | 284 | 196 | 186 |
| Temperature/K | 150(2) | 298 | 90 |
| Final $R_1$, $R_w$ ($I > 2s$) | 0.0411, 0.1043 | 0.0474, 0.1168 | 0.0396, 0.1048 |
| Final $R_1$, $R_w$ (all data) | 0.0715, 0.1197 | 0.0730, 0.1346 | 0.0439, 0.1212 |
| Goodness-of-fit | 1.050 | 0.965 | 1.413 |

Table 2. Selected bond lengths and angles for **1a**.

| Bond Lengths (Å) for 1a (150 K) | Bond Angles (°) for 1a (150 K) | |
| --- | --- | --- |
| Fe(1)–N(1): 2.234(9) | N(1)–Fe(1)–N(3): 175.9(3) | N(5)–Fe(1)–N(8): 90.0(3) |
| Fe(1)–N(3): 2.227(8) | N(1)–Fe(1)–N(5): 93.0(3) | N(6)–Fe(1)–N(7): 88.2(3) |
| Fe(1)–N(5): 2.161(8) | N(1)–Fe(1)–N(6): 94.5(3) | N(6)–Fe(1)–N(8): 176.4(3) |
| Fe(1)–N(6): 2.129(8) | N(1)–Fe(1)–N(7): 89.4(3) | N(7)–Fe(1)–N(8): 95.1(3) |
| Fe(1)–N(7): 2.121(8) | N(1)–Fe(1)–N(8): 84.0(3) | C(13)–N(5)–Fe(1): 167.4(8) |
| Fe(1)–N(8): 2.146(8) | N(3)–Fe(1)–N(5): 86.9(3) | C(14)–N(6)–Fe(1): 162.4(8) |
| Au(1)–C(13): 1.999(9) | N(3)–Fe(1)–N(6): 89.6(3) | C(15)–N(7)–Fe(1): 163.2(8) |
| Au(1)–C(16): 1.990(9) | N(3)–Fe(1)–N(7): 91.1(3) | C(16)–N(8)–Fe(1): 161.3(9) |
| Au(2)–C(14): 1.986(9) | N(3)–Fe(1)–N(8): 91.9(3) | C(13)–Au(1)–C(16): 178.0(4) |
| Au(2)–C(15): 1.981(10) | N(5)–Fe(1)–N(6): 86.8(3) | C(14)–Au(2)–C(15): 176.8(4) |
|  | N(5)–Fe(1)–N(7): 174.6(3) | |

## 3. Results and Discussion

*3.1. Crystal Structures*

3.1.1. Crystal Structure of Compound **1a** (T = 298 K)

The crystal structure of **1a** at 298 K crystallized in the orthorhombic centrosymmetric space group *Pbcn*. The asymmetric unit of the complex consisted of the Fe(ligand)$_2$[Au(CN)$_2$]$_2$ formula with a guest molecule (Figure 1a). This complex had one type of independent Fe$^{II}$ ion octahedrally coordinated by six N atoms. The axial Fe(1)–N$_{py}$ bond lengths (Fe(1)–N(1) = 2.234(9) Å, Fe(1)–N(3) = 2.227(8) Å) and Fe(1)–N$_{CN}$ bond lengths (Fe(1)–N(5) = 2.161(8) Å, Fe(1)–N(6) = 2.129(8) Å, Fe(1)–N(7) = 2.121(8) Å, Fe(1)–N(8) = 2.146(8) Å) were almost identical to that of **1b**. The equatorial positions were occupied by two quasilinear [Au$^I$(CN)$_2$]$^-$, which comprised a 2-D layer structure defined by square-shaped [Fe$^{II}$Au$^I$(CN)$_2$]$_4$ windows (Figure 1b). The layers interacted via pairs of defining bilayers (Figure 1c,d), in which strong aurophilic interactions held them together. The Au$\cdots$Au intermetallic distance was 3.1134(6) Å which was much shorter than that of **1b**. Although the former reported bilayer structures had no guest spaces, the interlayer space of **1a** formed one dimensional (1-D) channels parallel to the c axis, which were occupied by uncoordinated 4-CNpy. The guest molecules were disordered at two positions. The reason for enough space to include guest molecules was the steric effect from the 4-position substituent bulk. As shown in Figure 1c, the four position substituent was almost vertical to the layer. Thus, it would cause the pressure on the layer resulting in expanded interlayer space.

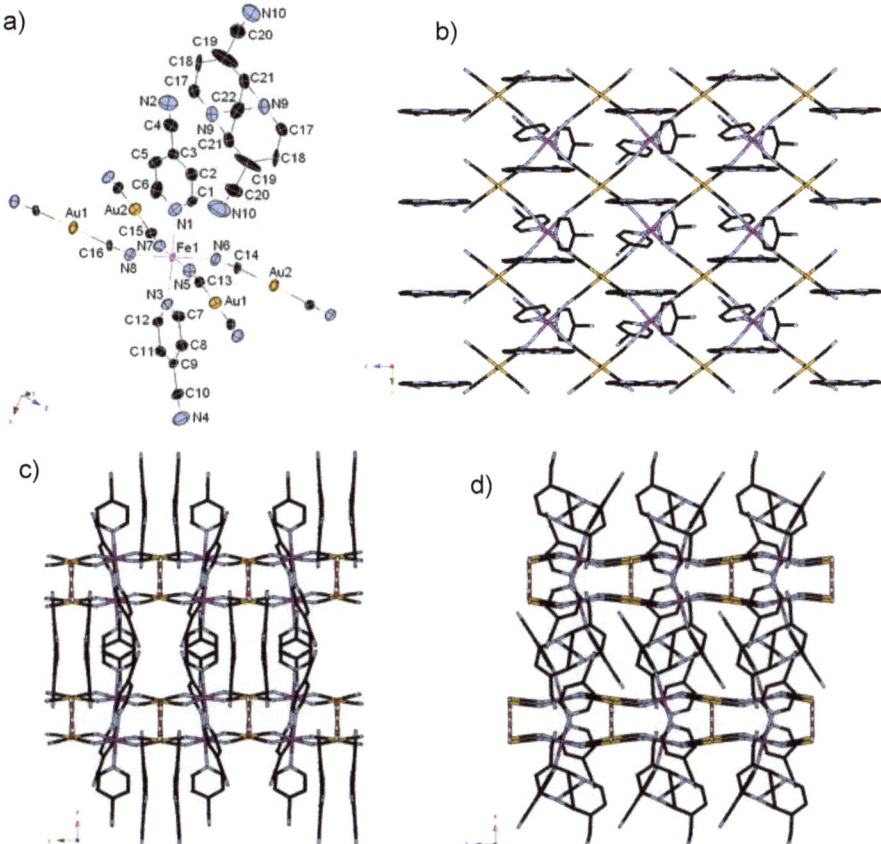

**Figure 1.** (**a**) Coordination structure of **1a** containing its asymmetric unit at 150 K; (**b**) view of the bilayer structure; (**c,d**) stacking of bilayers of **1a** along c (c) and b (d) axis involved in Au$\cdots$Au intermetallic interactions as indicated by red and white lines. In these pictures, hydrogen atoms are omitted for clarity.

The closely related Hofmann-like 2-D clathrate compound, {Fe[4-(3-pentyl)pyridine]$_2$[Au$^I$(CN)$_2$]$_2$·(guest)}$_n$ (guest = 4-(3-pentyl)pyridine), had been reported [19]. This compound formed a flat monolayer structure. 4-(3-pentyl) substituent was apparently of lager bulk than that of the 4-CN substituent. This much larger bulk caused bilayer interaction to break.

3.1.2. Crystal Structure of Compound **1b** (T = 298 K)

Compound **1b** at 298 K crystallized in the orthorhombic centrosymmetric space group Pnma. The asymmetric unit also consisted of the cyano bridged hetero-metal coordination (Figure 2a). There were two crystallographically different octahedral $Fe^{II}$ ions. Fe(1) was coordinated by six N atoms of two 4-CNpy ligands and four CN substituents from $[Au^I(CN)_2]^-$ units, which was similar to that of **1a**. The axial Fe(1)–$N_{py}$ bond lengths (Fe(1)–N(1) = 2.226(7) Å) were apparently longer than the Fe(1)–$N_{CN}$ bond lengths (Fe(1)–N(3) = 2.157(7) Å, Fe(1)–N(4) = 2.144(7) Å). On the other hand, the axial Fe(2)–O bond lengths (Fe(2)–O(1) = 2.078(13) Å, Fe(2)–O(2) = 2.158(9) Å) were close to the Fe(2)–$N_{CN}$ bond lengths (Fe(2)–N(5) = 2.130(8) Å, Fe(2)–N(6) = 2.187(8) Å). It is important to note that hydrogen bonds existed between N(2) of cyano group in 4-CNpy and O(1) (N(2)···O(1) = 2.905(15) Å). Due to the different coordination environment, the apical axis of Fe(1) was not parallel to the apical axis of Fe(2). Furthermore, the coordination geometry of the Fe(1) site was almost octahedron; the bond angle of the apical axis was 180(4)°. On the other hand, the Fe(2) site showed a quite distorted octahedron (O(1)–Fe(2)–O(2) = 169.5(6)° and N(5)–Fe(2)–N(6) = 172.3(3)°). Consequently, the bent bismonodentate $[Au^I(CN)_2]$ units and $Fe^{II}$ ions formed bent rectangular $[Fe^{II}Au^I(CN)_2]_4$ mesh network topology (Figure 2b). The rectangular moieties were penetrated by the other frameworks, which gave rise to a triply interpenetrated structure (Figure 2c). In addition, the closest approach between Au···Au suggested the presence of aurophilic interactions (Au(1)···Au(2) = 3.3705(5) Å and 3.3876(5) Å) which linked the other frameworks (Figure 2c,d). In previous works, isostructural compound {Mn(4-CNpy)$_2$[Ag(CN)$_2$]$_2$}-{Mn(H$_2$O)$_2$[Ag(CN)$_2$]$_2$} (**1b'**) had been reported [20]. This compound also showed similar hydrogen bonding and intermetallic Ag···Ag interactions.

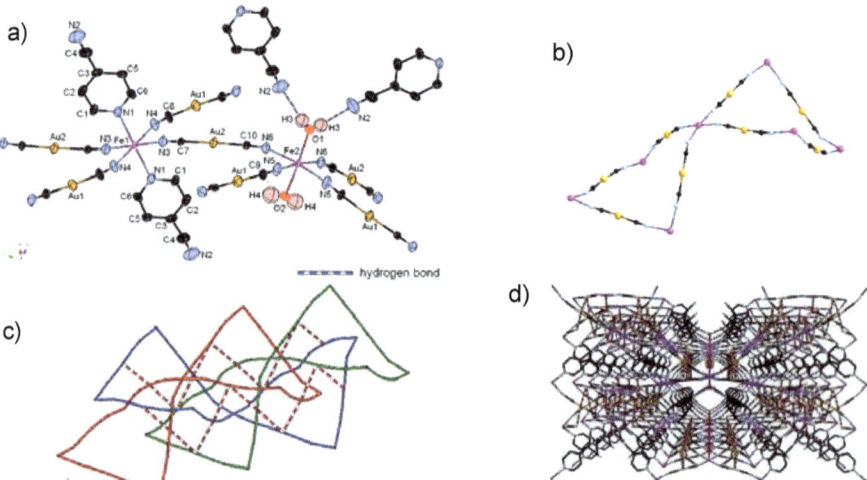

**Figure 2.** (**a**) Coordination structure of **1b** containing its asymmetric unit at 298 K involved in hydrogen bonding interactions as indicated by blue and white lines; (**b**) bent rectangular $[Fe^{II}Au^I(CN)_2]_4$ mesh structure; (**c**) cylinder drawing of triply interpenetrated 3D-networks of **1b** involved in hydrogen bonding interactions as indicated by red and white lines; (**d**) perspective view of the crystal structure along *a* axis. In these pictures, hydrogen atoms are omitted for clarity.

### 3.1.3. Structure of Compound 1b (T = 90 K)

The crystal structure of **1b** at 90 K was almost identical to that observed at 298 K. The Fe(1)–N$_{py}$ bond lengths (Fe(1)–N(1) = 1.998(7) Å) and Fe(1)–N$_{CN}$ bond lengths (Fe(1)–N(3) = 1.937(7) Å, Fe(1)–N(4) = 1.941(7) Å) corresponded quite well to those expected for LS state. On the other hand Fe(2)–O bond lengths (Fe(2)–O(1) = 2.095(9) Å, Fe(2)–O(2) = 2.112(9) Å) and Fe(2)–N$_{CN}$ bond lengths (Fe(2)–N(5) = 2.132(7) Å, Fe(2)–N(6) = 2.190(8) Å) were almost identical to that of 298 K. Hydrogen bonding and intermetallic interactions were slightly shorter than that of the HS state (Au(1)···Au(2) = 3.3462(4) Å, 3.3745(4) Å and N(2)···O(1) = 2.862(11) Å). The shortest Fe···Fe separation was 5.811(2) Å between Fe(2) centers which was bridged by the aurophilic interactions (see Figure S3). This was much shorter than that of the other Fe···Fe separation (Fe(1)···Fe(1) = 13.280(1) Å, Fe(1)···Fe(2) = 9.746(1) Å).

Table 3. Selected bond lengths and angles for **1b**.

| Bond Lengths (Å) for 1b (298 K) | Bond Angles (°) for 1b (298 K) | |
|---|---|---|
| Fe(1)–N(1): 2.226(7) | N(1)–Fe(1)–N(1): 180.0(3) | O(1)–Fe(2)–O(2): 169.5(6) |
| Fe(1)–N(3): 2.157(7) | N(1)–Fe(1)–N(3): 90.0(3) | O(1)–Fe(2)–N(5): 96.6(4) |
| Fe(1)–N(4): 2.144(7) | N(1)–Fe(1)–N(4): 89.5(3) | O(1)–Fe(2)–N(6): 89.9(3) |
| Fe(2)–O(1): 2.078(13) | N(3)–Fe(1)–N(3): 180.0(3) | O(2)–Fe(2)–N(5): 90.8(4) |
| Fe(2)–O(2): 2.158(9) | N(3)–Fe(1)–N(4): 91.3(3) | O(2)–Fe(2)–N(6): 83.3(3) |
| Fe(2)–N(5): 2.130(8) | N(4)–Fe(1)–N(4): 180.0(4) | N(5)–Fe(2)–N(5): 89.5(5) |
| Fe(2)–N(6): 2.187(8) | C(7)–N(3)–Fe(1): 167.1(8) | N(5)–Fe(2)–N(6): 85.6(3) |
| Au(1)–C(8): 1.985(9) | C(8)–N(4)–Fe(1): 166.6(8) | N(6)–Fe(2)–N(6): 98.7(4) |
| Au(1)–C(9): 1.978(9) | C(8)–Au(1)–C(9): 175.9(4) | N(5)–Fe(2)–N(6): 172.3(3) |
| Au(2)–C(7): 1.992(9) | C(7)–Au(2)–C(10): 176.6(3) | C(9)–N(5)–Fe(2): 169.6(8) |
| Au(2)–C(10): 1.992(9) | | C(10)–N(6)–Fe(2): 162.0(8) |
| **Bond Lengths (Å) for 1b (90 K)** | **Bond Angles (°) for 1b (90 K)** | |
| Fe(1)–N(1): 1.998(7) | N(1)–Fe(1)–N(1): 180.0(4) | O(1)–Fe(2)–O(2): 169.1(4) |
| Fe(1)–N(3): 1.937(7) | N(1)–Fe(1)–N(3): 90.5(3) | O(1)–Fe(2)–N(5): 96.2(3) |
| Fe(1)–N(4): 1.941(7) | N(1)–Fe(1)–N(4): 90.4(3) | O(1)–Fe(2)–N(6): 89.3(3) |
| Fe(2)–O(1): 2.095(9) | N(3)–Fe(1)–N(3): 180.0(2) | O(2)–Fe(2)–N(5): 91.5(3) |
| Fe(2)–O(2): 2.112(9) | N(3)–Fe(1)–N(4): 90.3(3) | O(2)–Fe(2)–N(6): 83.6(3) |
| Fe(2)–N(5): 2.132(7) | N(4)–Fe(1)–N(4): 180.0(4) | N(5)–Fe(2)–N(5): 90.6(4) |
| Fe(2)–N(6): 2.190(8) | C(7)–N(3)–Fe(1): 170.7(7) | N(5)–Fe(2)–N(6): 85.0(3) |
| Au(1)–C(8): 1.985(9) | C(8)–N(4)–Fe(1): 172.3(7) | N(6)–Fe(2)–N(6): 98.9(4) |
| Au(1)–C(9): 1.977(9) | C(8)–Au(1)–C(9): 174.8(3) | N(5)–Fe(2)–N(6): 173.4(3) |
| Au(2)–C(7): 1.983(9) | C(7)–Au(2)–C(10): 174.2(3) | C(9)–N(5)–Fe(2): 168.6(7) |
| Au(2)–C(10): 1.989(8) | | C(10)–N(6)–Fe(2): 161.7(7) |

## 3.2. Thermal Analysis

The thermal analysis of **1a** showed the three step weight loss between 380 K and 580 K corresponded to the loss of the two coordinated molecules of 4-CNpy and 0.5 solvent molecules (observed loss: 18.0% (first step) = ca. 1.5 molecules, 6.4% (second step) = ca. 0.5 molecule, 6.7% (third step) = ca. 0.5 molecule) (Figure S2). This result was consistent with the elemental analysis.

## 3.3. Magnetic Properties

### 3.3.1. Thermal Dependence Magnetic Behavior of Compound 1a

Figure 3a shows the thermal dependence of $\chi_M T$ for **1a** with $\chi_M$ being the molar magnetic susceptibility and T the temperature. At room temperature, $\chi_M T$ was 4.21 m$^3$·K·mol$^{-1}$. Upon cooling, $\chi_M T$ remained almost constant down to 120 K; below this temperature, $\chi_M T$ underwent a sharp decrease to around 50% conversion with approximately 1 K hysteresis loop (T$_c^{down}$ = 111 K, T$_c^{up}$ = 112 K). The decrease in the value of $\chi_M T$ at lower temperature was due to the typical behavior of zero-field splitting (ZFS).

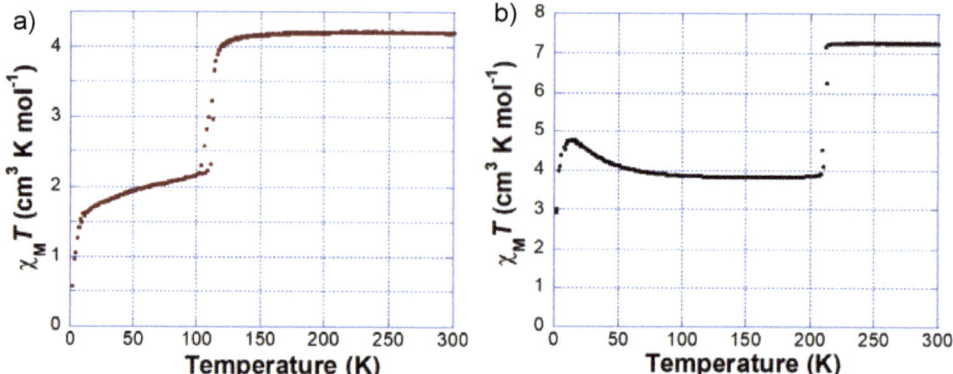

**Figure 3.** (a) Thermal dependence of $\chi_M T$ plot for **1a**; (b) thermal dependence of $\chi_M T$ plot for **1b**.

### 3.3.2. Thermal Dependence Magnetic Behavior of Compound 1b

$\chi_M T$ versus $T$ plotted for **1b** are shown in Figure 3b. At room temperature, the $\chi_M T$ value was 7.25 which indicated the characteristic of two Fe(II) ions in HS state. The value was slightly higher than that of a pure spin only system, whereas the value was similar to the values of other Hofmann-like SCO Fe(II) compounds. The $\chi_M T$ value was constant in the range 215–300 K. Below this temperature range, it displayed almost 50% abrupt spin transition ($T_c$ = 211 K). This half spin transition behavior agreed quite well with the different Fe-N bond lengths between Fe(1) and Fe(2) species at 90 K. Thus only Fe(1) species change the spin state. On the other hand, [Fe(2)N$_4$O$_2$] coordination environment must maintain the HS state at the full temperature range. After the transition, it then increased to a maximum of 4.77 cm$^3$·K·mol$^{-1}$ at around 20 K, indicative of a ferromagnetic interaction between the Fe(II) centers. On the other hand, the isostructural former reported that compound **1b'** showed weak antiferromagnetic interaction between two Mn$^{II}$ (HS state, S = 5/2). Although in a similar coordination environment, **1b** showed ferromagnetic interaction. In terms of the spin state, the X-ray structural analysis of **1b** at 90 K gave evidence of the arrangement of a···Fe(HS)–Fe(LS)···pair. On the other hand, **1b** showed the different arrangement of···Mn(HS)–Mn(HS)···. Thus, **1b** apparently much further distorted structure. Consequently, the magnetic structure of the residual HS site of Fe(2) could cause the different magnetic coupling. In fact Fe(2) ions of HS site were close to each other in the supramolecular networks. At even lower temperature, the decrease in the value of $\chi_M T$ was similar to **1a** due to ZFS effects.

## 4. Conclusions

The new supramolecular networks designed by the components of Hofmann-like frameworks with 4-CNpyridine were reported. These compounds showed unique multi-dimensional supramolecular networks involving hydrogen interactions and strong metallophilic interactions. Specifically, **1b** exhibited the two magnetic functions of a SCO and a ferromagnetic transition. The diversity of the self-assembly process offered both unexpectedly interesting structure and properties.

**Supplementary Materials:** The following are available online at http://www.mdpi.com/2073-4352/8/11/415/s1, Figure S1. X-ray powder diffraction data of **1a** and **1b** (red line: calculation, black line: experiment); Figure S2. Thermogravimetric analysis for complex **1a** was carried under nitrogen atmosphere at a heating rate of 10 K/min; Figure S3. Showing the closet approach between Fe(2) centers for **1b** at 90 K.

**Author Contributions:** Data curation, I.T.; Formal analysis, D.A. and T.S.; Investigation, T.K. and I.T.; Project administration, T.K.; Supervision, T.K.

**Funding:** This work was financially supported by KAKENHI (JSPS/15K05485 and 18K04964) and the Yashima Environment Technology Foundation. Part of this work was supported by the Ministry of Education, Culture,

Sports, Science, and Technology, Japan (MEXT)-Supported program for the Strategic Research Foundation at Private Universities 2012–2016.

**Conflicts of Interest:** The authors declare no conflict of interest.

## References

1. Real, J.A.; Andrés, E.; Munoz, M.C.; Julve, M.; Granier, T.; Bousseksou, A.; Varret, F. Spin crossover in a catenane supramolecular system. *Science* **1995**, *268*, 265–268. [CrossRef] [PubMed]
2. Miyamachi, T.; Gruber, M.; Davesne, V.; Bowen, M.; Boukari, S.; Joly, L.; Scheurer, F.; Rogez, G.; Yamada, T.K.; Ohresser, P.; et al. Robust spin crossover and memristance across a single molecule. *Nat. Commun.* **2012**, *3*, 938. [CrossRef] [PubMed]
3. Gentili, D.; Givaja, G.; Mas-Ballesté, R.; Azani, M.-R.; Shehu, A.; Leonardi, F.; Mateo-Martí, E.; Greco, P.; Zamora, F.; Cavallini, M. Patterned conductive nanostructures from reversible self-assembly of 1D coordination polymer. *Chem. Sci.* **2012**, *3*, 2047–2051. [CrossRef]
4. Gentili, D.; Demitri, N.; Schäfer, B.; Liscio, F.; Bergenti, I.; Ruani, G.; Ruben, M.; Cavallini, M. Multi-modal sensing in spin crossover compounds. *J. Mater. Chem. C* **2015**, *3*, 7836–7844. [CrossRef]
5. Gütlich, P.; Garcia, Y.; Goodwin, H.A. Spin crossover phenomena in Fe(ii) complexes. *Chem. Soc. Rev.* **2000**, *29*, 419–427. [CrossRef]
6. Chiruta, D.; Jureschi, C.-M.; Linares, J.; Garcia, Y.; Rotaru, A. Lattice architecture effect on the cooperativity of spin transition coordination polymers. *J. Appl. Phys.* **2014**, *115*, 053523. [CrossRef]
7. Kitazawa, T.; Gomi, Y.; Takahashi, M.; Takeda, M.; Enomoto, M.; Miyazaki, A.; Enoki, T. Spin-crossover behaviour of the coordination polymer $FeII(C_5H_5N)_2Ni^{II}(CN)_4$. *J. Mater. Chem.* **1996**, *6*, 119–121. [CrossRef]
8. Galet, A.; Muñoz, M.C.; Martinez, V.; Real, J.A. Supramolecular isomerism in spin crossover networks with aurophilic interactions. *Chem. Commun.* **2004**, 2268–2269. [CrossRef] [PubMed]
9. Muñoz, M.C.; Gaspar, A.B.; Galet, A.; Real, J.A. Spin-Crossover Behavior in Cyanide-Bridged Iron(II)-Silver(I) Bimetallic 2D Hofmann-like Metal-Organic Frameworks. *Inorg. Chem.* **2007**, *46*, 8182–8192. [CrossRef] [PubMed]
10. Agustí, G.; Muñoz, M.C.; Gaspar, A.B.; Real, J.A. Spin-Crossover Behavior in Cyanide-bridged Iron(II)−Gold(I) Bimetallic 2D Hofmann-like Metal−Organic Frameworks. *Inorg. Chem.* **2008**, *47*, 2552–2561. [CrossRef] [PubMed]
11. Kosone, T.; Kachi-Terajima, C.; Kanadani, C.; Saito, T.; Kitazawa, T. A two-step and hysteretic spin-crossover transition in new cyano-bridged hetero-metal $Fe^{II}Au^I$ 2-dimensional assemblage. *Chem. Lett.* **2008**, *37*, 422–423. [CrossRef]
12. Kosone, T.; Kanadani, C.; Saito, T.; Kitazawa, T. Synthesis, crystal structures, magnetic properties and fluorescent emissions of two-dimensional bimetallic coordination frameworks $Fe^{II}$(3-fluoropyridine)$_2$[$Au^I(CN)_2$]$_2$ and $Mn^{II}$(3-fluoropyridine)$_2$[$Au^I(CN)_2$]$_2$. *Polyhedron* **2009**, *28*, 1930–1934. [CrossRef]
13. Kosone, T.; Kanadani, C.; Saito, T.; Kitazawa, T. Spin crossover behavior in two-dimensional bimetallic coordination polymer $Fe^{II}$(3-bromo-4-picoline)$_2$[$Au^I(CN)_2$]$_2$: Synthesis, crystal structures, and magnetic properties. *Polyhedron* **2009**, *28*, 1991–1995. [CrossRef]
14. Kosone, T.; Tomori, I.; Kanadani, C.; Saito, T.; Mochida, T.; Kitazawa, T. Unprecedented three-step spin-crossover transition in new 2-dimensional coordination polymer {$Fe^{II}$(4-methylpyridine)$_2$[$Au^I(CN)_2$]$_2$}. *Dalton Trans.* **2010**, *39*, 1719–1721. [CrossRef] [PubMed]
15. Okabayashi, J.; Ueno, S.; Kawasaki, T.; Kitazawa, T. Ligand 4-X pyridine (X=Cl, Br, I) dependence in Hofmann-type spin crossover complexes: Fe(4-Xpyridine)$_2$[Au(CN)$_2$]$_2$. *Inorg. Chim. Acta* **2016**, *445*, 17–21. [CrossRef]
16. Kosone, T.; Kawasaki, T.; Tomori, I.; Okabayashi, J.; Kitazawa, T. Modification of Cooperativity and Critical Temperatures on a Hofmann-Like Template Structure by Modular Substituent. *Inorganics* **2017**, *5*, 55. [CrossRef]
17. Sheldrick, G.M. *SADABS, Program for Empirical Absorption Correction for Area Detector Data*; University of Göttingen: Göttingen, Germany, 1996.
18. Sheldrick, G.M. *SHELXL, Program for the Solution of Crystal Structures*; University of Göttingen: Göttingen, Germany, 1997.

19. Kosone, T.; Kitazawa, T. Guest-dependent spin transition with long range intermediate state for 2-dimensional Hofmann-like coordination polymer. *Inorg. Chim. Acta* **2016**, *439*, 159–163. [CrossRef]
20. Kawasaki, T.; Kachi-Terajima, C.; Saito, T.; Kitazawa, T. Triply Interpenetrated Structure of {$Mn^{II}$(L)$_2$[$Ag^I$(CN)$_2$]$_2$}{$Mn^{II}$(H$_2$O)$_2$[$Ag^I$(CN)$_2$]$_2$} (L = 4-CNpy or py-4-aldoxime). *Bull. Chem. Soc. Jpn.* **2008**, *81*, 268–273. [CrossRef]

© 2018 by the authors. Licensee MDPI, Basel, Switzerland. This article is an open access article distributed under the terms and conditions of the Creative Commons Attribution (CC BY) license (http://creativecommons.org/licenses/by/4.0/).

Article

# Impurity-Induced Spin-State Crossover in $La_{0.8}Sr_{0.2}Co_{1-x}Al_xO_3$

Ichiro Terasaki *, Masamichi Ikuta, Takafumi D. Yamamoto and Hiroki Taniguchi

Department of Physics, Nagoya University, Nagoya 464-8602, Japan; ikuta.masamichi@b.mbox.nagoya-u.ac.jp (M.I.); tdyamamoto@nagoya-u.ac.jp (T.D.Y.); hiroki_taniguchi@cc.nagoya-u.ac.jp (H.T.)
* Correspondence: terra@nagoya-u.jp; Tel./Fax: +81-52-789-5255

Received: 26 September 2018; Accepted: 29 October 2018; Published: 31 October 2018

**Abstract:** We have prepared a set of polycrystalline samples of $La_{0.8}Sr_{0.2}Co_{1-x}Al_xO_3$ ($0 \leq x \leq 0.2$), and have measured the magnetization as functions of temperature and magnetic field. We find that the average spin number per Co ion ($S_{Co}$) evaluated from the room-temperature susceptibility is around 1.2–1.3 and independent of $x$. However, we further find that $S_{Co}$ evaluated from the saturation magnetization at 2 K is around 0.3–0.7, and decreases dramatically with $x$. This naturally indicates that a significant fraction of the $Co^{3+}$ ions experience a spin-state crossover from the intermediate- to low-spin state with decreasing temperature in the Al-substituted samples. This spin-state crossover also explains the resistivity and the thermopower consistently. In particular, we find that the thermopower is anomalously enhanced by the Al substitution, which can be consistently explained in terms of an extended Heikes formula.

**Keywords:** cobalt oxide; spin polaron; impurity effect; spin-state crossover

---

## 1. Introduction

The spin state is one of the most fundamental concepts in transition-metal oxides/complexes [1]. The $d$ electrons in a transition-metal ion feel the Coulomb repulsion from the neighboring oxygen anions. In a transition-metal ion surrounded with octahedrally-coordinated oxygen anions, the five-fold degenerate $d$ orbitals in a vacuum are split into the triply degenerate $t_{2g}$ orbitals and the doubly degenerate $e_g$ orbitals. An energy gap from $t_{2g}$ to $e_g$ levels is called "ligand field splitting", and competes with the Hund coupling [2]. For a larger ligand field splitting, the $d$ electrons occupy the $t_{2g}$ levels first to minimize the total spin number $S$. On the other hand, for a larger Hund coupling, $S$ is maximized to satisfy Hund's rule. The former state is called the low-spin state, and the latter the high-spin state.

While the two spin states can be seen from $d^4$ to $d^7$ electron systems, $Co^{3+}$ is of particular importance in the sense that the low-spin state ($e_g^0 t_{2g}^6$, $S = 0$) and the high-spin state ($e_g^2 t_{2g}^4$, $S = 2$) are almost degenerate in energy, in which various external conditions such as temperature, pressure, and magnetic field can induce the spin-state transition/crossover [3]. $LaCoO_3$ [4–7] and $Sr_3YCo_4O_{10.5}$ [8–11] are prime examples.

In addition to the high- and low-spin states, the intermediate-spin state ($e_g^1 t_{2g}^5$, $S = 1$) has been theoretically proposed in the $Co^{3+}$ oxides, [12] and has been controversial for long time. Nakao et al. [13] have observed the characteristic resonant X-ray diffraction in a single-crystal sample of $Sr_3YCo_4O_{10.5}$ at low temperatures, indicating an $e_g$-like elongated electron density distribution around the $Co^{3+}$ ion. This naturally suggests that the $Co^{3+}$ ions are in the intermediate-spin state. In the case of $LaCoO_3$, however, no critical experiments have been done thus far, and it is still under debate whether or not the intermediate-spin state exists [14–20]. Recently, the intermediate-spin state has been regarded as a kind of excitonic insulator from the low-spin state, [21–23] and its excited properties have been reinvestigated [24].

In contrast to the undoped LaCoO$_3$, researchers in the community seem to reach a consensus that the Co$^{3+}$ ions exist as the intermediate-spin state in the doped LaCoO$_3$, and are responsible for the ferromagnetic metal state (FMM) in the heavily-doped samples [25–29]. Figure 1 shows an electronic phase diagram of La$_{1-x}$Sr$_x$CoO$_3$ simplified from the original one [30]. Note that the FMM phase shows cluster glass behavior, however, we do not further explore the glass nature of this phase in this paper. For $x \ll 0.05$, the system is insulating, and the magnetic Co$^{3+}$ ions experience the spin-state crossover denoted by the reverse triangles, showing the nonmagnetic ground state. According to this figure, the spin-state crossover is quickly suppressed with increasing Co$^{4+}$ content $x$. This has been explained in terms of spin polaron, [31] which is a spin cluster consisting of one Co$^{4+}$ ion in the low-spin state surrounded by several Co$^{3+}$ ions in the intermediate-spin state [32].

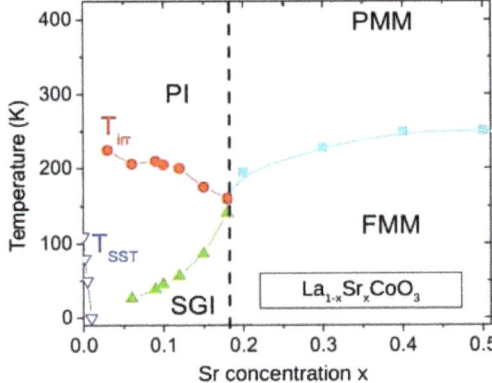

**Figure 1.** Phase diagram of La$_{1-x}$Sr$_x$CoO$_3$ simplified from the original one in Reference [30]. Depending on Sr content $x$ and temperature, the system takes the paramagnetic insulator (PI), the spin glass insulator (SGI), the paramagnetic metal (PMM), or ferromagnetic metal (FMM) state. FMM is often indistinguishable from cluster glass state. The reverse triangles show the spin-state crossover temperature $T_{SST}$. The other marks correspond to magnetic transition temperatures.

In this paper, we report on the effects of Al substitution for Co on the magnetic and transport properties of doped LaCoO$_3$. Impurity has been a powerful test for the essential properties of novel materials. Bardeen-Cooper-Schrieffer-type superconductors are robust against nonmagnetic impurities which is demonstrated in the Anderson theorem, but are very susceptible against magnetic impurity as explained by the Abrikosov-Gor'kov theory [33]. For unconventional superconductors, nonmagnetic impurity has revealed various anomalous properties in the high-temperature superconducting copper oxides, [34] and possible orbital-fluctuation-driven superconductivity in the Fe-based superconductors [35]. How the impurity effects deviate from simple dilution effects and/or percolation theories in various magnetic materials has been discussed. The Haldane chain compound exhibits anomalous $S = 1/2$ excitation on the Ni$^{2+}$ chain against Cu$^{2+}$ impurity doping, [36] and a transition from the dimered to uniform antiferromagnetic order is observed in Mg-doped CuGeO$_3$ [37]. In the case of LaCoO$_3$, Kyomen et al. [38] conducted a pioneering study focused on this, and they have found that the Rh substitution enhances the paramagnetism, while the Al and Ga substitution increases the spin-state crossover temperature to stabilize the low-spin state. Following their work, we studied the Co-site substitution effects in the doped and undoped LaCoO$_3$. In particuar, Asai et al. [39] found a weak ferromagnetism in LaCo$_{1-x}$Rh$_x$O$_3$, below around 15 K for $0.1 \leq x \leq 0.4$, which can be regarded as a ferromagnetism emerging from nonmagnetic end phases. As a natural extension, we conducted an extensive study of the Al-substitution effect in the doped LaCoO$_3$, where the carrier concentration is set to 0.2 per Co because metallic conduction is reported in this carrier concentration. Here we propose that spin-state crossover takes place in the doped LaCoO$_3$, and the substitution of Al for Co excellently highlights this crossover.

## 2. Experimental

Polycrystalline samples of La$_{0.8}$Sr$_{0.2}$Co$_{1-x}$Al$_x$O$_3$ ($x$ = 0, 0.05, 0.10, and 0.15) were prepared by a standard solid-state reaction. Stoichiometric amounts of La$_2$O$_3$, SrCO$_3$, Co$_3$O$_4$, and Al$_2$O$_3$ powders were mixed and calcined at 1273 K for 24 h in air. The calcined powder was ground and pressed into pellets, and sintered at 1473 K for 48 h in air.

The X-ray diffraction pattern was taken with a laboratory X-ray diffractometer (RINT-2200, Rigaku) in $\theta$-$2\theta$ scan mode with Cu K$\alpha$ as an X-ray source. The temperature dependence of the magnetization was measured with an SQUID susceptometer (MPMS, Quantum Design) in a field-cooling process in 0.1 T from 4 to 300 K. The magnetization–field curve was measured with the SQUID susceptometer from $-7$ to 7 T. The resistivity was measured with a four-probe technique with a homemade measurement station from 4 to 300 K in a liquid He cryostat. The thermopower was measured with a steady state and two-probe technique from 4 to 300 K in a liquid He cryostat. The sample bridged two separated copper heat baths, and the resistance heater made temperature differences between the two heat baths, which was monitored through a copper-constantan differential thermocouple. The contribution of the voltage leads were carefully subtracted.

## 3. Results

Figure 2 shows X-ray diffraction patterns of the prepared samples measured at room temperature. All the peaks are indexed as rhombohedral ($R\bar{3}c$), and no impurity phases are seen. The lattice parameter $a$ slightly decreased with increasing Al content $x$ ($a$ = 5.440, 5.434, 5.429, and 5.421 Å for $x$ = 0, 0.05, 0.10, and 0.15, respectively), and this indicates that the Al ion substituted well for Co in the whole set of the samples.

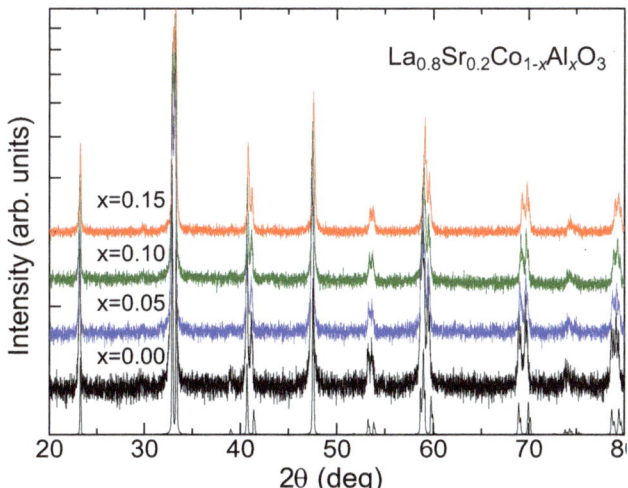

**Figure 2.** X-ray diffraction patterns of polycrystalline samples of La$_{0.8}$Sr$_{0.2}$Co$_{1-x}$Al$_x$O$_3$. The black curve plotted at the lowest position indicates the simulated diffraction patterns.

Figure 3a shows the temperature dependence of the magnetization in an external field of 0.1 T in the field-cooling process. All the samples showed a gradual increase in the magnetization below a certain temperature, indicating weak-ferromagnetic or cluster glass-like behavior. The data for $x = 0$ are consistent with preceding studies, [25–27] and the Al substitution systematically suppressed the low temperature magnetization and the transition temperature. As the Al ion is chemically stable as trivalent, it substitutes for $Co^{3+}$. Thus, the present results suggest that $Co^{3+}$ ions play a vital role in the FMM phase of the doped $LaCoO_3$, although FMM is driven by the doped holes (i.e., the $Co^{4+}$ ions). For $x > 0$, a broad maximum is seen in the magnetization as indicated by the arrows. The physical meaning of this maximum will be discussed later.

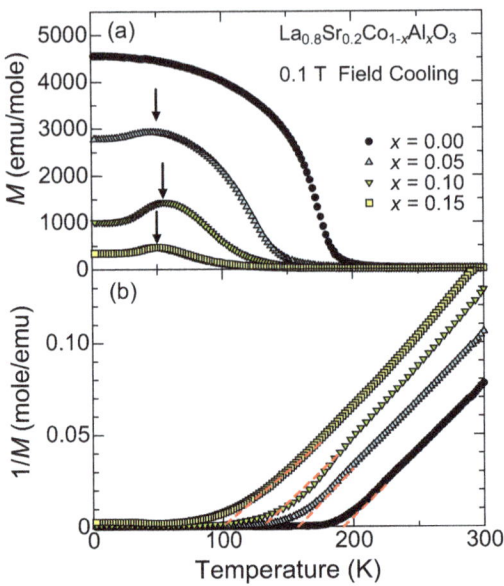

**Figure 3.** (a) Magnetization of polycrystalline samples of $La_{0.8}Sr_{0.2}Co_{1-x}Al_xO_3$ in an external field of 0.1 T in the field-cooling process. The arrows indicate a broad peak in the magnetization curve. (b) Inverse magnetization of the same data of (a). The black circles, blue triangles, green reverse triangles, and yellow squares represent $x = 0$, 0.05, 0.10 and 0.15, respectively. The dotted lines are guides to the eye for estimation of the transition temperature.

To see the transition temperature $T_c$ and the susceptibility at high temperatures clearly, Figure 3b shows the inverse magnetization for the same data. Although the transition is broadened with increasing $x$, $T_c$ decreases from 190 K for $x = 0$ down to 100 K for $x = 0.15$, as was estimated from the intersection points between the dotted lines and the temperature axis. This suppression is severer by a factor of three than the prediction from a simple dilution effect ($T_c(x)/T_c(0) \sim 1 - 3x$). This means that the Al substitution dramatically alters the ferromagnetic order. In contrast, the inverse magnetization above 250 K shows that the temperature slope is nearly the same for all the samples, and can be explained with a simple Curie–Weiss law in which the slope corresponds to the density of the magnetic moment. This indicates that the Al substitution left the whole number of magnetic moment almost unchanged. On one hand, this seems reasonable because $Al^{3+}$ is nonmagnetic, but on the other hand it is in contradiction with the strong suppression of the ferromagnetic order.

Figure 4a shows the magnetization hysteresis plotted as a function of external field for all the samples at 2 K. We show only the positive magnetization part just for clarity. All the curves tend to saturate at high magnetic fields, which is a hallmark of ferromagnetism. The saturation magnetization is around 1.3 $\mu_B$/f.u. for $x = 0$, while that for $x = 0.05$ it is around 1.0 $\mu_B$/f.u. This indicates that 5% substituted Al decreases the saturation moment of 0.3 $\mu_B$/f.u., i.e., one Al ion decreases a substantial value 6 $\mu_B$. For further substitution, the decrease in saturation magnetization is somewhat tempered, but still one Al ion suppresses roughly 4 $\mu_B$. Another notable feature is that the Al substitution makes the ferromagnetism harder; it increases the coercive field drastically. Owing to this hardening, the magnetization in 0.1 T shown in Figure 3a was measured as being smaller values than expected which underestimate the saturation magnetization.

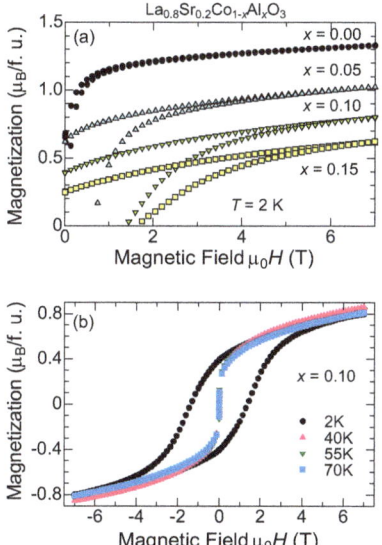

**Figure 4.** (a) Magnetization–field curve of polycrystalline samples of $La_{0.8}Sr_{0.2}Co_{1-x}Al_xO_3$ at 2 K. The positive magnetization part is shown for clarity. (b) The magnetization–field curves for $x = 0.10$ at various temperatures.

Figure 4b shows how the magnetic hysteresis evolves with temperature for $x = 0.10$. At 2 K, the magnetic hysteresis shows a clear loop with the coercive field of 1.5 T as was already mentioned above. With increasing temperature, the coercive field rapidly decreases above 40 K, and the magnetization does not show appreciable temperature dependence up to 70 K.

Let us evaluate the average spin number per Co ion quantitatively. We fit the magnetization above 250 K using the Curie–Weiss formula given by

$$\frac{M}{H} = \frac{C}{T - T_c}. \tag{1}$$

The parameter $C$ is the Curie constant which is further written by

$$C = \frac{N\mu_{\text{eff}}^2}{3k_B} = \frac{Ng^2\mu_B^2 S(S+1)}{3k_B}, \tag{2}$$

where $\mu_{\text{eff}}$ is the effective magnetic moment, $N$ the density of the effective moment, and $S$ the effective spin number. Figure 5 shows the thus evaluated $\mu_{\text{eff}}$ plotted as a function of the Al content $x$. Clearly $\mu_{\text{eff}}$ is almost independent of $x$ (slightly decreases with $x$), which is consistent with the fact that the temperature slope of the inverse magnetization is almost independent of $x$ as was already discussed in Figure 3b. In contrast, the saturation magnetization below $T_c$ rapidly decreases with $x$. We regard the magnetization at 7 T at 2 K ($M_{7T}$) as the saturation magnetization, and plot in the same figure.

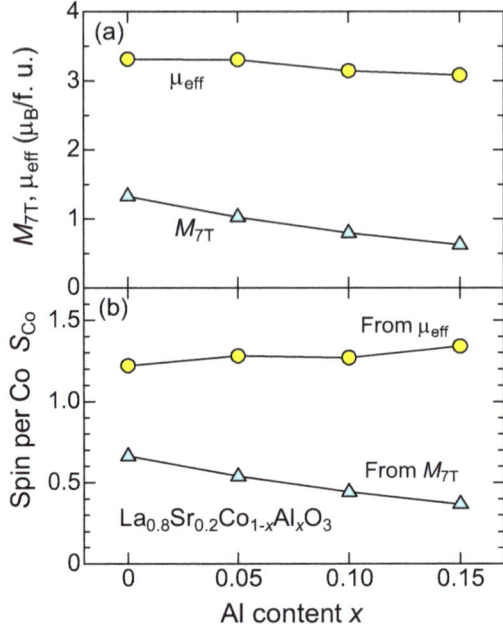

**Figure 5.** (a) Magnetization at 7 T ($M_{7T}$) evaluated from the magnetization–field curve at 2 K and the effective magnetic moment $\mu_{eff}$ evaluated from the Curie–Weiss fitting from 250 to 300 K. (b) The average spin number per Co ($S_{Co}$) evaluated from $M_{7T}$ and $\mu_{eff}$.

We evaluate the average spin number per Co ($S_{Co}$) from $\mu_{eff}$ and $M_{7T}$. In Figure 5b we plot the two kinds of $S_{Co}$, where we assume the g-factor to be 2. From $\mu_{eff}$, $S_{Co}$ is around 1.2–1.3 and almost independent of $x$. This suggests that most of the $Co^{3+}$ ions are in the intermediate state ($S = 1$). On the other hand, from $M_{7T}$, $S_{Co}$ is much smaller than unity and decreases with $x$. By combining these two, we are led to the conclusion that $S_{Co}$ decreases with decreasing temperature, and such tendency becomes more remarkable upon the Al substitution. This naturally indicates that some portion of the Co ions show a crossover to the low-spin state at low temperatures.

We think that the spin-state crossover is the origin for the broad maximum observed in the magnetization for $x > 0$ in Figure 3a, which looks similar to the drop of the magnetization in $Sr_3YCo_4O_{10.5}$ around 200 K [8,40]. Since the two $S_{Co}$ values do not coincide for $x = 0$, we propose that a small fraction of the $Co^{3+}$ ions may experience spin-state crossover even for $x = 0$. This conclusion is consistent with the magnetic-resonance experiment by Smith et al. [41], from which they propose that the spin-state transition survives around 100 K up to 15% doped $LaCoO_3$. In $La_{1-x}Sr_xCoO_3$, Rodoriguez and Goodenough [26] have reported $S_{Co}$ above $T_c$ is almost independent of Sr content, while $S_{Co}$ below $T_c$ is much smaller and depends on Sr content. Kriener et al. [29] have found that the saturation magnetization is smaller in $La_{1-x}Ca_xCoO_3$ than in $La_{1-x}Sr_xCoO_3$ for the same $x$, which cannot be explained by a simple combination of the low-spin state $Co^{4+}$ and the intermediate-spin state $Co^{3+}$. Prakash et al. [42] have also reported that the magnetization curves in $La_{1-x}Ca_xCoO_3$ are quantitatively different from those in $La_{1-x}Sr_xCoO_3$ in the form of nano-particles.

## 4. Discussion

### 4.1. Effects on Transport Properties

Let us discuss the effects of the Al substitution on the charge transport. Figure 6a shows the resistivity of the prepared samples plotted as a function of temperature. The resistivity for $x$

= 0 shows a low value of 1 mΩcm at room temperature, and weakly depends on temperatures above 100 K. With decreasing temperatures below 100 K, it turns into nonmetallic conduction, which has been regarded as the strong localization of carriers. This nonmetallic behavior happens when the double-exchange hopping between $Co^{3+}$ and $Co^{4+}$ ions is gradually suppressed below the magnetic transition temperature. The Al substitution increases the resistivity systematically at all temperatures, and makes the low-temperature upturn remarkable. We should note that the Al ions are expected to replace $Co^{3+}$ ions in the system, thereby leaving the carrier concentration (the $Co^{4+}$ concentration) intact. This is supported by the fact that the room-temperature resistivity remains in the order of mΩcm. Accordingly, we attribute the nonmetallic behavior to the reduced mobility arising from the localization of the spin polaron.

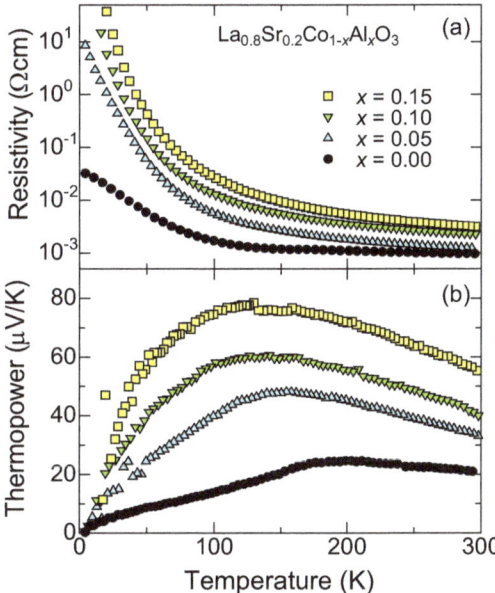

**Figure 6.** (a) Resistivity and (b) Thermopower of $La_{0.8}Sr_{0.2}Co_{1-x}Al_xO_3$.

Figure 6b shows the thermopower of the same set of the samples plotted as a function of temperature. The thermopower for $x = 0$ shows a small value of 20 μV/K at room temperature, and shows weak temperature dependence above the Curie temperature $T_c$ of 190 K. Then it shows a broad kink at $T_c$, and shows a temperature-linear dependence which is expected in the thermopower of degenerate semiconductors [43]. Since the diffusion term does not include scattering time directly, the $T$-linear thermopower is not in contradiction to the nonmetallic resistivity upturn as a result of the localization.

Most notably, the Al substitution systematically enhances the thermopower at all the temperatures. The increment in the thermopower reaches 60 μV/K at 100 K from $x = 0$ to 0.15, which is quite substantial. The $T$-linear thermopower at low temperatures indicates that their electronic states are essentially metallic without an energy gap in the density of states, being consistent with the picture of localized spin polaron. The samples for $x = 0.05, 0.10$, and $0.15$ show a broad peak, and the peak temperature decreases with $x$. The peak temperature seems to decrease with $x$ similarly to $T_c$, but the relationship between the two temperatures is yet to be explored. We should note once again that the Al ions are supposed to replace $Co^{3+}$ rather than $Co^{4+}$. Within a simple semiconductor physics, the thermopower depends mainly on the carrier concentration, not on the mobility. Thus, it is expected to be related to the Co$^{4+}$ concentration, but not to the Co$^{3+}$ concentration. Thus, the thermopower enhancement induced by the Al substitution is highly nontrivial.

The thermopower of the cobalt oxides has been discussed in terms of an extended Heikes formula proposed by Koshibae et al. [44]. The theoretical expression is given by

$$S = \frac{k_B}{e} \ln \frac{g_A}{g_B} \frac{p}{1-p}, \quad (3)$$

where $p$ is the concentration of the A ions, and $g_i$ ($i$ = A and B) is the internal electronic degeneracy due to the spin and orbital degrees of freedom. In the present case, $p$ equals 0.2 (the Co$^{4+}$ concentration) for all the samples. Consequently, $\ln g_{Co^{4+}}/g_{Co^{3+}}$ should determine the magnitude of the thermopower. We experimentally showed that the Co$^{3+}$ ions in the low-spin state can give a large value of $\ln g_{Co^{4+}}/g_{Co^{3+}}$ [45–47].

In this respect, we can associate the thermopower enhancement at low temperatures with the spin-state crossover induced by Al substitution. Although the spin-state crossover is suggested at low temperatues from the magnetic measurements, the thermopower already increases at room temperature. We do not yet understand this clearly. One possibility is that the substituted Al ions bring additional entropy into the system, and give additional thermopower in the Heikes formula ($g_A$ for Al ion and $g_B$ for Co ion in Equation (3)). Another possibility is that the formation of the spin polaron gradually starts from room temperature, and is suppressed by the Al substitution. Consequently, the Co$^{4+}$ ion is not fully surrounded with the intermediate-spin state Co$^{3+}$ at room temperature.

### 4.2. Comparison with Other Impurities

In this subsection, we review the Co-site substitution effects on LaCoO$_3$ and the related cobalt oxides. After the pioneering work of Kyomen et al. [38], we have studied the Co-site substitution effects in the doped and undoped LaCoO$_3$. Asai et al. [39] found that the Rh substitution induces weak ferromagnetism in LaCo$_{1-x}$Rh$_x$O$_3$ and further found that the Ga substitution in the weak-ferromagnetic LaCo$_{0.8}$Rh$_{0.2}$O$_3$ severely suppresses the effective moment in the Co site [48]. These two works suggest that the Rh substitution stabilizes the Co$^{3+}$ ion in the high-spin state, while the Ga substitution does the Co$^{3+}$ ion in the low-spin state [49]. Knizek et al. [50] have reported from GGA+U electronic structure calculations that the Rh impurity can stabilize the high-spin state of the neighboring Co$^{3+}$ ion. This tendency was further verified from optical reflectivity [51] and from X-ray absorption [52]. From the Curie–Weiss analysis, one substituted Rh ion leaves the value of $S_{Co}$ unchanged up to 50% substitution, while one substituted Ga ion reduces $S_{Co}$ by 4.6 $\mu_B$. Tomiyasu et al. [53] have recently shown that the doped magnetic impurity, such as Cr$^{3+}$, Fe$^{3+}$, Mn$^{4+}$, or Ni$^{2+}$, is surrounded by magnetic Co ions in LaCoO$_3$.

In the case of the doped LaCoO$_3$, Shibasaki et al. [54] found that the Rh substitution in La$_{0.8}$Sr$_{0.2}$Co$_{1-x}$Rh$_x$O$_3$ does not change $S_{Co}$ above $T_c$ while one Rh substitution effectively decreases $S_{Co}$ by 9 $\mu_B$ below $T_c$. Since their findings are qualitatively the same as those in the present work, we can also explain the big drop in $S_{Co}$ from room temperature down to 5 K in La$_{0.8}$Sr$_{0.2}$Co$_{1-x}$Rh$_x$O$_3$ in terms of spin-state crossover. As mentioned in the previous section, one Al substitution decreases $S_{Co}$ by 6 $\mu_B$ in La$_{0.8}$Sr$_{0.2}$Co$_{1-x}$Al$_x$O$_3$. It is not clear at present whether or not the difference between 9 $\mu_B$ for Rh and 6 $\mu_B$ for Al is intrinsic. One possibility is that the Rh ion destroy the spin polaron more seriously than the Al ion. While the 10% Rh substituted sample is nearly paramagnetic [54], the 10% Al substituted sample still shows ferromagnetic behavior.

### 4.3. An Origin of Impurity-Induced Spin-State Crossover

Let us discuss a possible mechanism of spin-state crossover induced by impurities. As already mentioned, the spin polaron is formed in the lightly-doped LaCoO$_3$. Podlesnyak et al. [31] have shown that the doped hole exists in the low-spin state Co$^{4+}$ ion and is surrounded by six octahedrally-coordinated $Co^{3+}$ ions in the intermediate spin state. The intermediate-spin state $Co^{3+}$ is believed to be induced by $Co^{4+}$ in order that the doped hole can gain kinetic energy through the double exchange interaction. This spin polaron acts as a super-spin of $S = 13/2$ via the double exchange interaction within the cluster. From the earlier magnetic measurement, Yamaguchi et al. [32] have found that the magnetization–field curve can be fitted with a large spin number of $S = 10$–$16$, which is consistent with the spin-polaron picture. The above experiments consistently show that the unit of magnetic moment extends to several unit cells.

Thus, it is naturally expected that the doped Al replaces one of the $Co^{3+}$ ions in the spin polaron, and suppresses the double exchange. In the strong coupling limit, a single Al ion can block the spin-polaron formation, and practically act as plural nonmagnetic ions. As already shown, the transition temperature is rapidly reduced with $x$ roughly expressed by $T_c(x)/T_c(0) \sim 1 - 3x$. The factor of three may come from the picture that one Al ion effectively makes around three Co ions be in the low-spin state at low temperatures. This is consistent with the fact that one Al ion suppress $6\,\mu_B$ in the saturation magnetization.

Since $S_{Co}$ is independent of $x$ above $T_c$, the spin polaron is likely formed around $T_c$ below which the effect of Al becomes remarkable. This is also suggested by the resistivity; the nonmetallic conduction becomes more remarkable below $T_c$. Since the spin polaron is strongly pinned or destroyed by the Al ion, it cannot act as a charge carrier in this system, where the isolated $Co^{4+}$ ion is strongly localized in the sea of the low-spin state $Co^{3+}$. At room temperature where the spin polaron is not yet formed, the resistivity is weakly dependent on $x$.

*4.4. Coercive Field Induced by Al Substitution*

As was already mentioned in the previous section, the weak ferromagnetism becomes harder with increasing Al concentration as shown in Figure 4. The coercive field $\mu_0 H_c$ is unusually high, and reaches near 2 T at 2 K for $x = 0.15$. This is difficult to understand from a conventional theory of ferromagnets, [55] in which the anisotropy energy and the pinning force determine $\mu_0 H_c$. Accordingly, $\mu_0 H_c$ increases roughly with $T_c$ in conventional hard ferromagnets. In the present case, however, *$\mu_0 H_c$ increases* and *$T_c$ decreases* with $x$. In particular, the coercive energy of $\mu_0 H_c g \mu_B S_{Co}$ is of 2 K/f.u. for $x = 0.15$, which is comparable with the ferromagnetic energy of $BH$/f.u. at 6 T. In the first place, polycrystalline samples are expected to show smaller coercive field than single crystal [56].

Such giant coercive fields have been overlooked thus far, possibly because they are regarded as coming from extrinsic origins such as pinning centers, dislocations, and particle size [57]. Actually nano-size ferromagnetic particles of $\varepsilon$-$Fe_2O_3$ exhibit a giant coercive field of 2 T, [58] and three-atom-width nanowires of Co metal show $\mu_0 H_c = 1.2$ T [59].

Aside from nano-magnetics, huge coercive fields have been reported in various oxide materials. For example, $\mu_0 H_c$ equals 5.9 T at 4.2 K for $Co_2(OH)_2(C_8H_4O_4)$ [60], 9 T at 4.2 K for $LuFe_2O_4$ [61], 12 T at 2 K for the hexagonal $Sr_5Ru_{5-x}O_{15}$ [62], and 2 T at 100 K for $BaIrO_3$ [63]. Although the origin of these huge $\mu_0 H_c$ is different from material to material, a strong spin-orbit interaction and Ising-like anisotropy seem to play vital roles.

We have discussed a possible spin-state crossover induced by Al substitution. In this context, we can associate the giant $\mu_0 H_c$ with magnetic-field induced spin-state crossover. As was discussed above, the substituted Al ion pins the spin polaron, and lets the neighboring $Co^{3+}$ ion be in the low-spin state. Since this pinning competes with the intra-polaron double exchange, we expect that the energy difference between the low- and intermediate-spin states is within the order of $k_B T$. In such a situation, an external field helps the intermediate-spin state to grow through the Zeeman effect, and the number of the intermediate spins gradually increases with increasing field. This is a kind of metamagnetic transition in the sense that the total spin number changes with external field, and accompanies a magnetic field hysteresis because the spin-state crossover causes a large magnetostriction [64]. The spin-state transition is induced by external field in $LaCoO_3$ and the related oxides, but the critical field is around 60 T for $LaCoO_3$ [65,66] and 53 T for $Sr_3YCo_4O_{10.5}$ [40]. It is not surprising that they are much higher than those in the present case, because the spin-state crossover in the pure cobalt oxides is caused by a cooperative interaction among the $Co^{3+}$ ions, while the spin-state energy and cooperative interaction are reduced near the Al impurity.

## 5. Summary

In summary, we prepared a set of polycrystalline samples of $La_{0.8}Sr_{0.2}Co_{1-x}Al_xO_3$ ($0 \leq x \leq 0.15$), and measured the X-ray diffraction, magnetization, resistivity, and thermopower. We obtained the average spin number per Co ion ($S_{Co}$) in two ways, and found that $S_{Co}$ is around 1.2–1.3 at room temperature, and rapidly reduces to 0.3–0.7 at 2 K. The reduction of $S_{Co}$ becomes more remarkable as $x$ becomes larger. We have ascribed this to a spin-state crossover and proposed that the Al substitution enhances this crossover by pinning the spin polaron. The spin-state crossover also semi-quantitatively explains the enhanced thermopower and the anomalously large coercive field induced by the substituted Al ion.

**Author Contributions:** I.T. organized the while project, found the analysis method, and wrote the first draft of the manuscript. M.I. made and characterized the samples, and he measured and analyzed the magnetization with T.D.Y. All the authors equally contributed to the discussion and understanding of the experimental results.

**Funding:** This research received no external funding and the APC was funded by the operating expenses grants of Nagoya University.

**Acknowledgments:** The authors would like to thank K. Tanabe for collaboration, and for S. Asai, R. Okazaki and Y. Yasui for collaboration for the early study of the Co-site subsitution effects.

**Conflicts of Interest:** The authors declare no conflict of interest.

## References

1. Gütlich, P.; Garcia, Y.; Goodwin, H.A. Spin crossover phenomena in Fe(II) complexes. *Chem. Soc. Rev.* **2000**, *29*, 419–427. [CrossRef]
2. Sugano, S.; Tanabe, Y.; Kamimura, H. *Multiplets of Transition-Metal Ions in Crystals*; Academic Press: New York, NY, USA, 1970.
3. Eder, R. Spin-state transition in LaCoO₃ by variational cluster approximation. *Phys. Rev. B* **2010**, *81*, 035101, doi:10.1103/PhysRevB.81.035101. [CrossRef]
4. Jonker, G.; Santen, J.V. Magnetic compounds wtth perovskite structure III. Ferromagnetic compounds of cobalt. *Physica* **1953**, *19*, 120–130. [CrossRef]
5. Raccah, P.M.; Goodenough, J.B. First-order localized-electron collective-electron transition in LaCoO₃. *Phys. Rev.* **1967**, *155*, 932–943. [CrossRef]
6. Asai, K.; Yoneda, A.; Yokokura, O.; Tranquada, J.M.; Shirane, G.; Kohn, K. Two spin-state transitions in LaCoO₃. *J. Phys. Soc. Jpn.* **1998**, *67*, 290–296. [CrossRef]
7. Vogt, T.; Hriljac, J.A.; Hyatt, N.C.; Woodward, P. Pressure-induced intermediate-to-low spin state transition in LaCoO₃. *Phys. Rev. B* **2003**, *67*, 140401. [CrossRef]
8. Kobayashi, W.; Ishiwata, S.; Terasaki, I.; Takano, M.; Grigoraviciute, I.; Yamauchi, H.; Karppinen, M. Room-temperature ferromagnetism in $Sr_{1-x}Y_xCoO_{3-\delta}$ ($0.2 \leq x \leq 0.25$). *Phys. Rev. B* **2005**, *72*, 104408. [CrossRef]
9. Kobayashi, W.; Yoshida, S.; Terasaki, I. High-temperature metallic state of room-temperature ferromagnet $Sr_{1-x}Y_xCoO_{3-\delta}$. *J. Phys. Soc. Jpn.* **2006**, *75*, 103702. [CrossRef]
10. Golosova, N.O.; Kozlenko, D.P.; Dubrovinsky, L.S.; Drozhzhin, O.A.; Istomin, S.Y.; Savenko, B.N. Spin state and magnetic transformations in $Sr_{0.7}Y_{0.3}CoO_{2.62}$ at high pressures. *Phys. Rev. B* **2009**, *79*, 104431. [CrossRef]
11. Sheptyakov, D.V.; Pomjakushin, V.Y.; Drozhzhin, O.A.; Istomin, S.Y.; Antipov, E.V.; Bobrikov, I.A.; Balagurov, A.M. Correlation of chemical coordination and magnetic ordering in $Sr_3YCo_4O_{10.5+\delta}$ ($\delta = 0.02$ and 0.26). *Phys. Rev. B* **2009**, *80*, 024409. [CrossRef]
12. Korotin, M.A.; Ezhov, S.Y.; Solovyev, I.V.; Anisimov, V.I.; Khomskii, D.I.; Sawatzky, G.A. Intermediate-spin state and properties of LaCoO₃. *Phys. Rev. B* **1996**, *54*, 5309–5316. [CrossRef]

13. Nakao, H.; Murata, T.; Bizen, D.; Murakami, Y.; Ohoyama, K.; Yamada, K.; Ishiwata, S.; Kobayashi, W.; Terasaki, I. Orbital ordering of intermediate-spin state of $Co^{3+}$ in $Sr_3YCo_4O_{10.5}$. *J. Phys. Soc. Jpn.* **2011**, *80*, 023711. [CrossRef]
14. Tokura, Y.; Okimoto, Y.; Yamaguchi, S.; Taniguchi, H.; Kimura, T.; Takagi, H. Thermally induced insulator-metal transition in $LaCoO_3$: A view based on the Mott transition. *Phys. Rev. B* **1998**, *58*, R1699–R1702. [CrossRef]
15. Noguchi, S.; Kawamata, S.; Okuda, K.; Nojiri, H.; Motokawa, M. Evidence for the excited triplet of $Co^{3+}$ in $LaCoO_3$. *Phys. Rev. B* **2002**, *66*, 094404. [CrossRef]
16. Maris, G.; Ren, Y.; Volotchaev, V.; Zobel, C.; Lorenz, T.; Palstra, T.T.M. Evidence for orbital ordering in $LaCoO_3$. *Phys. Rev. B* **2003**, *67*, 224423. [CrossRef]
17. Haverkort, M.W.; Hu, Z.; Cezar, J.C.; Burnus, T.; Hartmann, H.; Reuther, M.; Zobel, C.; Lorenz, T.; Tanaka, A.; Brookes, N.B.; et al. Spin state transition in $LaCoO_3$ studied using soft X-ray absorption spectroscopy and magnetic circular dichroism. *Phys. Rev. Lett.* **2006**, *97*, 176405. [CrossRef] [PubMed]
18. Klie, R.F.; Zheng, J.C.; Zhu, Y.; Varela, M.; Wu, J.; Leighton, C. Direct measurement of the low-temperature spin-state transition in $LaCoO_3$. *Phys. Rev. Lett.* **2007**, *99*, 047203. [CrossRef] [PubMed]
19. Chakrabarti, B.; Birol, T.; Haule, K. Role of entropy and structural parameters in the spin-state transition of $LaCoO_3$. *Phys. Rev. Mater.* **2017**, *1*, 064403. [CrossRef]
20. Shimizu, Y.; Takahashi, T.; Yamada, S.; Shimokata, A.; Jin-no, T.; Itoh, M. Symmetry preservation and critical fluctuations in a pseudospin crossover perovskite $LaCoO_3$. *Phys. Rev. Lett.* **2017**, *119*, 267203. [CrossRef] [PubMed]
21. Nasu, J.; Watanabe, T.; Naka, M.; Ishihara, S. Phase diagram and collective excitations in an excitonic insulator from an orbital physics viewpoint. *Phys. Rev. B* **2016**, *93*, 205136. [CrossRef]
22. Sotnikov, A.; Kuneš, J. Field-induced exciton condensation in $LaCoO_3$. *Sci. Rep.* **2016**, *6*, 30510. [CrossRef] [PubMed]
23. Afonso, J.F.; Kuneš, J. Excitonic magnetism in $d^6$ perovskites. *Phys. Rev. B* **2017**, *95*, 115131. [CrossRef]
24. Wang, R.P.; Hariki, A.; Sotnikov, A.; Frati, F.; Okamoto, J.; Huang, H.Y.; Singh, A.; Huang, D.J.; Tomiyasu, K.; Du, C.H.; et al. Excitonic dispersion of the intermediate spin state in $LaCoO_3$ revealed by resonant inelastic X-ray scattering. *Phys. Rev. B* **2018**, *98*, 035149. [CrossRef]
25. Itoh, M.; Natori, I.; Kubota, S.; Motoya, K. Spin-glass behavior and magnetic phase diagram of $La_{1-x}Sr_xCoO_3$ ($0 \leq x \leq 0.5$) studied by magnetization measurements. *J. Phys. Soc. Jpn.* **1994**, *63*, 1486–1493. [CrossRef]
26. Senarís-Rodríguez, M.A.; Goodenough, J.B. Magnetic and transport properties of the system $La_{1-x}Sr_xCoO_{3-\delta}$ ($0 < x \leq 0.50$). *J. Solid State Chem.* **1995**, *118*, 323–336. [CrossRef]
27. Masuda, H.; Fujita, T.; Miyashita, T.; Soda, M.; Yasui, Y.; Kobayashi, Y.; Sato, M. Transport and magnetic properties of $R_{1-x}A_xCoO_3$ (R = La, Pr and Nd; A = Ba, Sr and Ca). *J. Phys. Soc. Jpn.* **2003**, *72*, 873–878. [CrossRef]
28. Wu, J.; Leighton, C. Glassy ferromagnetism and magnetic phase separation in $La_{1-x}Sr_xCoO_3$. *Phys. Rev. B* **2003**, *67*, 174408. [CrossRef]
29. Kriener, M.; Zobel, C.; Reichl, A.; Baier, J.; Cwik, M.; Berggold, K.; Kierspel, H.; Zabara, O.; Freimuth, A.; Lorenz, T. Structure, magnetization, and resistivity of $La_{1-x}M_xCoO_3$ (M = Ca, Sr, and Ba). *Phys. Rev. B* **2004**, *69*, 094417. [CrossRef]
30. He, C.; Torija, M.A.; Wu, J.; Lynn, J.W.; Zheng, H.; Mitchell, J.F.; Leighton, C. Non-Griffiths-like clustered phase above the Curie temperature of the doped perovskite cobaltite $La_{1-x}Sr_xCoO_3$. *Phys. Rev. B* **2007**, *76*, 014401. [CrossRef]
31. Podlesnyak, A.; Russina, M.; Furrer, A.; Alfonsov, A.; Vavilova, E.; Kataev, V.; Büchner, B.; Strässle, T.; Pomjakushina, E.; Conder, K.; et al. Spin-state polarons in lightly-hole-doped $LaCoO_3$. *Phys. Rev. Lett.* **2008**, *101*, 247603. [CrossRef] [PubMed]
32. Yamaguchi, S.; Okimoto, Y.; Taniguchi, H.; Tokura, Y. Spin-state transition and high-spin polarons in $LaCoO_3$. *Phys. Rev. B* **1996**, *53*, R2926–R2929. [CrossRef]
33. Gor'kov, L.P., Theory of superconducitng alloys. In *The Physics of Superconductors*; Chapter 5; Springer: Berlin, Germany, 2003. [CrossRef]
34. Fukuzumi, Y.; Mizuhashi, K.; Takenaka, K.; Uchida, S. Universal superconductor-insulator transition and $T_c$ depression in Zn-substituted high-$T_c$ cuprates in the underdoped regime. *Phys. Rev. Lett.* **1996**, *76*, 684. [CrossRef] [PubMed]

35. Kontani, H.; Onari, S. Orbital-fluctuation-mediated superconductivity in iron pnictides: Analysis of the five-orbital Hubbard-Holstein model. *Phys. Rev. Lett.* **2010**, *104*, 157001. [CrossRef] [PubMed]
36. Hagiwara, M.; Katsumata, K.; Affleck, I.; Halperin, B.I.; Renard, J. Observation of S = 1/2 degrees of freedom in an S = 1 linear-chain Heisenberg antiferromagnet. *Phys. Rev. Lett.* **1990**, *65*, 3181. [CrossRef] [PubMed]
37. Masuda, T.; Fujioka, A.; Uchiyama, Y.; Tsukada, I.; Uchinokura, K. Phase transition between dimerized-antiferromagnetic and uniform-antiferromagnetic phases in the impurity-doped spin-Peierls cuprate $CuGeO_3$. *Phys. Rev. Lett.* **1998**, *80*, 4566. [CrossRef]
38. Kyômen, T.; Asaka, Y.; Itoh, M. Negative cooperative effect on the spin-state excitation in $LaCoO_3$. *Phys. Rev. B* **2003**, *67*, 144424. [CrossRef]
39. Asai, S.; Furuta, N.; Yasui, Y.; Terasaki, I. Weak Ferromagnetism in $LaCo_{1-x}Rh_xO_3$: Anomalous magnetism emerging between two nonmagnetic end phases. *J. Phys. Soc. Jpn.* **2011**, *80*, 104705. [CrossRef]
40. Kimura, S.; Maeda, Y.; Kashiwagi, T.; Yamaguchi, H.; Hagiwara, M.; Yoshida, S.; Terasaki, I.; Kindo, K. Field-induced spin-state transition in the perovskite cobalt oxide $Sr_{1-x}Y_xCoO_{3-\delta}$. *Phys. Rev. B* **2008**, *78*, 180403. [CrossRef]
41. Smith, R.X.; Hoch, M.J.R.; Moulton, W.G.; Kuhns, P.L.; Reyes, A.P.; Boebinger, G.S.; Zheng, H.; Mitchell, J.F. Evolution of the spin-state transition with doping in $La_{1-x}Sr_xCoO_3$. *Phys. Rev. B* **2012**, *86*, 054428. [CrossRef]
42. Prakash, R.; Shukla, R.; Nehla, P.; Dhaka, A.; Dhaka, R. Tuning ferromagnetism and spin state in $La_{1-x}A_xCoO_3$ (A = Sr, Ca) nanoparticles. *J. Alloys Comp.* **2018**, *764*, 379–386. [CrossRef]
43. Terasaki, I. Thermal conductivity and thermoelectric power of semiconductors. In *Reference Module in Materials Science and Materials Engineering*; Elsevier: Amsterdam, The Netherlands, 2016. [CrossRef][CrossRef]
44. Koshibae, W.; Tsutsui, K.; Maekawa, S. Thermopower in cobalt oxides. *Phys. Rev. B* **2000**, *62*, 6869–6872. [CrossRef]
45. Terasaki, I.; Sasago, Y.; Uchinokura, K. Large thermoelectric power in $NaCo_2O_4$ single crystals. *Phys. Rev. B* **1997**, *56*, R12685–R12687. [CrossRef]
46. Yoshida, S.; Kobayashi, W.; Nakano, T.; Terasaki, I.; Matsubayashi, K.; Uwatoko, Y.; Grigoraviciute, I.; Karppinen, M.; Yamauchi, H. Chemical and physical pressure effects on the magnetic and transport properties of the A-site ordered perovskite $Sr_3YCo_4O_{10.5}$. *J. Phys. Soc. Jpn.* **2009**, *78*, 094711. [CrossRef]
47. Terasaki, I.; Shibasaki, S.; Yoshida, S.; Kobayashi, W. Spin State Control of the Perovskite Rh/Co Oxides. *Materials* **2010**, *3*, 786–799. [CrossRef]
48. Asai, S.; Furuta, N.; Okazaki, R.; Yasui, Y.; Terasaki, I. $Ga^{3+}$ substitution effects in the weak ferromagnetic oxide $LaCo_{0.8}Rh_{0.2}O_3$. *Phys. Rev. B* **2012**, *86*, 014421. [CrossRef]
49. Asai, S.; Okazaki, R.; Terasaki, I.; Yasui, Y.; Kobayashi, W.; Nakao, A.; Kobayashi, K.; Kumai, R.; Nakao, H.; Murakami, Y.; et al. Spin State of $Co^{3+}$ in $LaCo_{1-x}Rh_xO_3$ Investigated by Structural Phenomena. *J. Phys. Soc. Jpn.* **2013**, *82*, 114606. [CrossRef]
50. Knizek, K.; Hejtmánek, J.; Marysko, M.; Jirák, Z.; Bursik, J. Stabilization of the high-spin state of $Co^{3+}$ in $LaCo_{1-x}Rh_xO_3$. *Phys. Rev. B* **2012**, *85*, 134401. [CrossRef]
51. Terasaki, I.; Asai, S.; Taniguchi, H.; Okazaki, R.; Yasui, Y.; Ikemoto, Y.; Moriwaki, T. Optical evidence for the spin-state disorder in $LaCo_{1-x}Rh_xO_3$. *J. Phys. Condens. Mat.* **2017**, *29*, 235802. [CrossRef] [PubMed]
52. Sudayama, T.; Nakao, H.; Yamasaki, Y.; Murakami, Y.; Asai, S.; Okazaki, R.; Yasui, Y.; Terasaki, I. Spin state of $Co^{3+}$ in $LaCo_{1-x}Rh_xO_3$ studied using X-ray absorption spectroscopy. *J. Phys. Soc. Jpn.* **2017**, *86*, 094701. [CrossRef]
53. Tomiyasu, K.; Kubota, Y.; Shimomura, S.; Onodera, M.; Koyama, S.I.; Nojima, T.; Ishihara, S.; Nakao, H.; Murakami, Y. Spin-state responses to light impurity substitution in low-spin perovskite $LaCoO_3$. *Phys. Rev. B* **2013**, *87*, 224409. [CrossRef]
54. Shibasaki, S.; Terasaki, I.; Nishibori, E.; Sawa, H.; Lybeck, J.; Yamauchi, H.; Karppinen, M. Magnetic and transport properties of the spin-state disordered oxide $La_{0.8}Sr_{0.2}Co_{1-x}Rh_xO_{3-\delta}$. *Phys. Rev. B* **2011**, *83*, 094405. [CrossRef]
55. Kittel, C. Physical theory of ferromagnetic domains. *Rev. Mod. Phys.* **1949**, *21*, 541. [CrossRef]
56. Kronmüller, H. Theory of the coercive field in amorphous ferromagnetic alloys. *J. Mag. Mag. Mater.* **1981**, *24*, 159–167. [CrossRef]
57. Kneller, E.F.; Luborsky, F.E. Particle size dependence of coercivity and remanence of single-domain particles. *J. Appl. Phys.* **1963**, *34*, 656–658. [CrossRef]

58. Jin, J.; Ohkoshi, S.i.; Hashimoto, K. Giant coercive field of nanometer-sized iron oxide. *Adv. Mater.* **2004**, *16*, 48–51. [CrossRef]
59. Gambardella, P.; Dallmeyer, A.; Maiti, K.; Malagoli, M.C.; Rusponi, S.; Ohresser, P.; Eberhardt, W.; Carbone, C.; Kern, K. Oscillatory magnetic anisotropy in one-dimensional atomic wires. *Phys. Rev. Lett.* **2004**, *93*, 077203. [CrossRef] [PubMed]
60. Huang, Z.L.; Drillon, M.; Masciocchi, N.; Sironi, A.; Zhao, J.T.; Rabu, P.; Panissod, P. Ab-initio XRPD crystal structure and giant hysteretic effect (H c= 5.9 T) of a new hybrid terephthalate-based cobalt (II) magnet. *Chem. Mater.* **2000**, *12*, 2805–2812. [CrossRef]
61. Wu, W.; Kiryukhin, V.; Noh, H.J.; Ko, K.T.; Park, J.H.; Ratcliff, W.; Sharma, P.A.; Harrison, N.; Choi, Y.J.; Horibe, Y.; et al. Formation of pancakelike ising domains and giant magnetic coercivity in ferrimagnetic $LuFe_2O_4$. *Phys. Rev. Lett.* **2008**, *101*, 137203. [CrossRef] [PubMed]
62. Yamamoto, A.; Hashizume, D.; Katori, H.A.; Sasaki, T.; Ohmichi, E.; Nishizaki, T.; Kobayashi, N.; Takagi, H. Ten layered hexagonal perovskite $Sr_5Ru_{5-x}O_{15}$ (x = 0.90), a weak ferromagnet with a giant coercive field $H_c \sim 12$ T. *Chem. Mater.* **2010**, *22*, 5712–5717. [CrossRef]
63. Kida, T.; Senda, A.; Yoshii, S.; Hagiwara, M.; Nakano, T.; Terasaki, I. Pressure effect on magnetic properties of a weak ferromagnet $BaIrO_3$. *J. Phys. Conf. Ser.* **2010**, *200*, 012084. [CrossRef]
64. Sato, K.; Bartashevich, M.I.; Goto, T.; Kobayashi, Y.; Suzuki, M.; Asai, K.; Matsuo, A.; Kindo, K. High-field magnetostriction of the spin-state transition compound $LaCoO_3$. *J. Phys. Soc. Jpn.* **2008**, *77*, 024601. [CrossRef]
65. Sato, K.; Matsuo, A.; Kindo, K.; Kobayashi, Y.; Asai, K. Field induced spin-state transition in $LaCoO_3$. *J. Phys. Soc. Jpn.* **2009**, *78*, 093702. [CrossRef]
66. Altarawneh, M.M.; Chern, G.W.; Harrison, N.; Batista, C.D.; Uchida, A.; Jaime, M.; Rickel, D.G.; Crooker, S.A.; Mielke, C.H.; Betts, J.B.; et al. Cascade of magnetic field induced spin transitions in $LaCoO_3$. *Phys. Rev. Lett.* **2012**, *109*, 037201. [CrossRef] [PubMed]

© 2018 by the authors. Licensee MDPI, Basel, Switzerland. This article is an open access article distributed under the terms and conditions of the Creative Commons Attribution (CC BY) license (http://creativecommons.org/licenses/by/4.0/).

*Article*

# Iron(II) Spin Crossover (SCO) Materials Based on Dipyridyl-N-Alkylamine

Taous Houari [1,2], Emmelyne Cuza [1], Dawid Pinkowicz [3], Mathieu Marchivie [4], Said Yefsah [2] and Smail Triki [1,*]

1. Univ Brest, CNRS, CEMCA, 6 Avenue Le Gorgeu, C.S. 93837-29238 Brest CEDEX 3, France; taous.houari@ummto.dz (T.H.); Emmelyne.Cuza@univ-brest.fr (E.C.)
2. Faculté des Sciences, Université Mouloud Mammeri, Tizi-Ouzou 15000, Algeria; said.yefsah@ummto.dz
3. Faculty of Chemistry, Jagiellonian University, Gronostajowa 2, 30-387 Kraków, Poland; dawid.pinkowicz@uj.edu.pl
4. CNRS, University of Bordeaux, ICMCB, UMR 5026, 87 Av. Doc. A. Schweitzer, F-33608 Pessac, France; Mathieu.Marchivie@icmcb.cnrs.fr
* Correspondence: smail.triki@univ-brest.fr; Tel.: +33-298-016-146

Received: 1 October 2018; Accepted: 21 October 2018; Published: 24 October 2018

**Abstract:** We present here a new series of spin crossover (SCO) Fe(II) complexes based on dipyridyl-N-alkylamine and thiocyanate ligands, with the chemical formulae [Fe(dpea)$_2$(NCS)$_2$] (**1**) (dpea = 2,2′-dipyridyl-N-ethylamine), I-[Fe(dppa)$_2$(NCS)$_2$], (**2**) II-[Fe(dppa)$_2$(NCS)$_2$], and (**2′**) (dppa = 2,2′-dipyridyl-N-propylamine). The three complexes displayed nearly identical discrete molecular structures, where two chelating ligands (dpea (**1**) and dppa (**2** and **2′**)) stand in the *cis*-positions, and two thiocyanato-κN ligands complete the coordination sphere in the two remaining *cis*-positions. Magnetic studies as a function of temperature revealed the presence of a complete high-spin (HS) to low-spin (LS) transition at $T_{1/2}$ = 229 K for **1**, while the two polymorphs I-[Fe(dppa)$_2$(NCS)$_2$] (**2**) and II-[Fe(dppa)$_2$(NCS)$_2$] (**2′**) displayed similar magnetic behaviors with lower transition temperatures ($T_{1/2}$ = 211 K for **2**; 212 K for **2′**). Intermolecular contacts in the three complexes indicated the absence of any significant interaction, in agreement with the gradual SCO behaviors revealed by the magnetic data. The higher transition temperature observed for complex **1** agrees well with the more pronounced linearity of the Fe–N–C angles recently evidenced by experimental and theoretical magnetostructural studies.

**Keywords:** Fe(II) complex; dipyridyl-N-alkylamine ligands; high spin (HS); low spin (LS); spin cross-over (SCO); magnetic transition

## 1. Introduction

The design of new coordination materials exhibiting the spin crossover (SCO) behavior is one of the most relevant challenge in the field of switchable materials [1–15]. In such materials, the spin state can be switched from a high-spin (HS) to a low-spin (LS) configuration through a number of external stimuli such as temperature, pressure, magnetic field, or light irradiation, for complexes involving transition metal ions of $d^4$–$d^7$ electronic configurations [3–14]. However, iron(II)-based SCO complexes, for which the transition takes place between the paramagnetic high-spin (HS) state ($t_{2g}^4 e_g^2$, $^5T_{2g}$, S = 2) and the diamagnetic low-spin (LS) state ($t_{2g}^6 e_g^0$, $^1A_{1g}$, S = 0) are, by far, the most studied switchable molecular materials [1–14]. From the synthetic point of view, one of the relevant strategies to design original SCO systems is based on the use of appropriate polydentate rigid nitrogen-based ligands and simple anionic entities acting as terminal ligands, such as NCX (X = S, Se, BH$_3$) anions [16–22] or the more sophisticated ones such as cyanocarbanions exhibiting terminal or poly-bridging coordination modes [4,7,23–27]. The latter are able to tune the ligand field energy and some SCO characteristics such as the transition temperature.

In the large families of polydentate molecules, the use of the polypyridine-based ligands of different denticities, such as 2,2'-dipyridylamine (dpa) [18,19], tris(2-pyridyl)methane (tpc) [20,28,29], and tris(2-pyridylmethyl)amine (tpma [21,23,30–33], has allowed the preparation of discrete and extended coordination compounds exhibiting original SCO transitions, allowing to understand more on the SCO phenomenon, such as the origin of cooperativity, the presence of complete or incomplete transitions, and the occurrence of one-step or multi-step behaviors and photo-induced effects. In this context, we have reported, in the last few years, a new series of dinuclear Fe(II) complexes based on the tetradentate tmpa ligand [23] and, more recently, a dinuclear complex and a one-dimensional coordination polymer, both based on the functionalized tris(2-pyridyl)methane (tpc) tripodal ligands and displaying unusual $FeN_5S$ coordination spheres. By experimental and theoretical magnetostructural studies, we have shown in both systems the crucial role of the linearity of the N-bound terminal thiocyanato ligand in the presence of the SCO transition. As a continuation of this research, we have pursued our investigations using the N-functionalized 2,2'-dipyridylamine (dpa) bidentate ligands (see Scheme 1). The two first Fe(II) SCO systems based on the dpa ligands were reported by J. A. Real et al. [18,19]. The first one, [Fe(dpa)$_2$(NCS)$_2$], containing two cis-thiocyanato-κN ligands, showed an incomplete SCO transition at 88 K, while the second one, Fe(dpa)(NCS)$_2$]$_2$bpym (bpym = 2,2'-bipyrimidine, acting as bis-chelating ligand), was reported as a dinuclear Fe(II) neutral complex with a very gradual SCO behavior at 245 K. Inspired by these observations, a few years later, S. Bonnet et al. prepared a new rigid ligand, N-(6-(6-(pyridin-2-ylamino)pyridin-2-yl)pyridin-2-yl)pyridin-2-amine (bapbpy, Scheme 1), composed by two directly linked dpa units, likely to induce stronger intermolecular interactions. The latter led to the new Fe(II) complex, [Fe(bapbpy)$_2$(NCS)$_2$], exhibiting a two-step SCO transition with an [HS–LS–LS] intermediate phase [22].

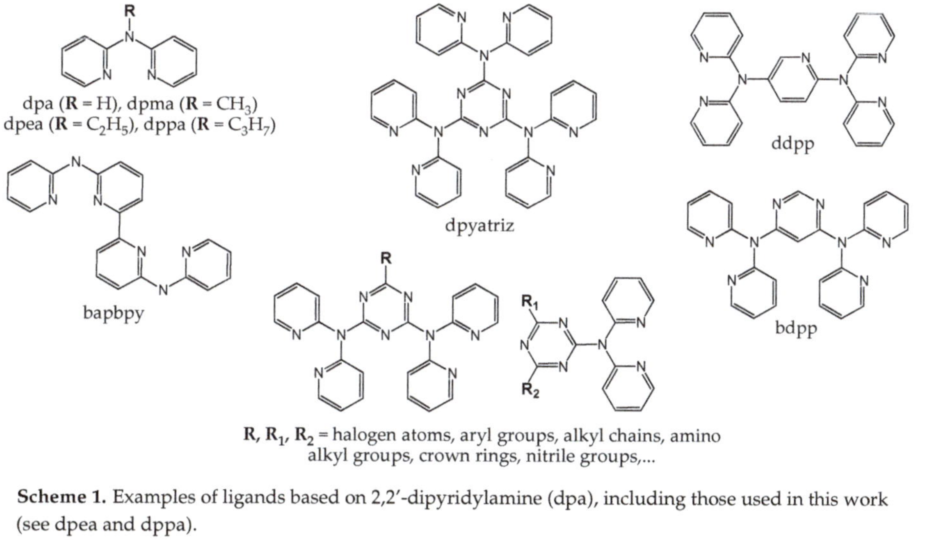

**Scheme 1.** Examples of ligands based on 2,2'-dipyridylamine (dpa), including those used in this work (see dpea and dppa).

With the same objectives, K.S. Murray et al. and P. Gamez et al. [34–50], separately designed triazines containing one, two, or three chelating dpa units and a variety of additional groups, such as halogen atoms, aryl groups, alkyl chains, aminoalkyl and nitriles units, as well as crown groups (see examples in Scheme 1). These sophisticated ligands have led to a variety of SCO materials exhibiting discrete structures generated by two chelating dpa units and two NCX (X = S, Se, BH$_3$)

acting as *cis*- or *trans*-terminal ligands [34–42], dinuclear complexes [43–45], or 1D coordination polymers in which the Fe(II) metal ions are connected through the central triazine group containing two or three dpa units (see dpyatriz ligand and some examples of its derivatives in Scheme 1) [45–50]. Magnetic investigations revealed various magnetic behaviors ranging from incomplete and gradual transitions to abrupt complete SCO transitions. However, since such sophisticated designed ligands did not result in significantly more cooperative SCO transitions than those obtained using simple dpa or bapbpy ligands [18,19,22], we have examined very recently the design of new Fe(II) SCO systems based on dpa ligands substituted by simple alkyl groups such dpma, dpea, and dppa (see Scheme 1) or by other rigid aryl functional groups such as luminophore units.

In this context, we report in the present work, the synthesis, crystal structures, and magnetic properties of a new series of spin crossover (SCO) Fe(II) complexes, based on dipyridyl-*N*-alkylamine and thiocyanate ligands, with the chemical formulae [Fe(dpea)$_2$(NCS)$_2$] (**1**) (dpea = 2,2'-dipyridyl-*N*-ethylamine), I-[Fe(dppa)$_2$(NCS)$_2$] (**2**), II-[Fe(dppa)$_2$(NCS)$_2$], and (**2'**) (dppa = 2,2'-dipyridyl-*N*-propylamine).

## 2. Results and Discussion

### 2.1. Synthesis

The compound 2,2'-dipyridyl-*N*-ethylamine (dpea) was prepared according to the procedure described in reference [51], while 2,2'-dipyridyl-*N*-propylamine (dppa) was prepared by using a slightly modified procedure, by replacing ethyl iodide by propyl iodide (see Figures S1–S8) [51]. The complexes, [Fe(dpea)$_2$(NCS)$_2$] (**1**), I-[Fe(dppa)$_2$(NCS)$_2$] (**2**), and II-[Fe(dppa)$_2$(NCS)$_2$] (**2'**), were prepared, as single crystals, using the slow-diffusion procedure in a fine glass tube (3.0 mm diameter). A solution resulting from the mixture of an aqueous solution of FeCl$_2$·4H$_2$O and of an ethanolic solution of dpea ligand was carefully layered onto an aqueous solution of potassium thiocyanate in a 1:2:2 ratio. The infrared spectra showed a strong absorption band pointed at 2049 cm$^{-1}$ for **1** and at 2057 cm$^{-1}$ for **2** and **2'**, which can be assigned to the asymmetric stretching vibration modes ($\nu$(CN)) of the thiocyanato-*N* coordination modes (see Figures S9–S11).

### 2.2. Crystal Structure Descriptions

Based on the conclusions derived from the thermal variation of the magnetic data, the crystal structures of the [Fe(dpea)$_2$(NCS)$_2$] (**1**) complex and of the two polymorphs I-[Fe(dppa)$_2$(NCS)$_2$] (**2**) and II-[Fe(dppa)$_2$(NCS)$_2$] (**2'**) were determined at 296 and 170 K. Complexes **1**, **2**, and **2'** crystallized in the *Pna2$_1$*, *Pccn*, and space *P*$\bar{1}$ space groups, respectively. The pertinent crystallographic data and selected bond lengths and bond angles for the three complexes are depicted in Table S1 and Table 1, respectively. The unit cell parameters of each complex (Table S1) revealed that there was no structural phase transition within the studied temperature range (170–296 K). The following structural descriptions of the molecular structures correspond to 296 K, and the structural modifications induced by cooling up to 170 k will be detailed in the paragraph dealing with structural and magnetic properties relationships. In Figure 1, the molecular structures of the complexes **1**, **2**, and **2'**, as well as the asymmetric units of each complex and the FeN$_6$ coordination environment of the iron (II) ions are depicted. Complexes **1** and **2'** display a similar asymmetric unit consisting of an iron metal ion, two 2,2'-dipyridyl-*N*-alkylamine molecules (dpea for **1** and dppa for **2'**), and two thiocyanate anions, while compound **2** exhibits an asymmetric unit involving one Fe(II) ion located on a special position, and a thiocyanate anion and a dppa molecule located on general positions. The molecular structures of the three complexes consist of discrete [FeL$_2$(NCS)$_2$] (L = dpea (**1**), dppa (**2** and **2'**) neutral units, where two chelating ligands (dpea (**1**), dppa (**2** and **2'**)) stand in the *cis*-positions, and two NCS$^-$ anions, acting as thiocyanato-κN ligands, complete the coordination sphere in the two remaining *cis*-positions (Figure 1). In each complex, the iron(II) metal ion exhibits a distorted FeN$_6$ polyhedron, arising from the coordination of the four pyridine nitrogen atoms (N3, N4, N5, N6 for **1** and **2'**; N3, N4, N3$^{(a)}$,

N4[(a)] for **2**) of the two 2,2′-dipyridyl-*N*-alkylamine chelating ligands and from the two nitrogen atoms (N1 and N2 for **1** and **2′**; N1, N1[(a)] for **2**) belonging to the two terminal thiocyanato-κN ligands. At room temperature (296 K), the four Fe–Npyr distances in the 2.151–2.204 Å range, are longer than the Fe–N distances corresponding to the terminal thiocyanato-κN ligands (2.102–2.150 Å), as observed in other Fe(II) complexes involving rigid pyridine-based ligands and terminal thiocyanato-κN groups [20,28,29]. The bond angles, depicted in Table 1, deviate considerably from the ideal values (80.05° to 95.22°), as demonstrated by the high values of the Σ distortion parameter [52] (Σ = 45.80° for **1**, 41.08° for **2** and 39.87° for **2′**) summarized in Table 1.

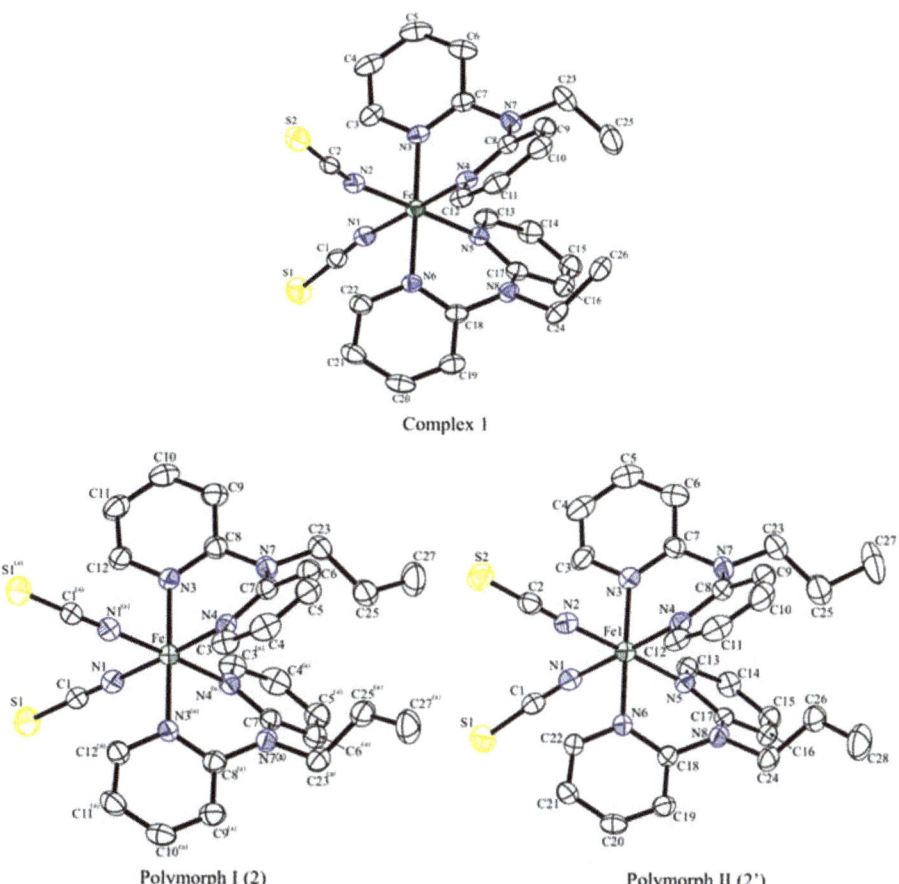

**Figure 1.** ORTEP drawings (50% probability ellipsoids) [53], showing the molecular structures at 170 K, the atom labelling schemes, and the coordination environments of the iron (II) ions for the three discrete complexes (**1** and the polymorphs **2** and **2′**). Codes of equivalent position: [(a)] $1/2 - x, 1/2 - y, z$.

**Table 1.** Selected bond lengths (Å) and bond angles (°) and the Σ distortion parameters for the complexes **1**, **2**, and **2'**.

| Complex | 1 | | 2' | | | 2 | |
|---|---|---|---|---|---|---|---|
| T/K | 296 | 170 | 296 | 170 | | 296 | 170 |
| Fe–N1 | 2.150(5) | 1.971(3) | 2.123(4) | 2.005(3) | Fe–N1 | 2.137(3) | 1.968(2) |
| Fe–N2 | 2.102(4) | 1.968(3) | 2.111(4) | 2.004(3) | Fe–N1 [a] | 2.137(3) | 1.968(2) |
| Fe–N3 | 2.151(5) | 1.983(3) | 2.175(3) | 2.032(3) | Fe–N3 | 2.184(3) | 1.990(2) |
| Fe–N4 | 2.198(4) | 1.986(3) | 2.184(3) | 2.017(3) | Fe–N4 | 2.204(2) | 1.989(2) |
| Fe–N5 | 2.179(4) | 1.976(2) | 2.174(3) | 2.019(3) | Fe–N4 [a] | 2.204(2) | 1.989(2) |
| Fe–N6 | 2.162(4) | 1.978(3) | 2.163(3) | 2.034(3) | Fe–N3 [a] | 2.184(3) | 1.990(2) |
| <$d_{(Fe-N)}$> | 2.157(5) | 1.977(3) | 2.155(4) | 2.018(3) | <$d_{(Fe-N)}$> | 2.175(3) | 1.982(2) |
| Fe–N1–C1 | 164.7(5) | 171.5(3) | 171.6(3) | 162.2(3) | Fe–N1–C1 | 174.8(3) | 174.9(2) |
| Fe–N2–C2 | 150.7(4) | 161.6(2) | 155.6(4) | 174.2(3) | Fe–N1 [a]–C1 [a] | 174.8(3) | 174.9(2) |
| N1–Fe–N2 | 94.10(17) | 93.36(11) | 90.98(15) | 89.70(12) | N1–Fe–N1 [a] | 91.21(17) | 90.15(12) |
| N1–Fe–N3 | 94.29(18) | 92.50(11) | 93.75(12) | 93.52(11) | N1–Fe–N3 | 92.75(11) | 91.65(8) |
| N1–Fe–N5 | 89.94(15) | 89.07(10) | 89.69(13) | 90.02(11) | N1–Fe–N4 [a] | 89.63(11) | 89.44(8) |
| N1–Fe–N6 | 89.42(17) | 86.98(11) | 90.37(12) | 87.23(11) | N1–Fe–N3 [a] | 91.74(11) | 87.92(8) |
| N2–Fe–N3 | 92.31(17) | 88.22(11) | 90.28(13) | 88.16(11) | N1 [a]–Fe–N3 | 91.74(11) | 87.92(8) |
| N2–Fe–N4 | 87.28(16) | 86.98(11) | 90.26(13) | 89.40(11) | N1 [a]–Fe–N4 | 89.63(11) | 89.44(8) |
| N2–Fe–N6 | 93.07(17) | 91.23(11) | 94.85(13) | 92.33(11) | N1 [a]–Fe–N3 [a] | 92.76(11) | 91.65(8) |
| N3–Fe–N4 | 81.24(17) | 86.42(11) | 80.52(11) | 85.91(11) | N3-Fe–N4 | 80.27(9) | 86.21(7) |
| N3–Fe–N5 | 94.29(17) | 93.83(11) | 94.36(12) | 93.73(10) | N3-Fe–N4 [a] | 95.15(9) | 94.22(7) |
| N4–Fe–N5 | 89.23(13) | 90.63(10) | 89.54(12) | 90.90(11) | N4–Fe–N4 [a] | 90.38(13) | 91.05(10) |
| N4–Fe–N6 | 94.90(15) | 94.10(11) | 95.22(12) | 93.34(10) | N3 [a]–Fe–N4 | 95.15(9) | 94.21(7) |
| N5–Fe–N6 | 80.05(18) | 86.74(11) | 80.45(12) | 85.78(10) | N3 [a]–Fe–N4 [a] | 80.27(9) | 86.21(7) |
| [b]Σ/° | 45.80 | 31.24 | 39.87 | 27.66 | [b]Σ/° | 41.08 | 25.79 |

Symmetry transformations used to generate equivalent atoms: [a] 1/2 − x, 1/2 − y, z. [b] Σ is the sum of the deviation from 90° of the 12 cis-angles of the FeN$_6$ octahedron [52].

Examination of the crystal packing in the three complexes did not reveal any strong intermolecular contacts. However, since the three complexes exhibit similar molecular structures, in particular the two polymorphs, a short description of the crystal packing for each compound should give the main differences between the complexes and show clearly that the two polymorphs display different crystal packing. In order to get a global view of the intermolecular interactions, Hirshfeld surface [54] was calculated for the three complexes, and the whole interaction map is displayed as fingerprints [55] in Figure 2. On fingerprints, $d_i$ and $d_e$ represent the distance to the surface of one atom respectively inside and outside the surface. Hirshfeld surfaces and fingerprints were drawn by using the crystalexplorer software [56]. In a first approximation, the fingerprints looked similar for the three complexes at room temperature. The main intermolecular interactions are thus of the same nature and consist of hydrogen-like contacts involving the sulfur atoms (corresponding to the couple $(d_i, d_e) \approx (1.7, 1.1$ Å$)$ on the fingerprints). The main differences between the three complexes involve H–H Van der Waals contacts corresponding to the broad peak at $(d_i, d_e)$ between $(1.0, 1.0$ Å$)$ for **2'** to $(1.2, 1.2$ Å$)$ for **2** on the fingerprints; consequently, the crystal structure of **2'** appeared slightly more compact than the others. At low temperature, the fingerprints looked very similar to the corresponding ones at room temperature but with lower $(d_i, d_e)$ couples.

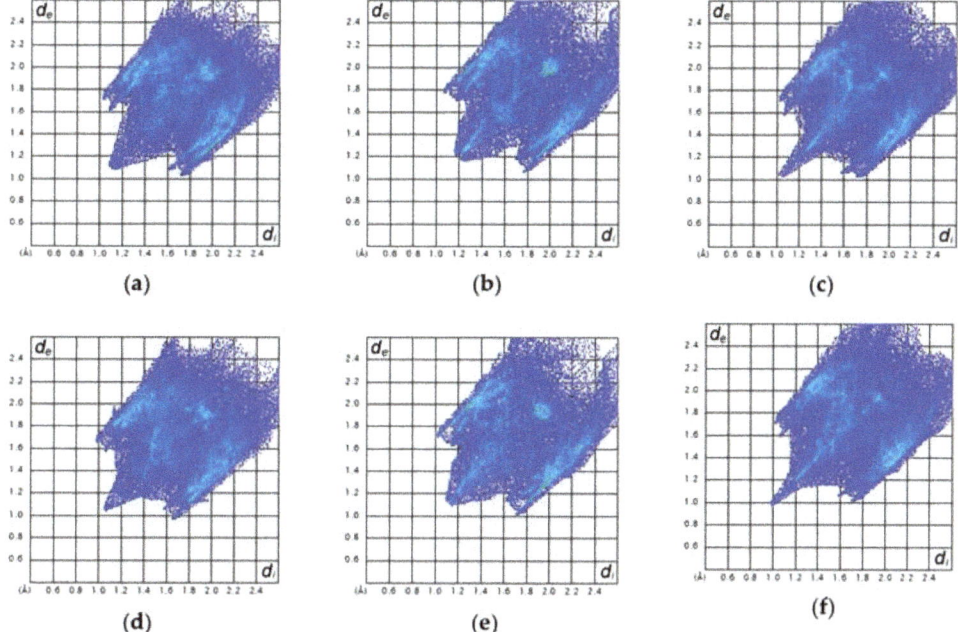

**Figure 2.** Fingerprints [55] of the intermolecular interactions for (**a**) **1**, (**b**) **2**, and (**c**) **2'** at room temperature (296 K) and for (**d**) **1**, (**e**) **2**, and (**f**) **2'** at 170 K (see text for definitions).

Thus, the intermolecular interactions are of the same nature but slightly shorter because of thermal contraction. This confirmed the absence of a structural transition associated to SCO for the three complexes. The main S··· H interactions were found in the three complexes between one sulfur atom and one aromatic hydrogen from the pyridine moiety in meta position to the N atom (corresponding to H6 and H19 for **1** and **2'**, and to H9 for **2**). According to the intermolecular S··· H distances (Table 2), which ranged between 2.873 and 3.105 Å, these interactions are weak comparing to those found in others SCO compound containing the NCS anion, such as in the [Fe(PM-L)$_2$(NCS)$_2$] series [52,57]. All these complexes should thus show a relatively low cooperativity, explaining the gradual spin conversions revealed by the magnetic data.

**Table 2.** Intermolecular S··· H (Å) and corresponding S··· C (Å) distances for compounds **1**, **2**, and **2'**.

|  | Compound 1 | | Compound 2 | | Compound 2' | |
| --- | --- | --- | --- | --- | --- | --- |
|  | d(S···C) | d(S···H) | d(S···C) | d(S···H) | d(S···C) | d(S···H) |
| S1···H6-C6 [(i)] | 3.755 | 2.999 | 3.769 | 3.105 | | |
| S1···H9-C9 [(ii)] | 3.692 | 3.037 | 3.811 | 2.970 | 3.782 | 2.898 |
| S2···H6-C6 [(iii)] | | | | | 3.803 | 2.886 |
| S2···H10-C10 [(iv)] | | | | | 3.677 | 3.028 |
| S2···H19-C19 [(v)] | 3.702 | 2.873 | | | | |

Symmetry codes: [(i)] $-3/2 - x, 1/2 + y, -1/2 + z$ for **1**; $1/2 - x, y, 1/2 + z$ for **2**; [(ii)] $1/2 + x, -1/2 - y, z$ for **1**; $-1/2 + x, -1/2 + y, 1 - z$ for **2**; $1 - x, 1 - y, 1 - z$ for **2'**; [(iii)] $1 - x, 1 - y, -z$; [(iv)] $1 + x, -1 + y, z$; [(v)] $-3/2 - x, -1/2 + y, 1/2 + z$.

## 2.3. Magnetic Properties

The susceptibility measurements were performed at 0.1 T magnetic field at variable temperatures in the 2–300 or 2–350 K range for the three complexes. The thermal dependences of the products of the molar magnetic susceptibility and the temperature ($\chi mT$) are shown in Figure 3 for complex **1** and in Figure 4 for the two polymorph complexes (**2** and **2′**). For compound **1**, the $\chi mT$ product of 3.205 cm³·K·mol⁻¹ at 300 K, slightly higher than the spin only value calculated for an isolated metal ion with S = 2 (3.0 emu·K·mol⁻¹), agrees well with the expected value for a magnetically isolated Fe(II) ion in the HS state (S = 2) (Figure 3) [17–20]. Upon cooling, the $\chi mT$ value decreased gradually until approximately 250 K and then sharply decreased, reaching a value of 0.024 cm³·K·mol⁻¹ at 2 K, indicating the presence of a complete and gradual HS to LS transition at $T_{1/2}$ = 229 K, as also revealed by the thermoschromism (yellow at 296 K and red at 150 K) observed on single crystals (see Figure 3). For the two polymorph complexes I-[Fe(dppa)₂(NCS)₂] (**2**) and II-[Fe(dppa)₂(NCS)₂] (**2′**), the thermal variation of the $\chi mT$ products depicted in Figure 4, showed clearly that the two polymorphs exhibited similar magnetic behaviors. For the polymorph **2**, the $\chi mT$ value at 300 K (3.377 cm³·K·mol⁻¹) was slightly lower than the corresponding value observed for the polymorph **2′** (3.462 cm³·K·mol⁻¹).

However, in both cases, these values are in agreement with the expected value for a magnetically isolated Fe(II) ion in the HS state [17–20] with g factors of 2.12 and 2.15, respectively. Upon cooling, the $\chi mT$ value decreased gradually, in both cases, until approximately 260 K and then sharply decreased reaching a value of 0.02 cm³·K·mol⁻¹ at 2 K, indicating the presence of a complete and gradual HS to LS transition which was accompanied, as expected, by a change of color observed for each single crystal (See Figure 4: orange to red for **2**, yellow to red for **2′**). The two magnetic behaviors were similar and agree well with the presence of complete spin cross-over transitions at almost similar transition temperatures ($T_{1/2}$ = 211 K for **2**; 212 K for **2′**). For the three complexes, the magnetic properties were measured in both cooling and warming modes, but no hysteretic effects were detected.

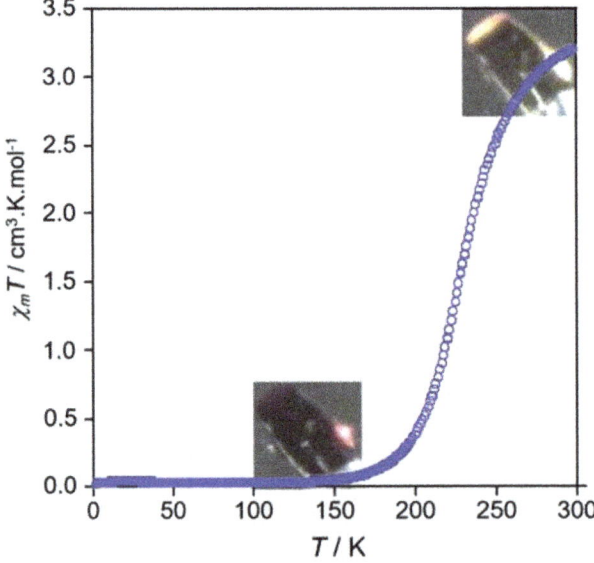

**Figure 3.** Thermal variation of the $\chi mT$ product for complex [Fe(dpea)₂(NCS)₂] (**1**).

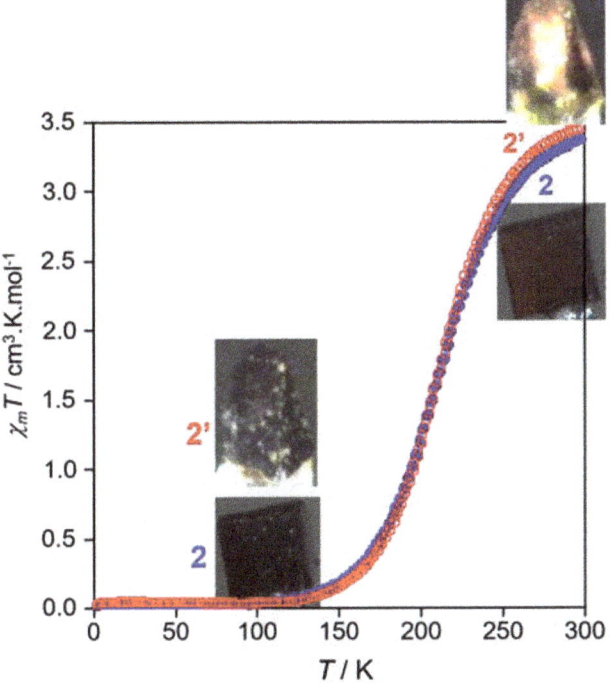

**Figure 4.** Thermal variation of the $\chi_m T$ product for the two polymorph complexes I-[Fe(dppa)$_2$(NCS)$_2$] (**2**) and II-[Fe(dppa)$_2$(NCS)$_2$] (**2'**).

## 2.4. Magneto-Structural Relationships

On the basis of the transition temperatures derived from the magnetic studies above, the crystal structures of **1**, **2**, and **2'** were determined at 170 K. Since the average value of the Fe–L distances (Fe–N) and the distortion parameter ($\Sigma$) are highly sensitive to the Fe(II) spin state, these structural parameters will be used in this section to assign the spin state on the Fe(II) centers. Table 1 lists the temperature evolution of the Fe–N bond lengths and selected bond angles (N–Fe–N and Fe–N–C) observed for each complex, as well as the values of the $\Sigma$ distortion parameter. At room temperature (296 K), the average value of the six Fe–N distances ($<d_{(Fe-N)}>$: 2.157(5) Å for **1**, 2.175(3) Å for **2**, and 2.155(4) Å for **2'**) are in good agreement with the corresponding values observed for the HS Fe(II) ion in a FeN$_6$ distorted octahedral environment [6,7]. As shown in Table 1, the Fe–N bond lengths constitute the first structural parameter at the origin of the distorted FeN$_6$ coordination spheres in the three complexes, since the four slightly different Fe–Npy distances (2.151–2.198 Å for **1**, 2.184–2.204 Å for **2**, and 2.163–2.184 Å for **2'**) are significantly longer than the two Fe–N distances corresponding to the terminal thiocyanato ligands (2.150(5) Å and 2.102(4) Å for **1**, 2.137(3) Å for **2**, 2.123(4) Å and 2.111(4) Å for **2'**). This metric distortion is strengthened by the values of the N–Fe–N *cis*-bond angles (see Table 1) which deviate considerably from ideal values (80.05° to 95.22°), as demonstrated by the relatively high values of the $\Sigma$ distortion parameter summarized in Table 1 for the three complexes, at room temperature. The crystal structures derived at 170 K for the three compounds revealed, as expected, significant changes since the six Fe–N distances are substantially smaller ($<d_{(Fe-N)}>$: 1.977(3) Å for **1**, 1.982(2) Å for **2** and 2.018(3) Å for **2'**) than the corresponding values observed for the HS state at room temperature, suggesting the presence of an LS state of the Fe(II) ion for the three complexes, as revealed by the magnetic data. However, in contrast to the crystallographic data observed at room temperature, the four Fe–Npy and the two Fe–N(NCS) distances did not show significant differences for the LS state (1.968–1.986 Å for **1**, 1.968–1.990 for **2**, and 2.004–2.034 for **2'**), suggesting less distorted FeN$_6$ environments, as demonstrated by the lower Σ distortion parameters (Table 1). It should be noted that the evolution of the Σ distortion parameter from the HS to the LS state (ΔΣ) for the three complexes was rather small (14.5° (**1**), 15.3° (**2**), and 12.2° (**2'**)) [20,52]. This may explain the absence of any photo-induced state in the three compounds. As clearly shown by the structural characterizations, the three complexes displayed a similar discrete mononuclear structure without significant intermolecular contacts, in agreement with gradual switching behaviors, suggesting the absence of any significant cooperative effects. This observation allows to expect almost similar transition temperatures for the three complexes. Effectively, the two polymorph complexes displayed, as expected, similar transition temperatures ($T_{1/2}$ = 211 K for **2**; 212 K for **2'**), while complex **1** exhibited a SCO transition at a higher temperature ($T_{1/2}$ = 229 K). This observation led us to examine other structural parameters within the molecular structure of the complexes, such as the Fe–N–CS bond angles. On the basis of previous experimental and theoretical magnetostructural studies in which some of us suggested that the bent N-bound terminal thiocyanato ligand promotes a weaker ligand field on the Fe(II) ion than the linear configuration [20,29], the examination of the Fe–N–CS angles, summarized in Table 1 for the three complexes, clearly showed that the linearity of the Fe–N–CS angles is more pronounced in complex **2**, exhibiting the highest transition temperature.

## 3. Experimental Section

### 3.1. Materials and Instrumentation

All the starting reagents were purchased from commercial sources (Sigma-Aldrich (Saint-Quentin Fallavier, Isère, France), Acros (Illkirch, Bas-Rhin, France)), and Alfa Aesar (Zeppelinstraße, Karlsruhe, Germany)) and used without further purification. Deuterated solvents were purchased from Sigma-Aldrich and Cambridge Isotope Laboratories. Elemental analyses were performed on a Perkin-Elmer Elemental Analyzer. Infrared (IR) spectra were collected in the range 4000–200 cm$^{-1}$ on a FT-IR BRUKER ATR VERTEX70 Spectrometer. $^1$H and $^{13}$C NMR spectra were recorded on Bruker AMX-400 and AMX-75 spectrometers, and the spectra were referenced internally using residual proton solvent resonances relative to tetramethylsilane (δ = 0 ppm). Magnetic measurements were performed with a Quantum Design MPMS3 SQUID magnetometer in the 2–350 K temperature range. Experimental susceptibility was corrected for the diamagnetism of the constituent atoms of the sample by using Pascal's tables and the diamagnetism of the sample holder.

### 3.2. Syntheses of the 2,2'-Dipyridyl-N-Alkylamine Ligands

**2,2'-Dipyridyl-N-ethylamine (dpea)** was prepared according to the procedure described in reference [51], with a yield of 1.777 g, 77%. IR data (vcm$^{-1}$): 3068 w, 3052 w, 3001 w, 2973 w, 2929 w, 2869 w, 1640 w, 1582 s, 1560 m, 1466 s, 1420 s, 1320 m, 1263 s, 1137 m, 1046 w, 984 m, 953 m, 922 w, 769 s, 736 m, 699 w, 637 w, 622 w, 573 w, 533 w, 494 w, 406 w. $^1$H NMR (400 MHz, CDCl$_3$ δ (ppm): 1.30 (3H, t, $^3J_{H-H}$ = 6.8 Hz); 4.30 (2H, q, $^3J_{H-H}$ = 7.2 Hz); 6.90 (2H, t, $^3J_{H-H}$ = 6 Hz); 7.08 (2H, d, $^3J_{H-H}$ = 8.4 Hz); 7.57 (2H, t, $^3J_{H-H}$ = 7.2 Hz); 8.37 (2H, d, $^3J_{H-H}$ = 4.3 Hz). $^{13}$C NMR (75 MHz, CDCl$_3$) δ (ppm): 13.66 (–CH$_3$, ethyl); 43.19 (N–CH$_2$–, ethyl); 114.87(C=C, aromatic); 116.94 (C=C, aromatic); 137.20 (C=C, aromatic); 148.44 (N=C, aromatic); 157.37 (C=C, aromatic, quat). **2,2'-Dipyridyl-N-propylamine (dppa)** was prepared using a similar procedure as reported for 2,2'-dipyridyl-N-ethylamine (dpea), by replacing the ethyl iodide by the propyl iodide [51]. Yield (0.935 g, 73%). IR data (v/cm$^{-1}$): 3420 br, 3068 w, 3052 w, 3007 m, 2961 m, 2872 m, 1640 w, 1582 s, 1558 s, 1529 m, 1465 s, 1419 s, 1377 s, 1321 m, 1276 s, 1236 m, 1142 m, 1106 m, 1050 m, 985 f, 957 m, 938 m, 890 s, 854 w, 768 s, 735 s, 638 w, 619 w, 605 m, 578 m, 531 m, 503 m, 466 w, 433 m. $^1$H NMR (400 MHz, CDCl$_3$) δ (ppm): 0.96 (3H, t, $^3J_{H-H}$ = 7.6 Hz), 1.74 (3H, q, $^3J_{H-H}$ = 7.6 Hz, $^3J_{H-H}$ = 7.9 Hz); 4.18 (2H, t, $^3J_{H-H}$ = 8 Hz); 6.91 (2H, t, $^3J_{H-H}$ = 5.6 Hz); 7.06 (2H, d, $^3J_{H-H}$ = 8.4 Hz); 7.58 (2H, t, $^3J_{H-H}$ = 7.6 Hz); 8.37 (2H, d, $^3J_{H-H}$ = 4.4 Hz). $^{13}$C NMR (75 MHz,

CDCl$_3$), δ (ppm): 21.62 (CH$_3$–CH$_2$–); 50.12 (N–CH$_2$–); 114.84 (C=C, aromatic); 116.91 (C=C, aromatic); 137.19 (C=C, aromatic); 148.57 (N=C, aromatic); 157.88 (C=C, aromatic, quat).

### 3.3. Preparation of the Fe(II) Complexes [Fe(dpea)$_2$(NCS)$_2$] (1) and [Fe(dppa)$_2$(NCS)$_2$] Polymorphs (2 and 2')

**[Fe(dpea)$_2$(NCS)$_2$] (1).** Single crystals of **1** were prepared using a slow diffusion procedure, in a fine glass tube (3.0 mm diameter): a solution of potassium thiocyanate (12.63 mg, 0.13 mmol) in 1.0 mL of H$_2$O was placed in the fine glass tube. A second solution (2 mL), containing a mixture of an aqueous solution (1.0 mL) of FeCl$_2$.4H$_2$O (13 mg, 0.065 mmol) and an ethanolic solution (1.0 mL) of dpea ligand (25.9 mg, 0.13 mmol), was then carefully added. After three days, yellow prismatic crystals of **1** were formed by slow diffusion at room temperature. CHN analysis: calculated for C$_{26}$H$_{26}$FeN$_8$S$_2$ (**1**): C, 54.7; N, 19.6; H, 4.6. Found: C, 54.9; N, 19.9; H, 4.6. IR data (ν/cm$^{-1}$): 3108 w, 3079 w, 3031 w, 2980 w, 2939 w, 2861 w, 2087 sh, 2049 s, 1595 s, 1573 s, 1489 w, 1461 s, 1435 s, 1335 s, 1294 m, 1233 m, 1164 m, 1073 m, 1056 m, 1012 m, 910 w, 773 s, 749 s, 642 w, 629 m, 572 m, 507 m, 478 m, 449 m, 421 s.

**[Fe(dppa)$_2$(NCS)$_2$] polymorphs (2 and 2').** Using a similar procedure as that described above for **1**, but replacing dpea with dppa (27.7 mg, 0.13 mmol), two single-crystal phases **2** (orange prisms) and **2'** (yellow prisms) formed after two weeks. CHN analysis: calculated for C$_{26}$H$_{26}$FeN$_8$S$_2$ (**2**): C, 56.2; N, 18.7; H, 5.0. Found: C, 56.4; N, 19.1; H, 4.9. IR data (ν/cm$^{-1}$) polymorph I (**2**): 3073 w, 3032 w, 2974 w, 2866 w, 2086 w, 2057 s, 1638 m, 1595 m, 1489 s, 1464 m, 1455 m, 1431 s, 1344 s, 1305 m, 1278 m, 1234 m, 1167 m, 1138 m, 1112 w, 1079 w, 1060 m, 1032 m, 1009 m, 965 w, 941 w, 914 m, 870 m, 819 s, 786 s, 775 s, 747 m, 643 m, 631 m, 602 m, 526 m, 505 m, 482 w, 472 w, 441 m, 421 m. CHN analysis: calculated for C$_{26}$H$_{26}$FeN$_8$S$_2$ (**2'**): C, 56.2; N, 18.7; H, 5.0. Found: C, 56.5; N, 19.0; H, 4.9. IR data (ν/cm$^{-1}$) polymorph II (**2'**): 3073 w, 3032 w, 2965 w, 2867 w, 2170 w, 2152 w, 2089 w, 2057 s, 1638 m, 1596 m, 1490 m, 1465 m, 1456 m, 1431 s, 1344 s, 1306 m, 1280 m, 1235 m, 1167 m, 1138 m, 1112 w, 1080 w, 1059 m, 1010 m, 964 w, 941 w, 903 w, 786 s, 775 s, 746 s, 747 m, 642 m, 631 m, 603 m, 528 m, 482 w, 472 w, 441 m, 422 m.

### 3.4. X-ray Crystallography

Crystallographic studies of compounds **1**, **2** and **2'** were performed at 296 and 170 K. The crystallographic data were collected on an Oxford Diffraction Xcalibur CCD diffractometer with Mo Kα radiation. For data collections, except for complex **2'**, similar single crystals were used at both temperatures: 0.20 × 0.18 × 0.13 mm$^3$ (**1**); 0.38 × 0.30 × 0.23 mm$^3$ (**2**); 0.14 × 0.12 × 0.10 mm$^3$ for **2'** at 296 K and 0.25 × 0.23 × 0.16 mm$^3$ for **2'** at 170 K. All the data collections were performed using 1° ω-scans with different exposure times (50 s and 40 s per frame for **1** at 296 and 170 K, respectively; 10 s per frame for **2** at 296 and 170 K; 50 s and 13 s per frame for **2'** at 296 and 170 K, respectively). The unit cell determinations and data reductions were performed using the CrysAlis program suite on the full set of data [58]. The crystal structures were solved by direct methods and successive Fourier difference syntheses with the Sir97 program [59] and refined on $F^2$ by weighted anisotropic full-matrix least-square methods using the SHELXL97 program [60]. All non-hydrogen atoms were refined anisotropically, while the hydrogen atoms were calculated and therefore included as isotropic fixed contributors to $F_c$. Crystallographic data including refinement parameters, bond lengths and bond angles, are given in Table S1 and Table 1, respectively.

## 4. Conclusions

We prepared a new series of spin crossover (SCO) Fe(II) materials based on dipyridyl-N-alkylamine and thiocyanate ligands, with the chemical formulae [Fe(dpea)$_2$(NCS)$_2$] (**1**) (dpea = 2,2'-dipyridyl-N-ethylamine), I-[Fe(dppa)$_2$(NCS)$_2$] (**2**), and II-[Fe(dppa)$_2$(NCS)$_2$] (**2'**) (dppa = 2,2'-dipyridyl-N-propylamine). All were structurally characterised by single-crystal X-ray diffraction at room temperature (296 K) and at 170 K and by magnetic studies as a function of temperature. Even if they displayed different crystallographic structures, as reflected by their different crystal packing, the three Fe(II) neutral complexes, exhibited almost similar molecular structures, which can be described as discrete mononuclear complexes of the general chemical formula

[FeL$_2$(NCS)$_2$], where two L chelating ligands (L = dpea (**1**), dppa (**2** and **2'**)) stand in the *cis*-positions, and the two thiocyanato-κN ligands complete the octahedral environment of the Fe(II) metal ions in the two remaining *cis*-positions. For complex **1**, the thermal variation of the $\chi_m T$ product showed a complete gradual HS–LS spin crossover transition at $T_{1/2}$ = 229 K, while the two polymorphs I-[Fe(dppa)$_2$(NCS)$_2$] (**2**) and II-[Fe(dppa)$_2$(NCS)$_2$] (**2'**) displayed similar magnetic behaviors at lower transition temperatures ($T_{1/2}$ = 211 K for **2**; 212 K for **2'**), which is in good agreement with the strong structural changes of the FeN$_6$ coordination spheres derived from the structural characterizations at room temperature and at 170 K. A careful examination of the intermolecular contacts in the three complexes did not reveal any significant intermolecular interaction, suggesting the absence of significant cooperative effects which agrees well the gradual behaviors shown by the magnetic data. However, complex **1** showed a transition temperature (229 K) clearly different from those observed for the two polymorph complexes ($T_{1/2}$ = 211 K for **2**; 212 K for **2'**). Such difference was ascribed to the more pronounced linearity of the Fe–N–CS angles observed for the two polymorphs **2** and **2'**.

**Supplementary Materials:** The following are available online at http://www.mdpi.com/2073-4352/8/11/401/s1. Crystallographic data for the structure reported in this paper were deposited in the Cambridge Crystallographic Data Centre as supplementary publication Nos. CCDC 1866637 (170 K) and 1866638 (296 K) for **1**; 1866639 (170 K) and 1866640 (296 K) for **2**; 1866641 (170 K) and 1866642 (296 K) for **2'**. A copy of the data can be obtained free of charge on application to CCDC, 12 Union Road, Cambridge CB21EZ, UK (Fax: +44-1223-336-033; E-Mail: deposit@ccdc.cam.ac.uk).

**Author Contributions:** The manuscript was written through contributions of all authors. T.H. and E.C. synthesized the ligand and the metal complexes and made the first experimental characterizations. M.M. and S.Y. analyzed the crystal data of the three complexes at room temperature (296 K) and at 170 K. D.P. performed and analyzed the magnetic studies as a function of temperature. S.T. supervised the experimental work and wrote the manuscript to which all the authors contributed.

**Funding:** This research was funded by the Université de Brest, the Centre National de la Recherche Scientifique (CNRS), Agence Nationale de la Recherche (project BISTA-MAT: ANR-12-BS07-0030-01) and by the Polish National Science Centre (Project 2016/22/E/ST5/00055).

**Acknowledgments:** The authors acknowledge the CNRS (Centre National de la Recherche Scientifique), the "Université de Brest", and the "Agence Nationale de la Recherche" (ANR project BISTA-MAT: ANR-12-BS07-0030-01). DP gratefully acknowledges the financial support of the Polish National Science Centre within the Sonata Bis 6 (2016/22/E/ST5/00055) research project.

**Conflicts of Interest:** The authors declare no conflict of interest.

## References

1. Halcrow, M.A. (Ed.) *Spin-Crossover Materials, Properties and Applications*; John Wiley & Sons Ltd.: Oxford, UK, 2013.
2. Gütlich, P.; Goodwin, H.A. (Eds.) *Topics in Current Chemistry*; Springer: Berlin/Heidelberg, Germany; New York, NY, USA, 2004; pp. 233–235.
3. Pittala, N.; Thétiot, F.; Triki, S.; Boukheddaden, K.; Chastanet, G.; Marchivie, M. Cooperative 1D Triazole-Based Spin Crossover Fe$^{II}$ Material with Exceptional Mechanical Resilience. *Chem. Mater.* **2017**, *29*, 490–494. [CrossRef]
4. Pittala, N.; Thétiot, T.; Charles, C.; Triki, S.; Boukheddaden, K.; Chastanet, G.; Marchivie, M. An unprecedented trinuclear Fe$^{II}$ triazole-based complex exhibiting a concerted and complete sharp spin transition above room temperature. *Chem. Commun.* **2017**, *53*, 8356–8359. [CrossRef] [PubMed]
5. Milin, E.; Patinec, V.; Triki, S.; Bendeif, E.-E.; Pillet, S.; Marchivie, M.; Chastanet, G.; Boukheddaden, K. Elastic Frustration Triggering Photoinduced Hidden Hysteresis and Multistability in a Two-Dimensional Photoswitchable Hofmann-Like Spin-Crossover Metal–Organic Framework. *Inorg. Chem.* **2016**, *55*, 11652–11661. [CrossRef] [PubMed]
6. Shatruk, M.; Phan, H.; Chrisostomo, B.A.; Suleimenova, A. Symmetry-breaking structural phase transitions in spin crossover complexes. *Coord. Chem. Rev.* **2015**, *289–290*, 62–73. [CrossRef]
7. Atmani, C.; El Hajj, F.; Benmansour, S.; Marchivie, M.; Triki, S.; Conan, F.; Patinec, V.; Handel, H.; Dupouy, G.; Gómez-García, C.J. Guidelines to design new spin crossover materials. *Coord. Chem. Rev.* **2010**, *254*, 1559–1569. [CrossRef]

8. Coronado, E.; Galán-Mascarós, J.R.; Monrabal-Capilla, M.; García-Martínez, J.; Pardo-Ibáñez, P. Bistable Spin-Crossover Nanoparticles Showing Magnetic Thermal Hysteresis near Room Temperature. *Adv. Mater.* **2007**, *19*, 1359–1361. [CrossRef]
9. Shalabaeva, V.; Ridier, K.; Rat, S.; Manrique-Juarez, M.D.; Salmon, L.; Séguy, I.; Rotaru, A.; Molnár, G.; Bousseksou, A. Room temperature current modulation in large area electronic junctions of spin crossover thin films. *Appl. Phys. Lett.* **2018**, *112*, 013301. [CrossRef]
10. Senthil Kumar, K.; Ruben, M. Emerging trends in spin crossover (SCO) based functional materials and devices. *Coord. Chem. Rev.* **2017**, *346*, 176–205. [CrossRef]
11. Dugay, J.; Giménez-Marqués, M.; Kozlova, T.; Zandbergen, H.W.; Coronado, E.; van der Zant, H.S.J. Spin switching in electronic devices based on 2D assemblies of spin-crossover nanoparticles. *Adv. Mater.* **2015**, *27*, 1288–1293. [CrossRef] [PubMed]
12. El Hajj, F.; Sebki, G.; Patinec, V.; Marchivie, M.; Triki, S.; Handel, H.; Yefsah, S.; Tripier, R.; Gomez-García, C.-J.; Coronado, E. Macrocycle-Based Spin-Crossover Materials. *Inorg. Chem.* **2009**, *48*, 10416–10423. [CrossRef] [PubMed]
13. Baadji, N.; Sanvito, S. Giant resistance change across the phase transition in spin-crossover molecules. *Phys. Rev. Lett.* **2012**, *108*, 217201. [CrossRef] [PubMed]
14. Prins, F.; Monrabal-Capilla, M.; Osorio, E.A.; Coronado, E.; van der Zant, H.S.J. Room-temperature electrical addressing of a bistable spin-crossover molecular system. *Adv. Mater.* **2011**, *23*, 1545–1549. [CrossRef] [PubMed]
15. Setifi, F.; Benmansour, S.; Marchivie, M.; Dupouy, G.; Triki, S.; Sala-Pala, J.; Salaün, J.-Y.; Gómez-García, C.J.; Pillet, S.; Lecomte, C.; et al. Magnetic bistability and thermochromism in molecular $Cu^{II}$ chain. *Inorg. Chem.* **2009**, *48*, 1269–1271. [CrossRef] [PubMed]
16. Arroyave, A.; Lennartson, A.; Dragulescu-Andrasi, A.; Pedersen, K.S.; Piligkos, S.; Stoian, S.A.; Greer, S.M.; Pak, C.; Hietsoi, O.; Phan, H.; et al. Spin Crossover in Fe(II) Complexes with $N_4S_2$ Coordination. *Inorg. Chem.* **2016**, *55*, 5904–5913. [CrossRef] [PubMed]
17. Matar, S.F.; Guionneau, P.; Guillaume Chastanet, G. Multiscale Experimental and Theoretical Investigations of Spin Crossover $Fe^{II}$ Complexes: Examples of [Fe(phen)$_2$(NCS)$_2$] and [Fe(PM-BiA)$_2$(NCS)$_2$]. *Int. J. Mol. Sci.* **2015**, *16*, 4007–4027. [CrossRef] [PubMed]
18. Gaspar, A.B.; Agustí, G.; Martínez, V.; Muñoz, M.C.; Levchenko, G.; Real, J.A. Spin crossover behaviour in the iron(II)-2,2-dipyridilamine system: Synthesis, X-ray structure and magnetic studies. *Inorg. Chim. Acta* **2005**, *358*, 4089–4094. [CrossRef]
19. Gaspar, A.B.; Ksenofontov, V.; Real, J.A.; Gütlich, P. Coexistence of spin-crossover and antiferromagnetic coupling phenomena in the novel dinuclear Fe(II) complex [Fe(dpa)(NCS)$_2$]$_2$bpym. *Chem. Phys. Lett.* **2003**, *373*, 385–391. [CrossRef]
20. Nebbali, K.; Mekuimemba, C.D.; Charles, C.; Yefsah, S.; Chastanet, G.; Mota, A.J.; Colacio, E.; Triki, S. One-dimensional Thiocyanato-Bridged Fe(II) Spin Crossover Cooperative Polymer With Unusual FeN$_5$S Coordination Sphere. *Inorg. Chem.* **2018**, *57*, 12338–12346. [CrossRef] [PubMed]
21. Paulsen, H.; Grünsteudel, H.; Meyer-Klaucke, W.; Gerdan, M.; Grünsteudel, H.F.; Chumakov, A.I.; Rüffer, R.; Winkler, H.; Toftlund, H.; Trautwein, A.X. The spin-crossover complex [Fe(tpa)(NCS)$_2$]. *Eur. Phys. J.* **2001**, *B23*, 463–472. [CrossRef]
22. Bonnet, S.; Siegler, M.A.; Sanchez Costa, J.; Molnar, G.; Bousseksou, A.; Spek, A.L.; Gamez, P.; Reedijk, J. A two-step spin crossover mononuclear iron(II) complex with a [HS–LS–LS] intermediate phase. *Chem. Commun.* **2008**, 5619–5621. [CrossRef] [PubMed]
23. Milin, E.; Belaïd, S.; Patinec, V.; Triki, S.; Chastanet, G.; Marchivie, M. Dinuclear Spin-Crossover Complexes Based on Tetradentate and Bridging Cyanocarbanion Ligands. *Inorg. Chem.* **2016**, *55*, 9038–9046. [CrossRef] [PubMed]
24. Dupouy, G.; Marchivie, M.; Triki, S.; Sala-Pala, J.; Salaün, C.J.Y.; Gómez-García, C.J.; Guionneau, P. The Key Role of the Intermolecular π-π Interactions on the Presence of Spin Crossover in Neutral [Fe(abpt)$_2$A$_2$] Complexes (A = terminal monoanion N-ligand). *Inorg. Chem.* **2008**, *47*, 8921–8931. [CrossRef] [PubMed]
25. Dupouy, G.; Marchivie, M.; Triki, S.; Sala-Pala, J.; Gomez-Garcia, C.J.; Pillet, S.; Lecomte, C.; Létard, J.-F. Photoinduced HS state in the first spin-crossover chain containing a cyanocarbanion as bridging ligand. *Chem. Commun.* **2009**, *23*, 3404–3406. [CrossRef] [PubMed]

26. Benmansour, S.; Atmani, C.; Setifi, F.; Triki, S.; Marchivie, M.; Gómez-García, C.J. Polynitrile anions as ligands: From magnetic polymeric architectures to spin crossover materials. *Coord. Chem. Rev.* **2010**, *254*, 1468–1478. [CrossRef]
27. Dupouy, G.; Triki, S.; Marchivie, M.; Cosquer, N.; Gómez-García, C.J.; Pillet, S.; Bendeif, E.-E.; Lecomte, C.; Asthana, S.; Létard, J.-F. Cyanocarbanion-based spin crossover materials: Photocrystallographic and photomagnetic studies of a new iron(II) neutral chain. *Inorg. Chem.* **2010**, *49*, 9358–9368. [CrossRef] [PubMed]
28. Yamasaki, M.; Ishida, T. First Iron(II) Spin-crossover Complex with an $N_5S$ Coordination Sphere. *Chem. Lett.* **2015**, *44*, 920–921. [CrossRef]
29. Mekuimemba, C.D.; Conan, F.; Mota, A.J.; Palacios, M.A.; Colacio, E.; Triki, S. On the Magnetic Coupling and Spin Crossover Behavior in Complexes Containing the Head-to-Tail [$Fe^{II}_2(\mu\text{-SCN})_2$] Bridging Unit: A Magnetostructural Experimental and Theoretical Study. *Inorg. Chem.* **2018**, *57*, 2184–2192. [CrossRef] [PubMed]
30. Park, J.G.; Jeon, I.-R.; Harris, T.D. Electronic Effects of Ligand Substitution on Spin Crossover in a Series of Diiminoquinonoid-Bridged $Fe^{II}_2$ Complexes. *Inorg. Chem.* **2015**, *54*, 359–369. [CrossRef] [PubMed]
31. Wei, R.-J.; Li, B.; Tao, J.; Huang, R.-B.; Zheng, L.-S.; Zheng, Z. Making Spin-Crossover Crystals by Successive Polymorphic Transformations. *Inorg. Chem.* **2011**, *50*, 1170–1172. [CrossRef] [PubMed]
32. Wei, R.-J.; Huo, Q.; Tao, J.; Huang, R.-B.; Zheng, L.-S. Spin-Crossover $Fe^{II}_4$ Squares: Two-Step Complete Spin Transition and Reversible Single-Crystal-to-Single-Crystal Transformation. *Angew. Chem. Int. Ed.* **2011**, *50*, 8940–8943. [CrossRef] [PubMed]
33. Li, B.; Wei, R.-J.; Tao, J.; Huang, R.-B.; Zheng, L.-S.; Zheng, Z. Solvent-Induced Transformation of Single Crystals of a Spin-Crossover (SCO) Compound to Single Crystals with Two Distinct SCO Centers. *J. Am. Chem. Soc.* **2010**, *132*, 1558–1566. [CrossRef] [PubMed]
34. Scott, H.S.; Ross, T.M.; Phonsri, W.; Moubaraki, B.; Chastanet, G.; Létard, J.-F.; Batten, S.R.; Murray, K.S. Discrete $Fe^{II}$ Spin-Crossover Complexes of 2,2′-Dipyridylamino-Substituted s-Triazine Ligands with Phenoxo, Cyanophenoxo and Dibenzylamino Functionalities. *Eur. J. Inorg. Chem.* **2015**, 763–777. [CrossRef]
35. Scott, H.S.; Moubaraki, B.; Paradis, N.; Chastanet, G.; Létard, J.-F.; Batten, S.R.; Murray, K.S. 2,2′-Dipyridylamino-based ligands with substituted alkyl chain groups and their mononuclear-M(II) spin crossover complexes. *J. Mater. Chem. C* **2015**, *3*, 7845–7857. [CrossRef]
36. Nassirinia, N.; Amani, S.; Teat, S.J.; Roubeau, O.; Gamez, P. Enhancement of spin-crossover cooperativity mediated by lone pair–p interactions and halogen bonding. *Chem. Commun.* **2014**, *50*, 1003–1005. [CrossRef] [PubMed]
37. Wannarit, N.; Roubeau, O.; Youngme, S.; Teatd, S.J.; Gamez, P. Influence of supramolecular bonding contacts on the spin crossover behaviour of iron(II) complexes from 2,2′-dipyridylamino/s-triazine ligands. *Dalton Trans.* **2013**, *42*, 7120–7130. [CrossRef] [PubMed]
38. Wannarit, N.; Roubeau, O.; Youngme, S.; Gamez, P. Subtlety of the Spin-Crossover Phenomenon Observed with Dipyridylamino-Substituted Triazine Ligands. *Eur. J. Inorg. Chem.* **2013**, 730–737. [CrossRef]
39. Scott, H.S.; Ross, T.M.; Batten, S.R.; Gass, I.A.; Moubaraki, B.; Neville, S.M.; Murray, K.S. Iron(II) Mononuclear Materials Containing Functionalised Dipyridylamino-Substituted Triazine Ligands: Structure, Magnetism and Spin Crossover. *Aust. J. Chem.* **2012**, *65*, 874–882. [CrossRef]
40. Ross, T.M.; Moubaraki, B.; Neville, S.M.; Batten, S.R.; Murray, K.S. Polymorphism and spin crossover in mononuclear $Fe^{II}$ species containing new dipyridylamino-substituted s-triazine ligands. *Dalton Trans.* **2012**, *41*, 1512–1523. [CrossRef] [PubMed]
41. Ross, T.M.; Moubaraki, B.; Wallwork, K.S.; Batten, S.R.; Murray, K.S. A temperature-dependent order-disorder and crystallographic phase transition in a 0D $Fe^{II}$ spin crossover compound and its non-spin crossover $Co^{II}$ isomorph. *Dalton Trans.* **2011**, *40*, 10147–10155. [CrossRef] [PubMed]
42. Quesada, M.; Monrabal, M.; Aromí, G.; de la Peña-O'Shea, V.A.; Gich, M.; Molins, E.; Roubeau, O.; Teat, S.J.; MacLean, E.J.; Gamez, P.; et al. Spin transition in a triazine-based Fe(II) complex: Variable-temperature structural, thermal, magnetic and spectroscopic studies. *J. Mater. Chem.* **2006**, *16*, 2669–2676. [CrossRef]
43. Amoore, J.J.M.; Kepert, C.J.; Cashion, J.D.; Moubaraki, B.; Neville, S.M.; Murray, K.S. Structural and Magnetic Resolution of a Two-Step Full Spin-Crossover Transition in a Dinuclear Iron(II) Pyridyl-Bridged Compound. *Chem. Eur. J.* **2006**, *12*, 8220–8227. [CrossRef] [PubMed]

44. Quesada, M.; de Hoog, P.; Gamez, P.; Roubeau, O.; Aromí, G.; Donnadieu, B.; Massera, C.; Lutz, M.; Spek, A.L.; Reedijk, J. Coordination Dependence of Magnetic Properties within a Family of Related [Fe$^{II}_2$] Complexes of a Triazine-Based Ligand. *Eur. J. Inorg. Chem.* **2006**, *2006*, 1353–1361. [CrossRef]
45. Neville, S.M.; Leita, B.A.; Offermann, D.A.; Duriska, M.B.; Moubaraki, B.; Chapman, K.W.; Halder, G.J.; Murray, K.S. Spin-Crossover Studies on a Series of 1D Chain and Dinuclear Iron(II) Triazine-Dipyridylamine Compounds. *Eur. J. Inorg. Chem.* **2007**, 1073–1085. [CrossRef]
46. Scott, H.S.; Ross, T.M.; Chilton, N.F.; Gass, I.A.; Moubaraki, B.; Chastanet, G.; Paradis, N.; Létard, J.-F.; Vignesh, K.R.; Rajaraman, G.; et al. Crown-linked dipyridylamino-triazine ligands and their spin-crossover iron(II) derivatives: Magnetism, photomagnetism and cooperativity. *Dalton Trans.* **2013**, *42*, 16494–16509. [CrossRef] [PubMed]
47. Ross, T.M.; Moubaraki, B.; Batten, S.R.; Murray, K.S. Spin crossover in polymeric and heterometallic Fe$^{II}$ species containing polytopic dipyridylamino-substituted-triazine ligands. *Dalton Trans.* **2012**, *41*, 2571–2581. [CrossRef] [PubMed]
48. Ross, T.M.; Moubaraki, B.; Turner, D.R.; Halder, G.J.; Chastanet, G.; Neville, S.M.; Cashion, J.D.; Létard, J.-F.; Batten, S.R.; Murray, K.S. Spin Crossover and Solvate Effects in 1D Fe$^{II}$ Chain Compounds Containing Bis(dipyridylamine)-Linked Triazine Ligands. *Eur. J. Inorg. Chem.* **2011**, 1395–1417. [CrossRef]
49. Neville, S.M.; Leita, B.A.; Halder, G.J.; Kepert, C.J.; Moubaraki, B.; Létard, J.-F.; Murray, K.S. Understanding the Two-Step Spin-Transition Phenomenon in Iron(II) 1D Chain Materials. *Chem. Eur. J.* **2008**, *14*, 10123–10133. [CrossRef] [PubMed]
50. Quesada, M.; de la Peña-O'Shea, V.A.; Aromí, G.; Geremia, S.; Massera, C.; Roubeau, O.; Gamez, P.; Reedijk, J. A Molecule-Based Nanoporous Material Showing Tuneable Spin-Crossover Behavior near Room Temperature. *Adv. Mater.* **2007**, *19*, 1397–1402. [CrossRef]
51. Rauterkus, M.J.; Fakih, S.; Mock, C.; Puscasu, I.; Krebs, B. Cisplatin analogues with 2,2-dipyridylamine ligands and their reactions with DNA model nucleobases. *Inorg. Chim. Acta* **2003**, *350*, 355–365. [CrossRef]
52. Guionneau, P.; Marchivie, M.; Bravic, G.; Létard, J.-F.; Chasseau, D. Structural Aspects of Spin Crossover. Example of the [Fe$^{II}$L$_n$(NCS)$_2$] Complexes. *Top. Curr. Chem.* **2004**, *234*, 97–128.
53. Farrugia, L.J. ORTEP-3 for Windows—A version of ORTEP-III with a Graphical User Interface (GUI). *J. Appl. Cryst.* **1997**, *30*, 565. [CrossRef]
54. Spackman, M.A.; Byrom, P.G. A novel definition of a molecule in a crystal. *Chem. Phys. Lett.* **1997**, *267*, 215–220. [CrossRef]
55. Spackman, M.A.; McKinnon, J.J. Fingerprinting intermolecular interactions in molecular crystals. *CrystEngComm* **2002**, *4*, 378–392. [CrossRef]
56. Turner, M.J.; McKinnon, J.J.; Wolff, S.K.; Grimwood, D.J.; Spackman, P.R.; Jayatilaka, D.; Spackman, M.A. *CrystalExplorer17*; University of Western Australia: Crawley, Australia, 2017.
57. Marchivie, M.; Guionneau, P.; Létard, J.-F.; Chasseau, D. Towards direct correlation between spin crossover and structural properties in iron II complexes. *Acta Cryst. B* **2003**, *59*, 479–486. [CrossRef]
58. Oxford Diffraction. *Xcalibur CCD/RED CrysAlis Software System*; Oxford Diffraction Ltd.: Abingdon, UK, 2006.
59. Altomare, A.; Burla, M.C.; Camalli, M.; Cascarano, C.; Giacovazzo, C.; Guagliardi, A.; Moliterni, A.G.G.; Polidori, G.; Spagna, R. *SIR97*: A new tool for crystal structure determination and refinement. *J. Appl. Cryst.* **1999**, *32*, 115–119. [CrossRef]
60. Sheldrick, G. Crystal structure refinement with SHELXL. *Acta Cryst. C* **2015**, *71*, 3–8. [CrossRef] [PubMed]

© 2018 by the authors. Licensee MDPI, Basel, Switzerland. This article is an open access article distributed under the terms and conditions of the Creative Commons Attribution (CC BY) license (http://creativecommons.org/licenses/by/4.0/).

Article

# Evolution of Spin-Crossover Transition in Hybrid Crystals Involving Cationic Iron Complexes [Fe(III)(3-OMesal$_2$-trien)]$^+$ and Anionic Gold Bis(dithiolene) Complexes Au(dmit)$_2$ and Au(dddt)$_2$

Nataliya G. Spitsyna [1], Yuri N. Shvachko [2,*], Denis V. Starichenko [2,*], Erkki Lahderanta [3], Anton A. Komlev [3], Leokadiya V. Zorina [4], Sergey V. Simonov [4], Maksim A. Blagov [1,5] and Eduard B. Yagubskii [1]

[1] Institute of Problems of Chemical Physics, RAS, Chernogolovka, MD 142432, Russia; spitsina@icp.ac.ru (N.G.S.); max-blagov@mail.ru (M.A.B.); yagubski@icp.ac.ru (E.B.Y.)
[2] M.N. Miheev Institute of Metal Physics, Ural Branch of Russian Academy of Sciences, S. Kovalevskaya str., 18, Yekaterinburg 620137, Russia
[3] Laboratory of Physics, Lappeenranta University of Technology, Box 20, FI-53851 Lappeenranta, Finland; erkki.lahderanta@lut.fi (E.L.); anton.komlev@lut.fi (A.A.K.)
[4] Institute of Solid State Physics, RAS, Chernogolovka, MD 142432, Russia; zorina@issp.ac.ru (L.V.Z.); simonovsv@rambler.ru (S.V.S.)
[5] Faculty of Fundamental Physical and Chemical Engineering MGU, Moscow, MOS 119991, Russia
* Correspondence: yuri.shvachko@gmail.com (Y.N.S.); starichenko@imp.uran.ru (D.V.S.); Tel.: +7(496)-522-83-86 (Y.N.S. & D.V.S.)

Received: 17 September 2018; Accepted: 27 September 2018; Published: 3 October 2018

**Abstract:** Hybrid ion-pair crystals involving hexadentate [Fe(III)(3-OMesal$_2$-trien)]$^+$ spin-crossover (SCO) cationic complexes and anionic gold complexes [Au(dmit)$_2$]$^-$ (**1**) (dmit = 4,5-dithiolato-1,3-dithiole-2-thione) and [Au(dddt)$_2$]$^-$ (**2**) (dddt = 5,6-dihydro-1,4-dithiin-2,3-dithiolate) were synthesized and studied by single-crystal X-ray diffraction, P-XRD, and SQUID magnetometry. Our study shows that both complexes have similar 1:1 stoichiometry but different symmetry and crystal packing. Complex **1** has a rigid structure in which the SCO cations are engaged in strong π-interplay with molecular surrounding and does not show SCO transition while **2** demonstrates a reversible transition at $T_{sco}$ = 118 K in a much "softer", hydrogen-bonded structure. A new structural indicator of spin state in [Fe(sal$_2$-trien)]$^+$ complexes based on conformational analysis has been proposed. Aging and thermocycling ruined the SCO transition increasing the residual HS fraction from 14 to 41%. Magnetic response of **1** is explained by the AFM coupled dimers S = 5/2 with $J_1$ = −0.18 cm$^{-1}$. Residual high-spin fraction of **2**, apart from a contribution of the weak dimers with $J_{12} = J_{34}$ = −0.29 cm$^{-1}$, is characterized by a stronger interdimer coupling of $J_{23}$ = −1.69 cm$^{-1}$, which is discussed in terms of possible involvement of neutral radicals [Au(dddt)$_2$].

**Keywords:** metal dithiolene complexes; [Au(dmit)$_2$]$^-$, [Au(dddt)$_2$]$^-$; ion-pair crystals; [Fe(III)(3-OMesal$_2$-trien)]$^+$; coordination complexes; crystal structure; magnetic properties; magnetic susceptibility; magnetization; spin-crossover transition

## 1. Introduction

Octahedral complexes with $d^4$-$d^7$ metal ions can demonstrate spin-crossover (SCO) transition under external factors such as temperature, pressure, and electromagnetic radiation [1–3]. As soon as the bistability between a high-spin (HS) and low-spin (LS) states of the metal ion embraces not only magnetic response but also the electronic structure of the entire complex, various endeavors have been undertaken to use the SCO compounds in magnetic memory or display devices, as well as multi-modal sensors [4–6]. Moreover, polyfunctional compounds combining the conductivity and the spin-crossover transition have been highlighted in connection with spin-dependent transport and single molecular switching [7–9]. Spin-lattice relaxation in the low-dimensional conducting networks of centrosymmetrical molecules is strongly suppressed. The hybrid structures implementing efficient exchange interaction between the SCO and conducting subsystems would facilitate the development of new molecular spintronic devices where the spin transport is controlled by spin-crossover complexes.

Metal bis-1,2-dithiolene complexes, as organic donors and acceptors, possess a delocalized electron system as a planar central core $M(C_2S_2)_2$ and present different formal oxidation states. This type of complex has been intensively studied as a component of molecular conductors [10–12]. In addition, metal dithiolene complexes have a rich variety of physical properties, such as Peierls instability of the low-dimensional systems and the quantum fluctuations [10,13]. Thus, both SCO and metal dithiolene complexes could undergo phase transition, and combining two components in one crystal structure might give rise to a novel molecular material with exotic phenomena.

The interplay between spin-crossover and conductivity was already observed in some of such materials [14–16]. The majority of conducting SCO compounds are represented by the Fe(III) cation complexes with $[M(dmit)_2]^{\delta-}$ anions [17–19]. As a first step in this way, Faulmann et al. have published a simple salt $[Fe(sal_2\text{-trien})][Ni(dmit)_2]$ showing a cooperative spin transition behavior with a wide hysteresis loop (30 K) [19]. These results convinced us that metal bis(dithiolene) anionic $[M(dmit)_2]$ and $[M(dddt)_2]$ complexes [12] might be promising candidates for enhancing cooperativity in SCO of cations based on mononuclear Fe(III) complexes with hexadentate Schiff base ligands. To achieve systems with molecular shape flexibility and rich in hydrogen bonding [20], we prepared the 3-methoxy derivative (3-OMesal$_2$-trien)$^{2-}$ and used it as a ligand (Figure 1).

In this paper, we report synthesis, crystal structure, and magnetic properties of $[Fe^{3+}(3\text{-OMesal}_2\text{-trien})][Au^{3+}(dmit)_2]$ (**1**) and the analogous compound $[Fe^{3+}(3\text{-OMesal}_2\text{-trien})][Au^{3+}(dddt)_2]\cdot CH_3CN$ (**2**) involving a nonmagnetic gold bis(dithiole) anions (Figure 1). The incomplete SCO conversion was observed in **2** and was not in the compounds **1**, where 100% of Fe(III) ions remained in HS state. Furthermore, while **1** exhibited weak antiferromagnetic (AFM) interactions between adjacent cations, the remnant HS fraction in **2** revealed an order of magnitude stronger AFM exchange coupling. We discuss a degradation of SCO phenomenon and a character of the exchange interactions in terms of aging and thermo cycling. Despite the degradation with aging, the structures and physical properties of fresh crystals from independent syntheses were consistent. We believe that the method of slow diffusion solution provides a reliable processability for other systems of that type.

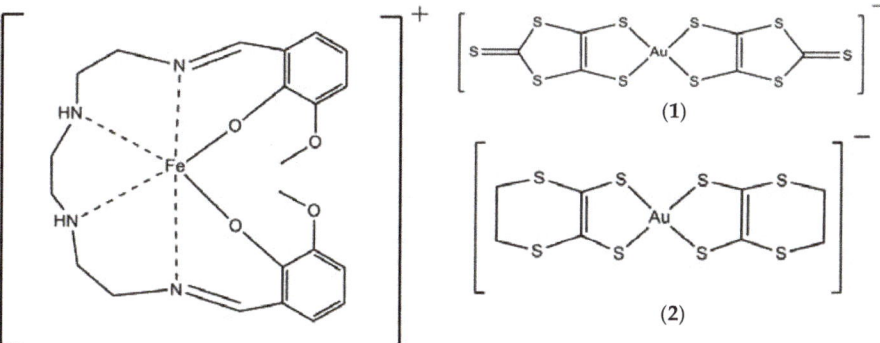

**Figure 1.** Scheme of the cation [Fe(3-OMesal$_2$-trien)]$^+$ and anions [Au(dmit)$_2$]$^-$ (**1**), [Au(dddt)$_2$]$^-$ (**2**).

## 2. Results and Discussion

### 2.1. Synthesis

Commercial solvents were used without further purification unless otherwise specified. Reactants commercially obtained: 3-methoxysalicylaldehyde (o-vanillin), 1,8-diamino-3,6,-diazaoctane (trien-oil), iron nitrate nonahydrate [Fe(NO$_3$)$_3$·9H$_2$O] and sodium methoxide (NaOCH$_3$) from Sigma-Aldrich (Saint Louis, MO, USA). Complexes [(Et)$_4$N][Au(dmit)$_2$], [(Et)$_4$N][Au(dddt)$_2$] and [Fe(3-OMe-sal$_2$-trien)]·NO$_3$·H$_2$O were prepared according to the literature [21–23].

#### 2.1.1. [Fe(3-OMesal$_2$-trien)][Au(dmit)$_2$] (1)

To the stirred solution of [(Et)$_4$N][Au(dmit)$_2$] (84 mg, 0.15 mmol) in acetonitrile (15 mL) was added dropwise a solution of [Fe(3-OMesal$_2$-trien)]·NO$_3$·H$_2$O (82 mg, 0.15 mmol)) in acetonitrile (5 mL). The reaction mixture was heated up to 60 °C for 10 min, and then cooled to −18 °C. The precipitate was filtered, washed with cooled acetonitrile, and dried in vacuo at room temperature. [Fe(3-OMesal$_2$-trien)][Au(dmit)$_2$] (1) was obtained as black polycrystalline powder with a total yield 78.14%. Elemental Anal. Found: C, 31.84; H, 2.73; N, 5.36%. Calcd for (Mw = 1057) C$_{28}$H$_{28}$FeN$_4$AuO$_4$S$_{10}$: C, 31.79; H, 2.65; N, 5.30%. For the results of thermogravimetric analysis see Figure S1.

By means the method of slow diffusion solution of [(Et)$_4$N][Au(dmit)$_2$] (6 mg, 0.01 mmol) in acetone (4 mL) into solution of [Fe(3-OMesal$_2$-trien)]·NO$_3$·H$_2$O (5.5 mg, 0.01 mmol) in acetonitrile (1 mL), after one week the black plate-like crystals of **1** suitable for X-ray diffraction was filtered and dried at ambient temperature. The electron-probe X-ray microanalysis (*EPMA*) information on the elements proportion in **1** is Fe:Au:S = 1:1:10. Powder diffractogram (P-XDR) of the samples taken for magnetic measurements is presented in Figure S2.

#### 2.1.2. [Fe(3-OMesal$_2$-trien)][Au(dddt)$_2$]·CH$_3$CN (2)

To the stirred solution of [(n-Bu)$_4$N][Au(dddt)$_2$] (120 mg, 0.15 mmol) in acetonitrile (10 mL) was added dropwise a solution of [Fe(3-OMe-sal$_2$-trien)]·NO$_3$·H$_2$O (82 mg, 0.15 mmol)) in acetonitrile (10 mL). The reaction mixture was heated up to 50 °C for 10 min, and then cooled to −18 °C. The precipitate was filtered, twice washed with ether, and dried at room temperature in air. [Fe(3-OMesal$_2$-trien)][Au(dddt)$_2$]·CH$_3$CN (2) was obtained as black polycrystalline powder with a total yield 76.13%. Elemental Anal. Found: C, 36.21; H, 3.73; N, 6.58%. Calcd for (Mw = 1066) C$_{32}$H$_{39}$FeN$_5$AuO$_4$S$_8$: C, 36.02; H, 3.66; N, 6.57%.

A weight loss of 3% was observed in the temperature range 145–200 °C which is assigned to the loose of one molecule of CH$_3$CN (3.85% by theory) from **2**. For the results of thermogravimetric analysis see Figure S3. The plot shows that weight loss occurs in two stages. A weight decrease of 3% corresponding to the loss of 0.8 solvate molecules is observed in the 145–200 °C range. The results indicate a relatively good agreement with the formulation deduced from elemental analysis and X-ray crystal structure determinations (see below). Decomposition of **2** begins above 225 °C. Although X-ray analysis before and after magnetic measurements did not show changes it seems likely that desolvation reveals itself at time exposures of months.

By means of slow diffusion method in U-shaped sell the solution of [(n-Bu)$_4$N][Au(dddt)$_2$] (8.0 mg, 0.01 mmol) in acetone (2 mL) was added into solution of [Fe(3-OMesal$_2$-trien)]·NO$_3$·H$_2$O (5.5 mg, 0.01 mmol) in acetonitrile (2 mL). Black shinny plate-like crystals of **2** with up to 3 mm length were obtained after one week. Crystals were filtered, washed with ether, and dried at ambient temperature. Suitable single crystals were chosen for X-ray diffraction. The *EPMA* information on the elements proportion in **2** is Fe:Au:S=1:1:8. Figure S4 displays the P-XDR of **2** in the 2θ range from 1.95° to 30°.

The purity of the complexes and their solvation state was verified by determining the carbon, hydrogen and nitrogen content, at the Vario MICRO Cube (Elementar GmbH, Langenselbold, Germany)

Analysis Service. Microphotos of single crystals of **1**, **2** are shown in Figure S5 left and right panels, respectively. Polycrystalline sample taken for magnetic measurements are shown in Figure S6.

*2.2. Crystal Structures*

2.2.1. [Fe(III)(3-OMesal$_2$-trien)] [Au(dmit)$_2$] (**1**)

Complex **1** crystallizes in the triclinic space group $P\bar{1}$ with one formula unit in the asymmetric volume (Figure 2). The refinement of the structure at 120 and 293 K (Table 1) showed identity of the structural characteristics at both temperatures. Below the details of the 120 K structure are described.

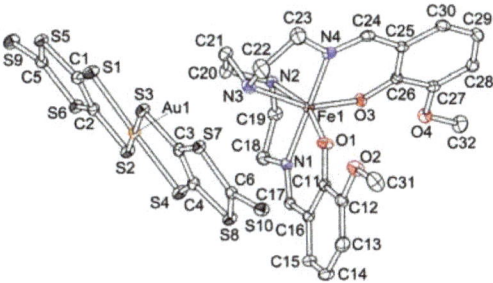

**Figure 2.** Asymmetric unit in **1** (120 K, ORTEP drawing with 50% probability ellipsoids).

**Table 1.** Crystal structure and refinement data.

|  | 1 | 1 | 2 | 2 |
|---|---|---|---|---|
| Chemical formula | C$_{28}$H$_{28}$AuFeN$_4$O$_4$S$_{10}$ | C$_{28}$H$_{28}$AuFeN$_4$O$_4$S$_{10}$ | C$_{32}$H$_{39}$AuFeN$_5$O$_4$S$_8$ | C$_{32}$H$_{39}$AuFeN$_5$O$_4$S$_8$ |
| Formula weight | 1057.96 | 1057.96 | 1066.98 | 1066.98 |
| Temperature (K) | 120 | 293 | 120 | 293 |
| Cell setting | triclinic | triclinic | monoclinic | monoclinic |
| Space group, Z | $P\bar{1}$, 2 | $P\bar{1}$, 2 | $I2/c$, 4 | $I2/c$, 4 |
| $a$ (Å) | 10.3711(1) | 10.4935(3) | 13.7776(2) | 13.9679(4) |
| $b$ (Å) | 10.9107(1) | 10.9970(3) | 12.9564(1) | 13.0757(4) |
| $c$ (Å) | 17.7081(2) | 17.7948(6) | 21.9490(2) | 22.1473(6) |
| $\alpha$ (°) | 75.5043(8) | 75.649(3) | 90 | 90 |
| $\beta$ (°) | 79.9758(8) | 79.768(3) | 98.4284(10) | 98.068(2) |
| $\gamma$ (°) | 68.9996(9) | 68.289(3) | 90 | 90 |
| Cell volume (Å$^3$) | 1803.23(3) | 1839.82(11) | 3875.77(7) | 4004.9(2) |
| Crystal size (mm) | 0.41 × 0.09 × 0.08 | 0.41 × 0.09 × 0.08 | 0.33 × 0.25 × 0.18 | 0.33 × 0.25 × 0.18 |
| $\rho$ (Mg/m$^3$) | 1.948 | 1.910 | 1.829 | 1.770 |
| $\mu$, cm$^{-1}$ | 50.86 | 49.85 | 46.30 | 44.81 |
| Refls collected/unique/observed with $I > 2\sigma(I)$ | 35,223/10,301/9875 | 19,892/10,236/8987 | 44,538/5889/5523 | 11,307/5471/4459 |
| $R_{int}$ | 0.0187 | 0.0223 | 0.0149 | 0.0305 |
| $\theta_{max}$ (°) | 31.06 | 31.07 | 31.17 | 30.84 |
| Parameters refined | 441 | 441 | 251 | 269 |
| Final $R_1$(obs), $wR_2$ (all) | 0.0148, 0.0338 | 0.0262, 0.0540 | 0.0140, 0.0333 | 0.0307, 0.0697 |
| Goodness-of-fit | 1.004 | 1.007 | 1.000 | 1.002 |
| Residual el. density (e Å$^{-3}$) | 0.606/−0.633 | 0.774/−0.756 | 0.807/−0.689 | 0.648/−0.561 |
| CCDC reference | 1868121 | 1868120 | 1868123 | 1868122 |

In the crystal structure the layers of [Au(dmit)$_2$]$^-$ anions alternate with the layers of [Fe(3-OMesal$_2$-trien)]$^+$ cations along the *c*-axis (Figure 3). The anion layer (Figure 4) consists of {[Au(dmit)$_2$]$_2$}$^{2-}$ dimers with the intradimer interplane distance of 3.56(3) Å and ring-over-atom overlap mode. Adjacent dimers are strongly shifted from each other along the long molecular axis and interdimer overlapping is absent (Figure 4b). Shortened S . . . S contacts are observed both inside the dimers [the S . . . S distances of 3.643(1) Å] and between them [3.218(1), 3.442(1), 3.508(1), 3.579(1) Å] and link the dimers into infinite ribbons along the [1 −1 0] direction. All the S . . . S distances between the anions from the neighbor ribbons exceed the sum of van-der Waals radii of 3.7 Å. Charge state of dmit in the [Au(III)(dmit)$_2$]$^-$ anion is 2$^-$. It is confirmed by short lengths of double C=C bonds in dmit which are sensitive to the charge: the average value of C=C bond lengths in **1** is 1.348(2) Å.

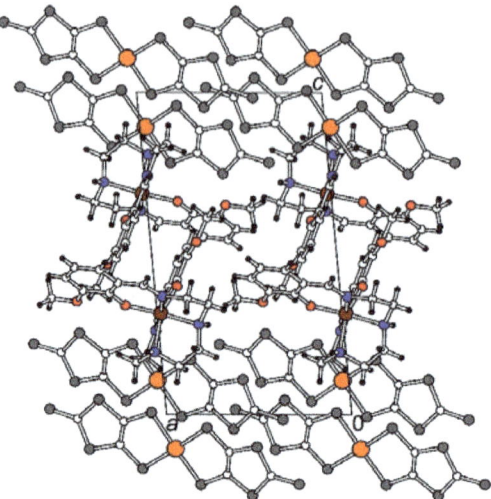

**Figure 3.** View of the structure **1** along $b$.

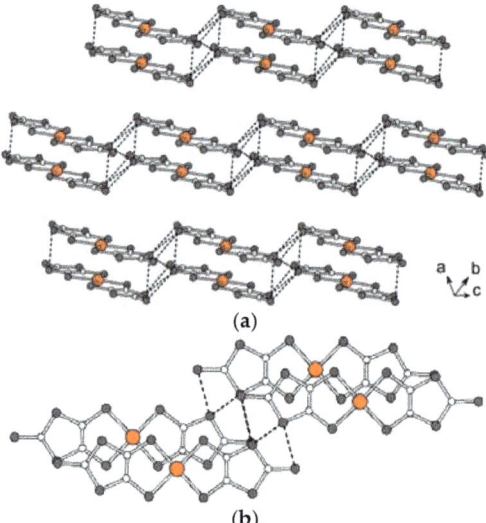

**Figure 4.** (**a**) The $ab$ layer of Au(dmit)$_2$. The S ... S contacts < 3.7 Å are shown by dashed lines. (**b**) Overlap mode inside and between the centrosymmetric [Au(dmit)$_2$]$_2$ dimers.

The magnetic layer consists of the chains of [Fe(3-OMesal$_2$-trien)]$^+$ cations, running along $a$ (Figure 5). The cations within the chain are paired by $\pi$-stacking. The distance between centroids of two interacting phenyl cycles in the pair is 3.467(2) Å; there is several intermolecular C ... C contacts in the range of 3.386(2)–3.507(3) Å which are shown by dashed lines in Figure 5. The Fe(III) ion in [Fe(3-OMesal$_2$-trien)]$^+$ is octahedrally coordinated by two O and four N atoms of the ligand. At 120 K Fe(III) is in a high-spin state, that is testified by lengthened Fe-N and Fe-O bonds (Table 2) as well as noticeable distortion of the octahedron: the O–Fe–N angles are 155.30(5) and 160.07(5)°. The other indicator of HS state is obtuse dihedral angle α between two salicylidene groups of 3-OMesal$_2$-trien ligand of 94.22(3)°. Very similar geometry of the cation, characteristic for the high-spin Fe(III) ion, is retained in the complex **1** at room temperature (Table 2).

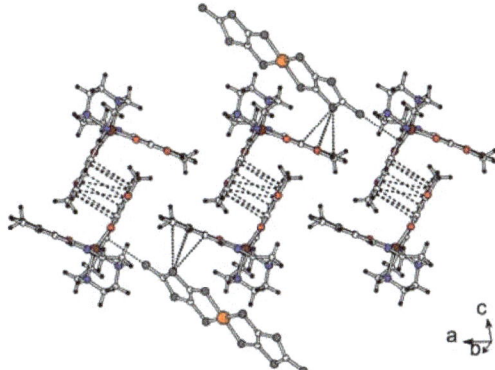

**Figure 5.** Ribbons of [Fe(3-OMesal$_2$-trien)]$^+$ in the *ab* plane. The shortened C ... C contacts < 3.6 Å in π-stacked pairs are shown by dashed lines. The anion ... cation S ... C contacts are shown by dotted lines.

**Table 2.** Selected bond lengths (Å) and angles (°) in [Fe(3-OMesal$_2$-trien)]$^+$.

| 1, 120 K (HS) | 1, 293 K (HS) | 2, 120 K * (LS) | 2, 293 K (HS) |
|---|---|---|---|
| Fe1 O1 1.9103(11) | 1.9056(17) | Fe1 O1 1.8714(9) | 1.8904(16) |
| Fe1 O3 1.9090(10) | 1.9090(17) | | |
| Fe1 N1 2.1233(13) | 2.123(2) | Fe1 N1 1.9358(10) | 2.082(2) |
| Fe1 N2 2.1761(14) | 2.178(2) | Fe1 N2 1.9987(11) | 2.152(2) |
| Fe1 N3 2.1797(14) | 2.182(3) | | |
| Fe1 N4 2.1055(13) | 2.103(2) | | |
| O1 Fe1 O3 105.34(5) | 105.52(8) | O1 Fe1 O1b 95.02(6) | 99.39(11) |
| O1 Fe1 N1 86.24(5) | 86.08(8) | O1 Fe1 N1 94.13(4) | 88.47(8) |
| O1 Fe1 N2 160.07(5) | 159.41(9) | O1 Fe1 N2 174.82(5) | 162.63(10) |
| O1 Fe1 N3 93.52(5) | 93.26(9) | O1 Fe1 N1b 88.71(4) | 93.94(7) |
| O1 Fe1 N4 88.72(5) | 89.55(8) | O1 Fe1 N2b 89.87(4) | 92.50(9) |
| O3 Fe1 N1 96.91(5) | 97.02(8) | | |
| O3 Fe1 N2 88.47(5) | 88.82(8) | | |
| O3 Fe1 N3 155.30(5) | 155.27(9) | | |
| O3 Fe1 N4 87.32(5) | 87.17(8) | | |
| N1 Fe1 N2 77.74(5) | 77.41(8) | N1 Fe1 N2 84.27(5) | 78.06(11) |
| N1 Fe1 N3 100.16(5) | 100.23(9) | N1 Fe1 N2b 92.64(5) | 99.02(11) |
| N1 Fe1 N4 174.14(5) | 174.62(8) | N1 Fe1 N1b 175.81(6) | 176.28(12) |
| N2 Fe1 N3 77.94(5) | 77.89(9) | N2 Fe1 N2b 85.30(7) | 79.04(15) |
| N2 Fe1 N4 106.52(5) | 106.15(9) | | |
| N3 Fe1 N4 77.14(5) | 76.87(9) | | |
| α 94.22(3) | 95.29(4) | 88.94(1) | 93.75(3) |
| ∠(i–ii)88.1(1) | 88.5(2) | ∠(i–ii)31.1(1) | 85.7(2) |
| ∠(ii–iii)84.3(1) | 83.7(2) | | |
| ∠(i–iii))35.4(1) | 35.7(2) | ∠(i–i(b))21.3(1) | 29.4(2) |

* Symmetry code: b ($-x, y, 0.5 - z$).

The cation and anion layers interact through hydrogen contacts of N−H...S (H ... S of 2.62 Å) and C−H...S types (H ... S of 2.81–2.98 Å). Besides, the S ... π interaction between the S(8) atom of [Au(dmit)$_2$]$^-$ and π-system of the C(11)-C(16) cycle of [Fe(3-OMesal$_2$-trien)]$^+$ is observed (Figures 2 and 5). The distance S(8) ... centroid C(11)-C(16) is 3.226(1) Å, the six S ... C contacts are in the range of 3.473(2)–3.566(2) Å (dotted lines in Figure 5). One more S ... C contact of 3.500(2) Å links terminal S(10) atom of the anion to the cation from the next cation pair in the ribbon. Figure 5 clearly demonstrates that both salicylidene arms of the [Fe(3-OMesal$_2$-trien)] cation have very strong $\pi$-interactions with surrounding molecules: one arm is $\pi$-stacked to the similar arm of another cation in the pair and linked to the terminal S atom of the anion, while the another arm has strong $\pi$ ... S connection to the anion. Thus, one can conclude that very strong cooperation (or interplay) of the [Fe(3-OMesal$_2$-trien)] cation with molecular environment should prevent its conformational switching needed for thermo-induced spin-crossover and in this reason the complex **1** retains the HS state at cooling.

2.2.2. [Fe(3-OMesal$_2$-trien)][Au(dddt)$_2$]·CH$_3$CN (**2**)

Complex **2** crystallizes in the monoclinic space group *I*2/*c*. Asymmetric unit includes a half of the [Au(dddt)$_2$]$^-$ anion with Au at an inversion center, a half of the [Fe(3-OMesal$_2$-trien)]$^+$ cation with Fe on a two-fold axis (Figure 6) and an acetonitrile molecule in a half-occupied general position near a two-fold axis. The structure of **2** was investigated at 120 and 293 K. Symmetry of the crystal is the same at both temperatures; however, some important structural changes are observed.

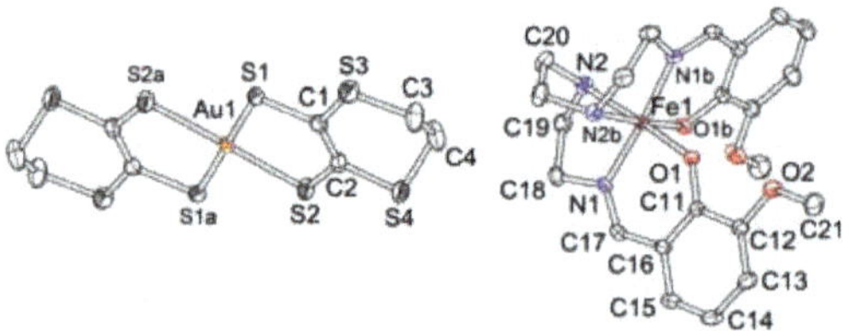

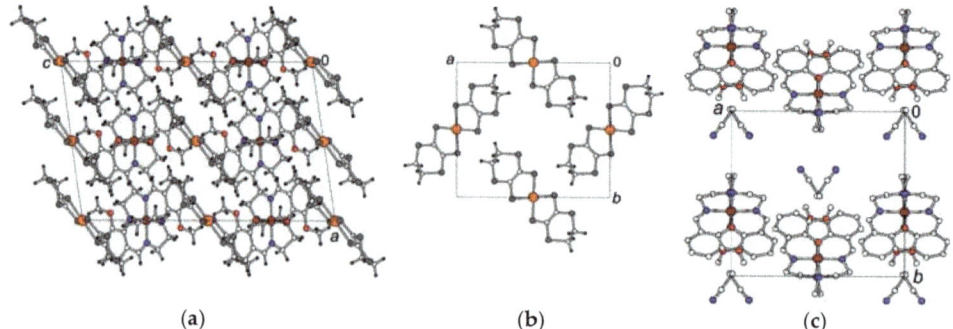

**Figure 6.** Independent molecules in **2** (120K, ORTEP drawing with 50% probability ellipsoids). Symmetry codes: a (1 − *x*, −*y*, −*z*), b (−*x*, *y*, 0.5 − *z*).

In the structure of **2** the layers of the [Au(dddt)$_2$]$^-$ anions alternate with the layers of the [Fe(3-OMesal$_2$-trien)]$^+$ cations along *c* (Figure 7a). The anion layer is shown in Figure 7b. Solvent molecules are sited in the cation layer (Figure 7c). The length of the double C=C bond in dddt is equal to 1.339(2) Å, i.e., the charge of the dddt part in the [Au(III)(dddt)$_2$]$^-$ anion is 2$^-$. There are no stacks in the anion layer because the anions are located near orthogonally to each other.

(a) (b) (c)

**Figure 7.** (**a**) Structure **2** (120 K) projected along *b*. Solvent molecules located in the cation layer are omitted for clarity. (**b**) View of the anion layer along *c*. (**c**) View of the cation layer along *c* (H-atoms are omitted).

In the cation layer (Figure 7c), $\pi$-stacking between the [Fe(3-OMesal$_2$-trien)]$^+$ cations is not observed. The disordered acetonitrile molecules are in the zigzag cavity of the layer, their –CH$_3$ groups being in between the salicylidene moieties of one cation and N atoms being near the middle ethylene group of the trien fragment of the next cation along *b*. The structure is supported by hydrogen bonding of cation ... cation type [C–H ... O contacts with H ... O distance of 2.49 Å) and cation ...

anion type [N–H ... S, H ... S 2.58 Å; C–H ... S, H ... S 2.81, 2.88 Å]. In the complex **2** the Fe(III) ions in [Fe(3-OMesal$_2$-trien)]$^+$ are in a low-spin state at 120 K. The LS geometry of central FeO$_2$N$_4$ octahedron is very different from the HS one found in **1** (Table 2): all Fe–N$_{im}$, Fe–N$_{am}$ and Fe–O bonds are noticeably shorter in LS whereas distortion of the octahedron decreases in comparison to the HS state, the O–Fe–N angle is 174.82(5)°. The dihedral angle α between two salicylidene groups of (3-OMesal$_2$-trien) ligand is 88.94(1)°, i.e., α < 90° that is a feature of LS complexes (Figure 7, middle). At room temperature the HS characteristics of Fe(III) coordinated octahedron similar to the complex **1** are found: extended coordination bonds Fe-O/Fe-N by 0.02/0.15 Å, decreased O–Fe–N angle by 12° and grown α angle by about 5° up to 93.75(3)° (Figure 8, right).

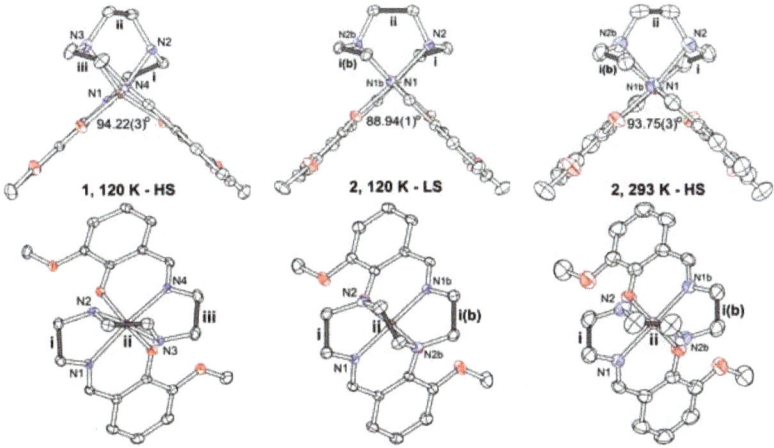

**Figure 8.** Molecular conformation of the [Fe(III)(3-OMesal$_2$-trien)]$^+$ cation in **1** at 120 K (left), in **2** at 120 K (middle) and 293 K (right), side (upper row) and top (bottom row) view. Symmetry code: b (−x, y, 0.5 − z). The values of α angles between the salicylidene groups are given in the figure. The angles between C–C lines of ethylene groups i, ii and iii are listed in Table 2.

We believe that it is important to also analyze a conformation of the ethylene groups –CH$_2$–CH$_2$– in the trien fragment of the cation. The ethylene groups i, ii and iii are denoted by grey-colored bonds in Figure 8. We found that the orientation of central ethylene group changes upon the spin-crossover whereas the orientations of ethylene groups in side cycles do not change significantly. In both **1** and **2** complexes central –CH$_2$–CH$_2$– bond ii is near normal to the side bonds i and iii in the HS state (Figure 8, left and right in bottom row). In the LS state the central ethylene group ii switches its conformation to be near parallel to the side bonds (Figure 8, middle in bottom row). The angles between lines i, ii and iii are given in Table 2. We have shown that different orientations of ethylene groups are energetically favorable for HS and LS complexes and therefore can be used as indicator of spin state of [Fe(III)(sal$_2$trien)]$^+$ cations along with usual characteristics such as bond lengths and valent angles in the FeO$_2$N$_4$ octahedron and α angle between salicylidene moieties [24,25]. The paper with detailed conformational analysis and density functional theory (DFT) calculations in the spin-crossover [Fe(III)(sal$_2$trien)]$^+$ complexes will be published soon [26].

*2.3. Conductivity.*

Electrical conductivity for **1** and **2** was measured in the conducting plane of single crystals at room temperature by a conventional four-probe method. Platinum wire contact (d = 10µm) was glued to the crystal using the DOTITE XC-12 graphite paste. Both compounds were insulators with $\sigma_1(300\ K) = 3 \times 10^{-9}\ \Omega^{-1}\ cm^{-1}$ (**1**) and $\sigma_2(300\ K) = 1 \times 10^{-9}\ \Omega^{-1}\ cm^{-1}$ (**2**). This is consistent with small transfer integrals between the anions compared to the known ones for the conducting salts of metal bis-dithiolene anion complexes family [7].

## 2.4. Magnetic Properties

Magnetic measurements were performed by using a Quantum Design MPMS-5-XL SQUID magnetometer. Static magnetic susceptibility $\chi(T)$ was measured on polycrystalline samples at the magnetic field $H$ = 100 Oe, while warming ($\uparrow$) and cooling ($\downarrow$) regimes. The temperature range of measurements was 2–300 K. The magnetization curves $M(H)$ were obtained during several loops over the field range +50 kOe to −50 kOe. Preliminary, the samples had been cooled down to 2.0 K in zero magnetic field and virgin curves were recorded.

Temperature dependences of the product $\chi T\downarrow$ for complexes **1** and **2** are depicted in Figure 9. The measurements were performed on fresh crystals while cooling in the range of 300→2 K, $H$ = 100 Oe. The $\chi T\downarrow$ value of **1** starts at 4.26 cm$^3$ K/mol, remains unchanged till 50 K, undergoes the inflection at $T_{i1}$ = 46 K, and then decreases continuously down to 2.79 cm$^3$ K/mol. Polycrystalline powder of **2** demonstrate a sharp SCO transition at $T_{sco}$ = 118 K, during which the $\chi T\downarrow$ value changes from 4.25 cm$^3$ K/mol at 300 K (HS) down to 1.07 cm$^3$ K/mol at the plateau in the vicinity of 60 K (LS). The shape of the transition curve was reproduced in several same-day subsequent measurements. Evolution at longer time exposures is discussed below. Hysteresis behavior was not observed between $\chi T\downarrow$ and $\chi T\uparrow$ curves in the transition range. In the vicinity of $T_{i2}$ = 55 K there was also a slightly pronounced inflection below which $\chi T$ decreases down to 0.51 cm$^3$ K/mol. The $\chi T$ values at 300 K of both compounds are consistent with the theoretical spin-only value of 4.375 cm$^3$ K/mol corresponding to 100% amount of HS Fe(III) ions, $S$ = 5/2, $g$ = 2.0. This is also in agreement with the assertion that both anions Au(dmit$_2$)$^-$ and Au(dddt$_2$)$^-$ do not have magnetic moment.

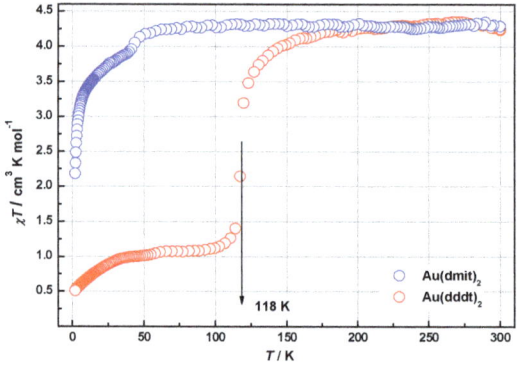

**Figure 9.** Temperature dependences of $\chi T\downarrow$ for **1** (○) and **2** (○).

Magnetic properties of both compounds changed with aging and thermal history, revealing an irreversible behavior of $\chi T$ while cooling and warming regimes. An uncomplicated behavior for **1** is shown in Figure 10. Irreversibility is observed below the inflection point, $T_{i1}$ = 46 K, in a form of a hysteresis loop, while above it the curves follow the same Curie-Weiss trend with $C$ = 4.3 cm$^3$ K/mol ($S$ = 5/2) and $\Theta$ = −0.5 K (see Figure 10, inset). This shape was reproduced in several cycles and it did not depend on the age of the sample. System **2** was found unstable. Both aging at ambient conditions and thermocycling led to a degradation of the SCO transition. The curves in Figure 11 present this evolution: the initial curve (black color, $\uparrow\downarrow$ regimes) was recorded on the crystals aged over a period of several months; the other data sets depicted in blue and purple colors show a character of SCO degradation after several subsequent cooling-warming cycles. A solid red line indicating the transition in freshly synthesized sample (see Figure 9) is given for reference. With aging the HS state was reached in a more gradual manner, so that the $\chi T$ value only got 3.8 cm$^3$K/mol at 300 K. The value $\chi T$ on the plateau below $T_{sco}$ corresponds to $\gamma_{LS}$ = 78% of $S$ = 1/2 and $\gamma_{HS}$ = 22% of $S$ = 5/2, where $\gamma_{LS}$ and $\gamma_{HS}$ are molar LS and HS fractions. We took 0.375 cm$^3$K/mol and 4.375 cm$^3$K/mol for 100 % of LS and HS fractions, respectively. The ratio $\gamma_{LS}/\gamma_{HS}$ did not change with time exposure, and varied from 3.6 (fresh) to 6.1 from sample to sample. However, thermocycling of seasoned crystals led to a quick spreading of the transition and to an irreversible growth of the residual HS fraction. In subsequent warming-cooling cycles $\gamma_{HS}$ changed from 14% to 32% and 41% respectively.

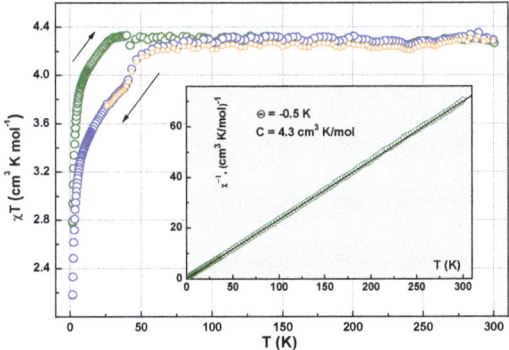

**Figure 10.** Temperature evolution of $\chi T$ in warming (○) and cooling (○, ○) regimes. Inset: Plot $\chi^{-1}(T)\uparrow$ vs. $T$ in **1**. Solid line is Curie-Weiss fit with $C$ = 4.3 cm$^3$ K/mol and $\Theta$ = −0.5 K.

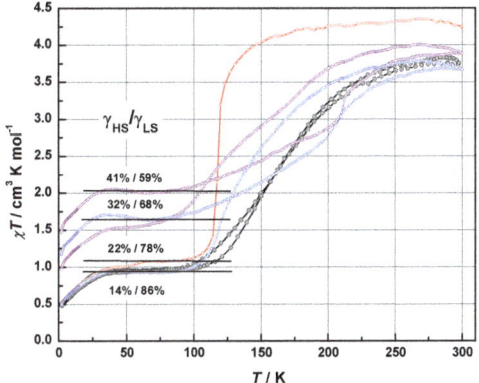

**Figure 11.** Temperature evolution of $\chi T$ for aged crystals **2**: initial curves $\chi(T)\uparrow\downarrow$ (○), $\gamma_{HS}$ = 14%, first warming-cooling cycle (○), $\gamma_{HS}$ = 32%, second cycle (○), $\gamma_{HS}$ = 41%. Solid red line is a reference curve for fresh crystals from Figure 1.

Magnetization curves, $M(H)$, are presented in Figures 12 and 13 for **1** and **2**, respectively. The data points in Figure 12 reveal a weak hysteresis with the coercive force $H_c$ = 13 Oe and the remnant magnetization value $M_r$ = 4 × 10$^{-3}$ $\mu_B$. Saturation at highest fields was not observed. However, a combination of the Brillouin function for $S$ = 5/2 with a magnitude factor 0.9 (black line), $M_s$ = 4.7 $\mu_B$, and linear function with the factor $k$ = 0.04 $\mu_B$/kOe gave a perfect fitting. Linear contribution is conceivable in terms of AFM interaction between HS Fe(III) cations. Magnetization curves for the AFM coupled spin system, including dimers, often show Brillouin dependences and a gradual growth proportional to the field strength when $|J_1| \ll kT$, $T$ = 2 K [27]. The magnetization curve of **2** depicted in Figure 13 also reveals hysteresis, $H_c$ = 50 Oe and $M_r$ = 3 × 10$^{-3}$ $\mu_B$. The value $M$ at 50 kOe reaches 1.14 $\mu_B$. Field dependence was perfectly fitted by Brillouin function for $S$ = 1/2 and magnitude factor 1.23 (solid line). The data points could not be fitted by a combination of Brillouin functions for $S$ = 1/2 and $S = 5/2$ regardless the weight factors. A surplus contribution of 0.23 $\mu_B$ remains unclear. Whereas the absence of contribution from HS Fe(III) cations could be explained by stronger AFM coupling, $|J_2| \gg kT$. Brillouin shape indicates that the local moments of LS Fe(III) ions remain non-interacting.

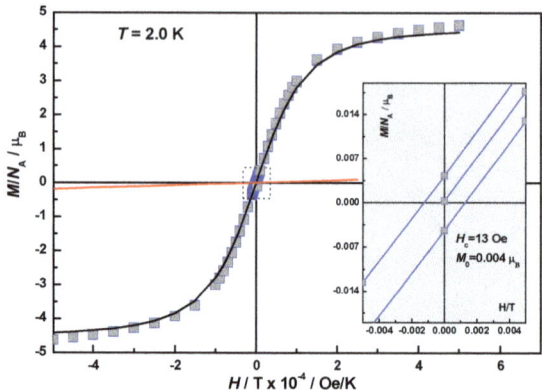

**Figure 12.** Field dependence of the magnetization, $M(H)$, in **1** at $T = 2.0$ K. Solid black line is a Brillouin function for $S = 5/2$ and magnitude factor 0.9. Solid red line fits the difference between experimental curve and Brillouin fitting. Inset: Near zero-field segment of the hysteresis loop with $H_c = 13$ Oe and $M_r = 4 \times 10^{-3}$ $\mu_B$.

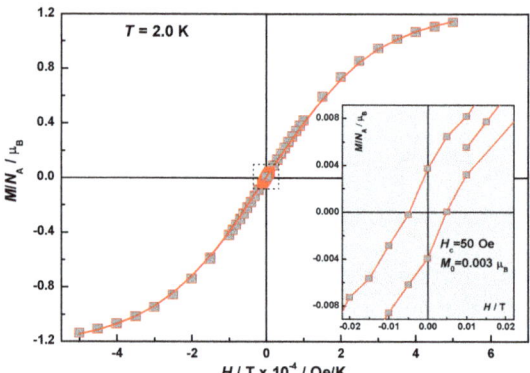

**Figure 13.** Field dependence of the magnetization, $M(H)$, in **2** at $T = 2.0$ K. Solid line is a Brillouin function for $S = 1/2$ and magnitude factor 1.23. Inset: Near zero-field segment of the hysteresis loop with $H_c = 50$ Oe and $M_r = 3 \times 10^{-3}$ $\mu_B$.

A comparison of the $\chi T$ evolution in **1** and **2** with their structural characteristics gives rise to several essential notions:

Results of our magnetic measurements agree well with the structural features of the compounds. Complex **1** has a rigid structure which is characterized by extensive $\pi$-interactions of both salicylidene moieties of the SCO cation with nearest neighbors, one with the cation and another with the anion. This makes changing conformation of [Fe(3-OMesal$_2$-trien)]$^+$ unit at the SCO impossible and explains absence of the latter in **1**. Similar situation was discussed for one of the polymorphic phases of [Fe(sal$_2$-trien)][Ni(dmit)$_2$] which also does not show SCO at cooling [28]. In complex **2**, the cation layer includes solvent molecules, $\pi$-stacking is absent and crystal packing is controlled mainly by weak hydrogen bonding allowing conformation freedom necessary for SCO. We believe this view is also consistent with the results of [28–30].

Although a significant loss of acetonitrile molecules was observed in the 145–200 °C range (Figure S3), the desolvation was evidently taking place at ambient conditions. These small changes significantly affect the configuration of the ligand and therefore the spin transition [25,31–33]. We believe, solvent vacancies in the seasoned crystals **2**, as well as the defects, provoke stabilization of the HS configuration in their vicinity. Thermocycling stimulates growth of the HS fraction by promoting a diffusion of the solvent molecules out of the bulk because the sample is placed in the chamber with helium atmosphere. In several cycles the remnant HS fraction in aged crystals **2** rose from 14% to 40% (Figure 11). It is worth noting that up to 20% HS fraction was found in the original fresh samples (depending on batch). This either relates to a presence of the HS conformers [26] or is likely deals with disproportion between [Fe(3-OMesal$_2$-trien)] and [Au(dddt)$_2$], when the metal dithiolene complex might become a neutral radical [Au(dddt)$_2$]$^\bullet$. The former was not confirmed by our structural data. Presence of neutral radical is possible since the dddt ligand can carry a various charge form [11]. The interaction of the local magnetic moment of the SCO complex with the radical in the anion counterpart was observed recently [14]. The additional spin contribution from the radicals (~7%) could hardly be distinguished on the top of the total magnetic response in SQUID measurements. However, a spin density on the dithiolene ligand may substantially enhance the superexchange coupling ($J_2$). This assumption will be verified in forthcoming electron paramagnetic resonance (EPR) experiments.

In both compounds the $\chi T$ curves revealed a characteristic inflection. Below the inflection points $T_{i1}$ = 46 K (**1**) and $T_{i2}$ = 55 K (**2**) $\chi T$ continuously declined. We fitted experimental data by using julX software [34]. The AFM coupled HS Fe(III) dimers were considered as a model system. The best fit parameters for **1** were $g_{i1} = g_{i2} = 2.0$, $D_{i1} = D_{i2} = 0$, $J_1 = -0.18$ cm$^{-1}$ (0.26 K). It is worth noting that even a very small ZFS factor such as for example $D_i = -0.1$ cm$^{-1}$ strongly distort the calculated $M(H)$ curve. We did not observe significant deviation from the Brillouin shape in Figure 12. Therefore, we took $D_i = 0$ in our fitting. The other interesting feature of the fitting was a magnitude factor of 0.9. This might mean that only ~90% of HS Fe(III) moments was effectively enough for the description of the $\chi T\downarrow$ curve below the inflection point. A noticeable portion of the moments Fe(III) ions "disappeared" out of the total magnetic response. This is a signature of weak AFM interactions.

The value $|J_1| = 0.26$ K for **1** is indeed lower than 2 K, at which the magnetization curves were measured. It is not surprising then that the calculated $M(H)$ line fits the Brillouin-like behavior in the experiment (Figure 12) and, at the same time, calculated $\chi(T)$ describes the decline of $\chi T\downarrow$ in experiment. Figure 14 shows a low temperature part of the experimental data set of **1** and two fitting lines: fitting by a Curie-Weiss law with $\Theta_1 = -1.7$ K and C = 4.0 cm$^3$ K/mol, and quantum calculations in the model of AFM coupled dimers with S = 5/2 and $J = J_1$. In cases when every magnetic ion has the same number and the same kinds of interactions, the Weiss temperature can be expressed via pair exchange constants $J_{i,j}$ in the usual form found in the literature:

$$\Theta = \frac{S(S+1)}{3k} \sum_{j=1(j\neq i)} z_{ij} J_{ij} = \frac{2S(S+1)}{3k} ZJ \quad (1)$$

where $z_{ij}$ is the number of $j$ neighbors of the $i$th atom. For **1** $\Theta$ and $J_{ij} = J_1$ match at $Z \approx 1.1$, which is close to the dimer case and may speak in favor of AFM interaction between the adjacent SCO complexes linked by the π–π bonding of their aromatic groups.

For the fresh crystals of **2** we estimated ~20% of the residual HS fraction. This contribution was extracted from the low temperature plateau (T < 70 K) of the experimental data set $\chi T\downarrow$ on Figure 9. Then we recovered its value to the virtual 100% HS phase, attributing a low temperature decline to the AFM coupling in this HS fraction. The data after the treatment are shown in Figure 14. The best fit curve by Curie-Weiss law, $\Theta_2 = -16$ K, is shown as the solid black line. Attempts of describing it within the uniform quantum model of AFM coupled moments failed. A satisfactory agreement was reached by using a model of two linked dimers and two coupling constants, $J$. A solid red line shows the best fit curve for $|J_{23}| = 2.46$ K and $|J_{12}| = |J_{34}| = 0.43$ K ($g_i = 2.0$, $D_i = 0$). Although there are no structural dimers in **2**, the magnetic dimer is a reasonable approximation of the short-range coupling. As soon as we do not have an idea about the morphology of the HS fraction this model might seem to a certain extent speculative. Nevertheless, it brings an understanding that the much stronger exchange interaction came to the stage. Therefore, in **2** additionally to a routine weak coupling between the adjacent iron complexes there comes a stronger coupling which apparently involves the gold dithiolene complexes as well. In other words, structural defects and, possibly, solvent vacancies may serve not only as a source of the residual HS fraction, but also as a promoter of stronger exchange interactions between respective HS complexes. If the defects cause a charge disproportionation, then the appearance of neutral radicals $[Au(dddt)_2]^\bullet$ would facilitate stronger exchange coupling via the anion sublattice. This also suggests an additional spin contribution from $Au(dddt)_2$, ~10% of $S = 1/2$ ($\chi T \approx 0.04$ cm$^3$ K/mol). Evidently, the inflections points $T_{i1} = 46$ K and $T_{i2} = 55$ K also reflect changes in metal dithiolene subsystem. A comparison of two above approaches by using (1) gives:

$$\Theta_2 = \frac{S(S+1)}{3k}(4J_{12} + 2ZJ_{23}) \qquad (2)$$

which matches at $Z = 0.9$. This means that one pair of stronger coupled $S = 5/2$ moments ($J_{23}$) and two linked pairs of weaker coupled ones ($J_{12}$, $J_{34}$) is enough to describe the decrease of $\chi T$ below the SCO transition in **2**. It is worth noting that anion driven modulation [11,33,35] may modify the intermolecular interactions but it cannot facilitate the exchange coupling via a network of hydrogen bonds.

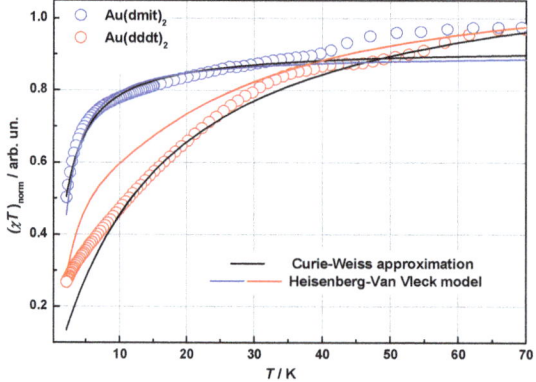

**Figure 14.** Temperature dependences of normalized $\chi T$ for HS fractions in **1** (○) and **2** (○). Solid lines are the best fit curves for the two models: Curie-Weiss law (black line) and Heisenberg-Van Vleck model (blue and red lines, see details in the text).

Of course, a model of two strongly coupled dimers plays a merely symbolic role linking together a weak coupling of the same nature as in **1**, and stronger coupling that is, in our understanding, a quantitative characteristic of the involvement of the electronic states of dithiolene into superexchange interactions between HS Fe(III) magnetic moments.

## 3. Conclusions

We have found that a structure organization of **1** with strong intermolecular π-interactions between aromatic rings of [Fe(III)(3-OMesal$_2$-trien)]$^+$ and molecular environment suppresses SCO. Such π-bonding is absent in **2** and the reversible SCO transition was observed at $T_{sco} = 118$ K. The conformational changes in [Fe(III)(3-OMesal$_2$-trien)]$^+$ at SCO were analyzed and it was shown at first time that reorientation of the central ethylene group of the trien occurs at the transition: it is near perpendicular to two side ethylene groups of the trien in the HS state and becomes near parallel to them in the LS state. This can be used as a very fast and visible indicator of the spin state in saltrien complexes [26]. It was also found that aging and subsequent thermocycling of the complex **2** led to a degradation of the spin-crossover transition. As a result, the residual HS fraction in our experiments increased to 41%, which was apparently associated with the release of acetonitrile out of the crystal bulk.

Total magnetic response of **1** indicates that 100% of magnetic moments in the system are the local magnetic moments of HS Fe(III) ions, $S = 5/2$. These moments form the AFM dimers with exchange constant $J_1 = -0.18$ cm$^{-1}$, which conceivably arise due to the strong $\pi-\pi$ bonding between terminal aromatic groups of adjacent iron complexes. Though there is no $\pi-\pi$ bonding in crystals **2**, they demonstrate surprisingly stronger AFM coupling, $J_{23} = -1.69$ cm$^{-1}$, within the residual HS fraction. A satisfactory agreement between experimental $\chi T$ vs. $T$ plot and model calculations was reached in the framework of two exchange coupled dimers, $|J_{12}| = |J_{34}| = 0.29$ cm$^{-1}$, linked via a stronger superexchange bridge $|J_{23}| \gg |J_{12}|, |J_{34}|$. We assume that such a bridge suggests involvement of the molecular orbitals of the [Au(dddt)$_2$] complexes, so that the distortions in the dithiolene sublattice not only stabilize the high-spin state of the iron complexes, but also facilitate the exchange interaction between them. If so, EPR experiments would shed the light on that. Such work is in progress.

**Supplementary Materials:** The following are available online at http://www.mdpi.com/2073-4352/8/10/382/s1, Figure S1: Thermogravimetric analysis of [Fe(3-OMe-sal$_2$-trien)][Au(dmit)$_2$] (**1**), Figure S2: Powder X-ray diffractogram for [Fe(3-OMe-sal$_2$-trien)][Au(dmit)$_2$]: experimental (line) and simulation (dots), Figure S3: The thermogravimetric analysis of [Fe(3-OMe-sal$_2$-trien)][Au(dddt)$_2$]·CH$_3$CN (**2**), Figure S4: Powder X-ray diffractogram for [Fe(3-OMe-sal$_2$-trien)][Au(dddt)$_2$]·CH$_3$CN: experimental (line) and simulation (dots), Figure S5: Single crystals of [Fe(3-OMesal$_2$-trien)][Au(dmit)$_2$] (left) and [Fe(3-OMesal$_2$-trien)][Au(dddt)$_2$] (right), Figure S6: Polycrystalline sample of [Fe(3-OMe-Sal$_2$-trien)][Au(dddt)$_2$]·CH$_3$CN.

**Author Contributions:** N.G.S., M.A.B., E.B.Y.—synthesis, chem. analyses; L.V.Z., S.V.S.—structure resolution; Y.N.S., D.V.S., E.L., A.A.K.—magnetic part.

**Acknowledgments:** This work was done on the topic of the State task (No. 0089-2014-0026) with using of the Computational and Analytical Center for Collective Use of the IPCP RAS tool base and was partially supported by the Program of Russian Academy of Sciences topic "Spin" №AAAA-A18-118020290104-2. The authors would like to thank L.I. Buravov for conductivity measurements.

**Conflicts of Interest:** The authors declare no conflict of interest. The founding sponsors had no role in the design of the study; in the collection, analyses, or interpretation of data; in the writing of the manuscript, and in the decision to publish the results.

## References

1. Gutlich, P.; Goodwin, A.H. *Spin Crossover in Transition Metal Compounds III*; Springerlink: Berlin, Germany, 2004; Volume 235, p. 1.
2. Brooker, S. Spin-crossover with thermal hysteresis: Practicalities and lessons learnt. *Chem. Soc. Rev.* **2015**, *44*, 2880–2892. [CrossRef] [PubMed]
3. Gutlich, P. Spin Crossover—Quo Vadis? *Eur. J. Inorg. Chem.* **2013**, 5–6, 581–591. [CrossRef]
4. Okai, M.; Takahashi, K.; Sakurai, T.; Ohta, H.; Yamamoto, T.; Einagad, Y. Novel Fe(II) spin crossover complexes involving a chalcogen-bond and π-stacking interactions with a paramagnetic and nonmagnetic M(dmit)$_2$ anion (M = Ni, Au; dmit = 4,5-dithiolato-1,3-dithiole-2-thione). *J. Mater. Chem. C* **2015**, *3*, 7858–7864. [CrossRef]
5. Bousseksou, A.; Molnár, G.; Salmon, L.; Nicolazzi, W. Molecular spin crossover phenomenon: Recent achievements and prospects. *Chem. Soc. Rev.* **2011**, *40*, 3313–3335. [CrossRef] [PubMed]
6. Gentili, D.; Demitri, N.; Schäfer, B.; Liscio, F.; Bergenti, I.; Ruani, G.; Ruben, M.; Cavallini, M. Multi-modal sensing in spin crossover compounds. *J. Mater. Chem. C* **2015**, *3*, 7836–7844. [CrossRef]
7. Sato, O.; Li, Z.-Y.; Yao, Z.-S.; Kang, S.; Kanegawa, S. Multifunctional materials combining spin-crossover with conductivity and magnetic ordering. In *Spin-Crossover Materials: Properties and Applications*; Halcrow, M., Ed.; John Wiley & Sons: Hoboken, NJ, USA, 2013; pp. 303–319.

8. Valade, L.; Malfant, I.; Faulmann, C. Toward bifunctional materials with conducting, photochromic, and spin crossover properties. In *Multifunctional Molecular Materials*; Ouahab, L., Ed.; Pan Stanford Publishing: Singapore, 2013; p. 149.
9. Huang, J.; Xie, R.; Wang, W.; Li, Q.; Yang, J. Coherent transport through spin-crossover magnet $Fe_2$ complexes. *Nanoscale* **2016**, *8*, 609–616. [CrossRef] [PubMed]
10. Kato, R. Conducting metal dithiolene complexes: Structural and electronic properties. *Chem. Rev.* **2004**, *104*, 5319–5346. [CrossRef] [PubMed]
11. Wang, H.H.; Fox, S.B.; Yagubskii, E.B.; Kushch, L.A.; Kotov, A.I.; Whangbo, M.-H. Direct observation of the electron-donating property of the 5,6-dihydro-1,4-dithiin-2,3-dithiolate (dddt) ligands in square planar $M(dddt)_2$ complexes (M = Ni, Pd, Pt, and Au). *J. Am. Chem. Soc.* **1997**, *119*, 7601–7602. [CrossRef]
12. Yagubski, E.B. Molecular conductors based on $M(dddt)_2$ bisdithiolene cation complexes. *J. Solid State Chem.* **2002**, *168*, 464–469. [CrossRef]
13. Fujiyama, S.; Kato, R. Algebraic charge dynamics of the quantum spin liquid β-$EtMe_3Sb[Pd(dmit)_2]_2$. *Phys. Rev. B* **2018**, *97*, 035131. [CrossRef]
14. Shvachko, Y.N.; Starichenko, D.V.; Korolyov, A.V.; Kotov, A.I.; Buravov, L.I.; Zverev, V.N.; Simonov, S.V.; Zorina, L.V.; Yagubskii, E.B. The highly conducting spin-crossover compound combining Fe(III) cation complex with TCNQ in a fractional reduction state. Synthesis, structure, electric and magnetic properties. *Magnetochemistry* **2017**, *3*, 9. [CrossRef]
15. Shvachko, Y.N.; Starichenko, D.V.; Korolyov, A.V.; Yagubskii, E.B.; Kotov, A.I.; Buravov, L.I.; Lyssenko, K.A.; Zverev, V.N.; Simonov, S.V.; Zorina, L.V.; et al. The conducting spin-crossover compound combining Fe(II) cation complex with TCNQ in a fractional reduction state. *Inorg. Chem.* **2016**, *55*, 9121–9130. [CrossRef] [PubMed]
16. Phan, H.; Benjamin, S.M.; Steven, E.; Brooks, J.S.; Shatruk, M. Photomagnetic response in highly conductive iron(II) spin-crossover complexes with TCNQ radicals. *Angew. Chem. Int. Ed.* **2015**, *127*, 837–841. [CrossRef]
17. Takahashi, K.; Cui, H.-B.; Okano, Y.; Kobayashi, H.; Einaga, Y.; Sato, O. Electrical conductivity modulation coupled to a high-spin–low-spin conversion in the molecular system $[Fe^{III}(qsal)_2][Ni(dmit)_2]_3$ $CH_3CN$ $H_2O$. *Inorg. Chem.* **2006**, *45*, 5739–5741. [CrossRef] [PubMed]
18. Takahashi, K.; Cui, H.-B.; Okano, Y.; Kobayashi, H.; Mori, H.; Tajima, H.; Einaga, Y.; Sato, O. Evidence of the chemical uniaxial strain effect on electrical conductivity in the spin-crossover conducting molecular system: $[Fe^{III}(qnal)_2][Pd(dmit)_2]_5$ acetone. *J. Amer. Chem. Soc.* **2008**, *130*, 6688–6689. [CrossRef] [PubMed]
19. Dorbes, S.; Valade, L.; Real, J.A.; Faulmann, C. $[Fe(sal_2\text{-trien})][Ni(dmit)_2]$: Towards switchable spin crossover molecular conductors. *Chem. Commun.* **2005**, *1*, 69–71. [CrossRef] [PubMed]
20. Coronado, E.; Gimenez-Lopez, M.C.; Gimenez-Saiz, C.; Romero, F.M. Spin crossover complexes as building units of hydrogen-bonded nanoporous structures. *CrystEngComm* **2009**, *11*, 2198–2203. [CrossRef]
21. Steimecke, G.; Sieler, H.-J.; Kirmse, R.; Houer, E. 1.3-dithiol-2-thion-4.5 ditholat aus schwefelkohlenstoff und alkalimetall. *Phosphorus Sulfur Silicon Relat. Elem.* **1979**, *7*, 49–55. [CrossRef]
22. Kirmse, R.; Stach, J.; Dietzsch, W.; Steimecke, G.; Hoyer, E. Single-crystal EPR studies on nickel(III), palladium(III), and platinum(III) dithiolene chelates containing the ligands isotrithionedithiolate, o-xylenedithiolate, and maleonitriledithiolate. *Inorg. Chem.* **1980**, *19*, 2679–2685. [CrossRef]
23. Nweedle, M.F.; Wilson, L.J. Variable spin iron(III) chelates with hexadentate ligands derived from triethylenetetramine and various salicylaldehydes. Synthesis, characterization, and solution state studies of a new 2T-6A spin equilibrium system. *J. Am. Chem. Soc.* **1976**, *98*, 4824–4834.
24. Halcrow, M.A. Structure: Functional relationship in molecular spin-crossover complexes. *Chem. Soc. Rev.* **2011**, *40*, 4119–4142. [CrossRef] [PubMed]
25. Pritchard, R.; Barrett, S.A.; Kilner, C.A.; Halcrow, M.A. The influence of ligand conformation on the thermal spin transitions in iron(III) saltrien complexes. *Dalton Trans.* **2008**, 3159–3168. [CrossRef] [PubMed]
26. Blagov, M.A.; Krapivin, V.B.; Simonov, S.V.; Spitsyna, N.G. Insights into the influence of ethylene groups orientation on iron (III) spin state in the spin crossover complex $[Fe^{III}(Sal_2\text{-trien})]^+$. *Dalton Trans.* **2018**, submitted.
27. Benelli, C.; Gatteschi, D. *Introduction to Molecular Magnetism: From Transition Metals to Lanthanides*; Wiley-VCH Verlag GmbH & Co.: Weinheim, Germany, 2015.

28. Faulmann, C.; Szilágyi, P.; Jacob, K.; Shahene, J.; Valade, L. Polymorphism and its effects on the magnetic behaviour of the [Fe(sal$_2$-trien)][Ni(dmit)$_2$] spin-crossover complex. *New J. Chem.* **2009**, *33*, 1268–1276. [CrossRef]
29. Faulmann, C.; Dorbes, S.; Real, J.; Valade, L. Electrical conductivity and spin crossover: Towards the first achievement with a metal bis dithiolene complex. *J. Low Temp. Phys.* **2006**, *142*, 265–270.
30. Faulmann, C.; Jacob, K.; Dorbes, S.; Lampert, S.; Malfant, I.; Doublet, M.-L.; Valade, L.; Real, J. Electrical conductivity and spin crossover: A new achievement with a metal bis dithiolene complex. *Inorg. Chem.* **2007**, *46*, 8548–8559. [CrossRef] [PubMed]
31. Yuan, J.; Liu, M.-J.; Liu, C.-M.; Kou, H.-Z. Iron(III) complexes of 2-methyl-6-(pyrimidin-2-yl-hydrazonomethyl)-phenol as spin-crossover molecular materials. *Dalton Trans.* **2017**, *46*, 16562–16569. [CrossRef] [PubMed]
32. Milin, E.; Benaicha, B.; El Hajj, F.; Patinec, V.; Triki, S.; Marchivie, M.; Gómez-García, C.J.; Pillet, S. Magnetic bistability in macrocycle-based Fe$^{II}$ spin-crossover complexes: Counter ion and solvent effects. *Eur. J. Inorg. Chem.* **2016**, *34*, 5305–5314. [CrossRef]
33. Vieira, B.J.C.; Dias, J.C.; Santos, I.C.; Pereira, L.C.J.; Vasco, D.G.; Waerenborgh, J.C. Thermal hysteresis in a spin-crossover Fe$^{III}$ quinolylsalicylaldimine complex, Fe$^{III}$(5-Br-qsal)$_2$Ni(dmit)$_2$·solv: Solvent effects. *Inorg. Chem.* **2014**, *54*, 1354–1362. [CrossRef] [PubMed]
34. Bill, E. *JULX 1.4.1 Simulation of Molecular Magnetic Data Software*; Matrix Diagonalization Was Realized with the Routine 'zheev' from the LAPACK Numerical Package; Max-Planck Institute for Bioinorganic Chemistry: Mulheim/Ruhr, Germany, May 2008.
35. Nemec, I.; Herchel, R.; Ruben, I.M.; Linert, W. Anion driven modulation of magnetic intermolecular interactions and spin crossover properties in an isomorphous series of mononuclear iron(III) complexes with a hexadentate Schiff baseligand. *CrystEngComm* **2012**, *14*, 7015–7024. [CrossRef]

 © 2018 by the authors. Licensee MDPI, Basel, Switzerland. This article is an open access article distributed under the terms and conditions of the Creative Commons Attribution (CC BY) license (http://creativecommons.org/licenses/by/4.0/).

Article

# Solvent Effects on the Spin-Transition in a Series of Fe(II) Dinuclear Triple Helicate Compounds

Alexander R. Craze [1], Mohan M. Bhadbhade [2], Cameron J. Kepert [3], Leonard F. Lindoy [3], Christopher E. Marjo [2] and Feng Li [1,*]

1. School of Science and Health, Western Sydney University, Penrith, Sydney, NSW 2751, Australia; 17717986@student.westernsydney.edu.au
2. Mark Wainwright Analytical Centre, University of New South Wales, Sydney, NSW 2052, Australia; m.bhadbhade@unsw.edu.au (M.M.B.); c.marjo@unsw.edu.au (C.E.M.)
3. School of Chemistry, University of Sydney, Sydney, NSW 2006, Australia; cameron.kepert@sydney.edu.au (C.J.K.); len.lindoy@sydney.edu.au (L.F.L.)
* Correspondence: Feng.Li@westernsydney.edu.au

Received: 11 September 2018; Accepted: 19 September 2018; Published: 23 September 2018

**Abstract:** This work explores the effect of lattice solvent on the observed solid-state spin-transition of a previously reported dinuclear Fe(II) triple helicate series **1–3** of the general form $[Fe^{II}_2L_3](BF_4)_4(CH_3CN)_n$, where **L** is the Schiff base condensation product of imidazole-4-carbaldehyde with 4,4-diaminodiphenylmethane ($L^1$), 4,4′-diaminodiphenyl sulfide ($L^2$) and 4,4′-diaminodiphenyl ether ($L^3$) respectively, and **1** is the complex when $L = L^1$, **2** when $L = L^2$ and **3** when $L = L^3$ (Craze, A.R.; Sciortino, N.F.; Bhadbhade, M.M.; Kepert, C.J.; Marjo, C.E.; Li, F. Investigation of the Spin Crossover Properties of Three Dinuclear Fe(II) Triple Helicates by Variation of the Steric Nature of the Ligand Type. *Inorganics.* **2017**, *5* (4), 62). Desolvation of **1** and **2** during measurement resulted not only in a decrease in $T_{1/2}$ and completeness of spin-crossover (SCO) but also a change in the number of steps in the spin-profile. Compounds **1** and **2** were observed to change from a two-step 70% complete transition when fully solvated, to a single-step half complete transition upon desolvation. The average $T_{1/2}$ value of the two-steps in the solvated materials was equivalent to the single $T_{1/2}$ of the desolvated sample. Upon solvent loss, the magnetic profile of **3** experienced a transformation from a gradual SCO or weak antiferromagnetic interaction to a single half-complete spin-transition. Variable temperature single-crystal structures are presented and the effects of solvent molecules are also explored crystallographically and via a Hirshfeld surface analysis. The spin-transition profiles of **1–3** may provide further insight into previous discrepancies in dinuclear triple helicate SCO research reported by the laboratories of Hannon and Gütlich on analogous systems (Tuna, F.; Lees, M. R.; Clarkson, G. J.; Hannon, M. J. Readily Prepared Metallo-Supramolecular Triple Helicates Designed to Exhibit Spin-Crossover Behaviour. *Chem. Eur. J.* **2004**, *10*, 5737–5750 and Garcia, Y.; Grunert, C. M.; Reiman, S.; van Campenhoudt, O.; Gütlich, P. The Two-Step Spin Conversion in a Supramolecular Triple Helicate Dinuclear Iron(II) Complex Studied by Mössbauer Spectroscopy. *Eur. J. Inorg. Chem.* **2006**, 3333–3339).

**Keywords:** spin-crossover; dinuclear triple helicate; Fe(II); solvent effects

---

## 1. Introduction

Spin-crossover (SCO) materials continue to attract a wide degree of multidisciplinary research effort [1–5]. For example, these materials have been demonstrated to show significant promise for use in molecular switches and sensors [6–9]. Such applications stem from the inherent bistability of SCO compounds and the many ways with which this bistability can be altered. For octahedral Fe(II)-based materials, a spin-transition may be induced by temperature, pressure or light between the paramagnetic $^5T_2$ HS (S = 2) state and the diamagnetic $^1A_1$ LS (S = 0) state [2,10,11].

The ability of the spin-transition to be readily altered can be a strong benefit in SCO research, in that factors such as the transition temperature ($T_{1/2}$), the number of steps, the completeness of SCO and its abruptness can often be subtly manipulated. On the other hand, such lack of stability can also serve as a distinct problem for these materials when it comes to finding real-world applications. That so many external influences can alter the spin-transition of a compound makes the precise nature of fabricated SCO materials very difficult to predict and control. The extensive and varied effects that solvent(s) of crystallisation have on the properties of SCO materials is a prime example of this [12,13]. Thus, it has been extensively shown that the spin-transition can be significantly affected by the intermolecular interactions induced by solvent molecules [14–26]. Solvent molecules of crystallisation can affect the transition temperature ($T_{1/2}$) [26–36], the nature of the spin-transition as well as the degree of cooperativity and thermal hysteresis that occurs [12,37–40]. Previous studies have indicated that the elastic interactions between the SCO active metal centres can be enhanced by the formation of hydrogen-bonding networks between solvent and/or anion molecules, leading to abrupt and hysteric spin-transitions [14,15]. Furthermore, solvent molecules can impose different crystal packing arrangements within the lattice, affecting the cooperative interactions between metal centres. Severe structural transformations such as single-crystal-to-single-crystal transformations arising from the exchange of solvent molecules of crystallisation have also been demonstrated [1,41–46]. In some examples, solvent effects can even be reversibly triggered by desolvation and resolvation of the solid material [47–50]. The onset of an apparent hysteresis, in which irreversible changes in the profile are observed, are often induced by the loss of solvent molecules, a process that tends to stabilise the high-spin (HS) state [24,25,51–53]. As a result, there is potential for thermally induced spin-transitions to be 'tuned' by the adsorption and desorption of solvent molecules of crystallisation, leading to a potential application as chemosensors [20,50].

The subtleties of the effects of solvent molecules on the spin-transition phenomenon are highlighted in a study performed by Kruger and co-workers who presented a pair of solvatomorphs of a Fe(II) dinuclear triple helicate structure that demonstrated a half-transition when incorporating water molecules of crystallisation and a full asymmetric spin-crossover when acetonitrile molecules were present in the crystal lattice [51,54]. Two independent studies were performed by Garcia et al. [55] and Neville et al. [56] on two pseudopolymorphs of the complex $[Fe^{II}_2(A)_5(cis\text{-}NCS)_4]n$MeOH, where n = 4 and 2 respectively and **A** = *N*-salicylidene 4-amino-1,2,4-triazole. The first presented a monoclinic structure with a $T_{1/2}$ of 155 K, while the latter yielded a triclinic structure with no SCO taking place. Despite the formula of these two compounds differing by only two MeOH solvent molecules, and both polymorphs exhibiting π-π stacking interactions, vastly dissimilar magnetic properties were obtained.

Previously, we reported the SCO of three desolvated Fe(II) dinuclear triple helicate compounds **1–3**, that displayed single-step spin-transitions of ca. 50% completion [57]. Herein we report the effects of acetonitrile solvent molecules on the spin-transitions in this series of dinuclear triple helicates, which have the general formula $[Fe^{II}_2L_3](BF_4)_4(CH_3CN)_n$ and which differ in the steric nature of **L** (Figure 1). For compound **1** **L** = **L$^1$**, for **2** **L** = **L$^2$** and for **3** **L** = **L$^3$**. These compounds exhibit a change in the degree of completeness of SCO as well as a transfer from a two-step to a single-step spin-transition upon desolvation in the case of **1** and **2**. Conversely, **3** exhibits an alteration from a gradual SCO (or possibly exhibits weak antiferromagnetic interactions) to a single-step incomplete SCO upon desolvation. Furthermore, the spin-transition profiles observed for **1–3** may provide an interesting insight into a previously reported discrepancy in the literature concerning Fe(II) dinuclear triple helicates presented by both teams of Hannon and Gütlich [58,59].

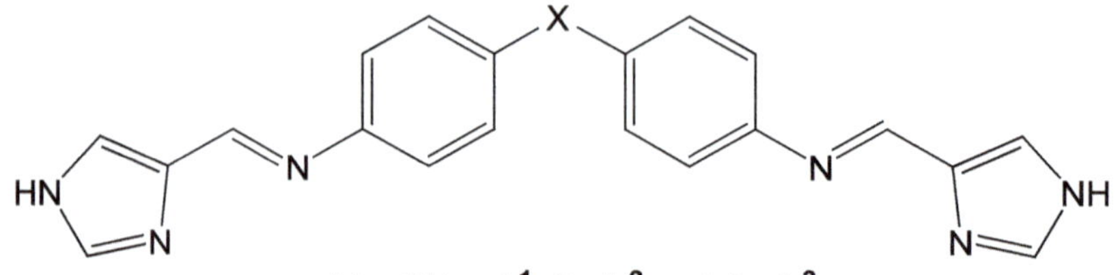

**Figure 1.** Schematic representation of $L^1$, $L^2$ and $L^3$ used to construct the dinuclear triple helicate architectures presented in this study.

## 2. Materials and Methods

All reagents and solvents were purchased from commercial sources (Sigma-Aldrich, Sydney, Australia and Alfa Aesar by Thermo Fisher Scientific, Australia and Tokyo Chemical Industry Co., Ltd., Adelaide, Australia), with no further purification being undertaken. Compounds **1–3** were prepared using the previously reported method [57]. Thermal gravimetric analysis (TGA) measurements were performed using a simultaneous thermal analysis (STA) 449 C Jupiter instrument (NETZSCH, Selb, Germany) using aluminium crucibles, nitrogen was used as both the protective and purge gas; the temperature range of 30–200 °C was cycled at a rate of 10 K·min$^{-1}$.

### 2.1. X-ray Crystallography

The X-ray crystallography experiments were performed on the MX1 beamline at the Australian Synchrotron (Clayton, Victoria, Australia.), using silicon double crystal monochromated radiation [60,61] or using a Bruker kappa-II CCD diffractometer, employing an IµS Incoatec Microfocus Source with Mo-Kα radiation (λ = 0.710723 Å). Data integration and reduction was undertaken with XDS [62] for synchrotron data and with APEX2. Ver. 2014.11-0 and SAINT. Ver. 8.34A (Bruker-AXS, Madison, WI, USA, 2014) [63] for the home source instrument. An empirical absorption correction was then applied using *SADABS* at the Australian Synchrotron [64]. The structures were solved by direct methods and the full-matrix least-squares refinements were carried out using a suite of *SHELX* programs (XT. Version 2014/4., XL. Version 2014/7.) [65,66] via the *OLEX2* graphical interface [67]. Non-hydrogen atoms were refined anisotropically. Carbon-bound hydrogen atoms were included in idealised positions and refined using a riding model. The crystallographic data in CIF format have been deposited at the Cambridge Crystallographic Data Centre with CCDC 1844792-1844794. It is available free of charge from the Cambridge Crystallographic Data Centre, 12 Union Road, Cambridge CB2 1 EZ, UK; fax: (+44) 1223-336-033; or e-mail: deposit@ccdc.cam.ac.uk. Specific refinement details and crystallographic data for each structure are presented below and in the supporting information.

Powder X-Ray diffraction measurements were conducted on a Bruker D8 ADVANCE diffractometer with a LynxEye position sensitive detector (PSD). The X-ray source was a Copper K-α1 at 1.54 Å at 40 kV and a current of 40 mA. The sample scan range was 5–55 degrees 2θ with a step size of 0.02° at a rate of 2 s per step. Data processing was conducted using Bruker's EVA software.

### 2.2. Magnetic Susceptibility Measurements

Samples of crystalline material, **1–3**, were measured under two separate conditions. First, a solvated sample was sealed in a plastic tube in the presence of mother liquid, with the temperature range reaching a maximum of 300 K to ensure as little solvent loss as possible during the experiment. Secondly, a filtered sample was placed in an unsealed magnetic sample holder, and the temperature ramped to a maximum of 350 K, allowing desolvation to occur. Data for magnetic susceptibility measurements were collected using a Quantum Design Versalab Measurement System

(Quantum Design, San Diego, CA, USA) with a vibrating-sample magnetometer (VSM) attachment. Measurements were taken continuously under an applied field of 0.5 T, at a heating rate of 4 K·min$^{-1}$.

*2.3. Hirshfeld Surfaces*

The Hirshfeld isosurfaces were calculated using the program Crystal Explorer v17.5 [68].

## 3. Results and Discussion

Magnetic susceptibility studies on polycrystalline samples of **1–3** between 50–350 K demonstrate that these compounds exhibit solvent-dependant SCO (Figure 2). The samples were measured under two separate conditions. First, the solvated solid sample was placed in a plastic tube which was then sealed with some mother liquid. The temperature was increased to a maximum of 300 K to ensure as little solvent loss as possible from the sample. Secondly, a filtered sample was placed in an unsealed magnetic sample holder and the temperature cycled to a maximum of 350 K, allowing solvent loss. TGA experiments were performed on filtered samples in order to provide evidence of solvent loss within the applied temperature range (30–200 °C) and compare the mass changes due to solvent loss. The filtered samples showed mass losses of 3.33%, 4.99% and 4.73% for **1**, **2** and **3** respectively when heated to 200 °C (Figures S4–S6). These mass changes demonstrate that the percentage of weight lost due to desolvation closely matches the mass percentage of acetonitrile present in the asymmetric unit (mass of acetonitrile/total mass × 100) at 100 K calculated from single crystal X-ray diffraction results (for the measurements TGA%/SCXRD%—**1**—3.33/3.92, **2**—4.99/4.95, **3**—4.73/5.01).

When sealed, **1** underwent a more complete two-step thermally induced spin-transition. The room temperature $\chi_M T$ value was 7.70 cm$^3$·K·mol$^{-1}$, corresponding to two uncoupled Fe(II) centres in the HS $^5T_2$ (S = 2) state. Upon cooling, the $\chi_M T$ value decreased steadily, reaching the first $T_{1/2\downarrow}$ at 205 K. After a minimum rate of change (first derivative) in the magnetic susceptibility at 138 K, the second $T_{1/2\downarrow}$ occurred at 100 K, after which the susceptibility dropped to a value of 2.13 cm$^3$·K·mol$^{-1}$ at 50 K. At this point, approximately 72% of Fe(II) ions have transitioned to the LS $^1A_1$ (S = 0) state.

On the other hand, when the sample was run in an unsealed magnetic sample holder and taken to 350 K, allowing the loss of solvent to occur, major changes in the transition profile were observed. In the heating mode, the room temperature magnetic susceptibility remained around 7.70 cm$^3$·K·mol$^{-1}$ and the minimum value was 4.55 cm$^3$·K·mol$^{-1}$ at 50 K, corresponding to only 40% of the Fe(II) metal centres undergoing a transition to the LS $^1A_1$ state at 50 K. The transition occurred in a single-step manner and the $T_{1/2\uparrow}$ was 180 K. In the cooling mode, loss of solvent experienced during the heating mode causes the transition to appear slightly more abrupt in manner, reaching a minimum $\chi_M T$ value of 4.40 cm$^3$·K·mol$^{-1}$ at 50 K with a $T_{1/2\downarrow}$ of 155 K.

Similarly, the sealed sample of **2** exhibited a two-step spin-transition, decreasing from a $\chi_M T$ value at 300 K of 7.55 cm$^3$·K·mol$^{-1}$ to 2.17 cm$^3$·K·mol$^{-1}$ at 50 K, with $T_{1/2\downarrow}$ of 219 and 135 K and $T_{1/2\uparrow}$ of 125 and 209 K upon thermal cycling. Again, around 72% of Fe(II) centres present in the material have undergone a spin-transition at this point. When the sample holder was not sealed, the spin transition profile exhibited a very similar trend to that of **1**, in which only 50% of the Fe(II) metal centres had undergone a transition to the LS state at 50 K in a single-step manner.

The sealed sample of **3** displayed a room temperature $\chi_M T$ value of 7.70 cm$^3$·K·mol$^{-1}$. This decreases in a steady fashion to 4.27 cm$^3$·K·mol$^{-1}$ at 50 K. The gradual, monotonic decrease in $\chi_M T$ may be interpreted as due to temperature-independent paramagnetism (TIP), weak antiferromagnetic coupling between HS Fe(II) centres or possibly very gradual spin-crossover [69]. On the other hand, the unsealed sample of **3** showed a similar shape to that of **1** and **2** under the same conditions. The heating mode exhibited a 50 K $\chi_M T$ value of 3.76 cm$^3$·K·mol$^{-1}$, which steadily increased over the range of 50 K to 350 K, at which point the magnetic susceptibility was 7.65 cm$^3$·K·mol$^{-1}$, demonstrating an inflexion point ($T_{1/2}$) at 182 K. At 50 K the spin-transition is incomplete, with ca. half of the Fe(II) centres remaining HS. The proceeding cooling mode demonstrated similar room temperature and 50 K

$\chi_M T$ values, although it occurred in a seemingly more abrupt fashion, again due to loss of solvent, with a $T_{1/2}$ value of 140 K, 42 K lower than that of the preceding heating mode.

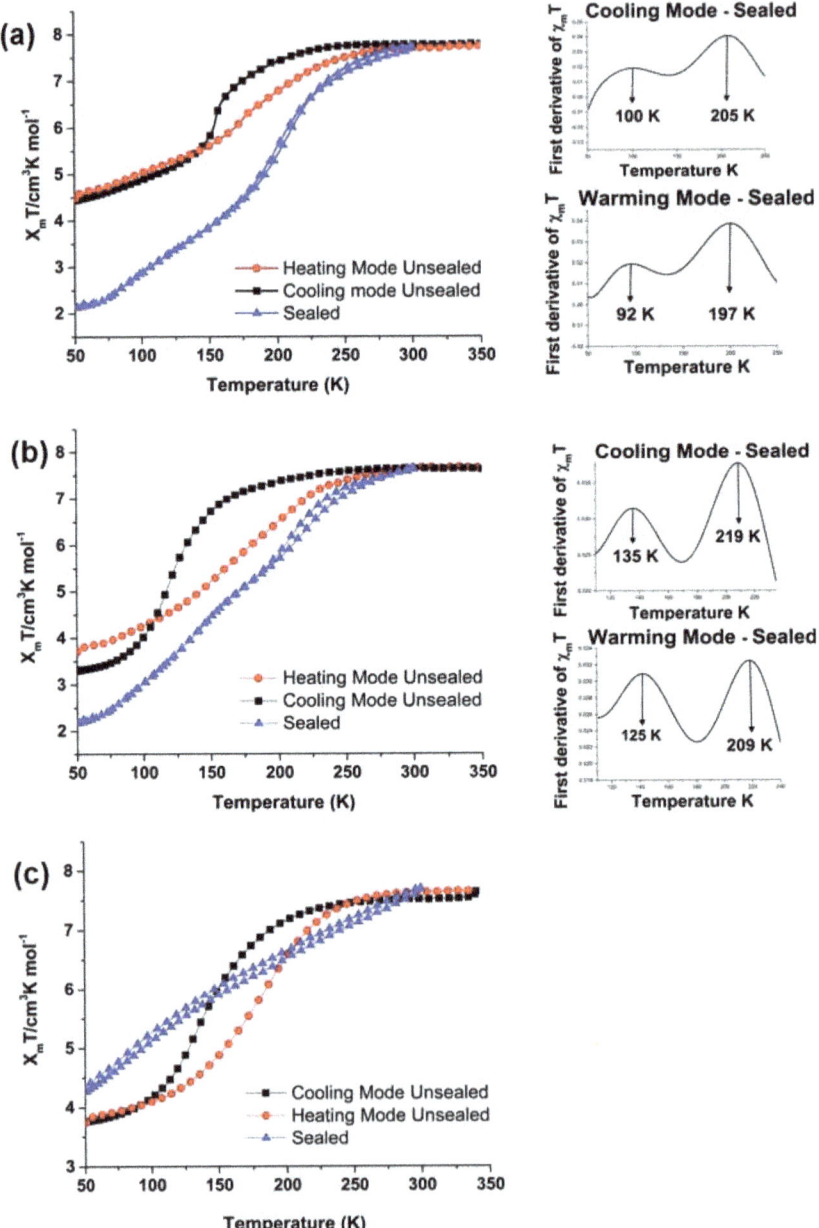

**Figure 2.** Magnetic susceptibility $\chi_M T$ versus $T$ plots for **1–3**, (**a–c**) respectively, at a scan rate of 4 K/min over the temperature range of 50–300 K (solvated) and 50–350 K (desolvated). Inserts for **1** and **2** display the rate of change (first derivative) of the $\chi_M T$ value in the sealed samples.

For **1** and **2**, the sealed samples both demonstrated a more complete (70%) two-step transition, whilst the unsealed samples of all three compounds demonstrated a one-step half-completed SCO. For all three compounds, the loss of solvent in the unsealed samples results in a shift in the $T_{1/2}$ to lower temperatures, which has also been the case in other solvent sensitive materials. A study performed by Kruger and co-workers [51] on a 2-positioned methylated imidazole donor helicate ($ClO_4^-$ salt), an architecture similar to **3**, showed an analogous trend to those obtained for unsealed samples of **1**, **2** and **3**. Here the $T_{1/2}$ moved to a lower temperature with solvent loss (water molecules in this case) and a more abrupt spin-transition occurred. Although on the other hand, the extent of the HS↔LS conversion (completeness) and the nature of the transition-profile (the number of steps) remained the same.

*Magneto-structural Correlations*

The single-crystal X-ray diffraction results at 100 K for **1–3** were presented in our previous report of the magnetic properties exhibited by completely dried samples [57]. Here, variable temperature single-crystal structures are presented in order to further explore the magneto-structural characteristics of these compounds. The structures of **1** and **3** at 298 K as well as **2** at 155 K were obtained in an attempt to further probe the spin-transitions occurring in these compounds. Desolvated samples were not suitable to be measured by single-crystal diffraction. Furthermore, powder X-ray diffraction (PXRD) patterns presented in the supporting information confirm that the synthesised materials retain high crystallinity at room temperature and show good correlation with single-crystal X-ray diffraction (SCXRD) data suggesting that phase purity is maintained at the compared temperatures (Figures S1–S3).

For a full structural description of **1** at 100 K, for which a solvatomorph was previously reported by Hannon and co-workers [58], see our previous study mentioned above [57]; a crystallographic summary of the variable temperature SCXRD analysis of **1–3** can be found in Table 1 below and further crystallographic details can be found in the supporting information (Tables S1–S3). At 298 K, **1** presents triclinic symmetry crystallising in the space group $P\bar{1}$. Hydrogen-bonding between $BF_4^-$ counter ions and the N-H donors of the non-coordinating 2,4-imidazole nitrogen atom is present for four of the six imidazole moieties of the helicate architecture. Similar to the 100 K structure, no supramolecular network of intermolecular interactions connects adjacent helicates within the crystal lattice. Crystallographic parameters at 298 K ($\Sigma$ = Fe01-84.84 and Fe02-84.90°, and average Fe-N = Fe01-2.21 and Fe02-2.19 Å) are in accord with the magnetic susceptibility measurements, indicating that the Fe(II) centres are present in the [HS-HS] state at room temperature. The two-step nature of the spin-transition for the solvated sample may be a result of the observed monoclinic to triclinic single-crystal-to-single-crystal symmetry breaking that occurs between 100 and 298 K.

Single-crystals of **2** were found to be of triclinic $P\bar{1}$ symmetry at 155 K, crystallising in the same space group as at 100 K. Hydrogen-bonding between imidazole N-H, $BF_4^-$ anions and acetonitrile solvent molecules connects adjacent helicates from both ends along the crystallographic *a*-axis, so as to arrange neighbouring helicates in a side-on manner (Figure 3). Both acetonitrile molecules present in the asymmetric unit interact through hydrogen-bonding ($CH_3CN \cdots$ H-N) with the non-coordinating nitrogen of the imidazole moiety, with contact lengths of 2.91 and 2.89 Å. The loss of such interactions in the desolvated material may, therefore, contribute to destabilisation of the [LS-LS] state. The measured octahedral distortions ($\Sigma$) of 60.7 and 92.68° for Fe01 and Fe02 respectively, in conjunction with average Fe-N coordination bond lengths of 2.00 and 2.18 Å for Fe01 and Fe02, confirm the results of magnetic susceptibility measurements for the solvated sample, and suggest that the helicates are present as the [LS-HS] spin-isomer at this temperature. Single-crystal measurements of **2** could not be obtained at higher temperatures. Room temperature PXRD indicates that at 155 K and 298 K the material is of the same phase, with no change in symmetry occurring despite the two-step magnetic susceptibility profile (Figure S2).

Alternatively, in contrast to the 100 K structure which crystallises in the triclinic space group $P\bar{1}$, at 298 K, **3** crystallises in the monoclinic space group $C2/c$. Tetrafluoroborate anions participate in hydrogen-bonding with each of the six helicate imidazole moieties, forming lengthwise end-to-end intermolecular contacts between adjacent helicates along the crystallographic c-axis (Figure 4). Acetonitrile solvent molecules of 0.5 and 0.25 site occupancy are present in the unit cell but do not participate in any intermolecular interactions. The degree of octahedral distortion was 87.81° and the average Fe-N distance was 2.20 Å, which in combination with magnetic measurements is in accord with the Fe(II) centres being in the [HS-HS] state at 298 K.

In the solvated samples of **1** and **2**, a more complete, two-step spin-transition occurs from the [HS-HS] state to the [LS-LS] state. Loss of solvent molecules from the crystal lattice was observed to destabilise the [LS-LS] state, trapping the material in a state of either [LS-HS] helicates or a 50:50 mixture of [HS-HS] and [LS-LS] compounds. This suggests that the solvent plays an integral role in the transition of Fe(II) centres from [HS-HS]↔[LS-LS] in these compounds. In contrast, the loss of acetonitrile from the lattice of **3** induces an incomplete spin-transition, while the solvated sample only gave a gradual and linear change in $\chi_M T$ with temperature, rather than the typical sigmoidal shape indicative of SCO. The desolvation of **3**, acts as a form of 'on-switch' for spin-crossover in this compound.

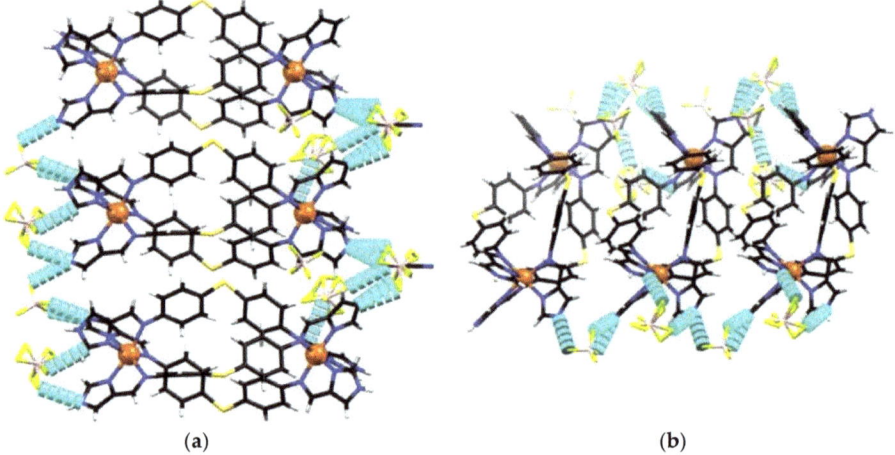

**Figure 3.** Schematic representations of the crystal packing and hydrogen-bonding (N-H··· $BF_4^-$) interactions present in the single-crystal structure of **2** at 155 K, demonstrating the manner with which adjacent helicates are connected in a side-on manner along the a-axis. (**a**) Projection down the crystallographic b-axis and (**b**) down the c-axis. Solvent molecules have been excluded for clarity. Hydrogen bonds are represented by thickened blue dotted lines.

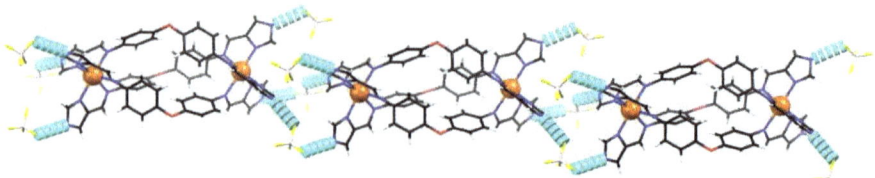

**Figure 4.** Schematic representation of the crystal packing arrangement of **3** at 298 K, demonstrating the network of hydrogen-bonding interactions that connect the helicates lengthwise along the crystallographic c-axis.

As illustrated in Figure 5, other parameters that have been used in this study to rationalise the role of intermolecular interactions and intramolecular ligand distortions in the SCO of these compounds, are the angles $\theta_{intermolecular}$ and $\theta_{intramolecular}$. In compounds **1–3**, there are three angles connecting the three interior imine nitrogen donors and the Fe(II) centre, and another three connecting the exterior imidazole nitrogen donors and the Fe(II) centre. The interior angles represent intramolecular distortions, while the exterior angles represent intermolecular distortions. These pairs of three exterior and interior angles (φ) are each individually subtracted from 90 and the absolute values are summed to give one distortion value for each pair ($\theta_{intermolecular}$ and $\theta_{intramolecular}$ in Table 2). These parameters were used to further document the effects of intermolecular interactions and intramolecular restraints on the distortion of the SCO coordination sphere.

In all three compounds, differences between the exterior distortions ($\Delta\ \theta_{intermolecular}$) of the HS and LS/MS Fe(II) centres are consistently more severe than those of the interior distortions ($\Delta\ \theta_{intramolecular}$). In other words, the largest geometric difference between the two centres of opposite spin is at the exterior of the helicate, indicating either that steric restraints imposed by the dinuclear triple helicate architecture are most severe at the interior of the molecule, or that intermolecular interactions may have an important role in accommodating the distortions required to reach the [LS-LS] state. Subsequently, the loss of solvent molecules upon desolvation may hinder the possibility of these necessary molecular rearrangements occurring, as observed in the stabilisation of the [LS-HS] state upon desolvation in both **1** and **2**.

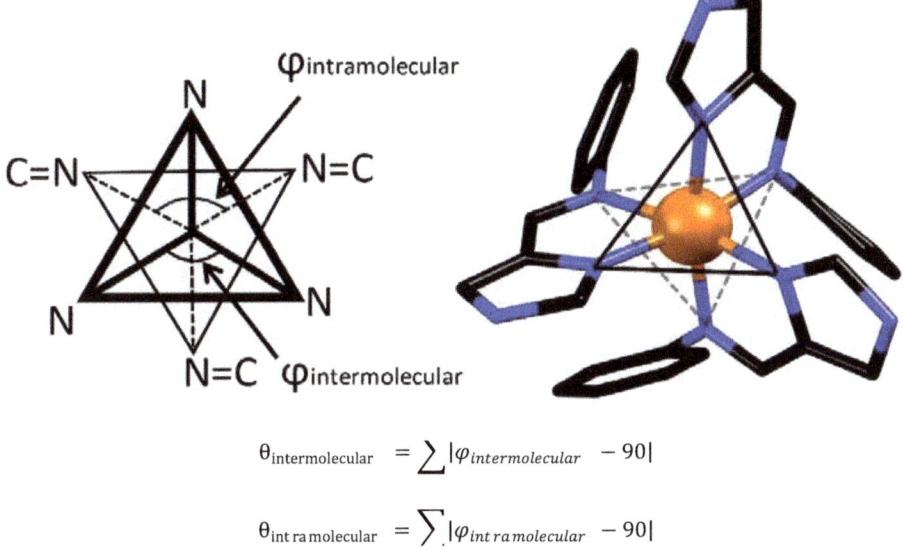

$$\theta_{intermolecular} = \sum |\varphi_{intermolecular} - 90|$$

$$\theta_{intramolecular} = \sum |\varphi_{intramolecular} - 90|$$

**Figure 5.** Schematic representation of the two geometric parameters $\theta_{intermolecular}$ and $\theta_{intramolecular}$ discussed in the text.

**Table 1.** Selected crystallographic parameters for **1**, **2** and **3**. MS represents a mixed HS/LS state population.

| | 1 | | 2 | | | 3 | |
|---|---|---|---|---|---|---|---|
| $\Sigma$ | 100 K<br>Fe01-76.3 | 298 K<br>Fe01-84.8<br>Fe02-84.9 | 100 K<br>Fe01-59.4<br>Fe02-90.3 | 155 K<br>Fe01-60.4<br>Fe02-92.7 | 100 K<br>Fe01-77.2<br>Fe02-85.2 | | 298 K<br>Fe01-87.5 |
| Average Fe(II)-N distance (Å) | Fe01-2.13 | Fe01-2.21<br>Fe02-2.19 | Fe01-2.00<br>Fe02-2.18 | Fe01-2.00<br>Fe02-2.18 | Fe01-2.10<br>Fe02-2.18 | | 2.20 |
| Spin state of Fe(II) | MS | HS-HS | LS-HS | LS-HS | HS-MS | | HS |
| $\theta_{intermolecular}/\theta_{intramolecular}$ 100 K | Fe01-15.4/15.3 | Fe01-23.8/20.6<br>Fe02-18.6/16.4 | Fe01-1.6/21.0<br>Fe02-6.1/23.9 | Fe01-1.75/23.407<br>Fe02-6.731/26.834 | Fe01-19.6/11.1<br>Fe02-14.8/13.1 | | 20.6/10.7 |
| Space group | C2/c | $P\bar{1}$ | $P\bar{1}$ | $P\bar{1}$ | $P\bar{1}$ | | C2/c |
| Intermolecular interactions | 6 × N-H ⋯ $BF_4^-$<br>$F_3BF \cdots H\text{-}CH_2C\text{-}N$<br>No supramolecular network present | 4 × N-H ⋯ $BF_4^-$<br>No supramolecular network present | 2 × N-H ⋯ $CH_3CN$<br>4 × N-H ⋯ $BF_4^-$<br>Form side-ways chain | 2 × N-H ⋯ $CH_3CN$<br>4 × N-H ⋯ $BF_4^-$<br>Form side-ways chain | 6 × N-H ⋯ $BF_4^-$<br>Form length-wise chain | | 6 × N-H ⋯ $BF_4^-$<br>Form length-wise chain |
| Number of intramolecular edge-to-face π interactions | 3 | 2 | 3 | 3 | 2 | | 2 |
| Number of acetonitrile solvent molecules | 1.5 | 1.25 | 2 | 2 | 2 | | 0.75 |
| C-X-C angle (where X = $CH_2$, S or O) | 113.6 | 115.4 | 104.9 | 105.1 | 115.8 | | 116.2 |
| Intrahelical-separation (Å) | 11.72 | 11.72 | 11.78 | 11.77 | 11.62 | | 11.72 |

Source: Note: the three 100 K structures have been reported previously [57].

**Table 2.** Comparison of the internal ($\theta_{intramolecular}$) and external ($\theta_{intermolecular}$) distortions of the Fe(II) coordination environment in the different spin-states observed for **1**, **2** and **3**. MS denotes a mixed HS/LS-state population. The $\Delta\theta$ values represent the differences between the LS (or MS) and HS $\theta$ values in each structure.

| Compound | Spin-State | $\theta_{intermolecular}$ | $\Delta\,\theta_{intermolecular}$ | $\theta_{intramolecular}$ | $\Delta\,\theta_{intramolecular}$ |
|---|---|---|---|---|---|
| 1 | MS | 15.439 | 8.347 | 15.252 | 5.418 |
|   | HS | 23.786 |       | 20.61  |       |
| 2 | LS | 1.57   | 4.554 | 21.047 | 2.847 |
|   | HS | 6.124  |       | 23.894 |       |
| 3 | MS | 19.55  | 4.8   | 11.05  | 2.079 |
|   | HS | 14.75  |       | 13.129 |       |

The role of these intermolecular interactions can be visualised utilising Hirshfeld surfaces [21,70]. Any electron density within the isosurface predominantly consists of the contribution of the considered molecule, while that outside the surface is dominated by the remainder of the crystal lattice. The parameter $d_{norm}$ is useful in visualising significant intermolecular interactions and is composed of two parameters that describe the distance an atom is from the isosurface, $d_i$, if the atom is inside the surface, and $d_e$ if the atom is outside [71]. The parameter $d_{norm}$ returns a zero value when the sum of $d_i$ and $d_e$ equates to the sum of the van der Waals radii of the atoms in question. Strong intermolecular interactions are represented as red areas on the Hirschfield surface, and signify regions in which the value of $d_{norm}$ is negative and the sum of $d_i$ and $d_e$ is less than the sum of the van der Waals radii. The Hirshfeld plots shown in Figure 6 are calculated for **2** at 100 K using the program Crystal Explorer [68]. The Hirshfeld surface of **2** displays red regions of the isosurface representing the hydrogen-bonding at the imidazole N-H donor site between acetonitrile solvent molecules and $BF_4^-$ anions respectively. When the isosurface is analysed with respect to the crystal packing of **2** (Figure 7), it can be observed that the anions and solvent molecules form a network of hydrogen-bonding along this row of helicates (*a*-axis). The helicates orientate themselves in the lattice so as to offset the regions of strongest intermolecular interactions relative to their neighbouring molecules. These surfaces demonstrate the role of such interactions in providing a network of intermolecular contacts between adjacent helicates, and, as a result, how the loss of the solvent molecules of crystallisation in the desolvated samples results in the destabilisation of the [LS-LS] state.

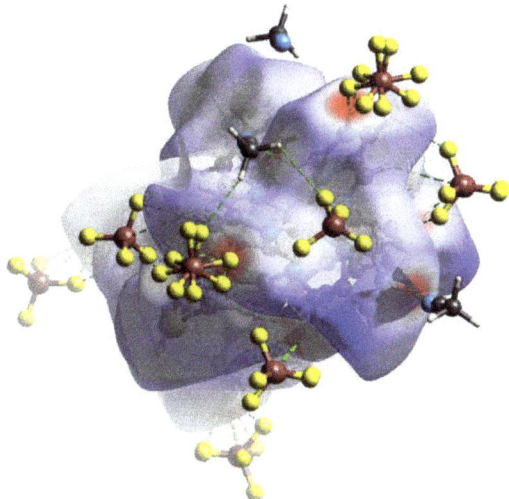

**Figure 6.** Hirshfeld surface of **2** at 100 K, showing the strongest intermolecular contacts in red located at the three external imidazole N-H groups at each end of the helicate, demonstrating the interactions between helicates, solvent and counterions.

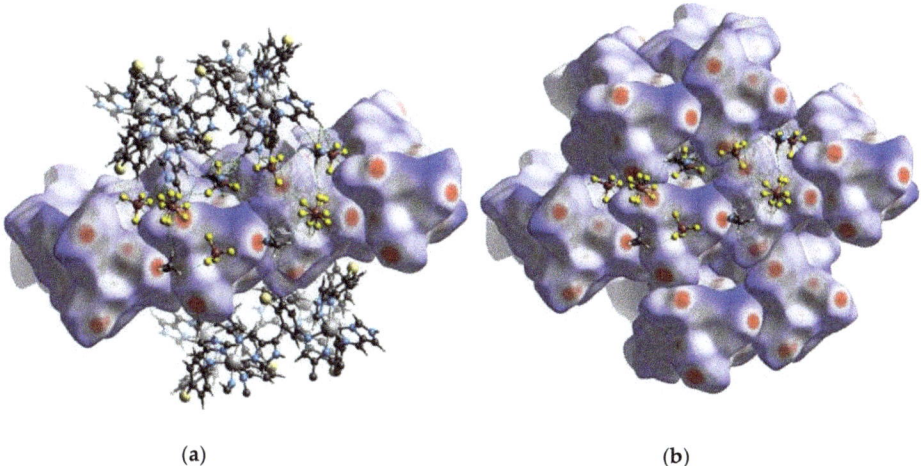

(a)            (b)

**Figure 7.** Hirshfeld surface within the crystal lattice of **2**, highlighting the interactions between the helicates, solvent and counter ions that connect adjacent helicates along the crystallographic *a*-axis. The strongest intermolecular contacts are denoted by surface regions shown in red. (**a**) reveals two rows of helicates with no isosurface (top and bottom) for clarity, while (**b**) depicts these surfaces.

The change in octahedral distortion ($\Sigma$) between the LS and HS centres in helicates **1**–**3** is around 30° (Table 1), and when considered over two Fe(II) centres in the one compound, this places quite a significant strain on each semi-rigid helicate architecture. The overlap figure below (Figure 8) shows **3** at 100 K and 298 K, and demonstrates the significant change in helicate architecture between the [HS-HS] and [LS-HS] structures, which may also help to rationalise the occurrence of the incomplete spin-transition observed. The majority of SCO dinuclear triple helicates reported previously exhibit such incompleteness, and, to the best of our knowledge, only the systems of Kruger and co-workers [54,72] as well as our laboratory [73] display a full transition of the two Fe(II) sites of the dinuclear triple helicate architecture. The distortion required of the two spin-centres in semi-rigid dinuclear triple helicates makes a complete [HS-HS]↔[LS-LS] transition difficult to achieve. That is, these compounds may be trapped in the [LS-HS] state by intramolecular steric constraints. Although, the presence of solvent molecules in **1** and **2** results in a two-step 70% complete spin-transition that does partially access the [LS-LS] state. Therefore, intermolecular interactions mediated by solvent molecules may influence the ability of **1** and **2** to achieve SCO at the second Fe(II) centre and to access a [LS-LS] state.

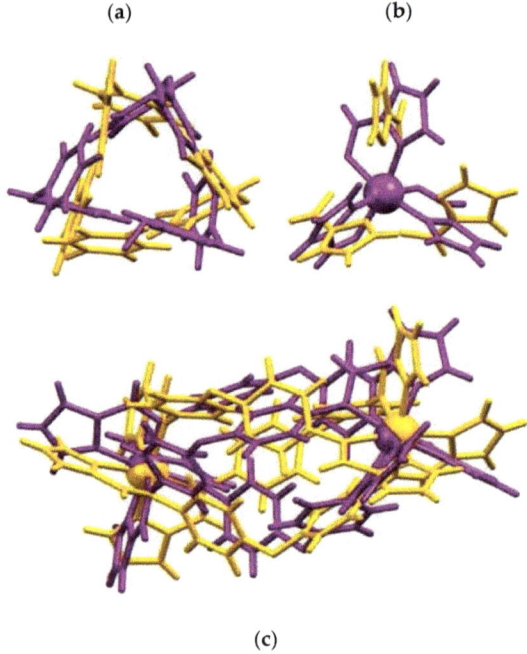

**Figure 8.** Schematic showing the overlapping representations of **3** at 100 K (purple) and 298 K (orange), illustrating the severity of distortion occurring between the [LS-HS] and [HS-HS] spin-isomers. (**a**) focuses on the change in the internal ring sections of the helicate ligands, (**b**) the external imidazoleimine groups, and (**c**) shows the overall change in conformation relative to the Fe⋯Fe positions.

Hannon and co-workers conducted a study of a complex using the same helicate architecture as **1**, in which they investigated the effect of different counter ions, namely $PF_6^-$, $BF_4^-$ and $ClO_4^-$, being present [58]. The $[Fe_2(L)_3][ClO_4]_4$ structure they obtained displayed a gradual, incomplete single-step transition like the desolvated samples in our previous study [57] and also the unsealed samples in this study. A later study by Gütlich and co-workers further explored the $ClO_4^-$ salt of this compound that led to an in-depth analysis of its magnetic behaviour and Mössbauer spectra [59]. Interestingly, these latter workers found a two-step spin-transition, where the average transition temperature of around 180 K corresponded to the $T_{1/2}$ of the gradual spin-transition presented by Hannon. The data presented by Gütlich and co-workers was cycled between 300–1.8 K, while that of Hannon and co-workers was cycled between 340–1.8 K, with acetonitrile being the solvent of crystallisation in each case. Heating to 300 K may not be sufficient for significant loss of solvent of crystallisation, whereas heating to 340 K would likely be sufficient, in accord with our TGA analysis of the $BF_4^-$ analogue, **1** (Figures S4–S6). In other words, it appears that the sample of Hannon and co-workers may have desolvated, while that of Gütlich remained solvated to a greater extent. In the present study of a $BF_4^-$ analogue of those compounds just mentioned, **1**, together with two similar helicates, **2** and **3**, differing in the steric nature of their ligands (**L**), we found that both **1** and **2** behaved in a similar manner to that described by Hannon and Gütlich in combination. That is, when the helicate samples are completely solvated, they exhibit a gradual two-step spin-crossover, while conversely, when each sample is heated and desolvated, each exhibits a gradual single-step transition where the average $T_{1/2}$ of the two-step transition is very close to the $T_{1/2}$ of the single-step transition present in the desolvated sample. This highlights the importance of solvent effects on SCO in the solid-state and further highlights the effects of sample preparation and experimental procedure on the observed spin-transition.

## 4. Conclusions

For all three compounds investigated, the partial desolvation of the sample during the magnetic measurements resulted in a decreased $T_{1/2}$ and a change in the completeness of transition. Compounds **1** and **2** demonstrated a two-step 70% complete profile when solvated and a single-step half-complete profile when desolvated. In this way, the solvent molecules of crystallisation may help to partially access the [LS-LS] state in **1** and **2**. Solvated samples of **3** displayed a monotonic decrease in magnetic susceptibility with temperature, indicative of either a gradual SCO or antiferromagnetic interactions between Fe(II) centres, while the desolvated samples exhibited an approximately 50% complete gradual spin-transition. The results may provide some insight into previous differences reported in the literature for dinuclear Fe(II) triple helicates by the groups of Hannon and Gütlich, as well as demonstrating the impact of solvent molecules of crystallisation on the SCO in these systems. The study also serves to further highlight the importance of sample preparation and experimental procedure when undertaking magnetic susceptibility measurements with solvent sensitive materials.

**Supplementary Materials:** The following are available online at http://www.mdpi.com/2073-4352/8/10/376/s1, Figure S1: PXRD pattern of 1, with the 298 K PXRD pattern in black on top and simulated spectrum from 298 K SCXRD data in red on the bottom; Figure S2: PXRD pattern of 2, with the 298 K PXRD pattern in black on top and simulated spectrum from 155 K SCXRD data in red on the bottom. No 298 K SCXRD structure could be obtained; Figure S3. PXRD pattern of 3, with the 298 K PXRD pattern in black on top and simulated spectrum from 298 K SCXRD data in red on the bottom; Figure S4. TGA analysis of filtered samples of 1; Figure S5. TGA analysis of filtered samples of 2; Figure S6. TGA analysis of filtered samples of 3; Table S1. Crystal data and structure refinement for 1 at 298 K; Table S2. Crystal data and structure refinement for 2 at 155 K; Table S3. Crystal data and structure refinement for 3 at 298 K.

**Author Contributions:** A.R.C. performed the synthetic, experimental, single crystal X-ray and magnetic studies. C.J.K. assisted in the magnetic studies. M.M.B. and C.E.M. assisted with crystallographic studies. A.R.C., L.F.L. and F.L. prepared the manuscript, and F.L. directed the work.

**Funding:** This research received no external funding.

**Acknowledgments:** The research described herein was supported by the Western Sydney University (WSU). The authors acknowledge Richard Wuhrer and the Western Sydney University Advanced Materials and Characterisation facility. The crystallographic data was undertaken on the MX1 beamline of the Australian Synchrotron, Clayton, Victoria, Australia. We thank the Australian Nuclear Science and Technology Organisation (ANSTO) and the Australian Synchrotron for travel support and their staff for assistance. Furthermore, A.R.C. acknowledges the Western Sydney University Masters of research scholarship and AINSE honours scholarship programs.

**Conflicts of Interest:** The authors declare no conflict of interest.

## References

1. Gütlich, P.; Hauser, A.; Spiering, H. Thermal and optical switching of Iron(II) complexes. *Angew. Chem. Int. Ed.* **1994**, *33*, 2024–2054. [CrossRef]
2. Bousseksou, A.; Molnár, G.; Salmon, L.; Nicolazzi, W. Molecular spin crossover phenomenon: recent achievements and prospects. *Chem. Soc. Rev.* **2011**, *40*, 3313–3335. [CrossRef] [PubMed]
3. Murray, K.S. The development of spin-crossover research. In *Spin Crossover Materials: Properties and Applications*; John Wiley & Sons Ltd.: Oxford, UK, 2013.
4. Halcrow, M.A. Structure:function relationships in molecular spin-crossover materials. In *Spin-Crossover Materials: Properties and Applications*; John Wiley & Sons Ltd.: Oxford, UK, 2013.

5. Kumar, K.S.; Ruben, M. Emerging trends in spin crossover (SCO) based functional materials and devices. *Coord. Chem. Rev.* **2017**, *346*, 176–205. [CrossRef]
6. Miyamachi, T.; Gruber, M.; Davesne, V.; Bowen, M.; Boukari, S.; Joly, L.; Scheurer, F.; Rogez, G.; Yamada, T.K.; Ohresser, P.; et al. Robust spin crossover and memristance across a single molecule. *Nat. Commun.* **2012**, *3*, 938. [CrossRef] [PubMed]
7. Lefter, C.; Davesne, V.; Salmon, L.; Molnár, G.; Demont, P.; Rotaru, A.; Bousseksou, A. Charge transport and electrical properties of spin crossover materials: Towards nanoelectronic and spintronic devices. *Magnetochemistry* **2016**, *2*, 18. [CrossRef]
8. Frisenda, R.; Harzmann, G.D.; Celis Gil, J.A.; Thijssen, J.M.; Mayor, M.; van der Zant, H.S.J. Stretching-induced conductance increase in a spin-crossover molecule. *Nano Lett.* **2016**, *16*, 4733–4737. [CrossRef] [PubMed]
9. Harzmann, G.D.; Frisenda, R.; van der Zant, H.S.J.; Mayor, M. Single-molecule spin switch based on voltage-triggered distortion of the coordination sphere. *Angew. Chem. Int. Ed.* **2015**, *54*, 13425–13430. [CrossRef] [PubMed]
10. Marchivie, M.; Guionneau, P.; Létard, J.-F.; Chasseau, D. Towards direct correlations between spin-crossover and structural features in Iron(II) complexes. *Acta Cryst. B* **2003**, *59*, 479–486. [CrossRef]
11. Gütlich, P.; Garcia, Y.; Goodwin, H.A. Spin crossover phenomena in Fe(II) complexes. *Chem. Soc. Rev.* **2000**, *29*, 419–427. [CrossRef]
12. Hostettler, M.; Törnroos, K.W.; Chernyshov, D.; Vangdal, B.; Bürgi, H.-B. Challenges in engineering spin crossover: structures and magnetic properties of six alcohol solvates of Iron(II) tris(2-picolylamine) dichloride. *Angew. Chem. Int. Ed.* **2004**, *43*, 4589–4594. [CrossRef] [PubMed]
13. Kahn, O.; Martinez, C.J. Spin-transition polymers: From molecular materials toward memory devices. *Science* **1998**, *279*, 44–48. [CrossRef]
14. Steinert, M.; Schneider, B.; Dechert, S.; Demeshko, S.; Meyer, F. Spin-state versatility in a series of $Fe_4$ [2 × 2] grid complexes: effects of counteranions, lattice solvent, and intramolecular cooperativity. *Inorg. Chem.* **2016**, *55*, 2363–2373. [CrossRef] [PubMed]
15. Lochenie, C.; Bauer, W.; Railliet, A.P.; Schlamp, S.; Garcia, Y.; Weber, B. Large Thermal hysteresis for Iron(II) spin crossover complexes with N-(Pyrid-4-Yl)Isonicotinamide. *Inorg. Chem.* **2014**, *53*, 11563–11572. [CrossRef] [PubMed]
16. Dankhoff, K.; Lochenie, C.; Puchtler, F.; Weber, B. Solvent influence on the magnetic properties of iron(ii) spin-crossover coordination compounds with 4,4'-dipyridylethyne as linker. *Eur. J. Inorg. Chem.* **2016**, *2016*, 2136–2143. [CrossRef]
17. Harding, D.J.; Phonsri, W.; Harding, P.; Gass, I.A.; Murray, K.S.; Moubaraki, B.; Cashion, J.D.; Liu, L.; Telfer, S.G. Abrupt spin crossover in an Iron(III) quinolylsalicylaldimine complex: Structural Insights and solvent effects. *Chem. Commun.* **2013**, *49*, 6340–6342. [CrossRef] [PubMed]
18. Duriska, M.B.; Neville, S.M.; Moubaraki, B.; Cashion, J.D.; Halder, G.J.; Chapman, K.W.; Balde, C.; Létard, J.-F.; Murray, K.S.; Kepert, C.J.; et al. A nanoscale molecular switch triggered by thermal, light, and guest perturbation. *Angew. Chem. Int. Ed.* **2009**, *48*, 2549–2552. [CrossRef] [PubMed]
19. Wu, X.-R.; Shi, H.-Y.; Wei, R.-J.; Li, J.; Zheng, L.-S.; Tao, J. Coligand and solvent effects on the architectures and spin-crossover properties of (4,4)-connected Iron(II) coordination polymers. *Inorg. Chem.* **2015**, *54*, 3773–3780. [CrossRef] [PubMed]
20. Fumanal, M.; Jiménez-Grávalos, F.; Ribas-Arino, J.; Vela, S. Lattice-solvent effects in the spin-crossover of an Fe(II)-based material. The key role of intermolecular interactions between solvent molecules. *Inorg. Chem.* **2017**, *56*, 4474–4483. [CrossRef] [PubMed]
21. Milin, E.; Benaicha, B.; El Hajj, F.; Patinec, V.; Triki, S.; Marchivie, M.; Gómez-García, C.J.; Pillet, S. Magnetic bistability in macrocycle-based FeII spin-crossover complexes: Counter ion and solvent effects. *Eur. J. Inorg. Chem.* **2016**, *2016*, 5282. [CrossRef]
22. Saureu, S.; de Graaf, C. On the role of solvent effects on the electronic transitions in Fe(II) and Ru(II) complexes. *Chem. Phys.* **2014**, *428*, 59–66. [CrossRef]
23. Liu, C.; Zhang, J.; Lawson Daku, L.M.; Gosztola, D.; Canton, S.E.; Zhang, X. Probing the impact of solvation on photoexcited spin crossover complexes with high-precision x-ray transient absorption spectroscopy. *J. Am. Chem. Soc.* **2017**, *139*, 17518–17524. [CrossRef] [PubMed]

24. Herold, C.F.; Shylin, S.I.; Rentschler, E. Solvent-dependent SCO behavior of dinuclear Iron(II) complexes with a 1,3,4-thiadiazole bridging ligand. *Inorg. Chem.* **2016**, *55*, 6414–6419. [CrossRef] [PubMed]
25. Phonsri, W.; Harding, P.; Liu, L.; Telfer, S.G.; Murray, K.S.; Moubaraki, B.; Ross, T.M.; Jameson, G.N.L.; Harding, D.J. Solvent modified spin crossover in an Iron(III) complex: phase changes and an exceptionally wide hysteresis. *Chem. Sci.* **2017**, *8*, 3949–3959. [CrossRef] [PubMed]
26. Vieira, B.J.C.; Dias, J.C.; Santos, I.C.; Pereira, L.C.J.; da Gama, V.; Waerenborgh, J.C. thermal hysteresis in a spin-crossover FeIII quinolylsalicylaldimine complex, $Fe^{III}(5-Br-Qsal)_2Ni(Dmit)_2 \cdot solv$: Solvent effects. *Inorg. Chem.* **2015**, *54*, 1354–1362. [CrossRef] [PubMed]
27. Bonnet, S.; Molnár, G.; Sanchez Costa, J.; Siegler, M.A.; Spek, A.L.; Bousseksou, A.; Fu, W.-T.; Gamez, P.; Reedijk, J. Influence of sample preparation, temperature, light, and pressure on the two-step spin crossover mononuclear compound $[Fe(Bapbpy)(NCS)_2]$. *Chem. Mater.* **2009**, *21*, 1123–1136. [CrossRef]
28. Craig, G.A.; Costa, J.S.; Roubeau, O.; Teat, S.J.; Aromí, G. Local coordination geometry and spin state in novel $Fe^{II}$ complexes with 2,6-Bis(Pyrazol-3-Yl)Pyridine-Type ligands as controlled by packing forces: Structural correlations. *Chem. Eur. J.* **2012**, *18*, 11703–11715. [CrossRef] [PubMed]
29. Wei, R.-J.; Tao, J.; Huang, R.-B.; Zheng, L.-S. Reversible and irreversible vapor-induced guest molecule exchange in spin-crossover compounds. *Inorg. Chem.* **2011**, *50*, 8553–8564. [CrossRef] [PubMed]
30. Zhang, W.; Zhao, F.; Liu, T.; Yuan, M.; Wang, Z.-M.; Gao, S. Spin crossover in a series of Iron(II) complexes of 2-(2-Alkyl-2H-Tetrazol-5-Yl)-1,10-Phenanthroline: Effects of alkyl side chain, solvent, and anion. *Inorg. Chem.* **2007**, *46*, 2541–2555. [CrossRef] [PubMed]
31. Amoore, J.J.M.; Kepert, C.J.; Cashion, J.D.; Moubaraki, B.; Neville, S.M.; Murray, K.S. Structural and magnetic resolution of a two-step full spin-crossover transition in a dinuclear Iron(II) pyridyl-bridged compound. *Chem. Eur. J.* **2006**, *12*, 8220–8227. [CrossRef] [PubMed]
32. Rajadurai, C.; Qu, Z.; Fuhr, O.; Gopalan, B.; Kruk, R.; Ghafari, M.; Ruben, M. Lattice-solvent controlled spin transitions in Iron(II) complexes. *Dalton Trans.* **2007**, 3531–3537. [CrossRef] [PubMed]
33. Bartel, M.; Absmeier, A.; Jameson, G.N.L.; Werner, F.; Kato, K.; Takata, M.; Boca, R.; Hasegawa, M.; Mereiter, K.; Caneschi, A.; et al. Modification of spin crossover behavior through solvent assisted formation and solvent inclusion in a triply interpenetrating three-dimensional network. *Inorg. Chem.* **2007**, *46*, 4220–4229. [CrossRef] [PubMed]
34. Chuang, Y.-C.; Liu, C.-T.; Sheu, C.-F.; Ho, W.-L.; Lee, G.-H.; Wang, C.-C.; Wang, Y. New Iron(II) spin crossover coordination polymers $[Fe(\mu\text{-Atrz})_3]X_2 \cdot 2H_2O$ ($X = ClO_4^-$, $BF_4^-$) and $[Fe(\mu\text{-Atrz})(\mu\text{-Pyz})(NCS)_2] \cdot 4H_2O$ with an interesting solvent effect. *Inorg. Chem.* **2012**, *51*, 4663–4671. [CrossRef] [PubMed]
35. Leita, B.A.; Neville, S.M.; Halder, G.J.; Moubaraki, B.; Kepert, C.J.; Létard, J.-F.; Murray, K.S. Anion−solvent dependence of bistability in a family of meridional N-Donor-Ligand-Containing Iron(II) spin crossover complexes. *Inorg. Chem.* **2007**, *46*, 8784–8795. [CrossRef] [PubMed]
36. Halder, G.J.; Kepert, C.J.; Moubaraki, B.; Murray, K.S.; Cashion, J.D. Guest-dependent spin crossover in a nanoporous molecular framework material. *Science* **2002**, *298*, 1762–1765. [CrossRef] [PubMed]
37. Weber, B.; Bauer, W.; Pfaffeneder, T.; Dîrtu, M.M.; Naik, A.D.; Rotaru, A.; Garcia, Y. Influence of hydrogen bonding on the hysteresis width in Iron(II) spin-crossover complexes. *Eur. J. Inorg. Chem.* **2011**, *2011*, 3193–3206. [CrossRef]
38. Greenaway, A.M.; Sinn, E. High-spin and low-spin.alpha.-picolylamine Iron(II) complexes. Effect of ligand reversal on spin state. *J. Am. Chem. Soc.* **1978**, *100*, 8080–8084. [CrossRef]
39. Ferguson, A.; Squire, M.A.; Siretanu, D.; Mitcov, D.; Mathonière, C.; Clérac, R.; Kruger, P.E. A face-capped $[Fe_4L_4]^{8+}$ spin crossover tetrahedral cage. *Chem. Commun.* **2013**, *49*, 1597–1599. [CrossRef] [PubMed]
40. Schulte, K.A.; Fiedler, S.R.; Shores, M.P. Solvent dependent spin-state behaviour via hydrogen bonding in neutral FeII diimine complexes. *Aust. J. Chem.* **2014**, *67*, 1595–1600. [CrossRef]
41. Dîrtu, M.M.; Neuhausen, C.; Naik, A.D.; Rotaru, A.; Spinu, L.; Garcia, Y. Insights into the origin of cooperative effects in the spin transition of $[Fe(NH_2trz)_3](NO_3)_2$: The role of supramolecular interactions evidenced in the crystal structure of $[Cu(NH_2trz)_3](NO_3)_2 \cdot H_2O$. *Inorg. Chem.* **2010**, *49*, 5723–5736. [CrossRef] [PubMed]
42. Costa, J.S.; Rodríguez-Jiménez, S.; Craig, G.A.; Barth, B.; Beavers, C.M.; Teat, S.J.; Aromí, G. Three-way crystal-to-crystal reversible transformation and controlled spin switching by a nonporous molecular material. *J. Am. Chem. Soc.* **2014**, *136*, 3869–3874. [CrossRef] [PubMed]
43. Gentili, D.; Demitri, N.; Schäfer, B.; Liscio, F.; Bergenti, I.; Ruani, G.; Ruben, M.; Cavallini, M. Multi-modal sensing in spin crossover compounds. *J. Mater. Chem. C* **2015**, *3*, 7836–7844. [CrossRef]

44. Barrios, L.A.; Bartual-Murgui, C.; Peyrecave-Lleixà, E.; Le Guennic, B.; Teat, S.J.; Roubeau, O.; Aromí, G. Homoleptic versus heteroleptic formation of mononuclear Fe(II) complexes with tris-imine ligands. *Inorg. Chem.* **2016**, *55*, 4110–4116. [CrossRef] [PubMed]
45. Huang, W.; Shen, F.; Zhang, M.; Wu, D.; Pan, F.; Sato, O. Room-temperature switching of magnetic hysteresis by reversible single-crystal-to-single-crystal solvent exchange in imidazole-inspired Fe(II) complexes. *Dalton Trans.* **2016**, *45*, 14911–14918. [CrossRef] [PubMed]
46. Galet, A.; Muñoz, M.C.; Real, J.A. Coordination polymers undergoing spin crossover and reversible ligand exchange in the solid. *Chem. Commun.* **2006**, *41*, 4321–4323. [CrossRef] [PubMed]
47. Wang, H.; Sinito, C.; Kaiba, A.; Costa, J.-S.; Desplanches, C.; Dagault, P.; Guionneau, P.; Létard, J.-F.; Negrier, P.; Mondieig, D. Unusual solvent dependence of a molecule-based FeII macrocyclic spin-crossover complex. *Eur. J. Inorg. Chem.* **2014**, *2014*, 4927–4933. [CrossRef]
48. Lennartson, A.; Southon, P.; Sciortino, N.F.; Kepert, C.J.; Frandsen, C.; Mørup, S.; Piligkos, S.; McKenzie, C.J. Reversible guest binding in a non-porous Fe(II) coordination polymer host toggles spin crossover. *Chem. Eur. J.* **2015**, *21*, 16066–16072. [CrossRef] [PubMed]
49. Steinert, M.; Schneider, B.; Dechert, S.; Demeshko, S.; Meyer, F. A Trinuclear Defect-Grid Iron(II) spin crossover complex with a large hysteresis loop that is readily silenced by solvent vapor. *Angew. Chem. Int. Ed.* **2014**, *53*, 6135–6139. [CrossRef] [PubMed]
50. Miller, R.G.; Brooker, S. Reversible quantitative guest sensing via spin crossover of an Iron(II) triazole. *Chem. Sci.* **2016**, *7*, 2501–2505. [CrossRef] [PubMed]
51. Archer, R.J.; Hawes, C.S.; Jameson, G.N.L.; McKee, V.; Moubaraki, B.; Chilton, N.F.; Murray, K.S.; Schmitt, W.; Kruger, P.E. Partial spin crossover behaviour in a dinuclear Iron(II) triple helicate. *Dalton Trans.* **2011**, *40*, 12368–12373. [CrossRef] [PubMed]
52. Šalitroš, I.; Pavlik, J.; Boča, R.; Fuhr, O.; Rajadurai, C.; Ruben, M. Supramolecular lattice-solvent control of Iron(II) spin transition parameters. *CrystEngComm* **2010**, *12*, 2361–2368. [CrossRef]
53. Ostermeier, M.; Berlin, M.-A.; Meudtner, R.M.; Demeshko, S.; Meyer, F.; Limberg, C.; Hecht, S. Complexes of click-derived bistriazolylpyridines: Remarkable electronic influence of remote substituents on thermodynamic stability as well as electronic and magnetic properties. *Chem. Eur. J.* **2010**, *16*, 10202–10213. [CrossRef] [PubMed]
54. Pelleteret, D.; Clérac, R.; Mathonière, C.; Harté, E.; Schmitt, W.; Kruger, P.E. Asymmetric spin crossover behaviour and evidence of light-induced excited spin state trapping in a dinuclear Iron(II) helicate. *Chem. Commun.* **2009**, *2*, 221–223. [CrossRef] [PubMed]
55. Garcia, Y.; Robert, F.; Naik, A.D.; Zhou, G.; Tinant, B.; Robeyns, K.; Michotte, S.; Piraux, L. Spin transition charted in a fluorophore-tagged thermochromic dinuclear Iron(II) complex. *J. Am. Chem. Soc.* **2011**, *133*, 15850–15853. [CrossRef] [PubMed]
56. Scott, H.S.; Ross, T.M.; Moubaraki, B.; Murray, K.S.; Neville, S.M. Spin crossover in polymeric materials using schiff base functionalized triazole ligands. *Eur. J. Inorg. Chem.* **2013**, *2013*, 803–812. [CrossRef]
57. Craze, A.R.; Sciortino, N.F.; Badbhade, M.M.; Kepert, C.J.; Marjo, C.E.; Li, F. Investigation of the spin crossover properties of three dinulear Fe(II) triple helicates by variation of the steric nature of the ligand type. *Inorganics* **2017**, *5*, 62. [CrossRef]
58. Tuna, F.; Lees, M.R.; Clarkson, G.J.; Hannon, M.J. Readily prepared metallo-supramolecular triple helicates designed to exhibit spin-crossover behaviour. *Chem. Eur. J.* **2004**, *10*, 5737–5750. [CrossRef] [PubMed]
59. Garcia, Y.; Grunert, C.M.; Reiman, S.; van Campenhoudt, O.; Gütlich, P. The two-step spin conversion in a supramolecular triple helicate dinuclear Iron(II) complex studied by mössbauer spectroscopy. *Eur. J. Inorg. Chem.* **2006**, *2006*, 3333–3339. [CrossRef]
60. Cowieson, N.P.; Aragao, D.; Clift, M.; Ericsson, D.J.; Gee, C.; Harrop, S.J.; Mudie, N.; Panjikar, S.; Price, J.R.; Riboldi-Tunnicliffe, A.; et al. MX1: A bending-magnet crystallography beamline serving both chemical and macromolecular crystallography communities at the Australian Synchrotron. *J. Synchrotron. Rad.* **2015**, *22*, 187–190. [CrossRef] [PubMed]
61. McPhillips, T.M.; McPhillips, S.E.; Chiu, H.J.; Cohen, A.E.; Deacon, A.M.; Ellis, P.J.; Garman, E.; Gonzalez, A.; Sauter, N.K.; Phizackerley, R.P.; et al. Blu-ice and the distributed control system: software for data acquisition and instrument control at macromolecular crystallography beamlines. *J. Synchrotron. Rad.* **2002**, *9*, 401–406. [CrossRef]

62. Kabsch, W. XDS. Automatic processing of rotation diffraction data from crystals of initially unknown symmetry and cell constants. *J. Appl. Crystallogr.* **1993**, *26*, 795–800. [CrossRef]
63. *APEX2*, version 2014; Bruker AXS Inc.: Madison, WI, USA, 2014. B) *SAINT*, version 8.34A; Bruker AXS Inc.: Madison, WI, USA, 2014.
64. *SADABS*, version 2014/5; Bruker AXS Inc.: Madison, WI, USA, 2001.
65. Sheldrick, G. *SHELX-2014: Programs for Crystal Structure Analysis*; University of Göttingen: Göttingen, Lower Saxony, Germany, 2014.
66. Sheldrick, G.M. Crystal structure refinement with SHELXL. *Acta. Cryst. C* **2015**, *71*, 3–8. [CrossRef] [PubMed]
67. Dolomanov, O.V.; Bourhis, L.J.; Gildea, R.J.; Howard, J.A.K.; Puschmann, H. OLEX2: A complete structure solution, refinement and analysis program. *J. Appl. Cryst.* **2009**, *42*, 339–341. [CrossRef]
68. Wolff, S.K.; Grimwood, D.J.; McKinnon, J.J.; Turner, M.J.; Jayatilaka, D.; Spackman, M.A. *CrystalExplorer, Version 3.1*; University of Western Australia: Perth, Western Australia, Australia, 2012.
69. Fatur, S.M.; Shepard, S.G.; Higgins, R.F.; Shores, M.P.; Damrauer, N.H. A synthetically tunable system to control mlct excited-state lifetimes and spin states in Iron(II) polypyridines. *J. Am. Chem. Soc.* **2017**, *139*, 4493–4505. [CrossRef] [PubMed]
70. Spackman, M.A.; Jayatilaka, D. Hirshfeld surface analysis. *CrystEngComm* **2009**, *11*, 19–32. [CrossRef]
71. McKinnon, J.J.; Jayatilaka, D.; Spackman, M.A. Towards quantitative analysis of intermolecular interactions with hirshfeld surfaces. *Chem. Commun.* **2007**, *37*, 3814–3816. [CrossRef]
72. Archer, R.J.; Scott, H.S.; Polson, M.I.J.; Williamson, B.E.; Mathonière, C.; Rouzières, M.; Clérac, R.; Kruger, P.E. Varied spin crossover behaviour in a family of dinuclear Fe(II) triple helicate complexes. *Dalton Trans.* **2018**, *47*, 7965–7974. [CrossRef] [PubMed]
73. Li, L.; Craze, A.R.; Akiyoshi, R.; Tsukiashi, A.; Hayami, S.; Mustonen, O.; Bhadbhade, M.; Bhattacharyya, S.; Marjo, C.E.; Wang, Y.; et al. Direct monitoring of spin transition at a high $T_{1/2}$ value in a dinuclear triple-stranded helicate Iron(II) complex through x-ray pho-toelectron spectroscopy. *Dalton Trans.* **2018**, *47*, 2543–2548. [CrossRef] [PubMed]

© 2018 by the authors. Licensee MDPI, Basel, Switzerland. This article is an open access article distributed under the terms and conditions of the Creative Commons Attribution (CC BY) license (http://creativecommons.org/licenses/by/4.0/).

*Article*

# Mosaicity of Spin-Crossover Crystals

Sabine Lakhloufi [1], Elodie Tailleur [1], Wenbin Guo [1], Frédéric Le Gac [1], Mathieu Marchivie [1], Marie-Hélène Lemée-Cailleau [2], Guillaume Chastanet [1] and Philippe Guionneau [1,*]

[1] CNRS, University Bordeaux, ICMCB, UMR 5026, 87 Avenue du Dr A. Schweitzer, F-33608 Pessac, France; sabine.lakhloufimathieu@gmail.com (S.L.); elodie.tailleur@icmcb.cnrs.fr (E.T.); guo0281@163.com (W.G.); le-gac.frederic@laposte.net (F.L.G.) mathieu.marchivie@icmcb.cnrs.fr (M.M.); guillaume.chastanet@icmcb.cnrs.fr (G.C.)
[2] Institut Laue-Langevin, 71 Avenue des Martyrs, BP 156, F-38042 Grenoble, France; lemee@ill.fr
* Correspondence: philippe.guionneau@icmcb.cnrs.fr; Tel.: +33-5-4000-2579

Received: 28 August 2018; Accepted: 11 September 2018; Published: 13 September 2018

**Abstract:** Real crystals are composed of a mosaic of domains whose misalignment is evaluated by their level of "mosaicity" using X-ray diffraction. In thermo-induced spin-crossover compounds, the crystal may be seen as a mixture of metal centres, some being in the high-spin (HS) state and others in the low spin (LS) state. Since the volume of HS and LS crystal packings are known to be very different, the assembly of domains within the crystal, i.e., its mosaicity, may be modified at the spin crossover. With little data available in the literature we propose an investigation into the temperature dependence of mosaicity in certain spin-crossover crystals. The study was preceded by the examination of instrumental factors, in order to establish a protocol for the measurement of mosaicity. The results show that crystal mosaicity appears to be strongly modified by thermal spin-crossover; however, the nature of the changes are probably sample dependent and driven, or masked, in most cases by the characteristics of the crystal (disorder, morphology ... ). No general relationship could be established between mosaicity and crystal properties. If, however, mosaicity studies in spin-crossover crystals are conducted and interpreted with great care, they could help to elucidate crucial crystal characteristics such as mechanical fatigability, and more generally to investigate systems where phase transition is associated with large volume changes.

**Keywords:** mosaicity; spin crossover; X-ray diffraction; fatigability; single crystal; phase transition; structural disorder

## 1. Introduction

It was as early as 1922 that C. G. Darwin, comparing theoretical expectations with experimental X-ray diffraction (XRD) observations, concluded that real crystals should be described as a conglomerate of perfect small blocks [1]. The model was further refined in 1926 by Bragg et al., who introduced the concept of "mosaicity"—the fact that crystals present a mosaic of blocks (now known as domains) each differing slightly in orientation [2]. A good example illustrating the mosaic model of real crystals is the assembly of bubbles on the surface of a soap solution [3]. Nowadays the single-crystal XRD experiments performed using modern diffractometers routinely start with a measurement of the mosaicity of the sample used, in order to assess the quality of the crystal prior to data collection. Values for mosaicity—hereafter "*M*"—are directly derived from the shape of the diffraction peaks which are, however, also related to factors other than mosaicity. These include microstrains affecting the distribution of unit cell parameters, local defects, the range and homogeneity of the domain sizes, and also the morphology of the crystal itself through the form factor. In addition, temperature can widen the foot of the Bragg peaks without modifying the sharpness of the peak, whilst reducing its maximum intensity [4]. Finally, and this is a central point, peak shape (and thus the value of $M$ obtained by automatic single-crystal XRD diffractometer protocols) strongly depends on instrumental factors: some of these are user-dependent e.g., the distance between crystal and detector); others are machine-dependent (e.g., the spatial and wavelength divergences of the X-ray beam). Reliable values for $M$ are thus difficult to obtain; they require precise protocols and high quality crystals. The protocols must also be adapted to the nature and size of the crystal [5]. However, even without ideal conditions, it is possible to examine and compare the effects of temperature on the mosaicity of crystals using identical protocols. This is the subject of this article.

It is known that spin-crossover behaviour (SCO) corresponds to a marked change in crystal volume. For instance, in iron (II) complexes based crystals, the unit cell volume may increase by over ten percent in the event of thermal-induced SCO with a transition from low-spin (LS) to high spin (HS) [6,7]. With such a large volume change, significant consequences could be expected for domain alignment within the crystals, considering especially that the crystal displays a combination of LS and HS domains in the course of thermal SCO. Mosaicity may therefore be strongly modified during thermal SCO and the value of $M$ may vary depending on whether the crystal is in a state of HS, LS, or a combination of the two. The modification of crystal mosaicity due to SCO is poorly documented in the literature, even though it is closely related to the important topic of the mechanical fatigability of the SCO process. We have previously demonstrated that a large number of thermal SCO cycles on the crystals of the molecular complex [Fe(PM-AzA)$_2$(NCS)$_2$] induces a strong increase in room-temperature (HS) mosaicity [8]. This increase corresponds to an irreversible rise in the misalignment of the crystal domains, clearly linked to the mechanical fatigability of the SCO phenomenon. The abrupt SCO effect observed in the [Fe(bntrz)$_3$][Pt(CN)$_4$] chain, on the other hand, is associated with strong mechanical resilience, since $M$ does not evolve despite the strong structural rearrangement of the crystal packing [9].

To understand the role of mosaicity change at the microstructural scale, there is a clear need for more systematic investigations into the evolution of mosaicity in the context of thermal spin crossover. In the present study we focused on the temperature-dependence of the mosaicity of [Fe$^{II}$(PM-L)$_2$(NCX)$_2$] compounds (including those mentioned above), with a view to shedding light on the relationship between mosaicity and spin state in crystals undergoing thermal SCO. Mosaicity is to be understood broadly, using $M$ as denoted above, covering coherent domain size effects/strain, defaults and domain misalignments.

An experimental protocol was first established, allowing $M$ values to be determined which were suitable for comparative purposes; a selection of diverse SCO behaviours was then examined using known compounds [Fe$^{II}$(PM-L)$_2$(NCX)$_2$] (Scheme 1). The relationship between mosaicity and gradual SCO was examined in [Fe(PM-AzA)$_2$(NCS)$_2$], **1**, [10], between mosaicity and incomplete SCO in [Fe(PM-TeA)$_2$(NCS)$_2$]·0.5MeOH [11], **2**, between mosaicity and dynamical disorder in [Fe(PM-TheA)$_2$(NCS)$_2$] [12], **3**, and between mosaicity and highly cooperative SCO in [Fe(PM-PeA)$_2$(NCSe)$_2$] [13], **4**, and [Fe(PM-BrA)$_2$(NCS)$_2$], **5**.

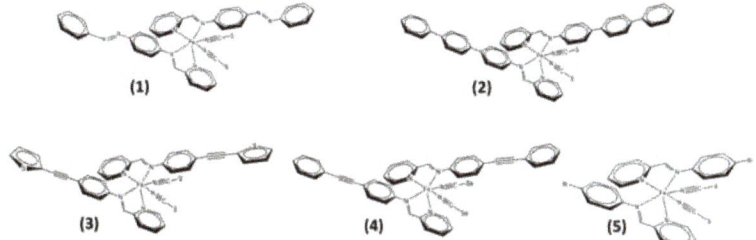

**Scheme 1.** schematic views of the molecular iron(II) complexes of the [Fe$^{II}$(PM-L)$_2$(NCX)$_2$] series studied in this paper, labelled **1** to **5**.

Note that all these compounds were in an HS state at room temperature or above when freshly synthetized and underwent SCO when cooling.

## 2. Materials and Methods

The measurement of mosaicity for the crystals **1** to **4** was performed on a Bruker-Nonius κ-CCD diffractometer (Bruker AXS, Karlsruhe, Germany) (MoKα radiation at 0.71073 Å) equipped with an Oxford Cryosystem $N_2$ open flow allowing a temperature range of [80–400 K]. The Denzo software programme (HKL Research, Inc., Charlottesville, USA) was used to calculate $M$ from the diffraction images [14]. The same global protocol and guidelines were used to determine the $M$ values. To define the protocol and clearly establish the limits for data interpretation, the influence of the experimental parameters on the $M$ values needed to be known. A series of tests was therefore carried out at room temperature providing numerous determinations of $M$, before investigating the temperature dependence of the SCO crystal. These tests were performed using the ammonium bitartrate κ-CCD (Bruker AXS, Karlsruhe, Germany) calibration crystal ($C_4H_9NO_6$; $oP2_12_12_1$ space group, almost spherical crystal of 0.20 mm diameter). For each test series, a single instrumental parameter was changed at a time, in order to identify its optimum value and range of use.

The following seven criteria were ultimately retained as potentially influential on the value of $M$ and were therefore carefully checked:

1. the crystal-detector distance
2. the $\omega$ angular range of the scans
3. the angular oscillation width
4. the resolution of the data collection
5. the duration of the image exposure
6. the diameter of the X-ray beam collimator
7. the temperature at which the experiment is performed

The variations of $M$ as a function of these experimental parameters are presented in Electronic Supplementary Information (ESI). The results are commented below and conclusions are set out in Section 3.1 below.

Compound **5** was investigated using a Bruker Apex-II (Bruker AXS, Karlsruhe, Germany) diffractometer. The experimental protocol defined from the above tests was adapted and applied.

The compounds **1–4** investigated are already known in the literature. The details and methodologies for the synthesis and crystallization processes can be found in the references below. In the present study, the single crystals were selected under microscope and full data collection was performed at room temperature before the temperature-dependence study; the crystal structures were fully refined to verify the initial structural quality of the samples and to check their correspondence to the reported compounds.

Compound **5** is new in this series and shows a first-order SCO transition; given its prolific polymorphism it is particularly difficult to investigate. A detailed study of the spin-crossover features will be provided elsewhere for this compound [15] and only the temperature-dependence mosaicity is commented below. Compound **5** was used in order to present $M$ values obtained on a different diffractometer.

Apart from the experimental protocol defined in Section 3.1 below, the specific characteristics for **1–5** were:

- For **1**: one small dark spearhead single crystal of approximate dimensions $0.10 \times 0.08 \times 0.08$ mm$^3$ investigated by cooling from 290 to 120 K with 10 K steps.
- For **2**: one needle-shaped crystal of dimensions $0.100 \times 0.125 \times 0.325$ mm$^3$; the full thermal SCO cycle was scanned by cooling from 300 to 90 K and then warmed to 300 K using 10 K steps.
- For **3**: one large single crystal of approximate dimensions $0.50 \times 0.50 \times 2.0$ mm$^3$ used to follow the full thermal cycles from 300 K to 100 K using 10 K steps.

- For **4**: one green dark crystal of approximate dimensions $0.2 \times 0.2 \times 0.4$ mm$^3$ investigated from 280 K to 100 K in 13 steps with a cooling speed of 12 K per hour. Almost full data collection (> 300° scanned) performed for each measurement.
- For **5**: a crystal of approximate dimensions $0.20 \times 0.08 \times 0.04$ mm$^3$ studied by cooling from 300 to 120 K and then warmed back to 300K.

## 3. Results

*3.1. Conclusions as Regards Obtaining Reliable Mosaicity Values*

The test results (see ESI) for the instrumental factors defined in paragraph 2 make it possible to draw conclusions on the $M$ measurement, most of which concord with common sense:

The $M$ value strongly depends on the crystal-to-detector distance (Figure S1). The distance must be fixed in order to observe a sufficient number of Bragg peaks, which also depends on the unit cell of the compound and the scan mode. Short distances increase the number of peaks obtained but also increase the risk of overlap between reflections, thus distorting the $M$ value. The best distance is probably that used for full data collection, i.e., between 30 and 50 mm for the Bruker–Nonius κ-CCD diffractometer with Mo/Kα radiation.

ω angular scanning ranges of 20° to 180° were tested. It became clear that the same angular range must be scanned throughout the investigation to obtain comparable $M$ values for a given crystal. In addition, smaller angular ranges lead to instable $M$ values, whilst $M$ values are reproducible above a certain scan range. In other words, the larger the angular range, the more consistent the $M$ value. The ideal case is probably to perform partial data collection on 60–90°.

The angular oscillation width determines reflection overlap and the number of images collected (Figure S2). The superposition of reflections is one of the main causes of unstable $M$ values; the angular oscillation width must therefore not be fixed at too high a value. Narrower scans provide less reliable intensities. The optimum oscillation width depends on the unit cell of the sample, but a value of 1.0° would be a good compromise for most molecular crystals with classic laboratory Mo/Kα radiation and beam size of ca. 300 μm.

The resolution of the data collected was tested from 0.40 to 1.64 Å (Figure S3). The lesson is that standard data collection resolutions (around 0.8 Å) must be used, while resolutions beyond 1.1 Å start to modify the $M$ value.

The duration of exposure was tested from 5 s up to 140 s for a 1-degree scan and was found to influence the $M$ value strongly (Figure S4). Above a certain value (40 s for the crystal tested), $M$ values are stable. Although the duration of exposure is sample dependent, the same value should be used for all the experiments conducted on a crystal.

The diameter of the X-ray beam collimator (0.25 and 0.35 mm in this case) is obviously important for the calculation of the $M$ value (Table S1). Large collimators are not the most favourable, for comparative purposes it is important, once again, to use the same collimator from one experiment to another.

Temperature in the range [110–300 K] barely modified the $M$ values for the crystal tested (Figure S5). Note that the crystal was well crystallized (weak mosaicity value) with no known thermal structural effects. The value of $M$ is therefore not significantly affected by temperature, in the absence of thermal effects on the sample. These conclusions may differ for samples displaying significant structural modifications induced by temperature, as will be illustrated below.

In summary, to obtain a series of $M$ values for comparison purposes, the parameters defined above should be respected; this corresponds to partial data collection and is far from the routine unit cell determination protocol proposed by default on diffractometers. It is essential that the same protocol be used for a series of measurements, as was done in this study. The values for the factors above are sample-dependent, but the experimental factors used in this study were very similar (since the crystals investigated resembled each other in terms of unit cell, symmetry, diffraction quality and morphologies): crystal-detector distance of about 40 mm, a total scan range >40°, exposition time >40 s/°, collimator of 0.35 mm, 1.0° oscillation and resolution of about 0.8 Å.

### 3.2. Mosaicity and Gradual SCO in [Fe(PM-AzA)$_2$(NCS)$_2$] (1)

The compound [Fe(PM-AzA)$_2$(NCS)$_2$] displays a gradual spin crossover from 280 K to 110 K with $T_{1/2}$ ~196 K. It has been studied thoroughly, including by crystallography [10,11]. The temperature dependence of $M$ for [Fe(PM-AzA)$_2$(NCS)$_2$] shows a strong modification at the HS to LS transition (Figure 1). The modification is characterised by three regimes, since $M$ initially increases slightly in the temperature range where the crystal is still essentially HS, from 280 to 240 K; it subsequently increases sharply to reach a top plateau in the range where the sample is approximately half HS and half LS. Finally $M$ decreases to a value at 120 K similar to its initial one. In the present case, the thermal SCO signature is visible in the mosaicity data.

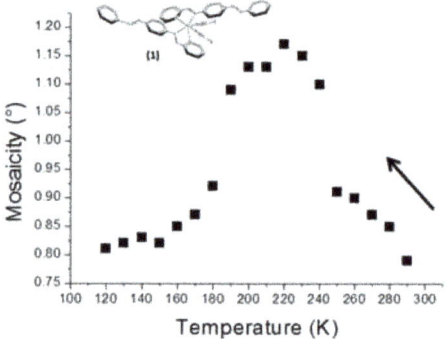

**Figure 1.** Temperature dependence of the mosaicity for a small crystal of [Fe(PM-AzA)$_2$(NCS)$_2$] in the cooling mode.

Clearly, in this case, mosaicity increases in the temperature region where the crystal is composed of a mix of metal centres in HS and LS states. The proximity of the $M$ values at 290 K and 120 K shows that when all the metal centres are in the same spin state, whatever it may be, the crystal perfection looks similar.

Note that room-temperature mosaicity as a function of the number of thermal SCO cycles has been measured in the past for [Fe(PM-AzA)$_2$(NCS)$_2$] [8]. A strong increase in mosaicity was seen after the first thermal cycles, even though the sample was fully HS at room temperature; this was interpreted in terms of mechanical fatigability. This may appear to contradict the present results, but this study has only examined the HS to LS conversion whilst the previous study measured mosaicity after complete HS-LS-HS cycles. Above all, the previous study was performed on very large crystals (>1.0 mm).

Sample size probably has a crucial effect on the mechanical behaviour of spin-crossover crystals, and therefore on mosaicity. In fact, the temperature dependence of the mosaicity of a very large crystal of [Fe(PM-AzA)$_2$(NCS)$_2$] does not reproduce the behaviour observed for the small crystal used in the present study. Although in large crystals, the change in $M$ from HS to LS are unclear, they globally tend to increase significantly. Since it is hard to achieve reliable and reproducible results for large crystals from one sample to another, these results are not discussed in detail here; it seems nevertheless clear that the effects of SCO (at the physical scale concerned by the mosaicity measurement) depend on the size of the crystal.

The value of the SCO temperature may be obtained supposing that maximum mosaicity is reached at temperatures where half of the entities in the crystal are HS and half are LS. In our study we obtained ~215 K; this can be compared with $T_{1/2}$ = 196 K from the magnetic measurements. These

## 3.3. Mosaicity and Incomplete SCO in [Fe(PM-TeA)$_2$(NCS)$_2$]·0.5MeOH (2)

The compound [Fe(PM-TeA)$_2$(NCS)$_2$]·0.5MeOH is known to undergo an incomplete (about 60%) and gradual SCO, taking place mainly in the range of 200–90 K [11]. The mosaicity of the crystal increases when temperature decreases in the range 190–100 K (Figure 2); this is coherent with the SCO temperature range. The disorder due to the combination of HS and LS entities within the crystal is again revealed by the increase in mosaicity, which is fully reversible when the sample is warmed. Maximum mosaicity is observed at about 110 K, the temperature at which magnetic measurements show that half of the entities are LS within the crystal.

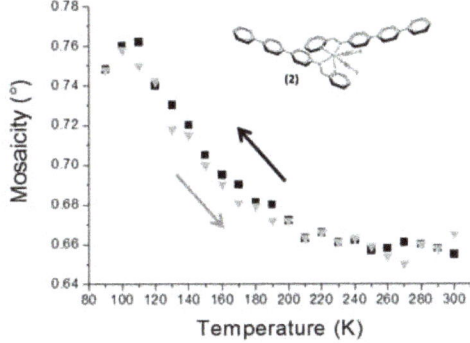

**Figure 2.** Temperature dependence of the mosaicity for a crystal of [Fe(PM-TeA)$_2$(NCS)$_2$]·0.5MeOH in the cooling (black squares) and warming (grey triangles) modes.

## 3.4. Mosaicity and Dynamical Disorder in [Fe(PM-TheA)$_2$(NCS)$_2$] (3)

The compound [Fe(PM-TheA)$_2$(NCS)$_2$] occurs in two polymorphs; the polymorph studied here undergoes a complete gradual SCO with $T_{1/2}$ = 243 K [12]. From the structural point of view, the peculiarity of this compound is to exhibit strong atomic disorder affecting the thiophene rings (Scheme 1). The detailed description of this disorder is not the subject of this paper. Suffice it to say that the disorder has two components, the one being statistical, the other dynamical. The dynamical contribution therefore significantly decreases with temperature, notably in the 300–100 K temperature range. The mosaicity of the crystal decreases almost continuously from 300 to 100 K (Figure 3), with no apparent sign of SCO. The decrease of mosaicity is also perfectly reversible by warming.

As explained above, the mosaicity in question comes from the measurement of $M$ that accounts for all kind of defects in the crystal which affect the width of the Bragg peaks. Any reduction in dynamical disorder leads to a drop in the $M$ value. Our results show that the change in mosaicity is low and that the part due to the SCO is totally masked by the drop in dynamical atomic disorder.

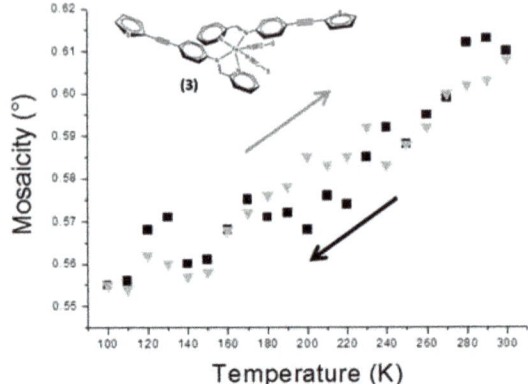

**Figure 3.** Temperature dependence of the mosaicity for a crystal of [Fe(PM-TheA)$_2$(NCS)$_2$]-II in the cooling (black squares) and warming (light-grey triangles) modes.

*3.5. Mosaicity and Mechanical Fatigability in [Fe(PM-PeA)$_2$(NCSe)$_2$] (4)*

The compound [Fe(PM-PeA)$_2$(NCSe)$_2$] has recently been reported; it is one of the very few compounds exhibiting SCO at room temperature together with a large hysteresis ($T_{1/2down}$ = 266 K and $T_{1/2up}$ = 307 K) [13]. As with sulphur, [Fe(PM-PeA)$_2$(NCS)$_2$] [11], this compound undergoes a highly cooperative phase transition associated with SCO and characterized by relatively distinct HS and LS crystal packings. Its mosaicity increases strongly and abruptly when cooling at about 260 K (Figure 4). The initial mosaicity (<0.7°) reflects the good crystalline quality of the sample in the HS state. In the LS state, below 250 K, its unusually high mosaicity (>4.0°) shows the marked degradation in crystal quality. There is no significant change in mosaicity upon further cooling from 250 to 100 K, clearly linking this loss of quality to the SCO. The loss of crystal quality is not reversible; there is no return of mosaicity to its initial value and there is even a tendency for the crystal to crack if warmed.

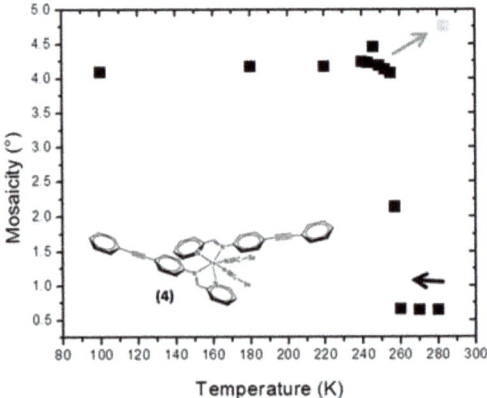

**Figure 4.** Temperature dependence of the mosaicity of [Fe(PM-PeA)$_2$(NCSe)$_2$] in the cooling mode (black squares), including the return values (light grey) after warming at room temperature.

Although reproducible, this phenomenon varies from one crystal to another; the drop in mosaicity and level of cracking depend on sample size. Cooling time may also be relevant; cooling was very slow in this study. These observations deserve further investigation. It is nevertheless clear that SCO degrades the sample through a form of mechanical fatigability, as already observed in other SCO materials [16,17]. It can be noted that there are no significant effects of this mechanical degradation on the magnetic characterization of the SCO, since the magnetic curve is not affected by repeated cycles [13]. It is interesting from the methodological point of view that SCO can also be tracked by the temperature dependence of the mosaicity in this compound.

*3.6. Pitfalls with the Mosaicity Approach: [Fe(PM-BrA)$_2$(NCS)$_2$] (5)*

The compound [Fe(PM-BrA)$_2$(NCS)$_2$] is a newly synthetized complex identified as a source of polymorphism; details on its behaviour are being published elsewhere [15]. Here we focus on the mosaicity of one polymorph undergoing abrupt SCO at $T_{1/2}$ ~210 K. Its mosaicity slightly decreases with temperature, the fact that the high temperature values are in the HS area would indicate better crystal quality after SCO (Figure 5). The improvement appears to be irreversible, since mosaicity remains lower with the return to room temperature (HS).

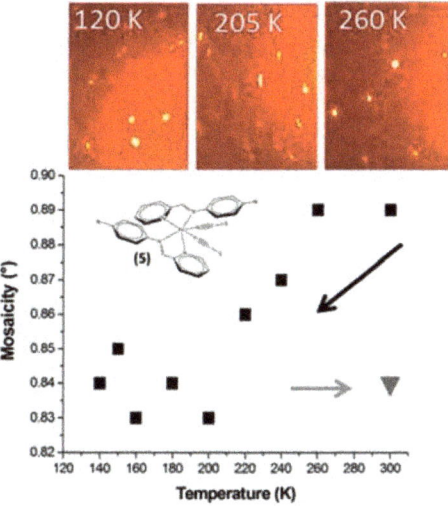

**Figure 5.** Temperature dependence of the mosaicity of *[Fe(PM-BrA)$_2$(NCS)$_2$]* in the cooling mode, with diffraction images at 120 K (LS), 205 K (mixture HS/LS) and 260 K (HS). Full squares mark the cooling mode and the grey triangle the value when returned to room temperature.

This is contradicted by the diffraction images however. The quality of the crystal seems to have been affected by the SCO and extra peaks appear at 120 K, suggesting a crack in the sample. Note that part of the sample is HS at 205 K and another part at LS, giving a number of double peaks which prevent any reliable measurement of $M$. At 120 K all the peaks may be indexed by the LS unit cell; there is no sign of HS residue, but the twin peaks indicate that the crystal has clearly suffered from the SCO probably due to cracks. These cracks are irreversible. After the SCO cycle the diffraction pattern at 300 K is similar to that at 120 K—i.e., worse than at 300 K before any temperature treatment.

The value of $M$ here does not reflect such degradation, since it is calculated for only one component, allowing a maximum number of peaks to be indexed. Other peaks are not taken into account in calculating the value of $M$. The mosaicity measured is therefore only true for one part of the sample and does not reflect its quality. The damage to the crystal occurs at an larger physical scale as it separates into different coherent domains. Within these coherent domains, however, crystal quality seems to improve slightly at low temperature. It can therefore be concluded that, for this compound, the measurement of mosaicity as we have defined it does not provide an efficient account of SCO-induced crystal degradation. This result illustrates one of the limits of the mosaicity investigation as reported here. Note also that we have no clear explanation on the difference of behavior between compounds **4** and **5** since they both show abrupt transitions but diverge in their mosaicity temperature dependence.

## 4. Discussion and Conclusions

The results described above for compounds **1–5** illustrate the diversity of situations encountered when dealing with spin crossover in molecular crystals. In all these studies the consistency of the series of values validates the experimental protocol.

Compounds **1** and **2** generally meet the initial expectations in term of temperature dependence of the mosaicity during a gradual SCO. In the temperature range where the sample shows both HS and LS, the significant increase in mosaicity was expected; it allows the SCO to be followed notably by determining the temperature at which half of the entities commute. Incidentally, this approach provides a new way to determine the so called $T_{1/2}$ experimentally.

Compound **3** exemplifies how a parameter other than mosaicity can affect the $M$ value, hiding the effect of SCO on mosaicity. This is also true for compound **2** (crystals larger than the crystal in our study do not behave in the same way) and of compound **4** (for which the amplitude of mosaicity variation is dependent on crystal size). This highlights the need for caution when interpreting temperature-dependent mosaicity in spin-crossover crystals.

In parallel, the study of compound **4** shows the potentiality of this approach since it clearly highlights the mechanical fatigability that can occur in crystals undergoing abrupt SCO. This is probably an important point for the development of mechanical devices based on spin-crossover crystals [18] and of molecular switches or bending crystals more generally [19].

The case of compound **5** confirms that mosaicity changes may occur at the limits of experimental sensitivity. More importantly, it shows that the measurement of mosaicity through $M$ does not necessarily capture all aspects of crystal quality. For instance, the true evolution of $M$ is more easily identified when SCO occurs without domain formation. The increase in $M$ should result from the distortion of the HS lattice by the presence of LS molecules, rather than through the misalignment of domains. We know, however, that, under SCO with domain formation, higher $M$ values can occur even if the domain shows no defect or misalignment, due to the partial overlap of Bragg peaks of both domains. In this case it is difficult to obtain the true value of $M$ for both domains unless both domains can be indexed separately.

The present study once again demonstrates that SCO on real crystals is a subtle and complex phenomenon involving a large number of parameters on different physical scales; some of these parameters are sample dependent. It would consequently be difficult, if not vain, to pronounce on general relationships between mosaicity and SCO behaviour. The temperature-dependence mosaicity approach developed in this paper nevertheless provides a tool which, if handled with care on a case-by-case basis, could shed new light on spin-crossover crystals, and on their potential mechanical fatigability in particular.

**Supplementary Materials:** The following are available online at http://www.mdpi.com/2073-4352/8/9/363/s1, Figure S1. Mosaicity of the crystal test as a function of the crystal-detector distance; Figure S2. Mosaicity of the crystal test as a function of the angular oscillation width; Figure S3. Mosaicity of the crystal test as a function of data collection resolution (Å); Figure S4. Mosaicity of the crystal test as a function of frame exposure time (s/°); Figure S5. Mosaicity of the crystal test as a function of temperature; Table S1. Mosaicity of the crystal test as a function of collimator diameter (mm).

**Author Contributions:** All authors contributed equally to the work.

**Acknowledgments:** S.L., F.L. and E.T. were funded by the French Ministry of Research through a PhD grant. W.G.'s PhD is funded by the China Scholarship Council (CSC). The authors gratefully acknowledge the Agence Nationale de la Recherche for its financial support under grant ANR FemtoMat 13-BS046002, and the Région Nouvelle Aquitaine. Robert CORNER (ILL) is thanked for his very meticulous correction of the English language.

**Conflicts of Interest:** The authors declare no conflict of interest.

## References

1. Darwin, C.G. The theory of X-ray reflection. *Philos. Mag.* **1922**, *43*, 800–829. [CrossRef]
2. Bragg, W.H. The intensity of reflection of X-rays by crystals. *Philos. Mag.* **1926**, *1*, 897–922. [CrossRef]
3. Bragg, W.L.; Nye, J.F. A dynamical model of a crystal structure. *Proc. R. Soc. Lond. A* **1947**, *190*, 474–481. [CrossRef]
4. Goeta, A.E.; Howard, J.A.K. Low temperature single crystal X-ray diffraction: Advantages, instrumentation and applications. *Chem. Soc. Rev.* **2004**, *33*, 490–500. [CrossRef] [PubMed]
5. Bellamy, H.D.; Snell, E.H.; Lovelace, J.; Pokross, M.; Borgstahlc, G.E.O. The high-mosaicity illusion: Revealing the true physical characteristics of macromolecular crystals. *Acta Crystallogr. Sect. D* **2000**, *56*, 986–995. [CrossRef]
6. Guionneau, P. Crystallography and spin-crossover. A view of breathing materials. *Dalton Trans.* **2014**, *43*, 382–393. [CrossRef] [PubMed]
7. Collet, E.; Guionneau, P. Structural analysis of spin-crossover materials: From molecules to materials. *C. R. Chim.* **2018**, in press, proof on line. [CrossRef]
8. Guionneau, P.; Lakhloufi, S.; Lemée-Cailleau, M.H.; Chastanet, G.; Rosa, P.; Mauriac, C.; Létard, J.F. Mosaicity and structural fatigability of a gradual spin-crossover single crystal. *Chem. Phys. Lett.* **2012**, *542*, 52–55. [CrossRef]
9. Pittala, M.; Thétiot, F.; Triki, S.; Boukheddaden, K.; Chastanet, G.; Marchivie, M. Cooperative 1D triazole-based spin crossover Fe$^{II}$ material with exceptional mechanical resilience. *Chem. Mater.* **2017**, *29*, 490–494. [CrossRef]
10. Lakhloufi, S.; Lemee-Cailleau, M.H.; Chastanet, G.; Rosa, P.; Daro, N.; Guionneau, P. Structural movies of the gradual spin-crossover in a molecular complex at various physical scales. *Phys. Chem. Chem. Phys.* **2016**, *18*, 28307–28315. [CrossRef] [PubMed]
11. Guionneau, P.; Létard, J.F.; Yuffit, D.S.; Chasseau, D.; Goeta Bravic, A.E.; Howard, J.A.K.; Kahn, O. Structural approach of the features of the spin crossover transition in iron(II) compounds. *J. Mater. Chem.* **1999**, *9*, 985–994. [CrossRef]
12. Marchivie, M. Approche structurale du phénomène de transition de spin par diffraction des rayons X sous contraintes (T, P, hv). Ph.D. Thesis, University of Bordeaux, Bordeaux, France, 25 November 2003.
13. Tailleur, E.; Marchivie, M.; Daro, N.; Chastanet, G.; Guionneau, P. Thermal spin-crossover with a large hysteresis spanning room temperature in a mononuclear complex. *Chem. Commun.* **2017**, *53*, 4763–4766. [CrossRef] [PubMed]
14. Otwinowski, Z.; Minor, W. Methods in Enzymology. In *Macromolecular Crystallography, Part A*; Carter, C.W., Sweet, R.M., Eds.; Academic Press: New York, NY, USA, 1997; Volume 276, pp. 307–326.
15. Guo, W.; Chastanet, G.; Guionneau, P. Polymorphism in the spin crossover [Fe(PM-BrA)$_2$(NCS)$_2$] compound. *J. Mater. Chem. C* **2018**. submitted.
16. Craig, G.A.; Sánchez Costa, J.; Roubeau, O.; Teat, S.J.; Aromí, G. An Fe$^{II}$ spin-crossover complex becomes increasingly cooperative with ageing. *Eur. J. Inorg. Chem.* **2013**, *5–6*, 745–752. [CrossRef]
17. Manrique-Juárez, M.D.; Suleimanov, I.; Hernández, E.M.; Salmon, L.; Molnár, G.; Bousseksou, A. In situ AFM imaging of microstructural changes associated with the spin transition in [Fe(htrz)$_2$(trz)](BF$_4$) nanoparticles. *Materials* **2016**, *9*, 537. [CrossRef] [PubMed]
18. Shepherd, H.J.; Gural'skiy, I.A.; Quintero, C.M.; Tricard, S.; Salmon, L.; Molnár, G.; Bousseksou, A. Molecular actuators driven by cooperative spin-state switching. *Nat. Commun.* **2013**, *4*, 2607. [CrossRef] [PubMed]
19. Naumov, P.; Chizhik, S.; Panda, M.K.; Nath, N.K.; Boldyreva, E. Mechanically responsive molecular crystals. *Chem. Rev.* **2015**, *115*, 12440–12490. [CrossRef] [PubMed]

© 2018 by the authors. Licensee MDPI, Basel, Switzerland. This article is an open access article distributed under the terms and conditions of the Creative Commons Attribution (CC BY) license (http://creativecommons.org/licenses/by/4.0/).

*Article*

# Cobalt(II) Terpyridin-4′-yl Nitroxide Complex as an Exchange-Coupled Spin-Crossover Material

**Akihiro Ondo and Takayuki Ishida \***

Department of Engineering Science, The University of Electro-Communications, Chofu, Tokyo 182-8585, Japan; ondo@ttf.pc.uec.ac.jp
\* Correspondence: takayuki.ishida@uec.ac.jp; Tel.: +81-42-443-5490; Fax: +81-42-443-5501

Received: 2 March 2018; Accepted: 29 March 2018; Published: 2 April 2018

**Abstract:** Spin-crossover (SCO) was studied in [Co(L)$_2$](CF$_3$SO$_3$)$_2$, where L stands for diamagnetic 2,2′:6′,2′′-terpyridine (tpy) and its paramagnetic derivative, 4′-{4-*tert*-butyl(*N*-oxy)aminophenyl}-substituted tpy (tpyphNO). The X-ray crystallographic analysis clarified the Co-N bond length change ($\Delta d$) in high- and low-temperature structures; $\Delta d_{\text{central}}$ = 0.12 and $\Delta d_{\text{distal}}$ = 0.05 Å between 90 and 400 K for L = tpy and $\Delta d_{\text{central}}$ = 0.11 and $\Delta d_{\text{distal}}$ = 0.06 Å between 90 and 300 K for L = tpyphNO. The low- and high-temperature structures can be assigned to approximate low- and high-spin states, respectively. The magnetic susceptibility measurements revealed that the $\chi_\text{m} T$ value of [Co(tpyphNO)$_2$](CF$_3$SO$_3$)$_2$ had a bias from that of [Co(tpy)$_2$](CF$_3$SO$_3$)$_2$ by the contribution of the two radical spins. The tpy compound showed a gradual SCO around 260 K and on cooling the $\chi_\text{m} T$ value displayed a plateau down to 2 K. On the other hand, the tpyphNO compound showed a relatively abrupt SCO at ca. 140 K together with a second decrease of the $\chi_\text{m} T$ value on further cooling below ca. 20 K. From the second decrease, Co-nitroxide exchange coupling was characterized as antiferromagnetic with $2J_{\text{Co-rad}}/k_\text{B}$ = −3.00(6) K in the spin-Hamiltonian $H = -2J_{\text{Co-rad}}(S_{\text{Co}} \cdot S_{\text{rad1}} + S_{\text{Co}} \cdot S_{\text{rad2}})$. The magnetic moment apparently switches double-stepwise as 1 $\mu_\text{B} \rightleftarrows$ 3 $\mu_\text{B} \rightleftarrows$ 5 $\mu_\text{B}$ by temperature stimulus.

**Keywords:** spin crossover; spin transition; cobalt(II) ion; paramagnetic ligand; aminoxyl; switch

## 1. Introduction

Spin crossover (SCO) is a reversible transition between low-spin (LS) and high-spin (HS) states by external stimuli like heat, light, pressure, or magnetic field [1–3]. A number of materials are studied toward application in sensors [4]. Much attention has been paid to develop multifunctional SCO materials; for example, mesophase or liquid crystal properties [5–7] and magnetic exchange coupling [8,9]. Iron(II) (3d$^6$) coordination compounds are the most developed materials among various SCO complexes [10–13], because SCO occurs between $S$ = 0 dia- and $S$ = 2 paramagnetic states [14], drastically exhibiting magnetic and chromic changes. However, if one develops SCO materials having exchange coupling from an adjacent paramagnetic center, iron(II) compounds are unsuitable. In this line, cobalt(II) (3d$^7$) SCO behavior between $S$ = 1/2 and $S$ = 3/2 states are promising because the LS state is still paramagnetic (Scheme 1a). The entropy and geometry changes in cobalt(II) SCO compounds are less pronounced than those of iron(II) and iron(III) ones and it generally leads to a gradual SCO profile as a function of temperature [15–18]. A possible scenario of exchange-coupled SCO materials involves multi-step magnetic property jumps as a function of temperature. Namely, in the $\chi_\text{m} T$ versus $T$ profile, a low-spin region has an additional spin-equilibrium regulated by magnetic exchange coupling. When a paramagnetic ligand (L) with $S_{\text{rad}}$ = 1/2 is available, a Co$^{2+}$(LS)/L = 1/1 compound would show a singlet-triplet equilibrium, or a Co$^{2+}$(LS)/L = 1/2 compound a doublet-quartet equilibrium.

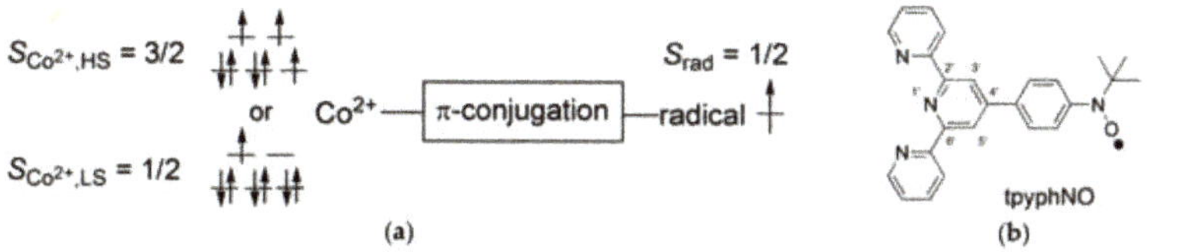

**Scheme 1.** (a) Schematic drawing of proposal for exchange-coupled spin-crossover materials; (b) Structural formula of tpyphNO.

The 2,2':6',2''-terpyridine (tpy) ligands are popular to the cobalt(II) SCO materials [19–22]; for example, [Co(tpy)$_2$](BF$_4$)$_2$ has been reported to exhibit SCO at 270 K [19] and [Co(tpy)$_2$](ClO$_4$)$_2$·0.5H$_2$O displayed SCO at 180 K [20]. There have been a number of reports on the valence tautomerism in cobalt(II)-radical coordination systems [23,24]. Various photomagnets have been developed from heterospin systems including prussian blue analogues [25,26]. Such charge transfer mechanism also works in cobalt(II) spin-crossover materials carrying a directly coordinated nitroxide ligand [8,9], often disturbing the analysis of exchange interaction. Thus, our hypothesis is as follows: the organic paramagnetic center should be remote from the SCO center. A ligand π-electron system is indispensable to maintain appreciable exchange coupling (Scheme 1a). As for a paramagnetic substituent, *tert*-butyl phenyl nitroxide radicals are sufficiently persistent under ambient conditions [27,28]. Therefore, we designed a novel paramagnetic ligand based on tpy, 4'-{4-*tert*-butyl(N-oxy)aminophenyl}-2,2':6',2''-terpyridine (tpyphNO) (Scheme 1b). Its cobalt(II) complexes would be a target to realize the present project.

## 2. Materials and Methods

### 2.1. Materials

The ligands tpy and 4'-bromo-2,2':6',2''-terpyridine (Br-tpy) are commercially available. The latter was subjected to the preparation of tpyphNO (Scheme 2), as follows. The counterpart 4-(N-*tert*-butyl-O-*tert*-butyldimethylsilylhydroxylamino)phenyl boronic acid (TBDMS-BA) was prepared according to the literature method [29].

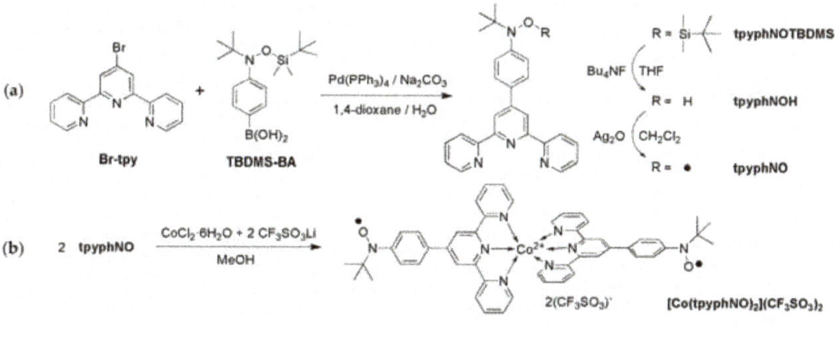

**Scheme 2.** Synthetic routes of (a) tpyphNO and (b) [Co(tpyphNO)$_2$](CF$_3$SO$_3$)$_2$.

A mixture of Br-tpy (730 mg; 2.34 mmol), TBDMS-BA (741 mg; 2.30 mmol), Pd(PPh$_3$)$_4$ (271 mg; 0.24 mmol), Na$_2$CO$_3$ (2.50 g; 14.3 mmol) in 50 mL of dioxane and 50 mL of water was heated at 100 °C for 72 h. The resultant mixture was extracted with dichloromethane, washed with brine and dried over anhydrous Na$_2$SO$_4$. The organic solution was filtered and concentrated under reduced pressure. The main product was separated through a basic alumina column eluted with dichloromethane. The concentration gave a yellow solid, which was characterized to be tpyphNOTBDMS (0.911 g; 1.79 mmol). Yield 85%. m.p:

212–223 °C. MS (ESI$^+$) $m/z$: 511.29 (M + H$^+$). $^1$H NMR (500 MHz CDCl$_3$): $\delta$ −0.084 (6H, br), 0.93 (9H, s), 1.13 (9H, s), 7.36 (2H, $J$ = 8.1 Hz, t), 7.36 (2H, $J$ = 8.6 Hz, d), 7.77 (2H, $J$ = 8.6 Hz, d), 7.88 (2H, $J$ = 8.1 Hz, t) 8.68 (2H, $J$ = 8.1 Hz, d), 8.73 (2H, s), 8.74 (2H, $J$ = 8.1 Hz, d). $^{13}$C NMR (126 MHz, CDCl$_3$): $\delta$ 156.34, 155.80, 152.14, 150.20, 149.12, 136.88, 134.56, 126.30, 125.46, 123.79, 121.34, 118.70, 61.44, 26.17, 26.12, 17.96, −4.64. IR (neat, attenuated total reflection (ATR)): 777, 857, 1466, 1061, 2853, 2928, 2957 cm$^{-1}$.

To a dry tetrahedrofuran (THF) solution (10 mL) containing tpyphNOTBDMS (0.911 g; 1.79 mmol) 2.5 mL (2.5 mmol) of tetrabutylammonium fluoride in a THF solution (1 mol L$^{-1}$) was added dropwise at 0 °C under nitrogen atmosphere. The mixture was stirred for further 1 h at room temperature. The resultant solution was extracted with dichloromethane after aqueous NaHCO$_3$ was added. The organic layer was dried over anhydrous Na$_2$SO$_4$ and filtered. After addition of a small amount of hexane, the deprotected product (tpyphNOH) was precipitated as a colorless solid (0.570 g; 1.43 mmol). Yield 85%. m.p.: 211–212 °C. MS (ESI$^+$) $m/z$: 397.20 (M + H$^+$), 419.17 (M + Na$^+$). $^1$H NMR (500 MHz CDCl$_3$): $\delta$ 1.20 (9H, s), 7.34 (2H, $J$ = 8.0 Hz, t), 7.38 (2H, $J$ = 8.6 Hz, d), 7.83 (2H, $J$ = 8.6 Hz, t) 7.88 (2H, $J$ = 8.0 Hz,d), 8.66 (2H, $J$ = 8.0 Hz, d), 8.72 (2H, s), 8.73 (2H, $J$ = 8.0 Hz, d). $^{13}$C NMR (126 MHz, CDCl$_3$): $\delta$ 156.36, 155.72, 150.84, 149.95, 149.16, 136.79, 135.09, 126.56, 124.56, 123.82, 121.22, 118.78, 60.97, 26.17. IR (neat, ATR): 730, 788, 1582, 2871, 2978, 3049, 3149, 3743 cm$^{-1}$.

After the above product (87 mg; 0.22 mmol) was dissolved in dichloromethane (20 mL), freshly prepared Ag$_2$O (510 mg; 2.2 mmol) was added and the resultant mixture was stirred at room temperature for 1 h. The solution portion was filtered and concentrated under reduced pressure. Crystallization from dichloromethane and hexane gave tpyphNO as a red solid (40 mg; 0.11 mmol). Yield 46%. M.p. 154–155 °C. MS (ESI$^+$) $m/z$: 396.18 (M + H$^+$), 418.16 (M + Na$^+$). IR (neat, ATR): 660, 778, 1193, 1250, 148, 1651, 2980 cm$^{-1}$. ESR (9.4 GHz, room temperature in toluene): $a_N$ = 1.165 mT, $a_{H(ortho)}$ = 0.212 mT (×2), $a_{H(meta)}$ = 0.094 mT (×2) at $g$ = 2.0065.

The target complexes were prepared as follows. A methanol solution (7 mL) involving tpyphNO (40 mg; 0.10 mmol), CoCl$_2$·6H$_2$O (12 mg; 0.050 mmol) and LiCF$_3$SO$_3$ (16 mg; 0.10 mmol) was allowed to stand at 0 °C under nitrogen atmosphere, to give [Co(tpyphNO)$_2$](CF$_3$SO$_3$)$_2$ as a dark red polycrystalline precipitation (27 mg; 0.024 mmol). Yield: 24%. m.p.: 286 °C (decomp.). The product was subjected to elemental, crystallographic and magnetic analyses without further purification. All data satisfied the formula of the target compound. IR (neat, ATR): 633, 790, 1030, 1136, 1260, 1603, 2935, 2978, 3083 cm$^{-1}$. Anal. Calcd. for C$_{52}$H$_{46}$Co$_1$F$_6$N$_8$O$_8$S$_2$: C, 54.40; H, 4.04; N, 9.76; S, 5.59%. Found: C, 54.34; H, 3.85; N, 9.81; S, 5.69%.

A similar method using tpy in place of tpyphNO gave [Co(tpy)$_2$](CF$_3$SO$_3$)$_2$ as orange polycrystals in an 85% yield. m.p.: 321 °C (decomp.). The product was subjected to elemental, crystallographic and magnetic analyses without further purification. All data satisfied the formula of the target compound. IR (neat, ATR): 513, 570, 632, 763, 1028, 1126, 1256, 1452, 1600, 3080 cm$^{-1}$. Anal. Calcd. for C$_{32}$H$_{22}$Co$_1$N$_6$O$_6$S$_2$: C, 46.67; H, 2.69; N, 10.20; S, 7.79%. Found: C, 46.29; H, 2.67; N, 10.08; S, 7.37%.

## 2.2. Crystallographic Analysis

X-ray diffraction data of [Co(L)$_2$](CF$_3$SO$_3$)$_2$ (L = tpyphNO, tpy) were collected on a Rigaku Saturn70 CCD diffractometer with graphite monochromated Mo K$\alpha$ radiation ($\lambda$ = 0.71073 Å). The structures were directly solved by a heavy-atom method and expanded using Fourier techniques in the CRYSTALSTRUCTURE [30]. Numerical absorption correction was used. Hydrogen atoms were located at calculated positions and their parameters were refined as "riding." The thermal displacement parameters of non-hydrogen atoms were refined anisotropically. Selected crystallographic data are given in Table 1 and selected bond distances and angles are listed in Tables 2 and 3. CCDC numbers 1826042, 1826043, 1826044 and 1826045 contain the crystallographic analysis details for [Co(tpyphNO)$_2$](CF$_3$SO$_3$)$_2$ at 90 and 300 K and [Co(tpy)$_2$](CF$_3$SO$_3$)$_2$ at 90 and 400 K, respectively. These data can be obtained free of charge via http://www.ccdc.cam.ac.uk/conts/retrieving.html.

**Table 1.** Selected crystallographic parameters of [Co(L)$_2$](CF$_3$SO$_3$)$_2$ (L = tpyphNO, tpy).

| L | tpyphNO | tpyphNO | tpy | tpy |
|---|---|---|---|---|
| T/K | 90 | 300 | 90 | 400 |
| Formula weight | 1148.03 | 1148.03 | 823.61 | 823.61 |
| Crystal system | monoclinic | monoclinic | orthorhombic | orthorhombic |
| Space group | $P2_1/c$ | $P2_1/c$ | $Pbcn$ | $Pbcn$ |
| a/Å | 19.6868(14) | 20.0956(19) | 16.554(4) | 16.884(2) |
| b/Å | 16.2129(14) | 16.5156(18) | 21.145(5) | 21.384(3) |
| c/Å | 16.3614(10) | 16.4648(14) | 9.0805(19) | 9.2858(13) |
| β/° | 107.774(4) | 108.753(4) | 90 | 90 |
| V/Å$^3$ | 4973.0(6) | 5174.4(9) | 3178.4(12) | 3352.7(8) |
| Z | 4 | 4 | 4 | 4 |
| $d_{calcd}$/g·cm$^{-3}$ | 1.533 | 1.474 | 1.721 | 1.632 |
| μ (MoKα)/mm$^{-1}$ | 0.517 | 0.497 | 0.765 | 0.725 |
| No. of unique reflections | 11373 | 10989 | 3516 | 3569 |
| $R(F)$ ($I > 2\sigma(I)$) [a] | 0.0639 | 0.0698 | 0.0567 | 0.0700 |
| $wR$ ($F^2$) (all reflections) [b] | 0.1597 | 0.1869 | 0.1426 | 0.1518 |
| Goodness-of-fit parameter | 1.036 | 1.011 | 0.951 | 1.031 |

[a] $R = \Sigma[|F_o| - |F_c|]/\Sigma|F_o|$; [b] $wR = [\Sigma w(F_o^2 - F_c^2)/\Sigma w F_o^4]^{1/2}$.

**Table 2.** Co–N bond distances ($d$) in Å for [Co(L)$_2$](CF$_3$SO$_3$)$_2$ (L = tpyphNO, tpy).

| L | tpyphNO | tpyphNO | tpy | tpy |
|---|---|---|---|---|
| T/K | 90 | 300 | 90 | 400 |
| $d$(O1–N4)/Å | 1.300(4) | 1.282(4) | - | - |
| $d$(O2–N8)/Å | 1.293(4) | 1.279(4) | - | - |
| N$_{central}$ | | | | |
| $d$(Co1–N2)/Å | 1.877(3) | 2.017(3) | 1.912(4) | 2.024(4) |
| $d$(Co1–N6/4)/Å | 1.941(3) | 2.025(3) | 1.886(4) | 2.023(4) |
| $d_{central,avg}$/Å | 1.91 | 2.02 | 1.90 | 2.02 |
| $\Delta d_{central}$/Å | | 0.11 | | 0.12 |
| N$_{distal}$ | | | | |
| $d$(Co1–N1)/Å | 2.011(3) | 2.144(3) | 2.143(3) | 2.134(3) |
| $d$(Co1–N3)/Å | 2.010(3) | 2.130(3) | 2.017(3) | 2.120(3) |
| $d$(Co1–N5)/Å | 2.170(3) | 2.154(3) | - | - |
| $d$(Co1–N7)/Å | 2.157(3) | 2.154(3) | - | - |
| $d_{distal,avg}$/Å | 2.09 | 2.15 | 2.08 | 2.13 |
| $\Delta d_{distal}$/Å | | 0.06 | | 0.05 |

**Table 3.** N–Co–N bond angles ($\phi$ in °) and distortion geometrical parameters Σ (in °) and CShM for [Co(L)$_2$](CF$_3$SO$_3$)$_2$ (L = tpyphNO, tpy).

| L | tpyphNO | tpyphNO | L [a] | tpy | tpy |
|---|---|---|---|---|---|
| T/K | 90 | 300 | T/K | 90 | 400 |
| $\phi$(N1–Co1–N2)/° | 80.99(11) | 76.99(11) | $\phi$(N1–Co1–N2)/° | 78.93(8) | 76.80(8) |
| $\phi$(N1–Co1–N3)/° | 161.09(11) | 152.95(12) | $\phi$(N1–Co1–N1 #)/° | 157.87(17) | 153.59(17) |
| $\phi$(N1–Co1–N5)/° | 97.53(10) | 100.25(11) | $\phi$(N1–Co1–N3)/° | 94.58(12) | 96.27(12) |
| $\phi$(N1–Co1–N6)/° | 94.64(11) | 95.43(11) | $\phi$(N1–Co1–N3 #)/° | 89.07(11) | 89.74(12) |
| $\phi$(N1–Co1–N7)/° | 89.03(10) | 89.93(11) | $\phi$(N1–Co1–N4)/° | 101.07(8) | 103.20(8) |
| $\phi$(N2–Co1–N3)/° | 80.14(10) | 76.20(11) | $\phi$(N1 #–Co1–N4)/° | 101.06(8) | 103.21(8) |
| $\phi$(N2–Co1–N5)/° | 101.31(11) | 103.23(12) | $\phi$(N1 #–Co1–N3)/° | 89.07(11) | 89.74(12) |
| $\phi$(N2–Co1–N6)/° | 175.55(11) | 172.30(11) | $\phi$(N1 #–Co1–N3 #)/° | 94.58(12) | 96.27(12) |
| $\phi$(N2–Co1–N7)/° | 102.63(11) | 104.88(12) | $\phi$(N2–Co1–N1 #)/° | 78.94(8) | 76.79(8) |
| $\phi$(N3–Co1–N5)/° | 87.43(10) | 89.28(11) | $\phi$(N2–Co1–N3)/° | 99.53(9) | 103.24(9) |
| $\phi$(N3–Co1–N6)/° | 104.24(11) | 111.44(11) | $\phi$(N2–Co1–N3 #)/° | 99.53(9) | 103.24(9) |
| $\phi$(N3–Co1–N7)/° | 93.83(10) | 93.53(11) | $\phi$(N2–Co1–N4)/° | 180.0 | 180.0 |
| $\phi$(N5–Co1–N6)/° | 78.32(11) | 76.55(12) | $\phi$(N3–Co1–N4)/° | 80.47(9) | 76.76(9) |
| $\phi$(N5–Co1–N7)/° | 155.89(10) | 151.61(12) | $\phi$(N3–Co1–N3 #)/° | 160.94(18) | 153.52(18) |
| $\phi$(N6–Co1–N7)/° | 78.04(11) | 76.14(12) | $\phi$(N4–Co1–N3 #)/° | 80.47(9) | 76.76(9) |
| Σ/° | 100.2 | 123.7 | Σ/° | 93.4 | 118.8 |
| CShM (Oh) | 2.824 | 4.282 | CShM (Oh) | 2.484 | 3.672 |

[a] Symmetry code for #: $1 - x, +y, 1/2 - z$.

*2.3. Magnetic Study*

Magnetic susceptibilities of [Co(L)$_2$](CF$_3$SO$_3$)$_2$ (L = tpyphNO, tpy) were measured on a Quantum Design MPMS-XL7 SQUID magnetometer with a static field of 0.5 T. The magnetic responses were corrected with diamagnetic blank data of the sample holder measured separately. The diamagnetic contribution of the sample itself was estimated from Pascal's constants [31].

## 3. Results and Discussion

*3.1. Preparation*

A new ligand tpyphNO was prepared via the Suzuki coupling reaction [32] from commercially available 4′-bromoterpyridine (Br-tpy) and a protected hydroxylaminophenyl boronic acid (TBDMS-BA) [29] (Scheme 2a). Paramagnetic tpyphNO was prepared after the deprotection of the above product with tetrabutylammonium fluoride followed by the oxidation with Ag$_2$O. The resultant nitroxide was isolated at room temperature under air and characterized as tpyphNO by means of spectroscopic methods including electron spin resonance (ESR) spectroscopy. A target complex [Co(tpyphNO)$_2$](CF$_3$SO$_3$)$_2$ was prepared by simply combining methanol solutions of the ligand and CoCl$_2$ in the presence of the counter anion CF$_3$SO$_3^-$ (Scheme 2b). As a reference complex, [Co(tpy)$_2$](CF$_3$SO$_3$)$_2$ was also prepared in a similar manner, using tpy in place of tpyphNO. The nitroxide-carrying derivative is dark red and the reference is orange at room temperature.

*3.2. Crystal Structures*

The X-ray crystallographic analysis on [Co(L)$_2$](CF$_3$SO$_3$)$_2$ (L = tpyphNO, tpy) was successful at 90 and 300 or 400 K (Table 1 and Figure 1). Though the crystal structure of [Co(tpy)$_2$](CF$_3$SO$_3$)$_2$ at 120 K has recently been reported [33], we measured them at 90 and 400 K to compare the LS and HS structures. The crystal structure of [Co(tpy)$_2$](CF$_3$SO$_3$)$_2$ possesses a relatively high symmetry orthorhombic *Pbcn*, which is kept between 90 and 400 K. A half molecule is crystallographically independent. Compound [Co(tpy)$_2$](ClO$_4$)$_2$·0.5H$_2$O is known to crystallize in a tetragonal cell [20] and the relatively low symmetry of CF$_3$SO$_3^-$ may cause the different crystal system. On the other hand, the crystal of [Co(tpyphNO)$_2$](CF$_3$SO$_3$)$_2$ belongs to monoclinic $P2_1/c$ and the whole molecule corresponds to an independent unit. The linear spin triad structure is unequivocally characterized. There is no solvent molecule in any crystal.

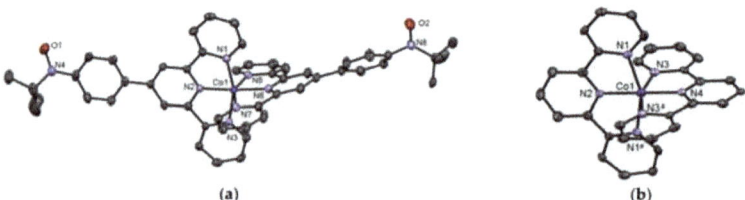

**Figure 1.** X-ray crystal structures of [Co(L)$_2$](CF$_3$SO$_3$)$_2$ (L = (**a**) tpyphNO; (**b**) tpy), measured at 90 K. The thermal ellipsoids are drawn at the 50% probability level. Hydrogen atoms and counter anions are omitted for clarity. The symmetry operation code for # is (1 − *x*, *y*, 1/2 − *z*) in (**b**).

The nitroxide group was characterized by the N–O bond lengths (1.300(4) Å N4–O1 and 1.293(4) Å for N8–O2 at 90 K) in a typical range of aryl *tert*-butyl nitroxides [34]. The two meridional chelate planes are arranged to be almost perpendicular with the dihedral angle of 94.99(6)°. The long molecular axis is somewhat bent at the metal center, as indicated with the N4...Co1...N8 angle of 159.98(3)° at 90 K, being considerably smaller than 180°. The 4-phenylpyridine core in each ligand is not coplanar.

The dihedral angles between the pyridine and adjacent phenyl rings are 10.4(1) and 33.8(1)° at 90 K with respect to the N2- and N6-pyridine sides, respectively.

The cell volume expansion of the tpy derivative is 4.0% from 90 to 300 K and that of the tpyphNO derivative 5.5% from 90 to 400 K. The considerable volume changes originate in the distance changes between the metal and the coordinated donor atom ($\Delta d$) accompanying SCO. Usually $\Delta d$ is not so large (0.07–0.11 Å) in cobalt(II) SCO complexes as those of the iron(II) complexes [15,19,20], because only one electron is transferred to the antibonding orbital upon SCO [16]. The present Co–N bond lengths are completely compatible with those of the previous SCO [Co(tpy)$_2$]$^{2+}$ compounds. For example, on the known SCO complex [Co(tpy)$_2$](BF$_4$)$_2$, Kilner et al. [19] reported that $d$(Co–N$_{central}$) of the HS state is longer than that of the LS state by 0.12 Å on the average. In our case $\Delta d_{central}$ = 0.11 and 0.12 Å for the complexes with L = tpyphNO and tpy, respectively (Table 2). As for the Co–N$_{distal}$ bond lengths, the HS state possesses longer distances than the LS state by 0.06 Å in [Co(tpy)$_2$](BF$_4$)$_2$ [19]. The present compounds showed $\Delta d_{distal}$ = 0.06 and 0.05 Å, respectively. These quite similar geometrical features strongly suggest that the low- and high-temperature structures can be assigned to approximate LS and HS states, respectively. This hypothesis is proven from the magnetic study (see below). The different sensitivity between $\Delta d_{central}$ and $\Delta d_{distal}$ is caused by the Jahn-Teller effect due to the LS e$_g$$^1$ state as well as the steric effect from the rigid ligand.

The HS states are known to favor distorted coordination geometry [35–38]. Among various geometrical parameters, $\Sigma$ and CShM seem to be sensitive and convenient metrics [38]. The $\Sigma$ values [39] were derived from the N–Co–N bond angles (Table 3), according to Equation (1). An ideal octahedron (Oh) possess $\Sigma$ = 0°. By using the SHAPE software [40], the continuous shape measures (CShM) are calculated with respect to an Oh. An ideal Oh returns null. The HS states possess relatively distorted Oh, as expected (4.282 at 300 K versus 2.824 at 90 K for [Co(tpyphNO)$_2$](CF$_3$SO$_3$)$_2$ and so on). The bite angle of the five-membered chelate ring seems to be responsible to the difference of $\Sigma$; namely, $\phi$ in the HS state tends to be smaller than that of the LS state (79.37° at 90 K versus 76.47° at 300 K on the average). Furthermore, the $\phi$ reduction is related to the elongation of the five-membered ring. In short, the Co–N distance regulates these distortion parameters.

$$\Sigma = \sum_{i=1}^{12} |\varphi(\angle cis\ N - Fe - N)_i - 90°| \tag{1}$$

We have to make a comment on the intermolecular interaction in particular in the crystal of [Co(tpyphNO)$_2$](CF$_3$SO$_3$)$_2$. The shortest interatomic distances with respect to the N–O groups are 5.205(4) Å for O4...O2' and 5.615(4) Å for N4...N8'' at 90 K [the symmetry operation codes for ' and '' are (1 + $x$, $y$, $z$) and (1 + $x$, 3/2 − $y$, 1/2 + $z$), respectively]. There hardly seems to be any exchange pathway. The tpy portions in the nearest neighboring molecules are arranged parallel with a separation of ca. 3.6 Å (Figure 2). The shortest Co...O(nitroxide) is found as 4.241(3) Å for Co1...O1* [the symmetry operation code for * is (1 − $x$, 1 − $y$, 1 − $z$)]. Two molecules are linked in a head-to-tail manner with two centrosymmetry-related Co1...O1* and Co1*...O1 distances. It is more likely that the intramolecular interaction through π-conjugation is dominant compared to the intermolecular through-space interaction but relatively short intermolecular distances cannot be neglected completely. In this case, the magnetic properties would be described as the sum of two Co...nitroxide pairs and two nitroxide doublets in every two molecules. This is another interpretation for exchange coupling in [Co(tpyphNO)$_2$](CF$_3$SO$_3$)$_2$. However, the motivation of this project never changes, because the cobalt(II) and nitroxide spins are exchange-coupled indeed, whether it works in an intra- or intermolecular fashion. By sharp contrast, such supramolecular contacts are absent from parent [Co(tpy)$_2$](CF$_3$SO$_3$)$_2$.

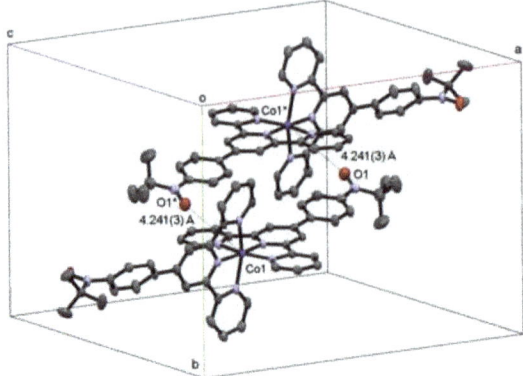

**Figure 2.** Molecular arrangement in the crystal of [Co(tpyphNO)$_2$](CF$_3$SO$_3$)$_2$. Only two molecules are shown. Hydrogen atoms and counter anions are omitted for clarity. Symmetry code for *: $1-x, 1-y, 1-z$.

### 3.3. Magnetic Properties

The magnetic susceptibilities of polycrystalline specimens of [Co(L)$_2$](CF$_3$SO$_3$)$_2$ (L = tpyphNO, tpy) were measured on a SQUID magnetometer in a temperature range of 1.8–300 K for the former and 1.8–400 K for the latter. As Figure 3 shows, the $\chi_m T$ values of [Co(tpy)$_2$](CF$_3$SO$_3$)$_2$ were 0.516 and 2.20 cm$^3$ K mol$^{-1}$ at 90 and 400 K, respectively. From the crystal structure analysis, the spin-state at 90 K is LS, namely, $S_{Co2+} = 1/2$ and accordingly the Landé factor $g_{Co2+,LS} = 2.35$. The spin state at 400 K is HS, $S_{Co2+} = 3/2$, which leads to $g_{Co2+,HS} = 2.15$. The latter involves a slight underestimation of $g_{Co2+,HS}$, because the $\chi_m T$ value still has a small positive slope at 400 K. The SCO temperature $T_{1/2}$ is defined as the temperature at which equimolar fractions of the HS and LS species are present. The gradual S-shaped curve in 150–400 K indicates $T_{1/2}$ = ca. 260 K for [Co(tpy)$_2$](CF$_3$SO$_3$)$_2$. The $\chi_m T$ value is ideally flat below 100 K. Note that practically no exchange coupling took place, especially illustrated with the constant $\chi_m T$ in a lowest-temperature region.

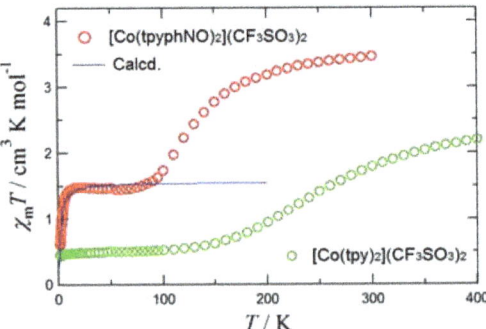

**Figure 3.** Temperature dependence of $\chi_m T$ for polycrystalline [Co(L)$_2$](CF$_3$SO$_3$)$_2$ (L = tpyphNO, tpy), measured at 5000 Oe. A solid line represents the calculated curve for a low temperature region. See the text for the equation and optimized parameters.

Referring to the results of [Co(tpy)$_2$](CF$_3$SO$_3$)$_2$, we can analogously analyze the data of [Co(tpyphNO)$_2$](CF$_3$SO$_3$)$_2$. The $\chi_m T$ versus $T$ profile of [Co(tpyphNO)$_2$](CF$_3$SO$_3$)$_2$ is apparently biased from that of [Co(tpy)$_2$](CF$_3$SO$_3$)$_2$, by the contribution of two radical spins (0.75 cm$^3$ K mol$^{-1}$). Another cause of this gap is the difference of the $g_{Co2+}$ values between the two compounds. The high-temperature $\chi_m T$ value of $[Co(tpyphNO)_2](CF_3SO_3)_2$ was 3.47 cm$^3$ K mol$^{-1}$ at 300 K. On cooling the $\chi_m T$ value decreased to draw an S-shaped profile in 250–100 K and reached a plateau at ca. 1.45 cm$^3$ K mol$^{-1}$ around 80 K. On further cooling, the $\chi_m T$ value again decreased to the smallest value 0.607 cm$^3$ K mol$^{-1}$ at 1.8 K (the base temperature of the apparatus available). The first drop is ascribable to Co$^{2+}$ SCO behavior with $T_{1/2}$ = ca. 140 K. The second drop is accordingly assigned to exchange coupling behavior among the LS Co$^{2+}$ spin and peripheral nitroxide spins.

The spin-Hamiltonian is defined as Equation (2), where $J_{Co\text{-}rad}$ stands for the exchange coupling constant. An approximation is introduced, where the spin centers are symmetrically arrayed in a linear manner and the interaction between the terminals is ignored. The fitting is performed only for analyzing the exchange behavior recorded in a low-spin region. The parameters were optimized according to the van Vleck equation, involving an averaged $g$ value [41,42], giving $g_{avg}$ = 2.352(9) and $2J_{Co\text{-}rad}/k_B$ = −3.63(12) K. Alternatively, the $g_{rad}$ and $g_{Co2+,LS}$ values can be separated with a more detailed van Vleck equation written as Equation (3) [43]. Assuming that the $g_{rad}$ value is frozen to 2.006 (from the ESR spectrum of tpyphNO), the optimization gave $g_{Co2+,LS}$ = 2.98(2) together with $2J_{Co\text{-}rad}/k_B$ = −3.00(6) K. The calculation curve is superposed in Figure 3.

$$H = -2J_{Co\text{-}rad}(S_{Co} \cdot S_{rad1} + S_{Co} \cdot S_{rad2}) \qquad (2)$$

$$\chi_m = \frac{N_A \mu_B^2}{4 k_B T} \frac{10 g_{3/2,1}^2 \exp(J_{Co-rad}/k_B T) + g_{1/2,0}^2 + g_{1/2,1}^2 \exp(-2 J_{Co-rad}/k_B T)}{2 \exp(J_{Co-rad}/k_B T) + 1 + \exp(-2 J_{Co-rad}/k_B T)} \qquad (3)$$

with

$g_{1/2,1} = (4 g_{rad} - g_{Co2+,LS})/3$
$g_{3/2,1} = (2 g_{rad} + g_{Co2+,LS})/3$
$g_{1/2,0} = g_{Co2+,LS}$

At the ground state, $S_{total}$ should be 1/2; on the other hand, three paramagnetic spins are present in the almost constant $\chi_m T$ region in ca. 20–80 K. Thanks to the different temperature regions where spin-crossover and exchange coupling effects are operative, the exchange coupling parameter is well resolved to give a precise evaluation. Furthermore, the $\chi_m T$ plateau clearly appeared. In total, $[Co(tpyphNO)_2](CF_3SO_3)_2$ can be regarded as a doubly switchable material showing 1 $\mu_B \rightleftarrows$ 3 $\mu_B \rightleftarrows$ 5 $\mu_B$ by temperature stimulus.

The $\chi_m T$ versus $T$ profile for $[Co(tpy)_2](CF_3SO_3)_2$ shows a very gradual SCO curve, whereas that of $[Co(tpyphNO)_2](CF_3SO_3)_2$ displays a relatively abrupt SCO curve (Figure 3). As described above (Figure 2), there are intermolecular interactions such as short Co...O(nitroxide) distances. The ligands in a neighboring molecule are centrosymmetry-related and planar portions are arranged in parallel with a separation of ca. 3.6 Å. Weak π-π stacking effects can be found in a dimeric structure as well as in interdimer relation. Owing to the spiro-type structure of the $[Co(tpyphNO)_2]^{2+}$ core, another parallel stacking motifs spread in the second direction, though the counter anion intervenes. The peripheral substituents like 4-tert-butyl(N-oxy)aminophenyl may serve additional intermolecular interaction, which may contribute cooperativity [44]. Such intermolecular interactions enhance an abrupt character of SCO [3,44,45].

Basically, the $e_g$ orbitals with σ-type symmetry possess no orbital overlap against π or π*-type orbitals of the ligand. This situation has been discussed when the nitroxide radical is directly coordinated to the metal ions [42,46,47] and in the present compound the ligating atom is a pyridine nitrogen atom. The 3d electron configuration of LS Co$^{2+}$ is $(t_{2g})^6(e_g)^1$ and the magnetic $e_g$ orbitals might lead to orthogonal geometry between the two magnetic orbitals (Figure 4a,b). However, the orthogonality is very sensitive to the coordination structure and out-of-plane deformation gives rise to loss of ferromagnetic coupling (Figure 4c,d) [14,34,42,48–51]. As the crystallographic analysis revealed, the long molecular axis is considerably bent (159.98(3)°) and the octahedral coordination sphere is largely distorted owing to the five-membered chelate ring. Therefore, the orthogonality is ready to breakdown.

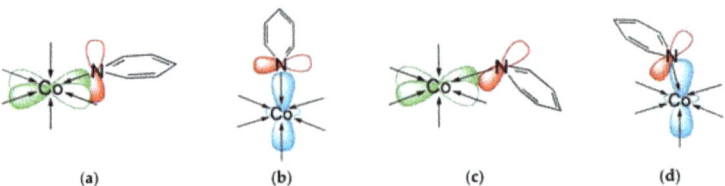

**Figure 4.** Schematic drawing of the absence or presence of orbital overlaps. (**a**,**b**) Geometries of (**a**) the Co $3d_{x^2-y^2}$ and N $2p_z$ orbitals or (**b**) Co $3d_{z^2}$ and N $2p_z$ orbitals with a negligible angular torsion. (**c**,**d**) With an appreciable angular torsion.

The magnitude of the exchange coupling is limited to be small ($2J_{Co\text{-rad}}/k_B$ = −3.00(6) K). It is comparable to several 3d-2p heterospin exchange coupling across a pyridine ring [52,53] and smaller than the 3d-3d exchange interaction found in the known dinuclear cobalt(II) SCO compound ($2J'/hc$ = 11.7 cm$^{-1}$) [54]. There is an intervening organic portion between the 2p and 3d spins in [Co(tpyphNO)$_2$](CF$_3$SO$_3$)$_2$. The spin-polarization mechanism is well documented with respect to the 2-, 3- and 4-pyridyl-substituted isomers [52,53]. As shown in Scheme 3, the 1,$p$-position of the 4-phenylpyridine core plays a role of a magnetic coupler and the ligating nitrogen atom has a positive spin density. As stated above, ferromagnetic coupling would be expected with an orthogonally placed $e_g$ spin. However, the spin-polarization is not so effective across a long distance. Moreover, a non-planar biaryl conformation brings about a reduction of the exchange interaction [55–57]. The dihedral angles between the pyridine and adjacent benzene rings are 10.4(1) and 33.8(1)° in [Co(tpyphNO)$_2$](CF$_3$SO$_3$)$_2$. A shorter ligand without a *para*-phenylene spacer—namely, *tert*-butyl 2,2′:6′,2″-terpyridin-4′-yl nitroxide—might be a promising exchange coupler to improve exchange interaction. Its cobalt(II) complexes will be a next target.

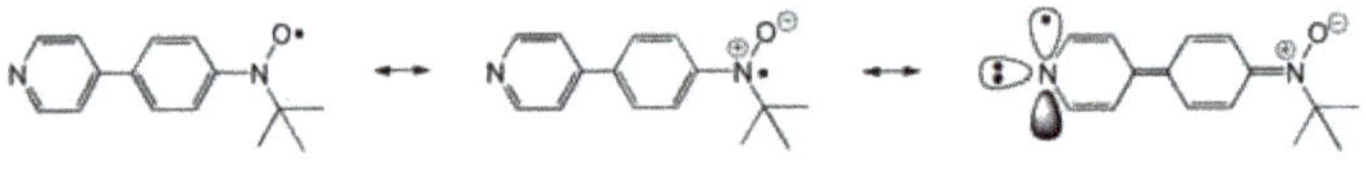

**Scheme 3.** Canonical structures of tpyphNO.

## 4. Conclusions

The SCO behavior was observed in [Co(tpy)$_2$](CF$_3$SO$_3$)$_2$ and [Co(tpyphNO)$_2$](CF$_3$SO$_3$)$_2$. The tpy compound showed a gradual SCO in 150–400 K. On the other hand, the tpyphNO derivative exhibited a relatively abrupt SCO in 100–250 K together with antiferromagnetic Co-nitroxide exchange coupling with $2J/k_B$ = −3.00(6) K. The comparison work has proven the coexistence of SCO and exchange coupling in a complex ion [Co(tpyphNO)$_2$]$^{2+}$. The d-π magnetic exchange coupling is rationalized with the pyridine π-conjugation system. Thanks to the different temperature regions where they are operative, the magnetic moment apparently switches double-stepwise as 1 $\mu_B$ ⇌ 3 $\mu_B$ ⇌ 5 $\mu_B$ by temperature stimulus. The present work can be regarded as a successful example of development of multifunctional SCO materials including additional magnetic exchange coupling.

**Acknowledgments:** This work was financially supported from KAKENHI (JSPS/15H03793).

**Author Contributions:** Akihiro Ondo participated in the preparation, X-ray structural analysis and magnetic study. Takayuki Ishida designed the study and wrote the manuscript.

**Conflicts of Interest:** The authors declare no conflict of interest.

**References**

1. Gütlich, P.; Goodwin, H.A. (Eds.) *Spin Crossover in Transition Metal Compounds I, II, and III*; Springer: Berlin, Germany, 2004.
2. Halcrow, M.A. *Spin-Crossover Materials: Properties and Applications*; John Wiley & Sons, Ltd.: Oxford, UK, 2013.
3. Kahn, O. Chapter 4, Low-Spin–High-Spin Transition. In *Molecular Magnetism*; VCH: Weinhein, Germany, 1993.
4. Gentili, D.; Demitri, N.; Schäfer, B.; Liscio, F.; Bergenti, I.; Ruani, G.; Ruben, M.; Cavallini, M. Multi-modal sensing in spin crossover compounds. *J. Mater. Chem. C* **2015**, *3*, 7836–7844. [CrossRef]
5. Galyametdinov, Y.; Ksenofontov, V.; Prosvirin, A.; Ovchinnikov, I.; Ivanova, G.; Gutlich, P.; Haase, W. First Example of Coexistence of Thermal Spin Transition and Liquid-Crystal Properties. *Angew. Chem. Int. Ed.* **2001**, *40*, 4269–4271. [CrossRef]
6. Oso, Y.; Ishida, T. Spin-crossover transition in a mesophase iron(II) thiocyanate complex chelated with 4-hexadecyl-N-(2-pyridylmethylene)aniline. *Chem. Lett.* **2009**, *38*, 604–605. [CrossRef]
7. Oso, Y.; Kanatsuki, D.; Saito, S.; Nogami, T.; Ishida, T. Spin-crossover transition coupled with another solid-solid phase transition for iron(II) thiocyanate complexes chelated with alkylated N-(di-2-pyridylmethylene)anilines. *Chem. Lett.* **2008**, *37*, 760–761. [CrossRef]
8. Gass, I.A.; Tewary, S.; Rajaraman, G.; Asadi, M.; Lupton, D.W.; Moubaraki, B.; Chastanet, G.; Létard, J.F.; Murray, K.S. Solvate-dependent spin crossover and exchange in cobalt(II) oxazolidine nitroxide chelates. *Inorg. Chem.* **2014**, *53*, 5055–5066. [CrossRef] [PubMed]
9. Gass, I.A.; Tewary, S.; Nafady, A.; Chilton, N.F.; Gartshore, C.J.; Asadi, M.; Lupton, D.W.; Moubaraki, B.; Bond, A.M.; Boas, J.F.; et al. Observation of ferromagnetic exchange, spin crossover, reductively induced oxidation, and field-induced slow magnetic relaxation in monomeric cobalt nitroxides. *Inorg. Chem.* **2013**, *52*, 7557–7572. [CrossRef] [PubMed]
10. Kitazawa, T.; Gomi, Y.; Takahashi, M.; Takeda, M.; Enomoto, M.; Miyazaki, A.; Enoki, T. Spin-crossover behaviour of the coordination polymer $Fe^{II}(C_5H_5N)_2Ni^{II}(CN)_4$. *J. Mater. Chem.* **1996**, *6*, 119–121. [CrossRef]
11. Letard, J.-F.; Guionneau, P.; Nguyen, O.; Costa, J.S.; Marcen, S.; Chastanet, G.; Marchivie, M.; Goux-Capes, L. A guideline to the design of molecular-based materials with long-lived photomagnetic lifetimes. *Chem. Eur. J.* **2005**, *11*, 4582–4589. [CrossRef] [PubMed]
12. Mochida, N.; Kimura, A.; Ishida, T. Spin-Crossover Hysteresis of $[Fe^{II}(L_H^{iPr})_2(NCS)_2]$ ($L_H^{iPr}$ = N-2-Pyridylmethylene-4-isopropylaniline Accompanied by Isopropyl Conformation Isomerism. *Magnetochemistry* **2015**, *1*, 17–27. [CrossRef]
13. Yamasaki, M.; Ishida, T. Heating-rate dependence of spin-crossover hysteresis observed in an iron(II) complex having tris(2-pyridyl)methanol. *J. Mater. Chem. C* **2015**, *3*, 7784–7787. [CrossRef]
14. Homma, Y.; Ishida, T. A New $S = 0 \rightleftarrows S = 2$ "Spin-crossover" Scenario Found in a Nickel(II) Bis(nitroxide) System. *Chem. Mater.* **2018**, *30*, 1835–1838. [CrossRef]
15. Hayami, S.; Komatsu, Y.; Shimizu, T.; Kamihata, H.; Lee, Y.H. Spin-crossover in cobalt(II) compounds containing terpyridine and its derivatives. *Coord. Chem. Rev.* **2011**, *255*, 1981–1990. [CrossRef]
16. Krivokapic, I.; Zerara, M.; Daku, M.L.; Vargas, A.; Enachescu, C.; Ambrus, C.; Tregenna-Piggott, P.; Amstutz, N.; Krausz, E.; Hauser, A. Spin-crossover in cobalt(II) imine complexes. *Coord. Chem. Rev.* **2007**, *251*, 364–378. [CrossRef]
17. Murray, K.S. Advances in Polynuclear Iron(II), Iron(III) and Cobalt(II) Spin-Crossover Compounds. *Eur. J. Inorg. Chem.* **2008**, 3101–3121. [CrossRef]
18. Brooker, S. Spin crossover with thermal hysteresis: Practicalities and lessons learnt. *Chem. Soc. Rev.* **2015**, *44*, 2880–2892. [CrossRef] [PubMed]
19. Kilner, C.A.; Halcrow, M.A. An unusual discontinuity in the thermal spin transition in $[Co(terpy)_2][BF_4]_2$. *Dalton Trans.* **2010**, *39*, 9008–9012. [CrossRef] [PubMed]
20. Oshio, H.; Spiering, H.; Ksenofontov, V.; Renz, F.; Gütlich, P. Electronic Relaxation Phenomena Following $^{57}Co$ (EC) $^{57}Fe$ Nuclear Decay in $[Mn^{II}(terpy)_2](ClO_4)_2 \cdot 1/2H_2O$ and in the Spin Crossover Complexes $[Co^{II}(terpy)_2]X_2 \cdot nH_2O$ (X = Cl and $ClO_4$): A Mössbauer Emission Spectroscopic Study. *Inorg. Chem.* **2001**, *40*, 1143–1150. [CrossRef] [PubMed]
21. Aroua, S.; Todorova, T.K.; Hommes, P.; Chamoreau, L.M.; Reissig, H.U.; Mougel, V.; Fontecave, M. Synthesis, Characterization, and DFT Analysis of Bis-Terpyridyl-Based Molecular Cobalt Complexes. *Inorg. Chem.* **2017**, *56*, 5930–5940. [CrossRef] [PubMed]

22. Enachescu, C.; Krivokapic, I.; Zerara, M.; Real, J.A.; Amstutz, N.; Hauser, A. Optical investigation of spin-crossover in cobalt(II) bis-terpy complexes. *Inorg. Chim. Acta* **2007**, *360*, 3945–3950. [CrossRef]
23. Adams, D.M.; Hendrickson, D.N. Pulsed laser photolysis and thermodynamics studies of intramolecular electron transfer in valence tautomeric cobalt o-quinone complexes. *J. Am. Chem. Soc.* **1996**, *118*, 11515–11528. [CrossRef]
24. Pierpont, C.G. Studies on charge distribution and valence tautomerism in transition metal complexes of catecholate and semiquinonate ligands. *Coord. Chem. Rev.* **2001**, *216*, 99–125. [CrossRef]
25. Escax, V.; Bleuzen, A.; Cartier dit Moulin, C.; Villain, F.; Goujon, A.; Varret, F.; Verdaguer, M. Photoinduced Ferrimagnetic Systems in Prussian Blue Analogues $CI_xCo_4[Fe(CN)_6]_y$ (CI = Alkali Cation). 3. Control of the Photo-and Thermally Induced Electron Transfer by the $[Fe(CN)_6]$ Vacancies in Cesium Derivatives. *J. Am. Chem. Soc.* **2001**, *123*, 12536–12543. [CrossRef] [PubMed]
26. Sato, O.; Cui, A.; Matsuda, R.; Tao, J.; Hayami, S. Photo-induced valence tautomerism in Co complexes. *Acc. Chem. Res.* **2007**, *40*, 361–369. [CrossRef] [PubMed]
27. Calder, A.; Forrester, A.R. Nitroxide radicals. Part VI. Stability of meta-and para-alkyl substituted phenyl-*t*-butylnitroxides. *J. Chem. Soc. C* **1969**, 1459–1464. [CrossRef]
28. Murata, H.; Lahti, P.M.; Aboaku, S. Molecular recognition in a heteromolecular radical pair system with complementary multipoint hydrogen-bonding. *Chem. Commun.* **2008**, 3441–3443. [CrossRef] [PubMed]
29. Lahti, P.M.; Liao, Y.; Julier, M.; Palacio, F. s-Triazine as an exchange linker in organic high-spin molecules. *Synth. Met.* **2001**, *122*, 485–493. [CrossRef]
30. *CrystalStructure, Version 4.2.1*; Rigaku/MSC: The Woodlands, TX, USA, 2015.
31. Kahn, O. Chapter 1.2, Diamagnetic and Paramagnetic Susceptibilities. In *Molecular Magnetism*; VCH: Weinhein, Germany, 1993.
32. Suzuki, A. Cross-coupling reactions of organoboranes: An easy way to construct C-C bonds (Nobel lecture). *Angew. Chem. Int. Ed.* **2011**, *50*, 6722–6737. [CrossRef] [PubMed]
33. Pavlov, A.A.; Denisov, G.L.; Kiskin, M.A.; Nelyubina, Y.V.; Novikov, V.V. Probing Spin Crossover in a Solution by Paramagnetic NMR Spectroscopy. *Inorg. Chem.* **2017**, *56*, 14759–14762. [CrossRef] [PubMed]
34. Okazawa, A.; Nagaichi, Y.; Nogami, T.; Ishida, T. Magneto-structure relationship in copper(II) and nickel(II) complexes chelated with stable *tert*-butyl 5-phenyl-2-pyridyl nitroxide and related radicals. *Inorg. Chem.* **2008**, *47*, 8859–8868. [CrossRef] [PubMed]
35. Wu, S.Q.; Wang, Y.T.; Cui, A.L.; Kou, H.Z. Toward Higher Nuclearity: Tetranuclear Cobalt(II) Metallogrid Exhibiting Spin Crossover. *Inorg. Chem.* **2014**, *53*, 2613–2618. [CrossRef] [PubMed]
36. Tiwary, S.K.; Vasudevan, S. Void geometry driven spin crossover in zeolite-encapsulated cobalt tris(bipyridyl) complex ion. *Inorg. Chem.* **1998**, *37*, 5239–5246. [CrossRef]
37. Halcrow, M.A. Structure: Function relationships in molecular spin-crossover complexes. *Chem. Soc. Rev.* **2011**, *40*, 4119–4142. [CrossRef] [PubMed]
38. Kimura, A.; Ishida, T. Pybox-Iron(II) Spin-Crossover Complexes with Substituent Effects from the 4-Position of the Pyridine Ring (Pybox = 2,6-Bis(oxazolin-2-yl)pyridine). *Inorganics* **2017**, *5*, 52.
39. Guionneau, P.; Marchivie, M.; Bravic, G.; Létard, J.-F.; Chasseau, D. Structural Aspects of Spin Crossover. Example of the $[Fe^{II}L_n(NCS)_2]$ Complexes. *Top. Curr. Chem.* **2004**, *234*, 97–128.
40. Lluncll, M.; Casanova, D.; Circra, J.; Bofill, J.M.; Alcmany, P.; Alvarez, S.; Pinsky, M.; Avnir, D. *SHAPE, v2.1*; University of Barcelona: Barcelona, Spain; The Hebrew University of Jerusalem: Jerusalem, Israel, 2005.
41. Gruber, S.J.; Harris, C.M.; Sinn, E. Metal complexes as ligands. VI. Antiferromagnetic interactions in trinuclear complexes containing similar and dissimilar metals. *J. Chem. Phys.* **1968**, *49*, 2183–2191. [CrossRef]
42. Okazawa, A.; Nogami, T.; Ishida, T. *tert*-Butyl 2-Pyridyl Nitroxide Available as a Paramagnetic Chelate Ligand for Strongly Exchange-Coupled Metal−Radical Compounds. *Chem. Mater.* **2007**, *19*, 2733–2735. [CrossRef]
43. Kahn, O. Chapter 10, Trinuclear Compounds and Compounds of Higher Nuclearity. In *Molecular Magnetism*; VCH: Weinhein, Germany, 1993.
44. Boca, R. *Theoretical Foundations of Molecular Magnetism: Current Methods in Inorganic Chemistry*; Elsevier: Amsterdam, The Netherlands, 1999; Volume 1.
45. Sorai, M.; Seki, S. Phonon Coupled Cooperative Low-Spin $^1A_1$ High-Spin $^5T_2$ transition in $[Fe(phen)_2(NCS)_2]$ and $[Fe(phen)_2(NCSe)_2]$ Crystals. *J. Phys. Chem. Solids* **1974**, *35*, 555–570. [CrossRef]

46. Luneau, D.; Rey, P.; Laugier, J.; Belorizky, E.; Conge, A. Ferromagnetic Behavior of Nickel(II)-Imino Nitroxide Derivatives. *Inorg. Chem.* **1992**, *31*, 3578–3584. [CrossRef]
47. Ondo, A.; Ishida, T. Structures and magnetic properties of transition metal complexes involving 2, 2′-bipyridin-6-yl nitroxide. *AIP Conf. Proc.* **2017**, *1807*, 020023.
48. Okazawa, A.; Nogami, T.; Ishida, T. Strong intramolecular ferromagnetic couplings in nickel(II) and copper(II) complexes chelated with *tert*-butyl 5-methoxy-2-pyridyl nitroxide. *Polyhedron* **2009**, *28*, 1917–1921. [CrossRef]
49. Okazawa, A. Magneto-Structural Relationship on Strong Exchange Interactions between Chelating Nitroxide Radical and Transition-Metal Spins. *IOP Conf. Ser. Mater. Sci. Eng.* **2017**, *202*, 012002. [CrossRef]
50. Okazawa, A.; Hashizume, D.; Ishida, T. Ferro-and antiferromagnetic coupling switch accompanied by twist deformation around the copper(II) and nitroxide coordination bond. *J. Am. Chem. Soc.* **2010**, *132*, 11516–11524. [CrossRef] [PubMed]
51. Okazawa, A.; Ishida, T. Spin-Transition-Like Behavior on One Side in a Nitroxide-Copper(II)-Nitroxide Triad System. *Inorg. Chem.* **2010**, *49*, 10144–10147. [CrossRef] [PubMed]
52. Shimada, T.; Ishida, T.; Nogami, T. Magnetic properties of transition metal complexes with 2, 2′-bipyridin-5-yl *t*-butyl nitroxide. *Polyhedron* **2005**, *24*, 2593–2598. [CrossRef]
53. Kumada, H.; Sakane, A.; Koga, N.; Iwamura, H. Through-bond magnetic interaction between the 2p spin of the aminoxyl radical and the 3d spin of the metal ions in the complexes of bis(hexafluoroacetylacetonato)manganese(II) and -copper(II) with 4-(*N*-*tert*-butyl-*N*-oxylamino)-2,2′-bipyridine. *J. Chem. Soc. Dalton Trans.* **2000**, 911–914. [CrossRef]
54. Brooker, S.; Plieger, P.G.; Moubaraki, B.; Murray, K.S. [Co$^{II}$$_2$L(NCS)$_2$(SCN)$_2$]: The First Cobalt Complex to Exhibit Both Exchange Coupling and Spin Crossover Effects. *Angew. Chem. Int. Ed.* **1999**, *38*, 408–410. [CrossRef]
55. Damrauer, N.H.; Boussie, T.R.; Devenney, M.; McCusker, J.K. Effects of intraligand electron delocalization, steric tuning, and excited-state vibronic coupling on the photophysics of aryl-substituted bipyridyl complexes of Ru(II). *J. Am. Chem. Soc.* **1997**, *119*, 8253–8268. [CrossRef]
56. Nishizawa, S.; Hasegawa, J.Y.; Matsuda, K. Theoretical investigation of the dependence of exchange interaction on dihedral angle between two aromatic rings in a wire unit. *Chem. Lett.* **2013**, *43*, 530–532. [CrossRef]
57. Ravat, P.; Baumgarten, M. "Tschitschibabin type biradicals": Benzenoid or quinoid? *Phys. Chem. Chem. Phys.* **2015**, *17*, 983–991. [CrossRef] [PubMed]

© 2018 by the authors. Licensee MDPI, Basel, Switzerland. This article is an open access article distributed under the terms and conditions of the Creative Commons Attribution (CC BY) license (http://creativecommons.org/licenses/by/4.0/).

MDPI  
St. Alban-Anlage 66  
4052 Basel  
Switzerland  
Tel. +41 61 683 77 34  
Fax +41 61 302 89 18  
www.mdpi.com

*Crystals* Editorial Office  
E-mail: crystals@mdpi.com  
www.mdpi.com/journal/crystals

www.ingramcontent.com/pod-product-compliance
Lightning Source LLC
LaVergne TN
LVHW071942080526
838202LV00064B/6651